Human Evolutionary Biology

Human Anatomy and Physiology from an Evolutionary Perspective

Arndt von Hippel

STONE AGE PRESS
Anchorage, Alaska

Human Evolutionary Biology
Human Anatomy and Physiology from an Evolutionary Perspective
by Arndt von Hippel

Published by:
Stone Age Press
1649 Bannister Drive
Anchorage, AK 99508
(907) 279-3740

Cover design: Leslie Newman
Book design and production: Newman Design/Illustration
Printer: R.R. Donnelley and Sons

Publisher's Cataloging in Publication Data

von Hippel, Arndt, 1932–
    Human evolutionary biology : human anatomy and physiology from an
evolutionary perspective / Arndt von Hippel.
        p. cm.
    Includes bibliographical references and index.
    Pre-assigned LCCN: 94-067336
    ISBN 0-9615808-2-8

    1. Human biology. 2. Evolution (Biology) 3. Natural selection.
I. Title
QP34.5.V65 1995   612
                                          QBI94-1508

Stone Age Books are selected and published with care by Stone Age Press of Alaska. It is the publisher's intent to present significant material in a sturdy, inexpensive and pleasing format that will assure Stone Age Press a good name and good will.

for

Marianne, Ted, Bill, Karin, Frank
and those yet to come

PERIODIC TABLE OF THE ELEMENTS

1 **H** 1.00797 Hydrogen																	2 **He** 4.0026 Helium
3 **Li** 6.939 Lithium	4 **Be** 9.0122 Beryllium											5 **B** 10.811 Boron	6 **C** 12.0111 Carbon	7 **N** 14.0067 Nitrogen	8 **O** 15.9994 Oxygen	9 **F** 18.9984 Fluorine	10 **Ne** 20.183 Neon
11 **Na** 22.9898 Sodium	12 **Mg** 24.312 Magnesium											13 **Al** 26.9815 Aluminum	14 **Si** 28.086 Silicon	15 **P** 30.9738 Phosphorus	16 **S** 32.064 Sulfur	17 **Cl** 35.453 Chlorine	18 **Ar** 39.948 Argon
19 **K** 39.102 Potassium	20 **Ca** 40.08 Calcium				25 **Mn** 54.938 Manganese	26 **Fe** 55.847 Iron	27 **Co** 58.933 Cobalt	28 **Ni** 58.71 Nickel	29 **Cu** 63.54 Copper	30 **Zn** 65.37 Zinc		31 **Ga** 69.72 Gallium	32 **Ge** 72.59 Germanium	33 **As** 74.922 Arsenic	34 **Se** 78.96 Selenium	35 **Br** 79.909 Bromine	36 **Kr** 83.80 Krypton

TABLE OF CONTENTS

PREFACE

Everyone is interested in human biology. Human anatomy and physiology permeate our everyday experiences from earliest childhood onward. Biology holds the key to our health and sickness, love and hate, reproduction and digestion, adolescent rebellion and death. Indeed, the study of human biology unlocks the door to an understanding of life itself.

Students enroll in Human Anatomy and Physiology to learn more about themselves. Too often they find the pathway toward useful knowledge blocked by endless lists of human anatomical parts bearing unpronounceable Latin names that must be memorized. Not surprisingly, many beginning biology students soon become discouraged and drop out. For those who persist just long enough to complete some minimal science requirement, recurrent nightmares about the final examination may be their most lasting memory. Yet little that is final will be found in any final examination, especially in biology where researchers are doubling the information content every five to seven years. Thus, beginning science courses are generally more helpful when they emphasize the "Big Picture" rather than requiring reliable regurgitation of arbitrarily selected details from some soon-to-be-revised illustrated catalog of facts. Furthermore, too many facts can easily confuse the student and thereby obscure both the beauty and incredibly rich organization of life.

Earth without life is difficult to imagine. Yet life itself is so strange that you never could imagine it, were it not already so familiar. When introducing strangers, one customarily includes information on what they do (function) and where they are from (ancestry). Our confidence increases when the stories of strangers contain no unseemly gaps, obvious discrepancies or troublesome disclosures. Thanks to many recent discoveries, we finally can introduce some of the stranger ancestral contributions to your own anatomy and physiology without encountering overwhelming gaps or obvious discrepancies—whether this human story contains troublesome disclosures is for you to decide.

Those with no background in science may find the first few chapters quite difficult. Nonetheless, a moderate effort invested here will reveal how physical and chemical processes set in motion by the Big Bang both constrained and enabled life's origin and progression. However, if physics and chemistry are inaccessible to you, do not despair. Just give it your best shot and keep moving, for later chapters can still be enjoyed without that preparation. But then you may have to take my word for how some things came about, rather than evaluating evidence of life's amazing advances for yourself. On the pages that follow, topics are developed in ordinary English, accepting the risk of appearing simplistic or even crude in the pursuit of clarity. Specialized terminology is brought in sparingly as needed, so that it becomes a useful tool or helpful mooring rather than an impediment. Each chapter is designed to correct and expand upon knowledge that you already possess about yourself and the world, without wasting your time. While individual chapters can stand alone, needless repetition has been avoided. Therefore you will find this material most digestible when taken in the order that it is served.

PROLOGUE TO CREATION

Overview;... All creatures bear evidence;... What is scientific evidence, and how can we be sure?... There are thousands of religions but just one science;... There is no scientific basis for any religion;... Religions lose credibility when they attack science;... How shall we recognize scientific truth when we see it?... A never-ending series of questions leads to new scientific discoveries;... Correlation, causation, anecdotal evidence and biased reporting;... So you must decide.

Overview

Nature is prolific. Resources are limited. Competition is fierce. Variation is inevitable. Only the most fit survive to reproduce. Thus natural selection brings about endless adaptive changes.

All Creatures Bear Evidence

Human anatomy and physiology are incredibly complex. Much is unknown—no one really understands the rest. And important advances in human biology are as likely to come from careful studies of yeast, bacteria, worms, fruit flies, corn, squid, toads or mice as from any direct examination of human function and malfunction. Indeed, such investigations regularly reconfirm how inextricably humans are entangled with the natural world from which they derive—which suggests that we really cannot know what we are without knowing how we got that way. But only in the past 150 years has the serious consideration of human origins finally moved from philosophical discussions on the Nature of Man and unforgiving religious disputes about Creation myths to a meticulous investigation of the copious evidence that surrounds us all. Tremendous advances in scientific instrumentation and understanding now uncover life's secrets at an ever-accelerating pace. Some of these findings have obvious immediate impacts, others more stealthily upset the way that we view ourselves. But old ideas rarely give way without protest. While much of this new and revolutionary information has yet to enter public awareness, already there is much dispute. So why are things moving so fast? And why all the dispute? Ought we not slow things down a bit until we can figure out what really is going on and what it all means? Well, fortunately or unfortunately, there is no stopping or going back. Returning to the good old days of continuous conflict over limited resources is impossible—furthermore it is pointless, since modern problems only respond to the most up-to-date and productive solutions.

Science is a cumulative adventure with each advance providing grounds for future research. No matter how inconvenient any new bit of information may be or how poorly it happens to fit the prevalent world view, such evidence can neither be ignored nor forgotten. Thus science brings a directionality to world history that is independent of all but the most catastrophic external changes. Nonetheless, the unexpected arrival of scientific discoveries without instructions for their use has always posed problems—for the public's understanding of science inevitably lags far behind the reality and even competent science teachers may lose their grip as copious new information displaces some of the old and alters the rest beyond recognition. Under such circumstances, the perplexed often seek reassurance and guidance from more traditional non-scientific authorities. These include many fundamentalist Christian leaders who are now trying to stem the rising tide of knowledge that threatens to wash away their own doctrine/following/life-style. Individuals most heavily invested in the status quo undoubtedly preached similar versions of hell-fire and damnation when our ancestors first learned to make fire, build a boat or plant a crop. But the ongoing information explosion can no longer be contained by any power on Earth. Nor can our human population explosion long continue without endangering all life on Earth.

Quite clearly, change is inevitable. Science merely accelerates that change and directs it into many new dimensions occupied by scientific concepts such as domesticated crops and animals, radios and televisions, automobiles and antibiotics, blood transfusions and birth control pills, computers and nuclear power, vitamins and pesticides. By creating winners and losers, change always generates dispute. Yet it takes religious involvement to make most disputes intractable—for religious fervor and finances flourish through the denial of any common factual basis for conflict resolution. The very human tendency toward outrage and passion—the "us against them" mind-set so easily stirred and inflamed by political and religious demagogues, has long brought power and profit to crusaders, invaders and more ordinary lynch mobs. But even though times have changed and peace now promotes productivity beyond the wildest dreams of our club-wielding ancestors, still we remain locked in costly and counterproductive struggles about tribal and territorial matters, human sexuality and population control, genetic studies and abortion, primarily because one or both sides remain sufficiently inflamed by religious (and other divisive) rhetoric to disregard obvious common interests. Fortunately, much new and revolutionary information about human biology is beginning to clarify many of the biological misconceptions and misdirected biological drives upon which such disputes are based. But even if (and as) additional facts become available that can help to settle these difficult matters objectively, how will you know whether information being provided is reliable? Well, before you accept statements made in this book or elsewhere, you should ask "Says who?" and "What is the evidence?". Let us therefore begin this brief discussion on the nature of scientific evidence by clarifying certain vital but widely misunderstood differences between science and religion.

There Are Thousands of Religions But Just One Science

A religion tells a story, detects a direction, reveals a plan, requires appropriate behavior in response to external authority, promises an outcome and offers certainty in an uncertain world. Because religions wrestle with eternal truths, they find it extraordinarily difficult to change with the times. Yet times always change, so in order to persist, any religion must evolve and adapt, often to the point where its original premises appear irrelevant and its previous manifestations would constitute outrageous or criminal conduct (e.g. human sacrifice, or stoning to death as punishment for adultery or for being disrespectful to one's parents). A study of history suggests that even the most powerful and bloodthirsty religious authorities cannot dominate everyone for long. So new and offshoot religions are as unavoidable as the ceaseless enmity within and between quite similar religions that still causes such endless grief and suffering.

Scientific inquiries seek order in the apparent randomness of nature while religious answers attempt to impose order. Scientists study what is. Religious authorities define what ought to be. But the fact that many people share an idea does not make that idea a fact. For a fact refers to something that actually exists, a value-free part of reality, a demonstrable truth. On the other hand, ideas, thoughts and concepts often deal with values. An idea may be helpful or interesting, noble and uplifting, merely crass or even despicable. But in the end, all ideas represent attempts at problem solving, efforts to organize facts in a more meaningful fashion. Of course, facts only have relevance when ideas organize them into coherent entities that can be manipulated for fun, profit and reproductive advantage. Yet all ideas remain mere approximations, subject to later modifications. And every idea must compete with contrary ideas until new evidence accumulates that further refines our understanding of ourselves and the Universe. Scientific progress means change. Although organized religions and science may overlap in their subject matter to some degree, they represent such entirely different approaches to life that they have nothing in common. This implies that neither is better than the other—they simply cannot be compared.

There Is No Scientific Basis For Any Religion

In their search for consistency and authority, newly organized religions commonly include current scientific thinking in their lore. As a result, many out-of-date scientific concepts remain embedded within the belief systems of long-established religions. There they are passed over as quaint anachronisms by later generations of believers. Yet even the most scientifically up-to-date religious answers, insights or predictions cannot contribute to scientific progress, for religious explanations are so all-encompassing that they can neither be verified nor falsified. Furthermore, the arbitrary distribution of life's rewards and punishments puzzles even the most religious, yet that is what religions presume to explain. Many of the devout therefore find a perverse confirmation of God's greatness in His apparent lack of respect for ordinary human concerns. Even if they cannot understand Him or affect His behavior, at least they can be on His team. But for others, the interposition of a random God between suffering humanity and a random Universe merely displaces the difficulty of understanding without resolving anything.

Some scientists view any claims of close communication with God as ordinary consumer fraud. Yet one could hardly expect a faithful following to invest hard-earned assets without some hint at inside information—unless an aspiring man-of-the-cloth chose to occupy the moral high ground with its great visibility and notoriously slippery slopes. Other scientists maintain strong personal religious beliefs by keeping science and religion "in separate parts of their brain". But regardless of how many prominent scientists may worship within any major or minor religion, their own religious beliefs can no more verify that particular religion than they falsify any confirmed scientific evidence. In addition, even famous scientists often lack an open mind and adhere religiously to some scientific *hypothesis* (testable prediction) as *dogma* (established opinion—not to be challenged) long after it has been discredited by new evidence. Thus outmoded scientific ideas may persist through various reincarnations until the affected generation of scientists has passed on. Meanwhile, the unstoppable avalanche of new information tends to leave busy scientists just as poorly informed as other educated laypersons on matters outside of their immediate area of expertise (in which they often learn more and more about less and less).

Nonetheless, uncertainty is uncomfortable and indecision can be unsafe so intelligent creatures feel a strong urge to study, organize and explain. Therefore religious and non-religious thinkers alike tend to accept the most plausible explanation once they have eliminated other alternatives that come readily to mind. But scientific ideas must forever be retested against any legitimate evidence. So while life endures and facts still accumulate, science will continue to evolve. In contrast, religions usually require ("or else!") acceptance of internal or authoritatively communicated evidence that is not equally accessible to all. For religiously acceptable evidence such as a vision cannot be evaluated/reviewed/measured by other competent objective observers. In other words, religious evidence usually is subjective and thus not falsifiable. Religious rewards also must be taken on faith, as they generally are collected in the hereafter rather than being distributed in the here-and-now where all can observe.

Religions Lose Credibility When They Attack Science

Controversies between religions are essential and unending. Religions differ and those differences must seem significant if a religion is to prosper. Indeed, the many religious regulations and disputes that isolate true believers also help them to keep the faith while attracting new adherents. Most religions recognize only one set of beliefs (fortunately their own) as true. Although religions regularly reinterpret scientific evidence to suit their own requirements, established religions gradually learn not to attack science directly. For no religion can retain long-term credibility if it is based upon obviously untrue claims about the world in which we all live—eventually every religion must come to grips with established reality. But major revelations do not come easily to conservative organizations—which explains the current Christian fundamentalist demand that U.S. public school students be taught how God created the Universe just a few thousand years ago. While such a quaint belief may have seemed plausible several hundred years ago, it clearly is refuted by every modern measurement. Nonetheless, the biblically based contention that Earth was created just over 5750 years ago is aggressively promoted as *creation science*. But creationism has no more to do with science than does Scientology or any other so-called Religious Science. Although it can be useful to group mutually incompatible terms in this fashion for their shock value or confusional impact, or to adopt respected terms such as *science* and *research* in order to share their hard-earned aura of truth and objectivity, such juxtapositions as "creation science" or "Christian Science" are as obviously oxymoronic as "Hell's Angels" or "Jews for Jesus".

Interestingly, creationists continue to proclaim biblical infallibility despite obvious inconsistencies (starting with the two incompatible versions of creation in

Genesis) and ample scholarly evidence for multiple authorship of the Hebrew Bible (Old Testament). Indeed, both Old and New Testaments have undergone repeated revisions over the centuries (for example, compare the varied tales of Matthew, Mark, Luke and John). Furthermore, while creationists rely heavily upon promises of Heaven and Hell as carrot and stick, neither Heaven nor Hell are even mentioned in the entire Hebrew Bible. Yet that sacred text dates back to a time when God supposedly roamed the Earth, helping His chosen few, smiting many others and even causing an occasional pregnancy. In addition, many Bible stories are quite obvious adaptations of older near-Eastern myths. But rather than utilize their God-given intelligence for questioning their own beliefs (which would be sinful and possibly dangerous), these fundamentalists tend to devote their intellects to snappy anecdotal proofs such as "I know God exists because I have a personal relationship with Him"—or when other arguments ring hollow, to simply pleading "Why not believe? What have you got to lose?" (other than the freedom to think and live as a rational human being).

How Shall We Recognize Scientific Truth When We See It?

Overwhelming scientific evidence (accessible to any interested observer) demonstrates that the entire Universe, including our Earth, is many billions of years old. Yet some creationists still try to explain away starlight that has traveled for billions of light years—or radiometric data and copious fossil evidence derived from ancient rocks—or uninterrupted sequential yearly deposits displayed in ice and sediment cores that date back several hundred thousand years—or even the annual growth rings from very old live bristlecone pines that overlap datable growth-rings of still older preserved pine logs to provide a continuous record of annual tree rings covering the past 8000 years—by alleging that when God set the whole thing in motion less than six thousand years ago, He also manipulated the evidence to make the Universe merely seem ancient. Comparable, perhaps, to the way modern manufacturers create antique furniture. But if some evidence provided by God is admittedly bogus, how can one have confidence in His word on other matters?

In the early 14th century, an English cleric, William of Ockham, provided a useful guideline for determining what is true. Ockham's Razor or the parsimony principle declares "That which can be done with less is done with more in vain". In other words, where a simple explanation will suffice, the more complex explanation generally has little merit. When Ockham's Razor is applied to the question of how long ago Earth formed, it suggests that other explanations are far less convincing than the simple idea that Earth really is as old as all scientific measurements would indicate. On the other hand, truly simple explanations that encompass all eventualities, such as "It was God's will" or "God must have been watching over me" or "God works in mysterious ways" are not scientifically useful since they are incomplete. Having no predictive power, they cannot be tested. And in the end, all answers that defer to God's judgment or imply Heavenly compensation for torment suffered here on Earth must fall short, for "No promise of Heaven can rectify Auschwitz".[1]

Still, one might reasonably argue that any of the thousands of major or minor religions within and about which people continue to bleed and die has thereby earned the right to be presented to students of religion. But surely there is no justification, especially under the Constitution of the United States, for presenting any specific religious belief such as creationism to students being educated in science at public expense. Creationism is no more the only reasonable alternative to science than it is the only reasonable alternative to sourdough bread. Religion and science are just not related.

A Never-ending Series Of Questions Leads To New Scientific Discoveries

This book presents biology as I understand it. When appropriate, I will refer to confirmatory evidence. Obviously, I, and you in turn, must accept a lot of incomplete evidence and work reported by others on faith. But I have tried to make only statements that seem consistent with scientific evidence and experiments currently being reported in reputable scientific journals. Those reports generally are sufficiently detailed so that other scientists can reevaluate the data presented or gather additional evidence to refute or confirm any of the conclusions advanced. But science is more an endlessly self-correcting process than a result, so much of the material presented in this book will eventually turn out to be approximate, dated or even incorrect as new evidence is unearthed and new relationships are discovered. After all, even advances worthy of the Nobel Prize soon become outmoded by new information. Nonetheless, that growing flood of new information does not imply that scientific studies leading to the automobile or airplane or "H" bomb—or to an approximate age for the Universe—have been discredited. Those autos will keep chugging, the planes will stay aloft and the bombs may explode regardless of any new and better theoretical understanding of thermodynamics. Clear and convincing evidence similarly assures us that Earth will remain billions of years old no matter how our ability to date her rocks, wrinkles and fossils may improve in their accuracy. In other words, uncertainty about tomorrow's weather does not make all eventualities equally probable—rain forests continue to generate and receive rain—deserts are likely to remain dry.

Disagreements between competent and often highly competitive scientists

[1] As quoted by J. Polkinghorne in Science and Providence SPCK Holy Trinity Church, London, 1989.

generally promote further research to resolve the dispute and thereby bring about new advances. The fact that scientists often differ on various aspects of evolutionary theory simply demonstrates the current vitality and progress in biological research. Although some fundamentalists cite this productive turmoil as evidence that many scientists actually are "closet creationists", the same faulty logic applied to the heated disputes between Martin Luther and his Pope would show that both men were "closet atheists".

Correlation, Causation, Anecdotal Evidence And Biased Reporting

Lastly, it may or may not be true that "Wheaties is the Breakfast of Champions". Even if that claim did prove correct, it is likely that eating a different breakfast cereal would not totally destroy a championship team. Nor ought such an advertisement be viewed as evidence that any bunch of losers could improve their game by ingesting Wheaties in small or large amounts, occasionally or even on a strict schedule. For any association of Wheaties with champions might also be explained by the free Wheaties being supplied to champions or by spoiled champions refusing to wait while their oatmeal is cooked. And things may appear to be associated without having any relationship at all. Many years ago, tobacco interests showed a close correlation between the increasing consumption of bananas in Britain and the rapidly rising incidence of lung cancer. Of course, this was just another attempt to blow smoke and thereby obscure the clearly incriminating relationship between cigarette smoking and lung cancer. Furthermore, repeated association may indicate that a real relationship exists without clarifying if A causes B, B causes A, or A and B are both the result of C, D and E. An example of this is our still developing understanding of interactions between kidney disease and high blood pressure. Either can cause the other or both may result from additional genetic and environmental causes. In other words, correlation does not prove causation.

Beware of anecdotal evidence as well. Scientists avoid using unusual individual experiences as evidence, for poorly documented unique experiences are too easily misinterpreted. To be scientifically plausible, an individual experience should generally be well-documented or reproducible at other times by other individuals. In addition, there may be hidden bias in any report of new evidence—for example, if positive results are more likely to be published. Consider a nearly drowned swimmer who reports that a porpoise saved her life by holding her up and pushing her toward shore. Does that report have statistical validity? Do similar events happen regularly? Is it possible that equal numbers of swimmers pushed away from shore and involuntarily submerged by porpoises have failed to report those less favorable findings?

So You Must Decide

During our everyday lives we are incessantly bombarded with credible and incredible information. Adolescent boys were once advised to smear Wildroot Creme Oil on their hair if they wish to have "a tough time keeping all the gals away" or that a cold bath would suppress lust. Cancer patients are still told that a regular vitamin shot, enema, back-cracking, laying on of the hands or other prayer session will soon restore their lost health. Farmers troubled by the increasing pests and ill health that often follow major pesticide applications are informed that more money spent upon bigger and better pesticide programs will save their crops without further endangering their families or customers. Indeed, liars abound. "I'm from the government and I'm here to help you". Or "Trust me! You can't possibly get pregnant the first time!" Or "When you are lonesome, our baby will love you and keep you company". Or "Congratulations! You just won a million dollars! Send $300.00 right away to ensure proper delivery of your prize". And especially in a dictatorship, one soon learns not to trust official propaganda. It obviously was inadvisable to "Tell us what you really think of Chairman Mao" in Beijing, or to "Start your own business here in the Soviet Union". Nor can anyone afford to sit back and expect a fair trial before the hanging.

People vary greatly in gullibility, but disbelief is the best defense against deception and misinformation. Indeed, only disbelief can limit deception by reducing its rewards. Quite clearly we cannot lead effective personal lives without disbelief. Furthermore, disbelief is essential to science. Scientists evaluate objective results in the light of their past education and experience. They seek new evidence to test statements of supposed fact. On the other hand, religions are subjective so they depend upon a suspension of disbelief. But the religious suspension of disbelief is not supposed to open one equally to all possibilities (indeed, it usually occurs under circumstances that deny equal access to competitive religious information or contrary scientific evidence)—more commonly, it strongly encourages the transfer of burdensome individual authority and responsibility to self-selected agents of a hypothetical father figure in hopes that such submission will deliver the magic, mystery and love one may not have experienced during childhood.

With so many dubious and conflicting religious beliefs and so many alternate routes to damnation or salvation, and with each of those mutually exclusive possibilities so vehemently affirmed by true believers, a reasonable person might justifiably disbelieve all religious teachings. But that would be rational, and religions deal mostly with the irrational aspects of life—which also explains why most religions view doubt as an evil to be suppressed or cast off. So while minor religions may benefit from freedom of worship, dominant religions enthusiastically justify the worst of humanity's inhumanities in the name of God and Country. Of course, until recently, the most abhorrent religion-justified behaviors against otherwise inoffensive infidels were at least evolutionarily correct in that they tended to eliminate infidel genes and wealth while similarly enhancing those of successful zealots—who paradoxically might lose religious fervor (as well as non-believers to persecute for fun and profit) if their own religious beliefs made more sense (were obviously and indisputably correct). Eventually, however, regardless of their apparent certitude, even the most religious must fall back upon the strength of shared beliefs for proof of their authenticity. Strongly held religious beliefs certainly bring solace to those who are troubled and afflicted but that merely confirms the dependence of religion upon the suspension of disbelief and once again shows how religions and science have nothing in common.

Yet we all rely upon individual, indirect and often inaccurate internal displays to confirm our views of ever-changing reality—whether during luncheon meetings, earthquakes or ordinary nightmares (see Nervous System). So it seems unlikely that any human will ever be truly free of magical beliefs or personal superstitions—for individual life seems far too precarious and precious to originate or end in a totally absent-minded fashion. Almost inevitably, thoughtful humans will continue to seek old and new ways to bring greater meaning into their seemingly random and inconsequential lives. Fortunately, it is now becoming apparent that the scientific study of life itself can lead us toward a healthier, more meaningful and rewarding existence right here on Earth. But enough on scientific evidence. Back now to the Big Bang....

Further Reading:

Speiser, E. A. 1964. *The Anchor Bible: Genesis; Introduction, Translation and Notes.* Garden City, N.Y.: Doubleday & Co. 379 pp.

Rosenberg, David and Bloom, Harold. 1990. *The Book of J.* New York: Grove Weidenfeld. 335 pp.

Wright, Robert. 1988. *Three Scientists and Their Gods; Looking for meaning in an age of information.* New York: Times Books. 324 pp.

Pagels, Heinz. 1988. *The Dreams of Reason; The computer and the rise of sciences of complexity.* New York: Simon and Schuster. 352 pp.

CHAPTER 1

THE BIG BANG

$E=mc^2$;... Overview;... The Moment Of Creation—A Big Bang;... Inflation Theory;... As Matter Cooled It Formed Simple Atoms, Some Of Which Were Unstable;... Both Nuclear Fission And Nuclear Fusion Can Convert A Little Matter Into A Lot Of Energy;... Proton Count Establishes Atom Identity While Electron Count Controls Chemistry;... Only Simple Atoms Resulted From The Big Bang, So Primordial Chemistry Was Simple;... Stars Utilize Gravitational Energy To Repackage Protons And Neutrons Within Heavier Elements;... Naturally Radioactive Isotopes Release Star Energy At Predictable Rates;... Summary.

$E=mc^2$

Einstein's famous equation declares energy and matter equivalent and establishes the going rate of exchange (E = energy, m = mass, c = speed of light).

Overview

About fifteen billion years ago, our Universe suddenly inflated from an incredibly tiny, hot and dense coincidence of subatomic particles. The resulting universal expansion soon cooled those fundamental particles until they could combine to form primordial hydrogen and helium. Many of the great outwardly moving hydrogen/helium clouds thus generated eventually underwent gravitational collapse to produce clusters of galaxies, individual galaxies and individual stars. Deep within each star-to-be, the primordial gas was progressively compressed and reheated by additional infalling mass until finally some hydrogen nuclei fused together. That nuclear fusion created new helium nuclei while also converting a tiny bit of the fused hydrogen's mass into a great deal of energy (being the smallest atomic nucleus and also three-fourths of all known mass in the Universe made hydrogen the primary fusion fuel for all stars). Soon the energy released during hydrogen fusion resisted further collapse of a star's outer layers. More massive stars had to put out far more energy to reach *hydrostatic equilibrium* so those extravagant beacons consumed their central hydrogen supplies most rapidly—heavy outer star layers then forced core temperatures and pressures ever higher until those recently created helium nuclei in turn fused to form carbon and even larger atomic nuclei. But the ever declining energy yield of such secondary fusion processes only briefly deferred the final collapse and explosion of those huge burnt-out stars.

As an ordinary more modest star, our sun converts about five million tons of hydrogen nuclei into energy every second—during almost five billion years of continuous nuclear fusion, the sun has consumed half of its central hydrogen supply. When that solar-core hydrogen finally runs out, further compression and heating will bring about the central fusion of helium into carbon while also warming overlying solar hydrogen layers to fusion temperatures. Being thus driven into its *red giant* phase, our life-giving sun will expand enormously to engulf Earth and the other inner planets while blowing away its outer hydrogen and helium layers as a ferociously beautiful planetary nebula. That pulsating red giant star will then slowly subside into an innocuous *white dwarf* (dense carbon/oxygen) star about the size of Earth, still being circled by scorched remnants of the inner solar system.

The entire solar system probably owes its existence to shock waves from some massive exploding star of long ago that stimulated collapse of our local hydrogen/helium cloud. But long before that seminal event, this gas cloud already had been littered by debris (those 90 heavier elements) from countless other gigantic stellar explosions—for many stars of all sorts had to die before Earth and your body could be constructed out of primordial hydrogen and recycled star trash.

The Moment of Creation—A Big Bang

The *Universe* encloses all known mass and energy, space and time. Its mass/energy so curves space/time that our Universe has neither an outside nor surroundings. Nevertheless, an endless number and variety of other universes could well exist out there, far beyond any possible interaction. *Galaxies* are enormous collections of *stars* held within the same region of space by mutual gravitational attraction. Our *solar system* (central sun plus distant circling planets) is located approximately 28,000 light years from the middle of a typical slowly rotating disc-shaped *barred* (central bulge elongated in one direction) spiral galaxy that measures about 100,000 light years in diameter. A *light year* is the distance (about 10^{13} kilometers or 6 trillion miles) that light will cross in one year while traveling at the *speed of light* (about 300,000 km/sec. or 186,000 miles/sec.). This Milky Way galaxy ("gala" means milk in Greek) is roughly 1,000 light years thick at our solar system. The sun takes 200 million years to complete one full galactic orbit, moving along at a brisk 220 kilometers/sec while Earth circles the sun at 30 kilometers/sec. In addition, the entire galaxy is being drawn toward the constellations Hydra and Centaurus at about 620 kilometers/sec (according to directional differences in the cosmic background radiation temperature—see below). The bright band of stars that can be seen arching across the sky on clear dark nights represents an edgewise view across a portion of our galaxy. Dark patches along that luminous (milky) band are caused by obscuring clouds of dust ejected by many huge stars of long ago. There are about 100 billion (10^{11}) stars in our galaxy and perhaps 100 billion galaxies in the observable Universe. Naturally we wonder where all this came from, how long it has existed, what it is made of, how it will end. And, of course, where do we fit in...?

While such questions still lack definitive answers, astronomers have gleaned an amazing amount of information through careful studies of starlight. Many of those findings led to predictions that could be tested and extended by further astronomical observations or in physics experiments performed right here on planet number three. One particularly exciting discovery was the repeatedly confirmed finding that light from distant galaxies is generally *red-shifted* (meaning the wave length of that light has been stretched with respect to comparable but brighter light from similar galaxies nearby). For just as police may detect your speeding automobile with radar, or you might estimate the speed of a train by noting how much its approaching whistle drops in pitch after passing on down the line, astronomers can determine from its light how fast a star or entire distant galaxy is moving toward or away from our solar system. Thus when light is higher in pitch or frequency (hence more violet) than usual, the light source must be moving quite swiftly toward us to so noticeably compress the wave length (raise the frequency). On the other hand, when light is reduced in pitch or frequency (with a correspondingly longer wave length—meaning its color has been red-shifted), the light source must have been moving away rather rapidly.[1] Such *Doppler* effects occur because the speeds at which light and sound travel are unaffected by the speed and direction of travel of their source. Instead, the standard speeds of light and sound are determined by the medium (air, water, vacuum, iron rails and so forth) through which their waves are passing (see Special Senses). However, *Einstein's Special Relativity Theory* shows that light must travel at the same *speed of light* with respect to all observers, regardless of how fast or in which direction those observers may themselves be traveling. On the

---

[1] Certain red shifts come about when light escaping from extremely strong gravitational sources loses part of its energy during that escape. And most of the red shift of light from distant galaxies represents stretching of the photon along with the expansion of space between galaxies—the so-called *cosmological* red shift discussed later in this chapter.

other hand, the apparent speed of sound in air will change with respect to an observer moving through that air—which allows certain airplanes to travel in silence by moving faster than the annoying sounds that they leave behind.

Each atomic element *absorbs* and *emits photons* (the quantized packets of *electromagnetic energy* given off as charged particles are accelerated) at a few characteristic frequencies. Elements emitting the photons of starlight can therefore be identified through the characteristic series of dark and bright lines formed when starlight is spread into an orderly *spectrum* (progression or rainbow of frequencies) by passing that light through a prism or else by allowing the light waves to bend and interfere at the many tiny parallel lines of a diffraction grating. The finding that light emitted by increasingly far-away galaxies is increasingly red-shifted (in comparison to starlight from similar but comparatively nearby and stationary sources) led to *Hubble's Law* that distant galaxies are generally receding (moving away) at a speed proportional to their distance from us. In other words, the space outside of local gravitationally held galaxies seems to be expanding quite briskly. One must keep in mind that light just now reaching Earth from very distant galaxies has been traveling for billions of years. So we see those galaxies as they were in their youth, which obviously influences their output of light as well. Indeed, the light from very distant *quasars* may have a red shift of 4 or more (meaning 400% or greater lengthening of those light waves). Such light signals therefore represent rather out-of-date news bulletins about strange objects and events from a time when the Universe was ⅕ its present size and ¹⁄₁₀ as old (about nine billion years before our Earth finally formed). Not surprisingly, it becomes more and more difficult to determine actual distances to really far-away galaxies by means other than their red shift, and so far it has not been possible to translate red shifts into exact distances.

Of course, the large red shifts in their light only provide evidence about how those distant objects were moving way back then. But one cannot calculate the rate at which gravity has since slowed the universal expansion without knowing the average mass density or gravitational pull exerted by the expanding Universe. Therefore many studies are underway to determine the Universe's mass density and establish a more accurate value for *Hubble's Constant* (the relationship of an average galaxy's recessional velocity to its distance from us). Even knowing the current value of that so-called constant would allow more accurate determinations of intergalactic distances as well as the age and expansion rate of the Universe. For now, it is estimated that an average galaxy one million light years away is probably receding from us at between 55,000 and 115,000 kilometers per hour (35,000 to 70,000 miles per hour). But galaxies also respond to the gravitational pull of nearby galaxies and clusters of galaxies, so the average or universal expansion rate is often modified by local interactions. Nonetheless, if galaxies twice as far away generally display a *cosmological red shift* that is twice as great, then those faraway galaxies must be receding from us twice as quickly. This suggests that all galaxies originated in the same place at the same time. Indeed if we reverse the present outward movement of galaxies in our minds (as if playing a movie backward) and also speed it up over time to compensate for gravitational slowing of that universal expansion rate, it appears that all parts of our expanding Universe originated explosively from a single source perhaps 15 (10–20) billion years ago. The inconceivably immense explosion that endowed the entire Universe with its ongoing outward momentum and hence cosmological red shift is commonly referred to as *The Big Bang*.

Inflation Theory

One theory suggests that immediately after that moment of creation (10^{-35} seconds after, to be more exact), the newborn Universe suddenly *inflated*. Thereafter it doubled in size at least 1000 times (until inflation ended about 10^{-32} seconds after creation) during the course of the *delayed phase transition* (as when supercooled water suddenly freezes) in which the *gravitational force*, suddenly supercooled by the ongoing universal expansion, finally separated out from the *other three* still-unified *fundamental forces* of our Universe. Those other forces are the *strong* and *weak* nuclear interactions and *electromagnetism*. These four forces were indistinguishable at the incredibly high energy densities of the very beginning (see below). This inflationary phase was probably associated with a firestorm of fundamental particles of matter (including *quarks, electrons* and *neutrinos*—also indistinguishable at those very high initial temperatures) coming into existence as positive energy to balance the *negative* gravitational energy invested in outward momentum. But how can gravitational energy be negative? The following simplistic analogies might reduce your discomfort with this mathematical concept.

Energy can be extracted from falling water or other large colliding masses because *as the gravitational force strengthens* (i.e. as Earth's and the water's separate gravitational forces are combined into a single stronger gravitational force), *the energy content* of that Earth-and-water system *diminishes*. Often we refer to the *potential* energy stored in an elevated object (some of the energy you expend to climb onto your roof, for example) and consider its release to represent kinetic energy (as object and Earth come together). But the first point here is simply that as large gravitationally attracted masses fall toward one another, their combined gravitational force rises while their energy content goes down (with you flat on the ground, for example). Consider also that an *absent* gravitational force can have *no* energy. That sort of reasoning allows one to view any non-zero gravitational force as stored negative energy. Not clear yet? It's all in the math.

Indeed, one of the more puzzling aspects of the Universe is why it operates so strictly in accordance with the rules of mathematics. There was no reason to expect that the Universe could be described so successfully by a human language of weights and measures. Of course, human thought processes must develop logically from experience in order to be consistent with the way things really are. And science would be impossible and irrelevant were it not for the common-sense relationship between cause and effect (which allows you to conclude "if this and that are so then the other is indisputable as well"). Perhaps a mathematically beautiful expression satisfies because it so elegantly expresses an underlying truth. In retrospect, rather than being surprised that the Universe operates in accordance with the particular rules of human mathematics, we might conclude that human limbs and brains brought greater reproductive advantage when they acted upon and patterned the real world more effectively. In other words, mathematics need not be viewed as a human invention that almost miraculously expresses basic universal principles. Rather, humans became successful by developing in accordance with universal principles—just as mathematics is an almost inevitable outcome of advanced thought processes.

For our purposes, it suffices to recognize that you are one of the many surprising and delightful returns on the initial gravitational investment made at the Big Bang. Since then, gravity has painstakingly or cataclysmically constructed the larger atoms essential to your being out of hydrogen. Furthermore, by stimulating stellar fusion, gravity has provided the flow of energy upon which life and all else of note has depended since the Big Bang. Gravity also provides Earth and you with the heaviness that keeps you together. So the next time you stand upon your bathroom scale, give thanks for the strength and persistence of gravity and for all of those wonderful opportunities to put on more weight that gravity brings. Indeed, the sudden expansion from nothing that brought the entire Universe into being represents the ultimate free lunch, for this enormously productive mechanism somehow arose from nowhere out of nothing.

In more detail, the so-called *New Inflationary Universe Theory* suggests that the entire known Universe was once just a few kilograms of extremely hot and dense something crammed into a space 100 billion times smaller than a proton. There is even a tiny likelihood that such a condition could arise by chance in previously "empty" space, since empty space never has an exactly zero energy level (according to the probabilistic rules of quantum mechanics—see below). As the energy represented by those initial few kilograms was then diluted and cooled by explosive expansion from its original temperatures of more than 10^{32} degrees *Kelvin*,[2] it almost instantly passed through the 10^{27}K° temperature at which gravity could first separate out from the other three fundamental forces. That energy content then became stored in a *Higgs*-like *Field* (a mathematical concept based upon energy-containing space as "scalar field matter" which helpfully provides the mechanism for the spontaneous symmetry breaking necessary to bring about separation of the gravitational force) rather than as ordinary subatomic particles. And according to calculations based upon Einstein's equations of *General Relativity* (the best theoretical description of gravity), that would allow gravity to briefly apply a repulsive (rather than its customary attractive) force.

This antigravity-like expansile effect is brought about by the impending *phase change* (supercooled gravity produces a negative energy density which is the same as negative mass). During its very rapid expansion, the ongoing *false vacuum state* would necessarily retain its same enormous (but very-lowest-possible under the circumstances) energy density that still permitted the gravitational force to remain unified with the other three—until that false vacuum eventually decayed when the markedly supercooled gravitational force finally separated out (still at the very beginning of that initial second of creation). In other words, while forced to expand under the influence of supercooled gravity, the already-at-minimum-energy-level false vacuum state had no option but to create more of itself, until eventually the false vacuum system broke down in a firestorm of fundamental particles (representing release of the enormous false vacuum energy density—somewhat comparable to the latent heat released when supercooled steam condenses or supercooled water finally freezes). This massive release of energy swiftly rewarmed the rapidly cooling Universe back toward 10^{27}K°. The irresistible expansile force of such an enormous energy density overwhelmed the gravitational self-attraction of all universal matter during this initial radiation-dominated phase. However, within a few thousand years the Universe became dominated by matter, for while the energy density of matter was diluted in proportion to volume (hence the universal radius cubed), the radiation energy intensity diminished even more rapidly (with universal radius to the fourth power) since the energy of each photon was simultaneously being redshifted by the great speed of that universal expansion.

---

2 Temperature °K—a temperature scale named after Lord Kelvin in which the degrees are the same size as those in the celsius or centigrade scale, but 273.16 is added—thus 0°K equals absolute zero, which is as cold as matter can get—no thermal motion.

To further complicate things, *matter* and *antimatter* seem to have been produced in equal amounts. Yet rather than simply recombining to vanish in a flash of energy and fundamental particles, the matter and antimatter of our Universe came together in an asymmetrical fashion that left an excess of matter to form our lovely Universe and you. That asymmetric interaction between matter and antimatter is somehow justified mathematically by another asymmetry of *time* between past and future (so we have gained our material existence in exchange for never becoming younger—but at least that obligatory one-way flow of time also requires cause to precede effect). The total energy content of our Universe remained about zero through all of this turmoil, so those strict but strange laws of *quantum physics* were not violated. Also by definition, some fundamental particles would have existed from the beginning, for a force has no meaning independent of particles that express such a force.

The falling temperatures and pressures associated with universal expansion and energy dilution encouraged the remaining three still-combined forces to separate from each other—soon conditions became appropriate for elementary particles to combine. The formation of as many *protons* (carrying a single positive charge) as *electrons* ($\frac{1}{2000}$ of a proton in mass and carrying a single negative charge) preserved the electroneutrality of the Universe (yet another example of how the entire Universe really amounts to nothing). *Neutrons* (proton-sized but carrying no net electrical charge) also formed at this time. Unlike the indivisible *point-sized* (no apparent volume at all) electrons, each proton and neutron had to be built up as a specific combination of three fundamental particles known as *quarks*—also point-sized and available in 6 varieties or flavors. Strangely enough, the interaction of such energetic non-entities gives rise to the effects we refer to as *mass* and *inertia* and *volume* and *life*.

As Matter Cooled It Formed Simple Atoms, Some Of Which Were Unstable

At these extremely high temperatures and pressures, newly created subatomic particles flew about in all directions at incredibly high *relativistic* speeds (a major fraction of the speed of light), smashing into one another constantly. The faster such particles traveled, the greater their total energy content and therefore mass. Electrons tended to avoid one another because all carried the same single electronegative charge. Protons similarly found each other electrically repulsive, but their comparatively huge size and greater inertia prevented rapid turns. Thus frequent impacts occurred between protons, and even more often between protons and uncharged neutrons and among neutrons (as long as independent neutrons still existed—see below). With a really strong impact, some particle momentum might be converted into matter (in the form of particle—antiparticle pairs) plus high-energy photons such as X-rays. At far lesser speeds, colliding particles could simply rebound. But intermediate strength collisions between proton and proton, or proton and neutron, allowed them to *fuse* (be welded together) as a result of coming within reach of the very-short-range *strong nuclear force* or *strong interaction* (so named because within the atomic nucleus it overcomes even the powerful electrical repulsion between protons). Each successful *nuclear fusion* of two energetically colliding subatomic particles was associated with a slight decrease in their total combined mass as an equivalent release of energy cemented the new relationship. Many unstable combinations of protons and neutrons also formed that soon flew apart again, releasing nuclear fragments and energy in a process known as *spontaneous nuclear fission* (or *radioactivity*). Of course, even comfortably fused particles were regularly disrupted when forceful impacts returned more energy than had been released during the prior nuclear fusion (thereby allowing such fused particles to buy out of that confining relationship).

When neutrons fuse in appropriate ratios with protons, such neutrons are very long lasting (and their protons last essentially forever). However, after just a few minutes, an *isolated* neutron usually disintegrates into three unequal pieces as one manifestation of the *weak nuclear force*, giving rise to a proton, an electron and an *antineutrino*. By converting one flavor of quark into another, that *weak nuclear interaction* can change protons to neutrons and vice versa. So if two protons are forcibly fused (held together by the strong interaction), one of them quickly decays into a neutron for that is a more stable arrangement. This decay is achieved by release of a positively charged electron (*positron* or *anti-electron*) plus a neutrino. Neutrinos and antineutrinos are tiny uncharged particles with little or no mass that therefore pass easily through matter or antimatter at about the speed of light. If neutrinos and antineutrinos turn out to have any mass at all, they also must move a bit slower than light. Otherwise their individual masses would become infinite at the speed of light. *Photons* are packets of electromagnetic energy of specific wave length that can travel at the speed of light because they have no mass *at rest* (which they never are). Despite its lack of rest mass, a photon's mass (in motion) and impact rises along with its frequency (energy content goes up as wave length goes down).

A proton fused to a neutron is called a *deuteron*. A proton fused with two neutrons is a *triton*. But regardless of whether a single proton is isolated or has fused with one or two neutrons, it remains a *hydrogen nucleus* carrying a single positive charge. And when a proton attracts a single (much smaller) electron into orbit around it, that electrically neutral (no net charge) proton/electron pair is called a *hydrogen atom*. As you might guess, the mass of a hydrogen atom is a tiny bit less than the total individual masses of an unencumbered electron and a free proton. One can consider such mass differences to represent freedoms given up within the constraints of these stable and restrictive relationships. By repaying that *binding energy* dowry, very high temperatures encourage protons and electrons to fly apart again (see plasma below). *Deuterium* and *tritium* atoms are heavier forms of the *element* hydrogen that combine one or two neutrons respectively into their single proton nucleus. Atoms of the same proton count that differ only in their neutron count and therefore *atomic weight* are classified as different *isotopes* of the same element. Thus the three possible isotopes of hydrogen correspond to a proton combined with zero, one or two neutrons ($_1^1H$, $_1^2H$ or $_1^3H$). One proton fused to two neutrons is a somewhat unstable arrangement so tritium is a *radioactive isotope* of hydrogen and about half of any collection of tritium nuclei will fly apart spontaneously within the following 12.4 years—such a short *half-life* means any tritium produced during the Big Bang has long since decayed.

These days, human-manufactured tritium is most likely to be encountered in the hollow core of the chemical-explosive-surrounded *plutonium pit* (fission trigger) of a thermonuclear fusion weapon. By increasing fission of the plutonium pit, tritium *boosting* allows the use of lighter plutonium triggers from which more energy can be released before the remaining plutonium is blown apart. But its natural radioactivity causes the tritium stored in a *hydrogen* or *H*-bomb to gradually disappear unless the H-bomb is detonated while still adequately boosted. In that detonation, the sudden release of tritium's stored nuclear energy and extra neutrons boosts plutonium fission, which in turn raises temperatures and pressures sufficiently to initiate fusion of the H-bomb's solid lithium deuteride fuel (see below). The main product of that explosive nuclear fusion is a nucleus containing two protons fused with one or two neutrons. Since it bears *2* protons, that newly formed and heavier-than-hydrogen *alpha particle* represents the nucleus of *element number 2*, also known as *helium* ($_2^4He$). Unlike an H-bomb, the technologically less advanced *fission* (or *atomic*) bomb achieves its impressive energy release by splitting certain heavy radioactive isotopes (such as the uranium 235 isotope of element *92* or the plutonium 242 isotope of element *94*) through neutron bombardment, for simply *splitting* a gram of uranium atoms releases about a million times more energy than the complete *chemical* combustion of one gram of carbon into CO_2.

Both Nuclear Fission And Nuclear Fusion Can Convert
A Little Matter Into A Lot Of Energy

A word about $E=mc^2$, the famous equation from Einstein's Theory of Relativity. Here E = energy, m = mass, and c stands for the speed of light. This less than obvious statement about the equivalence of energy and mass has now been amply confirmed. Examination of the equation shows that just a little mass can be converted into a great deal of energy, for any small bit of mass must be multiplied by a very large number—the velocity of light squared (which is about 300,000 kilometers/sec multiplied by itself)—to determine its equivalent energy. That is why total mass is only slightly reduced after either nuclear fusion or nuclear fission have released an awesome amount of energy. Actually just seven-tenths of 1% of the mass originally within two protons and two neutrons is released as energy when those actively moving independent particles are constrained within a single helium nucleus by nuclear fusion. It turns out that a single gram of mass entirely converted to energy (during either fusion or fission) becomes almost 22 trillion calories. Einstein's equation also indicates that if we add energy to any mass (thereby increasing its momentum) we simultaneously increase its mass (since c never changes)—but that enhancement of mass only becomes significant at relativistic speeds.

A reminder: except for one isotope of the element carbon (by definition), atomic weights are not exact whole numbers. In the first place, the listed atomic *weight* of any element is an average of naturally occurring isotopes in their commonly encountered proportions. Furthermore, every individual proton and neutron contributing to each isotope nucleus will have undergone an appropriate minor reduction in total weight as its element was formed by fusion—that mass converted into energy corresponds to the nuclear binding energy. Incidentally, atomic weight is usually the top left-hand number on the abbreviation that represents a specific element. The top right number refers to net electrical (ionic) charge (discussed below). The bottom left number usually reports the total proton count per nucleus—which is also the number assigned to that element in the Table of Elements (an element is equally well identified by its name or number). The bottom right number generally indicates how many of these atoms are included within a specific molecule (see below). Thus $_{1}^{2}\text{H}_{2}^{0}$ identifies a hydrogen molecule formed of two uncharged deuterium atoms.

Proton Count Establishes Atom Identity While
Electron Count Controls Chemistry

An atomic nucleus that contains just one proton must be hydrogen. Any two proton nucleus (usually with a couple of neutrons) has to be helium. Three protons always spell *lithium* (according to the Periodic Table of Elements) and so on. Whenever the number of *protons* in an atomic nucleus is altered by fusion or fission, that nucleus is automatically redefined as another atomic element. Electrons and neutrons may come and go, but as long as the total proton count of any nucleus is unchanged, that nucleus still represents the same element. Although an atom bearing its full complement of electrons carries no *net* electrical charge, every isolated atomic nucleus necessarily displays the total positive charge of all its protons. More generally, any atom holding more or fewer electrons than it has protons in the nucleus will carry a net negative or positive electrical charge. Such charged atoms are referred to as *ions*.

Single hydrogen atoms like to pair up at earthly temperatures while single helium atoms always remain *chemically inert* (refuse to bond with any other atom). For with two electrons in possession, helium has filled the closest possible electron shell to its nucleus—such a completed shell holds its electrons securely enough that usually they cannot be attracted away by another atom. But with only one electron in its innermost shell, a hydrogen atom tends to shed its single electron or else grab a second one (even if some other nucleus remains associated with that newly acquired electron). That is why hydrogen atoms pair up. Here again we can speak of binding energies, but such *inter*atomic electromagnetic attractions are far weaker than short-range *sub*atomic bonds based upon the strong nuclear force. Thus the combined weights of two hydrogen atoms sharing a pair of electrons is just barely below that of two solitary hydrogen atoms.

A *covalent* (shared-electron) bond between two atoms produces a *molecule* (any gathering of two or more atoms in which individual atoms are held snugly to the group by sturdy covalent bonds). Yet the *Coulomb repulsion* (named after Charles Coulomb, a French physicist who studied forces exerted by interacting electrical charges) between the positively charged protons in a hydrogen molecule simultaneously pushes those protons far enough apart so they remain well beyond reach of the strong nuclear force—thus "cold" nuclear fusion does not occur at earthly temperatures and pressures within even the most compact of molecules.

A *chemical reaction* alters the relationship between electron clouds of the atoms involved—those atoms may be solitary or part of one or more interacting molecules.

Note especially that chemistry has nothing to do with nuclear fusion or fission and that the same products result from a particular chemical reaction regardless of which isotope of any specific atomic element happens to be involved. Thus hydrogen, deuterium or tritium atoms all join in pairs using identical shared-electron covalent bonds despite the different neutron counts of their relatively remote nuclei.[3]

Only Simple Atoms Resulted From The Big Bang,
So Primordial Chemistry Was Simple.

It took Neils Bohr's quantum mechanics (which regards a system as fundamentally wavelike in nature, with energy, position and momentum being expressed mathematically as *probabilities* rather than definite values) to explain why negatively charged electrons orbiting a nucleus do not progressively slow until they can be captured by those positively charged nuclear protons deep down inside. For classical electrodynamic theory would have predicted that such charged particles in accelerated (circular orbital) motion should radiate electromagnetic waves, thereby losing energy until they spiral into the nucleus. But the electrons held in such orbits apparently accept and release energy only in specific size packets known as *quanta*. And the light photons absorbed or released by transitions between higher and lower electron energy levels must obey *Planck's Law* (frequency equals energy divided by Planck's constant—so the higher its frequency and shorter its wave length, the greater the energy carried by any photon). Since even subatomic (intranuclear) energy transfers are quantized, each element can only absorb and emit photons of certain energies. The specific frequencies of those photons account for the previously described spectral lines that not only can identify a *source* of photons but also specify which atomic element or perhaps molecular species may have *absorbed* (removed) photons of certain wave lengths en route to your telescope.

In any case, the least energetic electron still retains sufficient speed to orbit its nucleus forever. Furthermore, mathematical expressions for the wavelike nature of particles show that the wave functions for *position* and *momentum* are mutually exclusive—the more precisely one value is measured, the more uncertain the other value must remain. Indeed, an electron falling into the atomic nucleus would violate this (Heisenberg's) *Uncertainty Principle*. If that strikes you as implausible, it has repeatedly been shown that an electron is a point-like particle of no size at all, yet a single electron (or even atom) can sometimes act as an electromagnetic wave and pass through two side-by-side holes simultaneously, thereby producing an

---

[3] This last statement is only approximately correct, as certain isotopes may be slightly preferred or avoided in some processes or reactions, and there can be physical differences as well. For example, diamonds are pure crystalline carbon. Yet those diamonds containing only the Carbon-12 isotope are 50% better heat conductors than natural diamonds which incorporate one percent Carbon-13.

interference pattern with itself. But such an electron has to behave either as a wave or as a particle—both behaviors cannot be observed at the same moment—according to another principle of quantum mechanics known as *complementarity*. Fortunately, this is a book on biology, so we continue to skip lightly past such matters.

Within four minutes after the Big Bang, as protons and neutrons frantically produced and disrupted various isotopes of hydrogen and helium, a few larger nuclei also were created. When universal cooling (due to ongoing universal expansion) brought an end to primordial nuclear fusion, those nuclei carrying three protons plus some neutrons had become the third most common element. But with about 25% of all visible mass in the early Universe locked up inside helium nuclei (as $^4_2$He) and almost all the rest present as hydrogen ($^1_1$H), there were only traces of deuterium and $^3_2$He (at one part per 100,000), even less $^7_3$Li (at about one part in ten billion), and perhaps the tiniest bit of $^9_4$Be and $^{11}_5$B.

Interestingly, there is ample evidence for the existence of far more mass than can be accounted for by all visible matter in the Universe (including heavier elements created later during stellar fusion). For example, the average speed of galactic rotation appears sufficient to fling its outer stars away unless far more mass surrounds each galaxy than has yet been detected. Similarly, there is not enough visible mass to hold large clusters of galaxies together. Even the relatively slow rate of universal expansion implies more mass than meets the eye. Thus somewhere between 50% and 99% of all mass in the Universe seems to be present as some sort of invisible *dark matter*. Perhaps part of that dark matter is sequestered in abortive *brown dwarf stars* that (like large planets) are just too small to initiate fusion, and maybe some can be accounted for by badly faded white dwarf stars. Furthermore, it is still possible that those billions of neutrinos just now whizzing harmlessly through your body also have a tiny mass, or perhaps some other exotic and as yet undiscovered particles contribute enough gravity to keep the stars on their appointed galactic rounds.

In any case, the Universe of which we are aware apparently displays only a tiny fraction of whatever actually is out there and all around us. So it is difficult to determine the mass density of the Universe by direct observation, which in turn affects numerous other relationships and predictions including whether the Universe contains enough mass to eventually reverse its current expansion (into an inevitable *Big Crunch*) or whether the Universe will simply continue to spread out forever (into the *Big Chill*). This range of outcomes includes the theoretically plausible likelihood that the remaining kinetic energy of the Universe so exactly balances its gravitational attraction that the Universe is *flat* (an intermediate state between closed and open that also implies endlessly slowing expansion). Despite that wide range of possible outcomes, there are many clues as to how the observable Universe has developed so far, some dating back almost to the Big Bang.

During its first 300,000 years or so of turbulent expansion and cooling, our Universe consisted of an extremely hot dense *plasma* (flow of atomic nuclei and electrons) that continuously absorbed and emitted electromagnetic radiation. Only after reaching about 3000K° were those charged particles finally traveling slowly enough to come together as neutral hydrogen and helium atoms that no longer blocked all photons (for photons transfer electromagnetic forces between charged particles). As high-energy photons then traveled outward, they became markedly stretched in wave length along with the rapidly expanding space that each was crossing (hence their energy depletion represented a cosmological red-shift rather than being an ordinary Doppler effect). Big Bang theorists *(cosmologists)* correctly predicted that these primordial (but now massively red-shifted and cooled to 3°K) photons of the *cosmic microwave background radiation* would still be coming at Earth from all directions. Incidentally, this greatly stretched low-temperature electromagnetic radiation corresponds to a red shift of 1000 in that terrible first light released by the ultra-massive infant Universe undergoing very rapid early expansion. We have already seen how light from our Universe at one-tenth its present age has a red shift of about 4. So clearly red shifts need not be proportional to the time of light emission nor to distance or speed of its source. In fact, the exact distribution of that cosmic microwave radiation bears important tidings about conditions in that 300,000 year old Universe—especially regarding the scale of any inhomogeneities in mass distribution at those early times, since light emitted from more massive sectors was additionally red-shifted by energy losses during escape from that locally enhanced gravity. The previously mentioned slight directionality and other minor variations in this microwave radiation presumably reflect the movement of our solar system and the Milky Way galaxy through that background as well as the the orbital motion of Earth.

If an atomic nucleus carries 3 protons it must be *lithium* (atomic element #3 or ^7_3Li). Each electrically neutral lithium atom has three electrons orbiting its nucleus. Since only two of those electrons can enter the lowest orbit around the nucleus, lithium's third electron must fend for itself at the next higher orbital level. However, as is apparent from the Table of Elements, this next higher level comfortably seats eight electrons. So any atom with less than eight electrons in this second shell will be chemically reactive, urgently seeking some arrangement that more closely resembles a full outermost electron shell. Having just three positively charged protons in its nucleus, lithium could never hope to attract seven more electrons into its second electron shell. Therefore, just like hydrogen (also located in the first column of the Table of Elements), lithium is a confirmed and vigorous electron donor. For only by handing off that pesky outermost electron can lithium achieve serenity as an ion (bearing a single positive charge) with a completed outermost electron shell.

We have seen that our early Universe provided only hydrogen and helium or another lithium for lithium to pair with. Free lithium atoms or tamed Li^+ ions (a lithium atom that has donated that single outermost electron) tend to repulse one another and helium is unreactive so lithium preferentially paired with hydrogen ($_1^1H$) as lithium hydride or with deuterium ($_1^2H$) as in the solid lithium deuteride H-bomb fuel. Incidentally, the major and final thermonuclear fusion reaction of the H-bomb detonation mentioned earlier evolves in several energy releasing steps—a neutron plus $_3^7Li$ produces $_1^3H$ plus $_2^4He$ plus a neutron—then $_1^2H$ plus $_1^3H$ goes to $_2^4He$ plus more neutrons to make more tritium, while at the same time $_1^2H$ plus $_1^2H$ goes to $_2^3He$ plus extra neutrons that speed the reaction even further.

Normally, a hydrogen atom happily hands off its single electron in order to run around naked as a *proton* (also known as a *hydrogen ion* or H^+) carrying that single positive charge. However, lithium wants to be rid of its extra electron even more urgently (for one thing, lithium's third electron is farther from its nuclear protons). So any nearby hydrogen atom will be forced to accept lithium's third electron in transfer. The proton bearing two electrons that results is thereafter known as a *hydride* ion (H^-). Thus the relationship between the lithium-hydrogen duo is $Li^+ H^-$. These dissimilar atoms are attracted to one another by their equal and opposite charge in what is known as an *ionic bond*. Being based upon electrical attraction between ions of opposite electrical charge makes an ionic bond far weaker than a covalent (tightly shared pair of electrons) bond. Of course, ionically bound atoms can form very solid and stable crystalline structures such as rock salt and many other rocks of our Earth's crust. Yet ionic bonds do not a molecule make, for a molecule relies upon the stronger covalent commitments that keep each atom secured to at least one of its adjacent partner atoms. But regardless of whether they are neutral or ionized, alone or part of a molecule, atoms always repulse one another if their electron clouds approach too closely (within their so-called *van der Waals* limit). In view of such major differences between chemical and nuclear reactions, it is easy to see why ancient alchemists failed to transmute lead into gold, thereby not making a better living through chemistry. This raises the question of when and how our rapidly expanding and cooling hydrogen/helium (plus a trace of lithium) Universe went on to produce the 90-odd heavier elements that currently contribute to and complicate our lives.

Stars Utilize Gravitational Energy To Repackage Protons
And Neutrons Within Heavier Elements

Fortunately for Big Bang theories and those still working to improve them, the rapidly expanding clouds of molecular hydrogen plus atomic helium were not just homogeneously and progressively diluted. Instead, in some manner still unclear (possibly based upon early density disturbances consequent to the probabilistic nature of quantum phenomena applicable to the very small scale Universe initially inflating through atomic and molecular size), the swelling universal cloud broke up and became lumpy. That early non-uniformity then allowed gravitational attraction to draw the more dense protogalactic clouds together as they raced outward after creation. For despite its being far weaker than the weak and strong nuclear interactions or electromagnetism (indeed, the gravitational attraction between individual elementary particles is so weak that it has never even been detected), gravity does act over huge distances and attract everything. In comparison, the range of the weak interaction (which can scatter tiny neutrinos that otherwise would pass through matter unaffected) is about $\frac{1}{100}$ the diameter of an atomic nucleus. Even the strong interaction (responsible for the binding energy released as highly mobile hydrogen nuclei fuse to produce helium) only reaches the outer limits of atomic nuclei. And while electromagnetism cannot overpower the strong interaction within the nucleus, the photons that transfer electromagnetic energy do travel endlessly through space. Nonetheless, most electromagnetic interactions remain localized simply because positive and negative charges tend to cancel out on larger scales. So despite the electromagnetic force between charged particles being 10^{39} times stronger than their gravitational interaction, it was gravity that organized the Universe.

Furthermore, the gravitational energy being released as those cloud fragments destined to become individual stars gradually fell together (such a collapse process usually began at the center of a cloud before spreading outward) simply represented another tiny repayment on the huge gravitational investment in outward momentum at the Big Bang. As each growing protostar drew in ever more gas at higher speeds, the resulting heat and light boosted its early brightness up to 60 times above final steady-state star luminosity—but this early energy output remained trapped locally within the surrounding cloud of gas and dust. At the same time, increasingly energetic collisions deep inside each collapsing clump of primordial gas disrupted remaining molecules and atoms into free nuclei and electrons (producing another plasma of charged particles) as core temperatures rose toward 4000°K. That inner heat and gas pressure (generated by colliding nuclei and electrons not wishing to approach one another more closely) might resist the compressive burden of outer cloud layers for a while—until escaping photons and other high-speed particles brought about sufficient radiant cooling (removed enough energy) to allow further collapse, along with a gradual decline in star luminosity. If a potential star had sufficient mass, the temperature deep within that star-to-be eventually reached about one million degrees Kelvin, causing its small stores of deuterium (^2_1H) to fuse into helium. Heat released during that initial core-fusion reaction then expanded the warming protostar (still in hydrostatic equilibrium) to several times its final size. Star luminosity again rose as deuterium depletion allowed further star shrinkage and released additional gravitational energy.

Slowly the star contracted to its later, more stable size while significant quantities of infalling gas were blown away by the *protostellar wind* (a strong outflow of matter and energy from the heated star surface). Eventually internal temperatures within the increasingly compact star-to-be reached the 10 million degree Kelvin temperature at which hydrogen (^1_1H) fusion to helium could commence. The resulting copious new release of fusion energy enhanced star brightness and supported outer star layers while steadily flinging matter and energy across the heavens on the *stellar* wind. At this point, the young single or (more commonly) double star(s) would likely be rotating quite rapidly, having been wound up by any slight angular momentum present in the infalling gas cloud (just as a figure skater speeds the spin by bringing her arms in tightly). Such stars might also be surrounded by a rotating disc of material out of which planets could form. The bigger the star-to-be, the harder it fell together from the start, so the greater its ongoing central production of energy by fusion. But any star fused hydrogen once it acquired sufficient mass (8% or more of our solar mass). A rather massive star might burn most of its core hydrogen to helium within a couple of million decades. Then central cooling would again allow the gravitational collapse of unsupported outer star layers, which would further compress and heat its core until that helium fused (mostly) into carbon. Thereafter, any star more than four times as massive as our sun could sufficiently squeeze its carbon core to drive temperatures above 600 million °K, at which point even those carbon nuclei would begin to fuse. Helium nuclei (*alpha particles*) participated in many of the later fusion reactions, being absorbed or ejected during several of the steps necessary for production of heavier nuclei. However, none of these later fusion stages lasted long or provided anything like the increase in binding energy initially obtained when trapping light and free hydrogen nuclei within sedate helium nuclei. Incidentally, alpha particles were first detected during the radioactive decay of larger atoms—only later did it become apparent that they were helium nuclei. Indeed, helium was identified through its characteristic lines in the solar spectrum before it was recognized on Earth (Helios means "Sun" in Greek).

This stepwise fashioning of already bound clumps of protons and neutrons into ever heavier nuclei releases progressively less binding energy—but nuclear fusion remains an *exothermic* (heat producing) reaction *until* the most densely packed of the larger atomic nuclei has been formed. That nucleus with the highest binding energy or lowest residual energy per nuclear particle is *iron* ($^{56}_{26}$Fe) carrying 26 protons and 30 neutrons. Iron can only be produced within a star exceeding perhaps twenty solar masses that has reached a central temperature of about 3 billion °K. Deep inside such stars, the final complex series of nuclear fusions and revisions converts silicon nuclei into iron nuclei within a week. This progression is forced along because such ultra-high core temperatures encourage production of a great many neutrinos that leave the star at about the speed of light. In comparison, the electromagnetic photons produced at a star's core are so endlessly absorbed and reemitted by the surrounding stellar plasma that their energy may only arrive at the stellar surface a million years later.

Once its silicon fuel has been consumed, the huge and again collapsing star rapidly builds up sufficient heat and pressure within its Earth-size core to force further fusion of iron—but iron nuclei are as dense as atomic nuclei ever get. So that overwhelming heat and pressure rapidly crushes the protons and electrons of iron together to form neutrons instead. Since those neutrons occupy far less space than the electrically charged protons and electrons that preceded them, the star's iron core collapses into a single ball of neutrons perhaps 10 to 20 kilometers in diameter—which leaves the outer, still fusing and suddenly unsupported star layers plunging many hundred thousand kilometers toward their swiftly shrinking core. The inevitable outcome of this catastrophic collapse is an explosive release of gravitational energy with star parts being blown all over kingdom come. But for one brief and shining moment (a tenth of a second, actually), the energy released by that *supernova* (99% as neutrinos) can exceed the output of all other stars in the Universe combined.

That overwhelming release of gravitational energy almost instantly creates a great many heavier atomic nuclei as well, starting with cobalt and nickel. Because even these next heavier atoms beyond iron (see Table of Elements) have lower nuclear binding energies than iron, their construction through fusion of lighter nuclei must await that catastrophic moment of *endothermic* fusion during which they can *absorb* a great deal of energy while being produced. In this messy fashion, over the lifetime of our galaxy, about a quarter billion of those most massive extravagant exploding stars have enriched their surroundings in the Milky Way with elements heavier than helium (many of these elements were created suddenly through neutron bombardment and other nuclear collisions, the rest formed later during radioactive decays). The waves of energy, dust and debris that overwhelm the neighborhood at such times are thought to destabilize nearby molecular clouds and thereby encourage new star formation.

Naturally Radioactive Isotopes Release Star Energy At Predictable Rates

Given the slapdash manner in which those less densely packed nuclei are constructed, it is not surprising that many explosively created heavier elements are unstable (hence radioactive) isotopes. Indeed, all elements heavier than Bismuth ($^{209}_{83}$Bi) are radioactive. Although some newly created radioactive isotopes will reliably fly apart within moments or centuries, the half-life of others may be billions of years. Radioactive decay inside of Earth now produces only ⅓ as much heat as it did perhaps 3.5 billion years ago—however, nuclear energy being released by those radioactive elements still represents Big Bang gravitational energy crammed by fusion into atomic nuclei during mighty stellar explosions of long ago. An additional 20–40% of Earth's present internal heat production is due to another phase change—this time the comparatively low-temperature central solidification (freezing) of Earth's molten iron core. That heat release represents yet another delayed payback on the gravitational energy invested in compression and melting of infalling material during Earth's original accumulation. But were it not for ongoing heat production by spontaneous nuclear fission, Earth's present-day subsurface temperatures would have been reached within 20 to 40 million years rather than after 4.5 billion years. Indeed, before radioactive decay was discovered, Lord Kelvin insisted that both Earth and Sun could only have been cooling for about 25 million years. This dismayed Darwin and many geologists who were unable to imagine Earth's complexity arising within such a short time span by any processes then known.

It is evident that temperatures rise as one penetrates even a few hundred feet underground. Even the hardiest South African gold miner cannot tolerate the heat encountered at depths greater than about 3 miles. That Hades effect has been known for thousands of years. The 30 terawatts (30 trillion watts—remember watts are a measure of energy output *per second*) of heat currently being conducted and convected up through Earth's viscous mantle also shifts continents and drives ocean floor crustal plates outward from mid-ocean ridges at speeds of several centimeters per year (roughly the rate at which your fingernails grow), thereby producing mountain ranges, earthquakes, volcanoes and other changes in Earth's surface. The dynamo-like churning of its outer (still liquid) iron core also accounts for Earth's constantly changing (occasionally even reversed) magnetic field. These processes have slowed as Earth's radioactive heat output (mostly from the decay of uranium and thorium) has fallen. Although impressive, that central heating is dwarfed by the solar energy continuously bathing Earth's surface. For on an annual basis (assuming an average world-wide *albedo* or energy reflection back into space of over 30%) Alaska alone receives more heat from sunlight (at about 200 watts per square meter) than the entire Earth produces internally. Even the temperatures of permanently frozen northern Alaskan soils (*permafrost*) are dominated by solar energy inputs rather than the small fraction of Earth's total heat production rising from below. Nevertheless, the hot (up to 110°C) crude oil drawn from three thousand meters depth certainly flows more easily than its nearly congealed equivalent found in shallower Alaskan deposits.

The known half-lives of radioactive elements can be used to radiometrically date rocks in which those elements are found. For example, it has repeatedly been observed that a stable end product of the radioactive decay of a certain uranium isotope will be a certain lead isotope. So if a rock fragment being dated contains atoms of those specific uranium and lead isotopes in equal number, then that rock fragment must be at least as old as the known half-life of that particular uranium isotope. Such a rock might be far older, of course, as it could have been heated sufficiently within a volcano or deep inside Earth's crust to melt away the lead but not the uranium (just as gold is refined by driving off base metals in vapor form). But a partially lead and partially uranium rock cannot have been laid down that way initially since uranium and lead have such markedly different chemistries and melting points. Furthermore, any specific rock fragment being dated might have been displaced into an older or younger layer or formation, and there are many possible sources of error that can affect the radiometric dating of various ancient materials. But when different sorts of radiometric measurements and other geological and fossil evidence all provide the same age estimate for a certain rock or sedimentary layer, the evidence eventually becomes overwhelming. Interestingly, the oldest rocks yet found upon Earth's surface appear to have formed almost 4 billion years ago, while heat-resistant zircon inclusions within some ancient rocks may be up to 4.2 billion years old.

As for the age of our 4.6 billion year old solar system, that number comes from the careful radiometric dating of meteorites, many of which are samples of the pre-planetary disc of material that once rotated about our sun. Others amongst these meteorites (most readily collected from the Antarctic ice cap) were flung free during massive meteorite impacts on our moon and Mars, or by powerful collisions on or between the asteroids and inner planets (which can be confirmed in many ways, including viewing huge moon craters and direct comparisons with moon rocks returned by the Apollo space missions). Still other meteorites contain grains of silicon carbide and graphite that must have formed in the outer atmospheres of red giant stars (deduced from the ratios of various isotopes that these tiny star samples bear), as well as tiny diamonds created by the explosion of some giant star—perhaps the very same supernova whose shock waves so fatefully triggered the formation of our own solar system from a local interstellar gas cloud. And numerous meteorite analyses confirm the contamination of our prexisting gas cloud—as well as the newborn solar system—by fusion products from hundreds of giant stars over many eons.

Summary

The progressive red shift in light from more distant galaxies provided first evidence for the explosive origin of our Universe. Big Bang theorists probing the conditions of that hypothetical explosion were able to predict a diffuse low-temperature (2.7°K) cosmic microwave background radiation (those who initially discovered it were unaware of that published prediction) as well as the relative abundances of deuterium, helium-3, helium-4 and lithium-7 ($^2_1$H, $^3_2$He, $^4_2$He, $^7_3$Li) that would have been created by such an explosive event. Related nucleosynthesis arguments show that the number of families of neutrinos (based upon $^4_2$He abundance) should be three. All of these predictions have since been confirmed. In addition, very distant astronomical objects are detectably earlier along the pathway of galactic and stellar evolution than those nearby. Therefore, while Big Bang theories remain in ferment and subject to continuous revision, the general concept that the cosmological red shift signals ongoing expansion of space between galaxies seems securely based upon available evidence. And as we shall see, this complicated sequence of events provided all of the necessary materials (electrons, atomic nuclei and energy) for life's chemistry.

Further Reading:

The rapidly evolving topics of cosmology and astrophysics can be followed by reading articles in *Nature* and *Science* (weekly general science journals that include topic reviews, book reviews and editorial comments—expect difficult reading).

Science News (weekly) and *Scientific* American (monthly) provide helpful brief and article length reviews on many of the topics covered in this and subsequent chapters.

Cornell, James, ed. 1989. *Bubbles, Voids and Bumps in Time: The New Cosmology.* Cambridge: Cambridge University Press. 190 pp.

Lederman, Leon and Schramm, David. 1989. *From Quarks to the Cosmos.* New York: Scientific American Library. 242 pp.

Stanley, Steven. 1989. *Earth and Life Through Time.* 2d ed. New York: W.H. Freeman and Company. 689 pp. (useful for this and subsequent chapters)

Peebles P. J. E., Schramm D. N., Turner E. L. and Kron R. G. 1991. "The Case for the Relativistic Hot Big Bang Cosmology". *Nature* 352. pp 769–76.

Stahler, Steven W. July 1991. "The Early Life of Stars". *Scientific American.* pp 48–55.

Halliwell, Jonathan J. Dec. 1991. "Quantum Cosmology and the Creation of the Universe". *Scientific American.* pp 76–85.

Schwinger, Julian. 1986. *Einstein's Legacy.* New York: Scientific American Library. 239 pp.

Gribbin, John and Rees, Martin. 1990. *Cosmic Coincidences.* London: Blackswan. 291 pp.

Kaler, James B. 1992. *Stars.* New York: Scientific American Library. 260 pp.

Turner, Michael S. 1993. "Why is the Temperature of the Universe 2.726 Kelvin?" *Science* 262. pp 861–67.

Barrow, John D. 1992. *Pi in The Sky—Counting, Thinking and Being.* Boston: Little, Brown and Co. 297 pp.

Weinberg, Steven. 1992. *Dreams of a Final Theory.* New York: Pantheon Books. 315 pp.

CHAPTER 2

BEFORE LIFE WAS CHEMISTRY

Overview;... The Origin Of Life On Earth;... Prebiotic Earth;... Life Is Water Based;... Constant Molecular Collisions;... A Mole Is A Very Specific Number Of Molecules;... Water Is Liquid Because... Acid, Alkali And pH;... Oxidation And Reduction—Electron Donor Seeks Electron Grabber And Vice Versa;... First Life Was Anaerobic;... Carbon Chemistry Is The Chemistry Of Life;... Putting Out Fires;... Life Produces And Depends Upon An Unstable Atmosphere;... The Elements of Life.

Overview

Four and a half billion years ago the cooling crust of our planet was rocked by massive earthquakes, flooded by enormous lava flows, overturned by molten iron and other heavy elements sinking toward the center and pounded incessantly by falling star trash. The hot and wet atmosphere lacked free oxygen and was subject to violent electrical storms. Three and three-fourths billion years ago our water-based solute-laden solar-powered grease-coated little ancestors left the first unmistakeable fossil traces of their organized and productive existence. This chapter reviews certain interactions between atoms that undoubtedly played a crucial role in the origin of life on Earth.

The Origin Of Life On Earth

Life on Earth is generally viewed as inexplicable, highly irregular, possibly even miraculous. Although a miraculous origin for life would conveniently bypass the ordinary laws of physics and chemistry, there is absolutely no evidence that any miracle has ever taken place anywhere. Nor has the most intense investigation of life's chemistry revealed even one extraordinary reaction that violates any otherwise valid law of physics or chemistry. Indeed those seeking life's origins frequently replicate life's chemical processes *in vitro* (outside of the living entity) under controlled laboratory conditions in hopes of uncovering chemical events relevant to life's start.

Nevertheless, scientists remain a long way from understanding life, let alone constructing it from scratch within a test tube. And while common experience assures us that it takes life to beget life, common sense also tells us that it all started somewhere, somehow. Not surprisingly, there are many conflicting ideas about where and how life began. Certainly Earth's oldest rocks show no evidence of life being present at their formation and Earth's earliest fossil-bearing rocks carry only traces of the simplest life forms. Increasing structural complexity then becomes commonplace among plant and animal fossils from progressively less ancient strata until the present day when Earth is obviously infested with all forms of life from the uppermost reaches of the atmosphere to many kilometers underground. Evidently life on Earth began very long ago and has gradually become more specialized since that time.

Some scientists attribute life's origin to statistically improbable interactions among preexisting complex chemicals under drastic *prebiotic* (before life) conditions. This implies that even if we could exactly replicate those conditions in an unlimited number of test tubes, we still might have an indefinitely long wait before something stirred or crawled out (of course, if it was based upon almost imperceptible chemical changes and displayed little more vigor and personality than a rock, you might not notice that something underfoot today). Others expect life to appear via numerous alternate pathways wherever energy-acquiring self-replicating structures can survive. Thus they view life as an unavoidable universal process rather than an amazing outcome of some unique local event that requires very specific and convoluted explanations. And some insist that life never originated on Earth at all, that it simply saturates the entire Universe in primal spore or seed form, settling upon all possible locations as mold settles on fresh bread. Such a relegation of life's origins to incredibly ancient times and sufficiently far away places surely fuels speculation—indeed, it gives full freedom to imagine the most wonderful mechanisms and miracles. But with absolutely no evidence for life's extraterrestrial origins (or even of unearthly oversight or governance of our existence) we might as well seek life's beginnings right here on Earth. Indeed, a lot of evidence suggests that life really is indigenous to Earth, including the way it has developed and prospered with total dependence upon locally available rather than exotic materials and processes—which also implies that exotic life forms from elsewhere would (at least initially) be ill-suited for competition with native life forms (especially the bacteria) now flourishing on Earth.

Prebiotic Earth

In view of the drastic prebiotic conditions under which First Life originated and prospered, it would likely find Earth's present surface environment rather unbearable. Yet any or all of the prevailing conditions on prebiotic Earth may have been essential, helpful, irrelevant or directly harmful to First Life. Perhaps only the very unusual circumstances of some isolated microenvironment deep underground made life possible at all. More likely, however, prevalent conditions mattered a lot in the assembly and survival of First Life. So what were the chemical possibilities on our enlarging Earth as it swept up nearby matter from within the rotating solar disc? Well, Earth's continuous battering by huge numbers of gravitationally attracted meteorites of all sizes must have vaporized significant portions of those colliding rocks. That would have released gases such as CO, CO_2, SO_2, H_2O, H_2, N_2, H_2S, COS, CS_2 and HCN into Earth's growing atmosphere, while also producing various hydrocarbons and reactive carbon-containing molecules including ethanol (CH_3CH_2OH).[1] And Earth's developing atmosphere probably included many gases such as methane (CH_4) and ammonia (NH_3) that are still prevalent on the outer gas planets of our solar system as well as in interstellar space. Catastrophic collisions with larger comets and meteorites undoubtedly helped to heat Earth's surface toward its melting point (3000–3500 °K). One particularly cataclysmic impact with a planet-sized object apparently gave rise to the moon while also blowing away Earth's *original* atmosphere—which included far greater volumes of *noble* (chemically inert) gases such as helium and neon. The moon's persistent gravitational effect then stabilized Earth's spin axis (obliquity with respect to the sun) and thereby prevented Earth's *climate* from varying chaotically.

Our planet's cooling surface included about 90 different elements. Some of those atoms entered readily into an endless variety of molecular arrangements based upon shared electron bonds. Others were held together ionically, often in crystalline form, as part of Earth's rapidly forming rocky crust. In addition to lava and rocky debris, the weathering of early continents undoubtedly produced sand, silt and clay (but no *soil*, since that term implies live organic content). Innumerable bodies of standing or flowing water above and under the ground must have offered our early water-based ancestors access to every possible dissolved mineral, temperature and habitat. Certain of the endlessly varied and widely available clay or iron pyrite (fool's gold) surfaces may have provided essential templates for particular molecular syntheses and replications. Energy for production of life's complex molecules came from frequent electrical storms and intense solar ultraviolet light that

---

[1] Mukhin, L. et al. 1993. "Origin of Precursors or Organic Molecules During Evaporation of Meteorites and Mafic Terrestrial Rocks". *Nature* 340. pp 46–47.

penetrated Earth's thick clouds. Despite sunlight-induced disruption of water vapor in the upper atmosphere, the warm and wet lower atmosphere lacked free oxygen—therefore many fragile and complex molecules persisted that might have deteriorated in an oxygen-rich atmosphere. Perhaps life began almost at once. Possibly it took a long while to get started. Most likely it was extinguished repeatedly before it finally could endure. Indeed, the mere fact that life has continued against all apparent odds strongly suggests that young Earth was pregnant with possibilities for new life.

Life Is Water-Based

Life on Earth depends upon water (H_2O) and water accounts for about two-thirds of your body weight. Although the primordial protons and electrons of every hydrogen atom in the Universe were fully formed within four minutes after the Big Bang, water's oxygen atoms only entered production following first starlight. More specifically, Earth's oxygen atoms were individually created deep within countless overweight stars through a complex series of energy-releasing steps that allowed the strong nuclear interaction to fuse eight protons with a similar number of neutrons. Upon being blown out of their home star, those newly formed oxygen nuclei quickly attracted a full inner shell of two electrons, while gathering just six electrons for the *second* (outermost or *valence) shell* which comfortably seats eight. Therefore every electrically neutral oxygen atom avidly seeks two additional electrons in order to complete that second and outermost electron shell. When unable to find more generous partners, oxygen atoms will join in *covalently double-bonded* pairs as molecular oxygen (O_2 or $O{=}O$)—a relationship that requires each oxygen atom to place two of its precious electrons into joint custody so both partners can complete their second shell.

Covalently-bound atoms are stuck with each other. While they may twist and vibrate a bit within the confines of this arrangement, they hesitate to break it off because of the valuable property that they share (those sparkling electrons). Nonetheless, few relationships are eternal, so if some fast-traveling group of hydrogen molecules happens upon a bunch of oxygen molecules lusting for more electrons, it does not take much heat (a match perhaps, or a spark of static electricity) to boost local vibrations and collisions to a point where some highly reactive atoms can break free and change partners. Such a reaction will tend to continue in a self-sustaining and markedly *exothermic* (heat releasing) fashion—indeed, the molecular products of this chemical reaction are so much more stable than its starting reagents that H_2 gas combines explosively with O_2 gas to form water. Having thus filled their outermost electron shells and established a more stable relationship (at a lower energy state), each oxygen atom with its two covalently bound hydrogen partners becomes quite non-reactive. This rapid and complete chemical reaction (H_2 plus ½ O_2 becomes H_2O plus a lot of heat) explains why pilots prefer to fill their lighter-than-air dirigibles with non-reactive helium (^4_2He) even though that costly gas weighs twice as much as hydrogen ($^1_1H_2$).

Constant Molecular Collisions

Atoms and molecules in the gaseous state collide continuously—therefore they strike one another or the container wall or any intruding object such as your finger with the same average impact—which means lighter atoms and smaller molecules must move far more rapidly than their more massive cousins. Furthermore, average molecular velocity always rises with temperature (indeed, molecular velocity and temperature define each other) so hot air balloons go up because their enclosed air molecules hit harder and therefore occupy more space (hence weigh less per unit volume) than an equal number of cooler gas molecules outside. Despite making up three fourths of all visible mass in the Universe, free hydrogen molecules are rarely found within Earth's atmosphere—not only are those lightweight H_2 molecules very reactive, they also travel at extremely high average velocities. So if some nearby O_2 doesn't grab it first, several strong bumps in an outward direction are likely to kick an H_2 right out of Earth's dilute upper atmosphere—the *escape velocity* necessary for any molecule or spaceship to escape from Earth's gravitational pull diminishes with increasing altitude, being about 1.4 times the velocity required to maintain a circular orbit at that altitude (disregarding air resistance). From Earth's surface that escape velocity would be about 11 kilometers per second (7 miles per second or over 25,000 miles per hour). Escape velocity from our moon's surface is only 2.4 kilometers/sec (1.5 miles/sec)—close enough to the average speed of its atmospheric gas molecules so that the moon cannot long retain primordial gases still leaking from its interior or the other gases produced by radioactive decay of heavier atoms or during meteorite impacts.

An ordinary gas molecule at sea level in Earth's atmosphere must endure over ten thousand collisions during the exceedingly brief moment that it takes to travel one millimeter. It is even worse in liquid water, where an H_2O molecule will suffer literally millions of crashes along any one millimeter route at a staggering rate of 10^{13} (10 trillion) collisions per second. Such an immense number of collisions so encourages mixing and matching that nearby molecules in any water solution will surely meet and interact if suited for one another—and it only takes a *picosecond*

(10^{-12} second, or a trillionth of a second) to make or break a covalent molecular bond. When hot volcanic rock containing ferrous iron contacts sea water, each hot Fe^{++} will rip the oxygen heart right out of some nearby water molecule, thereby releasing its hydrogen. Solar photons similarly disrupt water molecules in the upper atmosphere. But the customary collisional recapture of free hydrogen molecules by atmospheric oxygen has allowed Earth to retain at least a third of its original H_2O allotment until now.

However, our sun is gradually becoming hotter and brighter as its helium production raises the density and thus temperature and fusion rate at the solar core. As Earth's *stratosphere* (upper atmosphere) then slowly warms and becomes more humid, solar ultraviolet light above the ozone layer will increasingly photo-dissociate that water vapor—the resulting escape of free hydrogen to outer space should eliminate Earth's remaining water within the next 2.5 billion years (a couple of billion years before our sun enters its red giant phase). Of course, Earth's higher life forms are unlikely to survive those rising temperatures for even another billion years unless Earth can be partially shielded from that enhanced sunlight.[2] But you ought not worry too much about that prediction since catastrophic cometary or asteroid impacts (plan on one every 500,000 years or so) will probably have eliminated another two thousand dominant life-forms or major civilizations by then. Furthermore, all of the above eventualities pose far less risk to you and your loved ones than your driving, your cooking, your slippery bathtub, your medicine cabinet and human overpopulation in general.

Microscopic particles of bacterial size (barely visible through an ordinary light microscope) are seen to vibrate or dance about when viewed in a water droplet. That *Brownian motion* is due to cumulative variations in the random impacts of millions of water molecules from all sides. Not knowing that Brown and others had already observed this phenomenon, Einstein predicted (in 1905) that if atoms and molecules were actual physical entities rather than mere convenient fictions, an investigation of the movements of tiny particles suspended in water should confirm both the existence of molecules and their calculated size. Perrin then measured those movements with such accuracy that he was able to calculate Avogadro's Number (see below). And modern techniques such as near-field optics, scanning tunneling electron microscopy and atomic force microscopy even allow the imaging and manipulation of individual atoms and molecules. In any case, at identical temperatures and pressures, a specific volume of oxygen gas (O_2) weighs 16 times more than an equal volume of hydrogen gas (H_2)—so hydrogen is far lighter than air. More generally, equal volumes of any atomic or molecular gas at equal tem-

2 Caldeira, Ken and Kasting, James F. 1992. "The Life Span of The Biosphere Revisited". *Nature* 360. pp 721–23.

peratures and pressures must contain equal numbers of independently moving atoms or molecules. That rule, known as *Avogadro's Law*, was proposed a century before Einstein's insight and Perrin's measurements.

A Mole Is A Very Specific Number Of Molecules

The output of any chemical reaction is limited by the number of molecules of each substance that are available to interact. This number of molecules can be determined from the weight of each substance involved but such calculations are cumbersome. It is far easier to deal in portions or multiples of the specific huge *Avogadro's* number (6.02×10^{23}) of molecules that defines *a mole*. A mole of any pure substance is *one gram molecular weight*—its molecular weight expressed in grams. In other words, add up the Table of Elements weights of every atom covalently bound within a single molecule of a pure substance, weigh out exactly that many grams of that substance and you will have 6.02×10^{23} molecules or one mole of that substance. And since the *gram atomic weight* of He is 4, we know that 4 grams of helium contains 6.02×10^{23} atoms. Similarly, because one mole of H_2 (2 grams) plus ½ mole of O_2 (16 grams) can react explosively to produce one mole of H_2O (18 grams), we know that 18 grams of H_2O contains that same huge number of molecules.

Water Is Liquid Because...

Life depends upon water being liquid at ordinary temperatures and pressures. Yet H_2O molecules are lighter than O_2 molecules (18 vs. 32 grams/mole) and O_2 is a gas—so why isn't H_2O a gas? It turns out that water molecules strongly attract each other (as we might have guessed from the way a water droplet rounds up as its molecules huddle together). This self-attraction or *surface tension* of water comes about because every water molecule is bent ($H \diagup O \diagdown H$) rather than straight (H—O—H). In addition, each oxygen atom has seven protons more than either of its hydrogen partners, so the covalently shared electrons of a water molecule spend more time around their O than about either H. That relative excess of electrons in current possession leaves the O mid-section of every H_2O somewhat negative while both H end-sections of every water molecule bear a slightly positive charge. Since both H's of each bent water molecule can be viewed as offset to the same side of the O, the entire water molecule has one sort-of-positive side (the H's) and a sort-of-negative side (the O). That slight separation of charges makes H_2O a *polar* molecule which will orient itself with respect to a strong electrical field. However, being electrically neutral overall (thus not an ion), a water molecule is not pulled toward a positive or a negative electrode in order to deposit or pick up an electron.

Now the interesting part. Put a bunch of water molecules together and every relatively negative O on one molecule will attract the relatively positive H ends of its neighboring water molecules while each of its own covalently-bound H's are similarly drawn to some nearby relatively-negative O. Thus despite all of its bopping about, a water molecule will on average be *hydrogen bonded* to 3.4 closely adjacent water molecules. Although covalent bonds are about twenty times stronger than such H-bonds, the many H-bonds between water molecules should make water quite stiff (and it surely felt stiff on your last belly-flop into the pool). But fortunately there seems just enough room for a fifth water molecule to intermittently squeeze in between its hydrogen-bonded brethren—thereby sharing and weakening one of their hydrogen bonds. So some hydrogen-bonded water molecules are always ready to rotate out of their current positions, which may explain why water flows so easily—and more rapidly under high pressure.

On the other hand, water stiffens when stretched around a *hydrophobic* (nonwettable) particle, due to diminished opportunities for the fifth molecule to crowd in. Similarly, each liquid water molecule moves increasingly slowly and becomes more securely hydrogen-bonded to four others during the formation of ice. Such an orderly inclusion of the fifth water molecule causes some expansion of water's volume with freezing. That is why ice floats on liquid water while other solids (e.g. Earth's solid iron core) tend to shrink and settle out beneath their melted phase as they cool. Life on Earth would surely be far different if ice sank—living things might be restricted to the tropics—or maybe Earth would be a frozen lifeless ball with permanently snowy surfaces reflecting incoming sunlight back into space and little atmospheric H_2O to help retain its heat (see below).

Water's hydrogen bonds also explain why it takes so much molecular velocity (heat) to fling each water-loving H_2O molecule out of liquid water as free gaseous H_2O. Thus evaporation causes cooling because any water molecule that does manage to evaporate must have been traveling far more rapidly than most of those remaining in liquid form—since temperature is defined by average molecular velocity, each high-speed departure of a vaporized H_2O molecule reduces the average velocity of the liquid water molecules remaining. Furthermore, water molecules that contain $^{16}_{8}O$ generally travel faster than those burdened with the two extra neutrons of the $^{18}_{8}O$ isotope. So during Ice Ages when more of Earth's water became tied up in snow and ice, sea water and sea life included a higher percentage of the less easily evaporated $^{18}_{8}O$ isotope. Water molecules are occasionally bumped directly out of ice in the process known as *sublimation*. Thus ice may evaporate gradually without ever passing through a visibly liquid water phase.

Acid, Alkali And pH

A liter of pure water weighs 1000 grams so it contains 55.5 moles of H_2O weighing 18 grams per mole. At any moment, a few of the water molecules in such a huge collection will have been split into H^+ and OH^- by an unusually violent intermolecular collision (note that the H^+ ion leaves without its electron). Those newly formed *hydrogen* and *hydroxyl ions* immediately attract a personal shell of nearby polar water molecules, for the somewhat-positive H ends of water molecules find OH^- exceedingly attractive while the relatively-negative O sides of nearby water molecules are naturally drawn to any free hydrogen ion (H^+). Although impermanent, such shells of water tend to separate dissolved ions from their original partners. This allows positive and negative ions to roam about in solution quite independently of each other. However, as concentrations of H^+ and OH^- rise, these ions are increasingly likely to meet and recombine as H_2O. Any liter of pure water therefore maintains steady low concentrations of H^+ and OH^- ions (10^{-7} or 0.0000001 moles of each, which defines a *neutral* solution). Adding more H^+ ions makes a neutral solution *acidic* while any excess of OH^- ions creates an *alkaline* or *basic* condition.

Furthermore, as the concentration of one ion rises, the concentration of the other must decline at an equal rate since the outnumbered ion is then more likely to encounter a suitable partner and interact to become water again. Therefore the molar concentration of H^+ ions multiplied by the molar concentration of OH^- ions is a constant number (10^{-14}). So if we know the H^+ concentration within any *aqueous solution* (meaning water with something dissolved in it), we can figure out the OH^- concentration without being told. For example, an H^+ concentration of 10^{-1} moles per liter (0.1 moles per liter) tells us that the OH^- concentration must be very small (10^{-13} moles per liter), so this is a very acidic solution. Similarly, an H^+ concentration of 10^{-11} means an OH^- concentration of 10^{-3} (0.001 moles of OH^- per liter), a very alkaline solution.

By custom, the term *pH* is used to express the hydrogen ion concentration of any solution with the ever-present "10" and its minus exponent sign left out for simplicity. Thus a 10^{-7} molar H^+ concentration has a pH of 7 which identifies a neutral solution, neither acidic nor basic. Any pH less than 7 represents an acidic solution while any pH greater than 7 refers to an alkaline solution. A pH of less than one is rarely encountered in biology. A pH of zero (10^0) would mean a 1 molar solution of H^+. A negative pH simply implies very concentrated acid. Effective acidity rises more rapidly than expected at minus pH's, possibly due to the absence of OH^- ions and reduced shielding of H^+ ions by water molecules as the relative H_2O concentration decreases. Negative pH's have no relevance to biology and will not be mentioned again.

A molecule is considered a *strong acid* if it ionizes in solution to release at least 1 mole (1 gram atom, actually) of H^+ per mole of that substance. Both HCl (hydrochloric acid) and H_2SO_4 (sulphuric acid) are examples of strong acids. HCl in solution separates completely into H^+ and Cl^-. Similarly H_2SO_4 in solution becomes H^+ and HSO_4^-. On the other hand, HSO_4^- is a *weak acid* and so is H_2CO_3 (carbonic acid) because a mole of either acid releases less than one mole of H^+ ion into the solution. In other words, some weak-acid molecules retain the original HSO_4^- or H_2CO_3 configuration in solution while others split to release H^+ and sulphate ($SO_4^=$) or H^+ and bicarbonate (HCO_3^-)—such partially dissociated weak-acid molecules have less effect on the pH of a solution than an equal number of fully dissociated strong-acid molecules.

To increase the hydrogen ion concentration (reduce the pH) of any solution, just add acid. The preexisting pH and quantity of acid added will determine whether that entire volume of solution then becomes acidic or merely less alkaline. Since certain molecules may act as proton donors at one pH, yet become proton acceptors under more acidic conditions, it is especially important to specify pH when referring to a weak acid or weak base. If pH is not mentioned during discussions of human biology, a *normal* (healthy human) *arterial blood* pH of about 7.4 is assumed. But regardless of pH, any molecule that releases a proton acts as an acid while any molecule that ties up a proton serves as a base.

Water can become *superheated* far underground when surrounding rock pressures prevent its turning to steam. The excessive violence of superheated high-pressure intermolecular collisions markedly enhances the total concentration of H^+ and OH^- ions. Because it reacts readily with hydrophobic molecules, superheated water may play an important role in the formation of coal and oil. And the cooling of solute-laden superheated water as it rises within Earth's fissures (cooling markedly reduces that water's ability to hold minerals in solution) has produced many valuable ore bodies. Fortunately life has come up with ways to utilize water as a solvent and reagent at temperatures and pressures more tolerable to you.

Oxidation And Reduction—Electron Donor Seeks Electron Grabber And Vice Versa

Atoms that need just one more electron to complete an outer electron shell can usually grab that electron from some nearby hydrogen atom, for all other atoms bear more protons (include more positive charges in their nucleus) than hydrogen. So hydrogen is considered a strong *reducing agent*. This means H usually emerges naked as an H^+ (a lone proton) from any electron-grabbing contest, while the atom that absconds with hydrogen's negative electron will have had its overall ionic charge *reduced* (made less positive). An atom is *oxidized* when it donates an electron and thereby acquires a *more positive* charge (as when H became H^+ in the above example). Oxygen atoms are especially *strong oxidizing agents* (avid *electron collectors*). Note that oxidation and reduction always go together and that a strong oxidizing agent tends to get itself reduced while a strong reducing agent is likely to become oxidized. In biological systems, any atom donating an electron must meet some other atom willing to take it, for otherwise electrons would pile up or need to disappear and reappear.

Spontaneous oxidation tends to be a self-sustaining exothermic reaction that proceeds to completion—which is another way of saying that the products of ordinary combustion are far more stable than the initial *reagents* (interacting molecules). However, simple oxidation can be reversed by reversing the flow of energy, as when we heat iron oxides (rust or iron ore) in a reducing atmosphere to recover metallic iron or when plants combine solar energy, water and CO_2 into carbohydrates (CO_2 plus H_2O plus solar energy becomes CH_2O plus O_2). In contrast to the plant production of carbohydrate, some photosynthetic bacteria utilize solar energy simply to create a proton gradient (separation of charges) across a membrane, thus making a minibattery of themselves in order to manufacture the high-energy phosphates they need.

In Chapter One we noted that lithium is one of the few reducing agents stronger than hydrogen. Indeed, an electrically neutral lithium atom remains highly reactive until another atom accepts its single electron from the second shell. Thereafter, that lithium atom becomes a peaceful Li^+ ion. When lithium hydride (LiH) is added to H_2O, the H^- (hydride ion) joins with an H^+ (hydrogen ion) to form H_2 gas, leaving Li^+OH^- in solution—this occurs simply because the products (output) of this reaction are more stable (less reactive or at a lower energy state) than the inputs. Incidentally, lithium salts are useful in the treatment of manic depressive states, and the hydride ion is a safe and convenient way to package a couple of electrons for intracellular energy transfers (see Metabolism).

First Life Was Anaerobic

The atmosphere of prebiotic Earth included no free O_2 since that electron-grabbing element readily entered into more stable covalent relationships such as H_2O, CO, CO_2, SiO_2, FeO, SO_2, $SO_4^=$ and all the other oxides of the *inorganic* (non-living) world. So First Life living *anaerobically* could directly extract only one-fourteenth as much stored solar energy from the covalent chemical bonds of a glucose molecule as modern aerobic life does by dumping its *food electrons* onto molecular oxygen via the iron-containing electron transfer chain. For the double-bonded oxygen molecules ($O=O$) discarded by plants now make up almost 21% of Earth's atmosphere while triple-bonded nitrogen ($^{14}_{7}N_2$ or $N\equiv N$) accounts for another 78%—indeed, those very inert N_2 molecules are among the few gases able to coexist with oxygen on a long-term basis. Or to put it another way, oxygen and nitrogen atoms hold onto their outer electrons more avidly than do the atoms of other elements. The last 1% of Earth's present atmosphere is $^{40}_{18}$ Argon—another unreactive (bearing a full outer electron shell in the uncharged state) noble gas produced by radioactive decay of the rare $^{40}_{19}$ potassium isotope. Earth's atmosphere also includes about 0.035% carbon dioxide—up from about 0.030% CO_2 in 1920. The prebiotic world of four billion years ago had a far more reducing atmosphere with CO_2 levels possibly 1000 times higher. In those days the sun had a cooler less-dense core and only 70% of its current energy output so those high atmospheric CO_2 levels may have provided the necessary *greenhouse effect* that prevented Earth's waters from freezing. For in addition to being *transparent* (allowing free entry to the visible light photons bringing in solar energy), gas molecules such as H_2O, CO_2, CH_4 (methane), NH_3 (ammonia) and O_3 (ozone) help to preserve Earth's warmth by blocking the escape of infra-red photons (all gas molecules composed of 3 or more covalently bound atoms absorb infra-red in this fashion—nowadays Earth's relatively high concentration of atmospheric water vapor provides nearly twice as much greenhouse effect as all other greenhouse gases combined).

Long-lasting chlorofluorocarbons (CFC's) are very stable non-toxic molecules (used as refrigerants, fire extinguishers and cleaning fluids) whose atmospheric levels were rising at up to 4% per year until recently. Besides being effective greenhouse gases, CFC's also reduce stratospheric ozone levels when their eventual breakdown releases ozone-destructive chlorine (oceanic and volcanic chlorides and bromides usually cannot reach the stratosphere before encountering water vapor and being returned as precipitation). While the net effect of chlorofluorocarbons on Earth's heat balance remains uncertain, similar gases have been suggested for possible production and release on Mars (which is farther than Earth from the sun) in order to warm the Martian surface and melt its frozen CO_2 layer. Release of that CO_2 along with water vapor and other greenhouse gases might then warm Mars sufficiently allow survival of photosynthetic bacteria and perhaps plants shipped in from Earth.[3] Such interplanetary speculation may eventually bear fruit (or at least bring germs)

3 McKay, C. P. et al. 1991. "Making Mars Habitable". *Nature* 352. pp 489-96.

but even on Earth there are too many known and unknown interacting variables and feedbacks for anyone to be sure how Earth's rising atmospheric CO_2 levels (due to combustion of hydrocarbon fossil fuels) will play out. For example, while it seems likely that industrial and volcanic aerosols (dusts, sulphates, etc.) can cool Earth by reflecting incoming sunlight, those aerosols also serve as water-condensation nuclei that increase cloud formation, which can both warm Earth by reducing its heat losses and cool Earth by reflecting sunlight—but warmer oceans in winter would build up atmospheric water and so increase snowfall at higher latitudes (especially above 65°N latitude)—and any increase in northern winter snowfalls would require additional summer warmth to allow melt-off—yet ice and snow reflect sunlight so they encourage cooling and reduced melting, especially at higher elevations. Furthermore, the current tilt of Earth's axis with respect to the sun ought to result in cooler Northern Hemisphere summers anyhow.

In addition, any change in sea level consequent to major ice sheet formation or melting could markedly alter continental erosion and nutrient release, which might significantly affect coral production and oceanic plant uptake or release of CO_2 as well as methane release from both land and sea (see below). Even the changing positions of Earth's drifting continents and the associated location, growth and orientation of major mountain ranges can have great impact on world climate. Thus the Rocky Mountains, the Andes and the high Tibetan Plateau significantly alter atmospheric heat transfers and global precipitation patterns. And while warm Gulf Stream waters presently are cooled and concentrated by evaporation in the North Atlantic until eventually they sink and flow back south at deeper levels, this circulation may be subject to rather sudden changes. So if North Atlantic surface waters become sufficiently cooled and freshened by a long term enhancement of river water inflows due to additional precipitation and melting of glacier ice (which seems to be the current trend), this important tropics-to-high-latitudes heat transfer process could stop as deep ocean currents enter the Atlantic from Antarctica. During times of marked continental uplift consequent to collisions between Earth's crustal plates, new mountainous areas undergo accelerated erosion and sedimentation. High atmospheric CO_2 levels increase rainfall acidity which hastens the dissolution of surface silicates (rocks). Calcium (and magnesium) ions released by such weathering tend to coprecipitate with CO_2 as follows: $CaSiO_3$ (rock) plus CO_2 goes to $CaCO_3$ (calcium carbonate or limestone) plus SiO_2 (sand).

Increased release of rock-based plant nutrients and high CO_2 levels encourage the production of more organic material. Enhanced erosion and sedimentation tend to bury more organic and inorganic carbon compounds, thereby reducing atmospheric CO_2 levels. High sea-bottom water pressures tend to dissolve sea-floor carbonates although shallow seas usually are saturated in carbonates—indeed the oceans contain 60 times more inorganic carbon than the atmosphere. Therefore climate-related changes in the production of organic molecules or carbonates can affect oceanic pH and alter atmospheric CO_2 levels (see Respiration). Deep-ocean crustal plates submerging below the edges of continents carry along copious sea floor deposits of carbon compounds and also huge quantities of Na^+ and K^+ bound up as silicates—that *subduction* (burial and heating) eventually results in metamorphic creation of $CaSiO_3$ (rock again) as well as release of much CO_2 from continental rim volcanoes. Note that our quick glance at known or postulated greenhouse-related effects and their complex cycles and feedbacks does not even consider any so-far unpredictable but possibly major variations in total solar radiation (which could overwhelm all of these secondary processes and effects). While many of these specific and reiterated incremental changes may turn out to have little impact upon Earth's complex and chaotic climate, the records of past climates now being extracted from undisturbed sediment samples and ice cores suggest that similar gradual changes have occasionally boosted Earth's complex heat-transfer system over into an entirely different stable state (such as a new Ice Age).

Earth's original oceans contained plenty of partially oxidised *ferrous* (or Fe^{++}) iron ions. But as free oxygen became a significant fraction of Earth's atmosphere about two billion years ago, that dissolved ferrous iron was oxidized to far less soluble *ferric* (Fe^{+++}) iron which then precipitated out in those valuable *banded iron formations* (somewhat comparable to persistent bathtub stains) along the margins of ancient seas. So modern oceans located far from eroding continental margins generally contain just traces of dissolved iron, and newly dissolved ferrous iron soon precipitates out of solution adjacent to its deep-ocean-floor hydrothermal-vent source (thereby depleting dissolved O_2 in that locale as well). Iron depletion may limit the biological productivity of the Southern and equatorial Pacific Oceans and other remote deep-sea areas where any available iron generally arrives as windborne dust from the arid lands of Africa and Asia. Some have speculated that a tanker-load of iron might markedly enhance the bioproductivity of those seas (at least temporarily) but possible adverse outcomes of such world-scale interventions include unrestricted algal blooms (which can have direct toxic effects on other sea life while also depleting dissolved O_2 during oxidative decay of all that additional plant matter). Many other unforeseen environmental problems could result from such an endeavor since ocean-based life processes are still so poorly understood. Interestingly, the Southern Ocean accounts for about 15% of Earth's annual production of carbon compounds. An estimated 25% of those 3.5 *giga*tons (billions of tons) of biomolecules are consumed by birds and mammals at the top of that food chain.

Carbon Chemistry Is The Chemistry Of Life

You will note in the Periodic Table of Elements that carbon ($^{12}_{6}C$) has an outermost shell containing 4 electrons. So does Silicon ($^{28}_{14}Si$), an important dietary trace element for mammals. But while Si resembles C in its ability to form all sorts of interesting bonds with many different atoms (including other Si atoms), carbon has a major advantage in being smaller so it can fit comfortably between more different atoms. Furthermore, carbon compounds are far more stable in water than similar constructs of silicon. Carbon also has a unique ability to form durable linear and far more complex 3-dimensional structures (including hollow balls of 60 or more carbon atoms and extended carbon tubes) based upon carbon-to-carbon bonds. Many ring-like carbon molecules (as well as the above-mentioned more complex structures known as Buckminster Fullerenes) are sufficiently stable to accumulate in the soot resulting from incomplete oxidation of hydrocarbons. Carbon atoms readily form four single covalent bonds in order to complete their second (outermost) electron shell. Alternatively, a carbon atom can share two or even three pairs of electrons with another carbon atom, thereby creating double or triple covalent bonds. Although single covalent bonds tend to be flexible and mobile, double or triple shared-electron bonds become rigid parts of a molecule. All of which helps to explain how life came to be carbon-based and why the complex carbon-containing molecules of life are referred to as *organic* molecules. Indeed, carbon and hydrogen atoms are such essential participants in organic chemistry that they are often left out to simplify molecular diagrams—the reader is supposed to add hydrogen atoms until the outer ring of each carbon atom is complete (bears eight electrons). That means a molecule written as $C-C$ really is H_3C-CH_3 (ethane), while $C=C$ means $H_2C=CH_2$ (ethylene) and $C\equiv C$ really is $HC\equiv CH$ (acetylene). And carbons are routinely left out of ring or longer linear formulas, being merely indicated ritually by corners, as in ⬡ which is shorthand for a benzene ring—a very stable form of carbon that resists combustion and thereby raises the octane rating of gasolines. So whether a formula is ring-like or linear, you can assume that every corner stands for another carbon atom unless some other element has been written in.

Putting Out fires

The hydrogens of H_2O and the carbon in CO_2 (O=C=O) are fully oxidized. That makes them valuable for putting out fires—as is aluminum hydroxide powder, which releases water and becomes aluminum oxide when heated [$2Al(OH)_3 \rightarrow Al_2O_3$ plus $3H_2O$]. Water has a high heat capacity and it absorbs an additional large amount of heat when turned into steam. But H_2O is dangerous on electrical fires because it conducts electricity and thereby increases the risk of electrical shock for those nearby. Carbon dioxide expands rapidly upon release from pressurized storage in a fire extinguisher at room temperature. That expansion to atmospheric pressure swiftly dilutes its available molecular energy so newly expanded CO_2 is exceedingly cold. Refrigerators compress CFC's or other gases far from their cool compartments and then reexpand those gases adjacent to those cool compartments in order to remove heat. You utilize the same principle when you blow air through your pursed lips so it will expand and cool the soup, or breathe out through your wide-open mouth to warm your cold hands. A mole of CO_2 weighs 44 grams which is heavy for a gas. That cold dense just-expanded CO_2 gas therefore settles down nicely and smothers a fire by preventing oxygen access while also cooling the flames until easily oxidized *hydrocarbon gases* (vaporized molecules in which carbon is covalently bonded to hydrogen) no longer leave the fuel to support ongoing combustion. Since ordinary fires require continuous access to free atmospheric oxygen concentrations above 15%, we can be certain that no fires raged on prebiotic Earth. On the other hand, Lovelock suggests that atmospheric oxygen concentrations above 25% would sufficiently boost wildfires to eliminate forests.

Life Produces And Depends Upon An Unstable Atmosphere

Despite their heavy weight, constant molecular collisions cause individual CO_2 molecules to fly randomly about the atmosphere without settling down. Many CO_2 molecules eventually reach the upper atmosphere where they can block infrared radiation escaping from sun-warmed surfaces on Earth. Thus atmospheric CO_2 (currently increasing in concentration at 0.5% per year) serves in the same way as glass in a greenhouse—allowing light to pass in either direction but trapping heat. Of course, greenhouse glass also prevents convective heat losses due to mixing of warm inside and cooler outside air. Ozone (O_3) is formed by the action of solar ultraviolet light upon O_2 in the upper atmosphere. Ozone is also produced in electrical discharges (such as lightning) and by chemical reactions within that photochemical brew known as *smog*. Although *tropospheric* (lower atmospheric) ozone is a strong oxidizing agent that damages leaves and lungs, stratospheric ozone is a fine sun blocker, protecting life on Earth from harmful high-energy photons (solar ultraviolet radiation). Both high-temperature fossil-fuel combustion (in motor vehicles and industry) and biomass combustion (grass and forest fires) contribute to smog and thus to tropospheric ozone production by encouraging the incidental oxidation of atmospheric nitrogen into various nitrogen oxides near Earth's surface. Nitrous oxide (N_2O) is another important greenhouse molecule whose atmospheric levels rising at 0.25% per year—it is generated not only through fossil fuel and biomass combustion but also during breakdown of organic nitrates and nitrites (NO_3^- and NO_2^-) by anaerobic bacteria. And methane (CH_4) is released in huge quantities when anaerobic bacteria metabolize the fermentation products and dead bodies of other bacteria under low-oxygen conditions (e.g. on the sea floor, in rice paddies, swamps, water-logged tundra, land fills, termite mounds and the intestinal tracts of cows). About 2% of life's carbon compounds reach low-oxygen sediments and 95% of that carbon is eventually returned as methane. So only one in every thousand plant-produced organic molecules stays deeply buried (in markedly modified form)—but that is still a very large amount.

Methane concentrations in the atmosphere (now 1.7 parts per million) have been rising for 300 years—currently at almost 1% per year. Methane is quite readily oxidized (subjected to electron extraction) in your gas-burning furnace or by short-lived but highly reactive (because the oxygen needs just one more electron) *neutral hydroxyl* ($^\cdot OH$) *radicals* in the atmosphere, or in well-illuminated soils where plant photosynthesis provides excess oxygen. Furthermore, methane is a far more effective greenhouse gas than CO_2. Early life may have helped Earth to maintain a temperature appropriate for life by simultaneously producing a methane-based upper atmospheric smog as it removed CO_2 from the atmosphere. Even current low levels of atmospheric methane contribute about a fourth as much greenhouse warming as atmospheric CO_2. Since the breakdown of atmospheric methane releases CO_2 and H_2O (both greenhouse gases), the net effect of that particular oxidation on Earth's solar heat balance remains unclear. In comparison, Venus, our neighboring planet on the sunward side, has an out-of-control greenhouse atmosphere with high CO_2 levels and a total atmospheric pressure 90 times that on Earth. Interestingly, Venus and Earth have similar levels of atmospheric nitrogen (N_2) but Venus long ago lost the H_2 of its water into space—allowing the residual O_2 to combine with reduced minerals exposed through volcanic action. Venus lost its oceans by orbiting closer to the sun and lacking photosynthetic life while Earth's oceans (currently equivalent to 257 Earth atmospheres composed solely of water vapor) were saved through photosynthetically released oxygen. Incidentally, the oceans contain 97% of our world's water and the atmosphere includes only 0.001% of that water. The rest is located on or under the land as solid or liquid H_2O—an additional unknown amount (possibly equivalent to several world oceans) is chemically combined within Earth's minerals. Since liquid water absorbs a great deal of CO_2, sea life has been able to deposit about 100 atmospheres worth of CO_2 into Earth's sedimentary rocks. So while Venus and Earth had similar initial endowments, only life's early onset made Earth fit to live on.

At high pressures and low temperatures, methane and water combine to form the solid ice-like *gas hydrates* found under sea floor sediments, under Arctic permafrost and within high-pressure natural gas pipelines. That is why gas pipeline companies must dry their natural gas or add antifreeze before shipping it. A single liter of pressurized gas hydrate (water-caged methane) at the sea floor expands to become 160 liters of methane at the sea surface. It is quite possible that much of the ocean floor deeper than about 500 meters is underlain by a thick layer of gas hydrates. Apparently this methane is produced by bacterial breakdown of organic molecules in an environment totally depleted of O_2 as an electron acceptor. The resident bacteria therefore must reduce nitrate, sulphate, ferric iron and oxidized manganese. Where none of these molecules are available, bacteria can only gain access to the reducing power (stored solar energy) within organic materials by acquiring oxygen and carbon from CO_2 and eliminating extra hydrogen and electrons as methane. Those solid ocean-floor gas hydrates may block the escape of warmer gaseous methane from deeper underground where it is too warm for stable hydrates to form. If heated or mechanically disturbed by an earthquake or oil drilling rig, such a gaseous layer could liquify sea floor sediments and also allow downslope slippage of the overlying hydrates—in the process perhaps toppling an oil rig or releasing a great deal of trapped gas in a potentially explosive belch. Similarly, any reduction in sea floor depth consequent to the massive storage of solid water on land during an Ice Age might sufficiently reduce sea floor water pressures (and thereby destabilize enough sea floor gas hydrates) to help end that Ice Age via the greenhouse effect of newly released methane. These truly enormous undersea stores of gas may include several times the solar energy (reducing power) represented by all known gas, coal and oil reserves on land.

Clearly life has pushed Earth's atmosphere far from equilibrium through removal of CO_2, addition of O_2, N_2 and CH_4, and recapture of H_2 within H_2O. Since Earth's oxidizing atmosphere destroys free methane within about ten years, both the production and destruction of atmospheric methane can be attributed to life. And O_3 can only be present in significant amounts if O_2 already is available. As Lovelock points out, Earth's flourishing life keeps Earth's life flourishing. Indeed, the equilibrium (or rundown) state of the Martian atmosphere with hardly any detectable water or oxygen led Lovelock to predict the absence of life on Mars long before the Viking expeditions were able to land there without further clarifying that particular issue.

Combustion of sulphur-containing fossil fuels gives rise to sulphur dioxide (SO_2) along with oxides of nitrogen. These products produce the sulphuric and nitric acids that contribute to acid rain. Reduced sulphur (H_2S) and oxidized sulphur (SO_2) are released in large quantities from mid-ocean hot spot (e.g. Hawaii) and continental-edge (e.g. Pacific Rim) volcanoes. In addition, a significant percentage of all atmospheric sulphuric acid droplets originate from the dimethyl sulphide released by dying oceanic algae (the same fresh fish odor is detectable on the breath of humans with liver failure). Although sulphuric acid does not persist in the atmosphere (since it tends to serve as a nucleus for water droplet formation), it does encourage the buildup of thick reflective clouds (for example, in the wake of ships). Thus it is possible that a marked reduction in fossil fuel utilization or a switch to low sulphur fuels could temporarily increase greenhouse heating of the Earth by reducing the reflection of incoming solar radiation (from sulphur-nucleated water clouds) back into space. Still, that might be less disruptive than a continued build-up of SO_2 in the industrialized Northern Hemisphere, for eventually this could disturb atmospheric and oceanic interactions between Northern and Southern Hemispheres. But natural levels of SO_2 in the lower atmosphere are essential for returning oceanic sulphur to the land where it is needed by all living things—which again shows how life encourages life.

The Elements Of Life

Viewing the Table of Elements once again, we can make a few additional comments about elements that are essential for life. The only gaseous element we utilize directly is molecular oxygen which must first enter solution in our lungs. Noble gases (elements #2, 10, 18, 36, and 54) clearly are useless in biology—too inert: But take note of their proton counts. Although 2, 10, and 18 are the usual full or noble capacities of the first, second and third electron shells in smaller atoms, the rules of this concentric electron shell game become increasingly complex with heavier elements (those complex rules we henceforth ignore—except see phosphorus below). Two other gas molecules, chlorine ($^{35}_{17}CL_2$) and iodine ($^{127}_{53}I_2$)—as crystals or gas—are far too reactive for biological use until they have filled their incomplete outer ring by stealing an electron to reduce themselves while oxidizing some other atom—the chloride (Cl^-) and iodide (I^-) that result are the nicest non-toxic ions you could ever hope to meet. Nitrogen ($^{14}_{7}N_2$) is a biologically essential gaseous element that remains inaccessible to most organisms or plants until *fixed* (made available by certain bacteria in dissolved, oxidized or reduced form). The principal soluble inorganic forms of biologically useful nitrogen are nitrate (KNO_3, the K^+ salt of nitrate, also known as *saltpetre*) and nitrite (NO_2^-) or ammonium (NH_4^+) ions. Nitric oxide (NO) and nitrous oxide (N_2O) also have biological activity. Although an intact H_2 molecule is very easily oxidized, the triple-bonded gaseous nitrogen molecule is one of the most chemically stable molecules known ($N\equiv N$). Thus all life depends upon the skills of a few specialized sorts of nitrogen-fixing bacteria that can convert atmospheric nitrogen into biologically useful form. The extreme stability of N_2 gas implies that any sudden return to that gas state can release a lot of energy. So nitrogen compounds play an important role in explosives such as *nitroglycerine* (in addition to the chemical energy being released, 4 moles of liquid nitroglycerine turn into 29 moles of gas for a further immediate 1200-fold volume expansion), *hydrazine* rocket fuel and *TNT* (tri-nitro-toluene).

No living thing could regularly incorporate radioactive atoms without suffering reproductively disadvantageous damage to its information-bearing molecules. Since all elements heavier than bismuth ($^{209}_{83}Bi$) are radioactive, they certainly cannot be essential to life. In fact, biologically essential elements tend to come from the light end of the Table of Elements. Those smaller atoms also fit more comfortably within life's complex molecules. Among metals essential to life are sodium ($^{23}_{11}Na$) and potassium ($^{39}_{19}K$). Partially oxidized iron ($^{56}_{26}Fe^{++}$) plays an important role in oxygen storage and transport as well as the transfer of electrons. And certain other metals *catalyse* specific inorganic reactions. That is, they greatly increase the speed of those chemical reactions without themselves being consumed or altered into reaction products. This they do by stabilizing particular molecules in a highly reactive intermediate form that makes the specific reaction very likely to proceed. For example, the catalytic converter of your automobile exhaust system may expose a platinum surface that completes the combustion of passing partially oxidised hydrocarbons without itself being altered or used up. Quite a few different metals play essential biological roles in tiny (trace) quantities. Usually these metal atoms stabilize proteins or serve at the active site of metalloprotein enzymes. But with or without such metal atoms, many proteins and some RNA molecules can serve as exquisitely specific *enzymes* (organic catalysts that are not consumed in the reaction that they expedite—some RNA enzymes also can bring about changes in themselves—see next chapter).

Like lithium, both sodium (Na) and potassium (K) are highly reactive metals. Elemental H, Li, Na and K atoms all carry a single electron in their outermost shell and Li, Na and K are even more powerful reducing agents than hydrogen. So in water, metallic Na or K atoms force their unwanted electron on hydrogen, causing the release of hydrogen from water. In an oxygen-containing atmosphere, the heat released during that oxidation of Na or K with reduction of those H's encourages immediate further oxidation of that newly released hydrogen gas back to water ($2Na$ plus $2H_2O \rightarrow 2Na^+$ plus $2OH^-$ plus H_2 plus heat; followed by H_2 plus $\frac{1}{2}O_2 \rightarrow H_2O$). So the explosive release of energy that occurs when Na or K enter water, along with the rapid reduction/reoxidation of all those hydrogens, simply demonstrates the extreme chemical reactivity of metallic sodium and potassium. But compare that fiery introduction of Na or K atoms to water with the mere salty flavor that results when a comparable number of chemically inert, singly charged, biologically essential Na^+ or K^+ *cations* (meaning positively charged ions that are therefore attracted to any negatively charged electrode or *cathode)* are added to water. Since H^+, Li^+, Na^+ and K^+ differ primarily in size and overlap in their chemical characteristics, they can often be interchanged usefully in biological systems. Presumably the medical benefits of Li^+ relate in some fashion to that element's intermediate position between H^+ and Na^+ in the Table of Elements.

Calcium ($^{40}_{20}Ca$) and magnesium ($^{24}_{12}Mg$) are biologically valuable metals but only in their soluble *divalent* (two charges) cation form. Ca^{++} and Mg^{++} ions also tend to form more durable relationships with *anions* (negatively charged ions that are attracted to a positively charged electrode known as an *anode)*. Thus Ca^{++} and Mg^{++} combine with some multivalent anions to form relatively insoluble crystalline *salts* (ionic solids that can readily produce rigid structures of alternating positive and negative ions such as rocks and bones). Their smaller shells of water molecules allow *mono*valent ions (those with single charges such as Na^+, K^+ and Cl^-) to move through water solutions more easily and independently than divalent ions do.

Organic molecules commonly include carbon, hydrogen, oxygen, nitrogen, phosphorus and sulphur atoms (CHONPS). Elemental phosphorus ($^{31}_{15}P$) is exceedingly reactive. White phosphorus (the most reactive form) burns upon exposure to air, making it handy for smoke screens and antipersonnel weapons. The fully oxidized pentavalent phosphate ion, which is of crucial importance in your biological energy exchanges, is completely ionized at pH 7.4, so it is available as

$$
\begin{array}{ccc}
& O & & O \\
& \parallel & & \parallel \\
O^- - P - O^- & \text{rather than} & HO - P - OH \\
& | & & | \\
& O^- & & OH
\end{array}
$$

Here each of its *5* original outer shell electrons has paired up and P exposes an outer ring of 10 (rather than 8) electrons. Sulphur (also known as *brimstone*) is highly reactive ($^{32}_{16}S$). With just 8 more protons (one electron shell larger) than $^{16}_{8}O$, sulphur can be oxidized as far as biologically important sulphate (SO_4^-) or reduced to hydrogen sulphide (H_2S) which is toxic for humans. Sulphur is present in your body as dissolved sulphate ions and also within certain amino acids—indeed, polypeptide chains (composed of many amino acids) are often held next to each other by covalent disulphide (—S—S—) cross-linkages. In the next chapter we will see how molecules that are useful for life can be built from these atomic elements.

Further Reading:

Stryer, Lubert. 1988. *Biochemistry.* 3rd ed. New York: Freeman. 1089 pp. (useful for this and subsequent chapters).

Lovelock, James. 1988. *The Ages of Gaia.* New York: Bantam Books. 244 pp.

McPhee, John. 1993. *Assembling California.* New York: The Noonday Press. 304 pp.

Levin, Harold L. 1994. *The Earth Through Time.* 4th ed. New York: Harcourt Brace. 651 pp. (useful for this and subsequent chapters)

CHAPTER 3

THE GATHERING INFORMATION STORM

*There Were No Guidelines For Prebiotic Molecules;... Natural Selection
Chooses Without Caring;... Alive, Not Alive Or Dead;... Life's Limits;...
Life Is A Bacteria-based Show;... Sterility;... Inheritance;... Naked DNA
As Free Advice;... Protein-covered Advice;... Parasitic Opinions And
Afterthoughts;... The Rising Information Tide Lifts All Life Before It;...
Advice Comes From All Directions;... The Path Not Taken;... Mind Over
Living Matter And Vice Versa;... Diffusion Of Information;... Gaia And
Homeostasis;... The Thermodynamics Of Life;... Life Is A Four Letter
Word;... Complementary DNA Strands Fit Together;... RNA Utilizes U
Instead Of T;... Those 100,000 Genes Are The Real You;... How The
Modular Tools And Processes Of Life Evolve;... 100,000 Genes Are
Not Enough.*

There Were No Guidelines For Prebiotic Molecules

Far out in interstellar space, cosmic rays ionized H_2 molecules to H_2^+ which
then combined with H to form H_3^+. Oxygen, carbon and the many other elements
making up interstellar dust attracted H_3^+ electrically and utilized its protons. Progressive chemical reactions at those frigid interstellar temperatures led to endless
varieties of complex organic molecules that rained down through prebiotic Earth's
protective reducing atmosphere in countless tons to join far greater numbers of
delicate organic molecules being randomly created under the influence of incoming solar radiation as well as the dozens of new molecular species produced wherever
superheated water or steam met hot carbon-and-mineral-bearing rocks. Innumerable unique clay and iron pyrite surfaces repetitively catalysed formation of yet
again subtly different molecules. And all of these molecules collided continuously,
—joining, breaking up and rejoining under the always different conditions that
could be found within Earth's prebiotic waters.

Natural Selection Chooses Without Caring

Over long ages such complex molecules formed and interacted chaotically. The initial advantage always went to molecules best suited for self-replication under locally unique circumstances, but only rarely could such molecules integrate usefully within the latest self-organizing complex. As countless moles of the least likely molecules gradually accumulated, the most improbable chemical interactions (having perhaps one chance in a trillion of ever taking place) eventually became inevitable, since even a single mole included almost 10^{24} or a trillion trillion such molecules open to all possibilities. At the same time, Earth's hydrocarbons regularly self-assembled tiny bubble-like structures that indiscriminately included some of the self-replicating molecular systems becoming so prevalent within Earth's warm and agitated waters—there was always a faint chance that these enclosed systems might prosper, grow and divide at the expense of molecules that still remained exposed. And radical recombinations of self-replicating molecules were encouraged when enclosed units collided and their membranes fused. Every sort of surface and internalized membrane provided a site for electrochemical storage of solar energy as ordinary physical and chemical processes led to the separation of positive and negative ions.

Molecular complexes able to enhance and utilize such energy stores gained great advantage. Certain *ribonucleic acid* sequences proved so successful at storing information, transferring energy and self-replication that eventually RNA ruled over all organic molecules. But as information became ever more complex and molecular interactions grew increasingly organized, the occasional RNA copying error caused greater difficulties for RNA-dominated mechanisms. In contrast, being a thousand times less prone to error allowed the more stable *complementary* (matching) structure of DNA (a slightly reduced RNA with one oxygen removed from the ribose—hence *deoxy*ribonucleic acid) to overwhelm the competition. Furthermore the unlimited variability that could be achieved when constructing *enzymes* out of twenty different amino acids made *proteins* far more versatile than *ribozymes* (RNA enzymes built using only four sorts of nucleotides) at carrying out life's complex chemistry. Eventually RNA, life's founder molecule, was retained mostly for its central role in life's energy exchanges and its tediously acquired ability to assemble proteins according to DNA-encoded instructions.

So it is nucleotide-inscribed information (as DNA or RNA) that permeates and regulates the living world. All life is based upon and replicates its own unique DNA in endless copies and all life is subject to relentless assault by foreign DNA, both directly and via all of the wild and wonderful competing life forms that DNA may happen to designate. Bits of DNA information spread mindlessly wherever they can gain access or bring advantage. And every living exemplar of inherited DNA instructions becomes yet another fleeting contender in the desperate never-ending battle to achieve reproductive success and thereby define the next generation. Although the battle for survival and control of limited resources inevitably gives rise to "Nature, red in tooth and claw", no harm is intended. Indeed, there is no intent. It simply happens that some DNA sequences designate those whose ancestors succeeded by killing and eating while other DNA sequences inform the primary producers whose ancestors flourished despite such risks. If wolf instructions only persist by victimizing those bearing moose or hare DNA sequences, that has no special significance at the ever-reshuffled DNA level. The meaning and worth of each new DNA message is for the marketplace to determine. But at least the rules never change. "Adapt to available resources. Cooperate and compete to maximize your advantage. Be fruitful, multiply and fill the Earth with your seed, until eventually your particular brand/information-packet/DNA recipe loses shelf space to some more competitive, better-selling variety". By its unique existence in the face of incredible odds, each life form shows how DNA copying errors very occasionally can combine in fruitful fashion with intruding foreign DNA. For every living thing simply represents the most recent manifestation of ancient and endlessly revised information-bearing nucleotide sequences. The overwhelming multitude of others who fell by the wayside without issue were somehow less fortunate. Of course, we only equate reproductive success with good fortune because that is the most fundamental message borne within our own competitively inherited DNA sequences.

Alive, Not Alive Or Dead

A person may require mechanical breathing assistance, yet still be alive. The same holds true for someone supported by a heart-lung machine. But often we conclude that a person is dead when there is no further physical or electrical evidence of brain function. At that moment, many body parts of the newly dead person might still continue to function usefully under appropriate conditions. Clearly being dead is entirely different from not being alive (like a rock). And only rarely is death associated with early total destruction of all organic molecules in that individual. More commonly, death releases DNA carrying useful information out into the environment where it may affect others or persist for a very long while. Indeed, fossil weevil DNA sequences have allegedly been recovered from 120 million year old amber and species-specific DNA fragments from 30 million year old amber-preserved insects as well as from 18 million year old fossil magnolia leaves initially preserved within low-oxygen lake bottom sediments.

My sourdough bread starter represents an uneasy partnership between yeast and bacteria. When I ignore it for too long, it develops a terminal odor and will no longer cause bread to rise. If I happen to detect this condition while it still seems reversible and administer fresh-flour resuscitation, good function and odor may resume. Or it may not, in which case my sourdough goes onto the mulch pile and I borrow a healthy sample. But if I had a better understanding of badly depleted yeast and bacteria, I might be able to resurrect my starter at even that advanced mulch-pile stage of decay. What if I could artificially substitute for all but two or three of their essential life processes and thereby bring those yeast and bacteria back to active bread making? Would they still be considered alive? Undoubtedly, my bread would still rise if I found it necessary and practical to replace every single chemical process in both yeast and bacteria, but at that point the sourdough starter surely would neither be alive nor likely ever to return to life. Perhaps it is easier to recognize and define non-life than life.

On the one hand, it seems that we might be talked into accepting almost any simple self-replicating interactive combination of organic chemical reactions as First Life. On the other hand, many exceedingly complex seeds and *spores* (the especially durable bacterial or cellular forms that remain capable of giving rise to a new individual) may show no detectable signs of life for decades, centuries or even millenia. Yet some relatively ancient seeds and spores still retain the ability to come alive under appropriate circumstances while others that remain unresponsive may really be dead after all. Often we cannot determine if a seed or spore is alive without taking it apart, or even then—for it might someday have come alive had we not killed it with our curiosity. Apparently life's definition lies in terms other than "simple" or "complex", even when we take care to define complexity by the variety of molecular interactions rather than their total number.

Life's Limits

To further confuse this life and death issue, is it possible that very different, perhaps simpler or more diffuse life forms might exist? Could they even now be avoiding our detection simply by being too small, too large, too slow, too fast, or even too unusual? After all, "new" varieties of microorganisms, insects and worms are still being discovered regularly. So why can't we find something more exotic in our ongoing search for life? Are we searching incorrectly or are there significant physical limitations, some real rules that restrict life to the sort of thing with which we already are familiar? The answer seems to be "Yes, life on Earth truly has limits, sort of". We have seen that seeds and spores sometimes exceed any common-sense expectations for life in the slow lane as long as they are isolated from attack by bacteria, fungi, worms, mice and birds. A short discussion of bacteria should allow us to determine how small or fast-living an organism can be. Then we can consider restrictions on largeness and how these can be circumvented by using molecular diffusion as a focus. But only Gaia comes readily to mind in the extremely large and possibly unusual category.

Life Is A Bacteria-based Show

Let us start with bacteria, more commonly known as germs, *microbes* (a category that includes any microscopic life-form likely to trouble you) or even "the bug", as in "I see you've got the bug that is going around" (that expression might also refer to viral or parasitic infections). Bacteria are the smallest of living organisms. They are by far the most numerous, prevalent and versatile in their metabolism (eventually they eat everything, including ear wax and eye lashes which otherwise would carpet the Earth). Indeed, bacteria (a category that probably includes the first well-organized living things) are fundamental to all life. For cows rely upon bacteria (and single-cell organisms known as *protozoa)* to digest their grassy lunch while cowboys and cockroaches rely on bacteria for essential metabolic supplements—furthermore, such larger life forms are both covered with and constructed from bacteria (See Cell). Bacteria recycle all of life's chemicals and deposit a variety of Earth's mineral resources including some uranium ores—one bacterium-based accumulation of $^{235}_{92}U$ went critical about 1.8 billion years ago and thereafter produced kilowatts of power for millions of years in Gabon, West Africa. This natural reactor would speed up when water was present (water slows emitted neutrons and thus makes them more effective) but if it got too hot the water boiled off and things cooled down again. Some bacteria break down rocks, others provide us with essential vitamins or *fix* (hence make available to other life forms) that most unreactive N_2 molecule by converting atmospheric nitrogen into inorganic nitrogen compounds and essential amino acids. Certain *symbiotic* bacteria and their products you cannot live without, other *pathogenic* bacteria you can hardly live with. In the meanwhile, that vast silent majority of *saprophytic* bacteria goes about its own affairs with little direct effect on you.

The earliest bacteria may have acquired sufficient solar energy to persist and multiply by utilizing the organic molecules still accumulating under Earth's reducing atmosphere. Or possibly they supported their energy requirements by oxidising simple inorganic sulphides and other materials at elevated temperatures deep under land or sea. But as Earth's available organic accumulations declined, the *phototrophic* bacteria (sunlight-driven producers of essential high-energy RNA-phosphate compounds) came to dominate. Some phototrophs required organic materials as their carbon source, others could use CO_2. Various types also depended upon H_2, H_2S, S, H_2O or perhaps even Fe^{++} as their source of electrons. Those *cyanobacteria* breaking down H_2O to obtain hydrogen atoms released O_2 gas as a waste product—and their descendents still reside within all photosynthetic plant cells as the *chloroplasts* (or *plastids*) that do the actual work of photosynthesis. Some bacteria prefer to munch sewage or oxidize sulphur into sulphuric acid which may then leach from coal mines to acidify nearby waters. Others are only found near ocean floor hot water vents where they tolerate and even require extreme heat and pressure, or in volcanic hot springs. Still others produce H_2S in three-kilometer-deep oil wells (where they happily reduce sulphur and corrode pipes at 110°C and 350 atmospheres pressure) or live in maximally concentrated salt water (36% NaCl). The Great Salt Lake of Utah is 25% salt by weight, the Dead Sea between Israel and Jordan is about 27% salt (only one-quarter of which is NaCl), Earth's oceans contain about 3% NaCl by weight—all of these waters are infested by bacteria. Bacteria that pass off waste electrons to specific metal ions create many valuable metal ore deposits including placer gold. There are even bacteria that flourish at the bottoms of permanently frozen Antarctic lakes. Of course, as the environment becomes more extreme, fewer varieties of bacteria tend to view that spot as home.

Sterility

Thus there are endless varieties of bacteria, some of which will be found under almost any circumstance eating just about anything. Indeed, bacteria are so pervasive that appropriate populations appear wherever they can flourish—from the bacteriologist's perspective "the environment chooses". So we cannot hope to sterilize (kill all bacteria on) living skin with our antiseptic scrub before making an injection or surgical incision—we just plan to reduce the density of harmful bacteria. Nor do we expect to kill all resident germs by pasteurizing milk and beer—we merely intend to suppress the sorts that could endanger our health. The bacteria that remain will eventually rot our milk with their ongoing excretions, but rotten milk is obviously unpalatable whereas unpasteurized milk may be both tasty and toxic. Since we can neither avoid airborne bacterial contamination of a newly opened jar of food nor of a fresh surgical incision, we try instead to reduce the likelihood of rotten food or postoperative infection. Therefore we refrigerate opened food jars, for cold happens to selectively inhibit most pathogenic bacteria that trouble warm-blooded folks (cold also reduces their production of toxic wastes). And we try to minimize tissue injury at surgery as well as limit our warm patient's exposure to potential pathogens. But even cold-stored food eventually goes bad as less dangerous bacteria and fungi help themselves to left-over portions forgotten at the back of your refrigerator. Incidentally, most bacteria in the environment cannot even be cultured by current laboratory techniques (which were developed for growing the bacteria that like to grow on you). So those other relative unknowns must often be identified by their characteristic appearance on electron microscopy or by specific molecules that they express. High concentrations of salt or sugar preserve food by depriving bacteria and fungi of water for their metabolism. Food can also be dehydrated directly or by cooking and storage in fat (non-polar hydrocarbons). Food in a sealed container can be heated sufficiently (cooked) with infra-red photons to kill all resident bacteria, or sterilized with high-energy photons (X-rays, gamma rays and so on) that may have less impact upon consistency and flavor. In addition, the low pH of sauerkraut, yoghurt and pickles inhibit the growth of most bacteria.

Inheritance

Bacteria may interact and even hunt cooperatively while still competing vigorously for nutrients. Often that competition requires speedy growth and reproduction while times are good. Indeed, some bacteria are born, grow and divide to be born again as two identical descendents within 20 minutes—note that successful bacterial replication generally proceeds without sexual intercourse. In order to reproduce at such competitive speeds, bacteria have trimmed their molecular machinery as much as possible and eliminated all but the most essential information on how to be a bacterium from their circular *chromosome* (a single long DNA molecule joined end-to-end that is duplicated prior to each cell division). Many bacteria multiply at their maximum rate while food and water last, then produce tough dormant spore forms stable enough to wait even millenia for the good times to return to their salt deposit or other burial site. Stabilization of bacterial DNA by close-fitting proteins during spore formation markedly reduces the sensitivity of that DNA to damage by ultraviolet light. Apparently those particular proteins prevent production of harmful covalent cross-links between DNA thymidines (see below)—just another reason that bacterial spores are so hard to kill by radiation, boiling and so on.

Bacteria differ in the limited instructions that they inherit. So as foods and other environmental conditions change, new varieties of bacteria come to predominate while those formerly prevalent may no longer be detectable. Some bacteria do best on tough diets or during tough times. Then the pressure is not so much on rapid reproduction as on how to approach or avoid the out-of-ordinary chemical lunch or other hazardous situation. Bacterial diameters generally do not get very much below one *micron* (which is 10^{-6} meters or a millionth of a meter). Apparently one-fifth of a micron is about as small as any living organism can become and still include all essential molecular machinery. And since the speed of bacterial reproduction has been a matter of life or death for several billion years, a twenty minute start-to-finish bacterium presumably approaches the minimum time required for completing those millions of sequential molecular interactions and revisions involved in nourishment, growth and reproduction.

Naked DNA As Free Advice

Plasmids are circular double-stranded much-shorter-than-chromosome DNA or RNA molecules that commonly lead a relatively independent parasitic existence within bacteria. Plasmids tend to be present only in low concentrations when they bring their bacterial host no advantage, and parasitized bacteria often die out along with their useless plasmids during hard times. Certain plasmids can cause otherwise non-toxic bacteria to become virulent. Other plasmids reduce virulence by weakening their bacterial hosts, which may actually help those bacteria to persist as an ongoing infection—for were they not so weakened, they might kill the current plant or animal host too quickly at a time when other susceptible hosts were not readily available to be infected by subsequent bacterial generations (you may view an infection as an ongoing entity but an ordinary bacterium living life in the fast lane only experiences a moment of that prolonged process since bacterial generations follow one another the way individual treads of a traveling bulldozer pass a particular spot). Because so many plasmids are flourishing within such enormous populations of the most highly successful bacteria, a few plasmids will occasionally (by happy accident and sloppy packing) take along some useful molecular advice as they depart from their current host. That useful molecular advice is thereafter replicated right along with those normally unwanted instructions on how to make many more of those pesky plasmids. The lucky few bacteria that become infected by helpful plasmid information then flourish, multiply madly and their descendents soon predominate until some new problem arises that in turn eliminates most of them along with their plasmids.

Since a bacterium must minimize its own DNA load to maintain a competitive replication rate, plasmids provide at least some chance to access essential new information—e.g. how to cope with a community-wide problem such as antibiotics that are killing off all susceptible bacteria. Plasmids that bear transmissible *F* (fertility) factors may induce their host bacterium to extend a long indecent-looking probe that establishes contact with nearby uninfected bacteria as a preliminary to the transfer of plasmid information. This sort of proto-sexual information transfer can occur between individuals from different species of bacteria and has nothing to do with reproduction—which strongly suggests that *sex* and *reproduction* began as entirely separate activities. But while sex may initially have served to transfer infective plasmid DNA, it was inevitable that some host DNA information would occasionally be included by accident, and if a little information exchange was sometimes good, why not exchange more and even enjoy it (see Reproduction). Individual bacteria often coordinate their production of some toxic chemical in a way that maximizes its impact. Some bacterial colonies show slow pulsations on time-lapse photographs as well as other cooperative behavior. Many bacteria travel alone or in groups over wet land or sea, and their resistant spore forms waft easily through the grossly unsterile air (or water) that surrounds you. Bacteria bearing whip-like flagellae are able to swim voluntarily toward nutritious objects or move in unison out of toxic neighborhoods. Other bacteria kill and digest nearby bacteria, and any bacterium may "catch the bug going around" if that bug happens to be an appropriate variety of *bacteriophage* (viruses that attack certain bacteria).

Protein-covered Advice

A virus is quite similar to (and no more alive than) a plasmid or other free-floating sequence of DNA. However, unlike the naked DNA or RNA of plasmids, viral DNA or RNA suggestions come encased within a precise protein attack mechanism that adheres to specific surface molecules of particular target cells in order to insert its viral information or stimulate the host cell to engulf it. The various several-thousand-nucleotide sequences of viral information have different ways in which they disturb or dominate a bacterium or cell upon entry—successfully inserted polio virus RNA is converted directly into protein by the invaded cell's own protein-producing machinery while the *minus* strand RNA of rabies or influenza virus must first be copied onto a mirror-image *plus* strand before its translation can commence. In contrast, the single-RNA-strand human immunodeficiency virus that causes AIDS usually transcribes its information into double-stranded DNA for more or less untargeted inclusion within the host cell's own genetic information (these RNA viruses ordinarily bring along their own *reverse transcriptase*—a protein enzyme that makes multiple DNA versions of the attacking viral RNA information. Note that enzyme names are often quite descriptive and tend to end in *ase*). Viruses that bear double-stranded DNA information (e.g. herpes or smallpox) generally include enzymes to help make multiple new copies of their DNA within the host cell. And as with successful plasmids, DNA or RNA viral advice always includes "make a lot more of me" (this virus).

Viruses are widely distributed and huge numbers occupy every drop of seawater, but most of those viruses are quite specifically released by and also targeted toward local bacteria and algae. That allows you to swim in waters unpolluted by human waste with little fear of viral attack. Some viruses begin replication immediately upon entry by commandeering their host's living machinery. Other viruses may only become active after numerous cell generations have passed, perhaps following slight structural modifications by ultraviolet light or some chemical exposure. But a great many apparently successful viral insertions simply remain inactive, frequently because of minor mistakes in their own nucleotide sequences. Indeed, construction errors are so common that only 10% or less of newly produced viral particles are likely to be infectious. When many ineffective viral insertions stimulate specific host immune responses, there is growing pressure upon that virus to come up with new and improved information—otherwise it risks becoming extinct during the course of progressively less effective attacks upon moderately resistant hosts.

Viruses sometimes remain *latent* (have no apparent effect) if positioned within an inactive portion of the host's DNA or within cells no longer able to divide such as adult nerve cells. On the other hand, viral advice that lands within an active host gene may disrupt the function of that gene, whether or not such viral advice achieves construction of more virus. While disrupting a single gene within an entire complex bacterium or cell might not seem very important, any damage to certain critical controls can cause cell malfunction or death, or even give rise to *malignant cells* not subject to normal restrictions on cell growth and duplication. The resulting *tumor* or cancerous growth then endangers the entire multicellular organism. Obviously, a huge number of viral particles infecting a great many cells will have endless opportunities for such mischief. Although virus particles must adhere to appropriate cell-surface molecules on particular host bacteria or cells, a great many other factors affect host susceptibility or resistance to viruses and bacteria—these include age, overall condition, cellular defenses and antibody production (see Immunity). Many other known and unknown influences help to determine whether viral expression will be immediate, delayed or never. Nonetheless a typical viral infection of a susceptible host depends upon a lot of virus particles getting their information into numerous bacteria or cells. That unwanted viral information then causes the host to produce numerous self-assembling copies of the same viral information plus protein covering. When those virus-packed bacteria or cells then rupture, each releases many thousand newly minted viral particles to infect other similar bacteria or cells. But the viruses themselves are subject to attack by parasitic information borne on *satellite* molecules that once again might modify a viral attack—perhaps allowing the infection persists in a milder form that brings reproductive advantage to both virus and satellite molecule.

Parasitic Opinions and Afterthoughts

Successful viruses, bacteria and other parasites often force their host to adopt new and strange behaviors in order to infect others. Thus infectious diarrhea is common but infectious constipation could not spread so it does not happen—even if it did occur occasionally, it would not persist. Similarly, coughing and sneezing effectively clear your airways while also distributing your latest bacterial or viral infection to your relatives, friends and co-workers. An equally *virulent* (nasty in its effects) bacterial or viral airway infection that suppressed coughing or sneezing would be a non-starter. And a mildly irritating genital infection often encourages sexual intercourse which further disseminates that infective information. Consider also *rabies*, a virus that can change its previously sensible human or other victim into a raging, biting, scratching maniac who effectively transmits the rabies virus onward before dying. Or the parasite that causes a small clam to reverse direction and dig upward toward the light so that a shore bird will eat that clam and become infected—thereby allowing that parasite to complete its life cycle at the expense of clam and bird. Or the wasp that introduces an immune-system-disrupting virus as it lays eggs within a moth. The only logic in all of these strange interactions is competition within and between species for reproductive advantage—which also explains ordinary instinctive behavior. For example, migration deprives local predators while simultaneously providing access to lush seasonal resources elsewhere. But animals and birds continue to migrate only because that behavior brought reproductive success to their parents. So while certain migration patterns, routes or destinations may seem (or actually could be) unnecessarily arduous, strange or suboptimal, their persistence suggests that they once had advantages in comparison with other adequately tested options. Although ongoing success tends to entrench certain migratory patterns, there will always be lazy or otherwise misguided or befuddled individuals inadvertently testing other alternatives that might rapidly become dominant if circumstances were to change.

The Rising Information Tide Lifts All Life Before It

Thus far we have barely considered other major sources of environmental DNA such as all that DNA routinely released by living or dying creatures. Clearly we are submerged in a sea of free and often infectious DNA-based advice on how to run or ruin our lives. But only the tiniest part of that information could ever benefit the recipient. Among the more effective bacterial defenses against takeover by outside information are individualized *restriction enzymes* that make cuts through very specific segments of foreign DNA. Comparable genetic sites within the bacterium's own chromosome are shielded from that restriction enzyme by attachment of methyl ($-CH_3$) or other groups of atoms. *Methylation* is also used by many different life forms to activate or deactivate particular genes (see below and also *imprinting* in Growth and Development). Because practically all of life's DNA-based suggestions are written in a single standard code, a particular DNA message can be *transcribed* into RNA and then *translated* into an identical protein within a bacterium, a birch tree, a young beagle or an old buzzard. Thus it is hardly surprising that various versions of your own critically important blood hemoglobin molecules are found in many other life forms—wherever hemoglobin might bring reproductive advantage. And regardless of whether an early alder bush acquired its hemoglobin gene from one of your dead ancestors (or vice versa) or both caught it from some other source, the alder apparently finds hemoglobin handy for storing oxygen in those root nodules where atmospheric nitrogen undergoes fixation into nitrogenous compounds by oxygen-sensitive symbiotic bacteria. Hemoglobin also serves as scuba gear for certain aerobic bacteria that commonly dine deep within cow dung since deep dung oxygen concentrations often drop dangerously with so many happy campers at work.

Infectious suggestions carried by bacteria, plasmids, viruses and satellite molecules often appear to have been appropriated from the chromosome of some previous host. For example, the mosquito-borne sindbis virus which shares various traits with other groups of viruses, bears many copies of a protein that closely resembles human pancreatic chymotrypsin (a protein-digesting enzyme). That makes sindbis virus yet another obvious hodge-podge of parts from hither and yon. So viruses may pick up and deliver additional tag-along DNA or RNA messages as they infect one host after another, just as plasmids do. Indeed, plasmids and viruses may originate as escaped bacterial or cellular information that then uses "inside information" to attack its former employer. Those enormous numbers of plasmids, viruses and bacteria eventually provide natural selection with every sort of handy as well as worthless combination and recombination of any available information. Indeed, most genetic innovations probably originate and develop at this micro level (see next chapter) since a few well-adapted survivors can rapidly regenerate an overwhelming new population of microorganisms. Even within your own body, just a few days of antibiotic treatment might suppress or eliminate all but the fittest (most antibiotic-resistant) members of one hundred consecutive bacterial generations. The intent of that antibiotic treatment would be to allow you sufficient time for identification, response and eradication of any remaining pathogenic bacteria—whether antibiotic resistant or not (see Immunity). One could also hope that any pathogenic antibiotic-resistant bacteria released into the environment would soon be overwhelmed by their unmodified wild cousins (perhaps better-suited to more ordinary bacterial circumstances). There is less likelihood of antibiotic resistance developing among wild bacterial populations when antibiotics are used sparingly and then only in fully effective doses. Such sparing use also reduces the reproductive advantage for bacteria able to resist this particular antibiotic. Unfortunately, bacteria often become resistant to several antibiotics at the same time.

Advice Comes From All Directions

Organisms inherit DNA-based advice on how to be, grow and reproduce from their parent(s). Such *vertically transmitted* DNA is supplemented by nucleotide-borne information attacking from the environment, but very little of that *horizontally transmitted* parasitic information reaches the reproductive cells of other life forms (for otherwise ants might have tomato cousins and elephant grandparents). Furthermore, most horizontally acquired changes ought to be detrimental to the next generation and therefore selected against. But despite the low error or *mutation* rate in ordinary DNA replication and the vanishingly small opportunity for successful inheritance of any horizontally derived DNA, individuals vary a great deal and all life forms *evolve* as those genes providing current reproductive advantage become more prevalent. Presumably any variation outside of previously observed limits for that species (e.g. beagle or long-hair chihuahua versus ancestral wolf) either represents the interaction of new or altered genes (thus a new *genotype)* or occurs in response to new environmental pressures on the old genotype (such altered expression of those usual genes thereby producing a new *phenotype)*. But no matter what the source of variation, any increase in cat speed tends to enhance the reproductive success of fast mice and vice versa. And if lengthening of a moth's nectar-sucker means that only the deepest orchid will successfully brush pollen onto that moth for distribution, then moths with long suckers will tend to help deepest orchids prosper and vice versa. In this fashion, the huge number of semi-independently changing variables inherited by all living things encourages an appropriate response by a lucky few to any new selection pressures.

The Path Not Taken

Early biologists commented that God must especially have loved beetles of which He made so many (over a million) different species. More recent information indicates that beetles rely upon live-in metabolically essential bacterial companions. So if a new and useful bacterial variety successfully displaces the old, every newly infested beetle involuntarily becomes reproductively isolated from otherwise attractive beetles still sharing assets with the old bacterial crowd. Thereafter, the lunches, lifestyles and loves of these beetles (who initially differed only in their inadvertent choice of symbiotic bacterial comrades) will inevitably diverge as new traits or characteristics appear by chance within one group or the other. The *sexual isolation* of (meaning lack of DNA sharing between) nearly identical groups of beetles is just another of the rather endless interactions possible between sex, infection and reproduction—that most fundamental conflict among many interests. A uniform ancestral population may also diverge into two reproductively isolated species as a result of geographic separation or even diet. For example, if certain flies reproduce on only one sort of fruit while others of their cohort prosper by utilizing later-developing fruit, those two exclusive groups of flies will soon display different characteristics—demonstrating once again how evolution is simultaneously adaptive and aimless.

Mind Over Living Matter And Vice Versa

Human epidemics can be caused by viruses, bacteria or other parasites, or by the poverty and malnutrition often associated with misgovernment, or by the use of alcohol, tobacco and other harmful drugs, or by floods, famine, drought, over-population and environmental degradation, or by insufficient education or inadequate regulation of the hazardous chemicals and other toxic products of our increasingly complex and contaminated world—even your well-balanced affluent diet poses dangers. And all such epidemics undoubtedly exert selective pressures by altering reproductive success. But most humans still enjoy reproductive success aplenty and modern humans usually respond intellectually rather than genetically to new environmental challenges. So rather than waiting passively for a slow upgrade in information processing to come about as less competitive humans die out, humans have enhanced their inherited ability to store and process information by inventing numbers and writing, the abacus, the slide rule and the computer. The delayed decline in birth rate that follows each improvement in human health and longevity still allows human overpopulation to increasingly stress other animal, plant and even soil bacteria populations. Indeed, human-caused changes in our common environment have far outpaced the ability of most long-lived creatures to adapt on their own. There simply is no time for natural selection to drive adaptive change by repeatedly choosing among subsequent generations—only the smaller life forms that reproduce rapidly in great multitudes can still accomodate to our intellectual whims or resist our often inadvertent attacks with their own genetically diverse responses. So while humans often win major battles with specific rodents, insects, worms and microorganisms, the enormous variety of "simple" life forms under such direct attack have generally been productive and versatile enough to at least endure—and often they flourish in exceedingly large numbers. Of course, these life forms must simultaneously compete relentlessly amongst themselves and with their own predators and parasites for any reproductive advantage. It would appear that life has been forced to enlarge, shrink, speed up, slow down and specialize about as much as possible. The unchanging rules remain simple, severe and sufficient—adapt to available resources, respond appropriately to current challenges or die.

Diffusion Of Information

The size of a living cell determines the time required for internal distribution of molecules by *diffusion*. You will recall that air molecules are incessantly slamming into one another at very high speeds and that molecules dissolved in water are constantly colliding with water molecules and each other. As a result, local odor molecules in the air or unevenly distributed solutes in water will gradually spread out even in the absence of *convection* (stirring or currents). But molecules spreading solely by diffusion make rather slow headway, since diffusion depends entirely upon randomly directed molecular collisions driving more molecules away from than toward their zones of highest concentration. Thus light from exploding fireworks reaches you at about 300,000 kilometers (186,000 miles) per second—and the associated sound waves that alternately compress and expand the intervening air will approach at around 1100 kilometers or 700 miles per hour—but the odor of burning powder only becomes apparent if the breeze is toward you. In absolutely calm air, diffusion by repeated molecular collisions only brings odors to those nearest the source unless a very concentrated odor has been released over a prolonged time. Indeed, noses provide reproductive advantage precisely because smells usually reflect nearby conditions and events. And the perfumes that can drive men mad (or at least make them sneeze) hardly spread by diffusion at all—rather they travel with the air being stirred by body movements, body heat or other causes. When the most vigorous surface waves at sea cannot mix large volumes of fresh water with underlying heavier salt water at any appreciable rate, the far slower process of diffusion can add little. So fresh Amazon River water is still tasted hundreds of miles offshore and warm water layers over more dense cold water in a persistent *thermocline* and any relatively warm air overlying colder ground-level air during a northern winter creates a stable *inversion* (so-called because dense atmospheric air near Earth's surface ordinarily cools as it rises and expands) that must await convective stirring by wind to achieve significant mixing.

However, diffusion does encourage rapid molecular distribution on a microscopic scale—distributing many essential ions and molecules inside bacteria and cells. Since *eucaryotes* (cells with nuclei) are almost always significantly larger than *prokaryotes* (bacteria), most eucaryotes include intracellular transport systems for the delivery of certain essential molecules. But even eucaryotic cells rarely exceed 20 microns in diameter, probably because diffusion delays become noticeable at about 20 microns, causing larger cells to lose real-time contact with their surroundings—an obvious disadvantage. Despite all of that theory, *prochloron*, a giant green photosynthetic bacterium often found existing in symbiotic fashion with sea squirts, seems to prosper although it ranges in size from 9-30 microns. And the intestinal tracts of surgeon fish harbor mobile bacteria over one half mm (500 microns) in length—large enough to be seen with the naked eye. Cells of certain green algae even form cylinders up to 5 centimeters long. However, in this case, the living cytoplasmic contents are mostly displaced to the cell periphery by a large central *vacuole* (storage chamber) and that rim of cytoplasm circulates at rates of up to 4.5 mm per minute (see Muscle). Furthermore, many of your own skeletal muscle fibers are centimeters in length and the osteoclasts that drill holes through your bones may exceed 100 microns (0.1 mm) in diameter (see Bones). Ordinarily, however, rather than expand individual cells much beyond 20 microns in diameter, most living things grow in size by adding more cells. And in addition to *intracellular* (inside of cell) active-transport systems, all but the simplest multicellular life forms utilize dedicated *extracellular* (outside of cells) passageways that can be pumped to encourage the transfer of gases, messages, nutrients and wastes.

Gaia And Homeostasis

As for life forms large or unusual enough to have escaped our attention, Gaia is the only example currently under consideration. This inclusive name for the inorganic Earth and its *biosphere* (all life on Earth) was proposed by Lovelock and Margulis to emphasize that self-stabilizing feedback systems are as important to the whole Earth as to yourself. Just as every creature (and its wastes) represents opportunity for others in this intensively recycled world, so the present build-up of atmospheric CO_2 by fossil fuel combustion encourages corals to deposit more calcium carbonate while also enhancing photosynthesis on land and sea. Other large scale stabilizations attributed to Gaia include the preservation of water on Earth, the redistribution of essential nutrients such as sulphur and the many other soil, oceanic and atmospheric modifications critical to life (see preceding chapter). The idea here is that any *negative* or complex feedback generally reduces the opportunity for a dangerous runaway result (Earth smothering in life's waste products or overheating markedly, for example). On the other hand, a simple *positive* feedback (More! More!) can swiftly send matters out of control, as when an exothermic campfire becomes a raging forest fire. Under that sort of positive feedback, runaway oxidation soon involves all accessible fuel. As you might expect, negative feedback is characteristic of living systems and apparently of Gaia as well. For negative feedback maintains *homeostasis* (the status quo or essential inner stability of all living organisms) in the face of highly variable inputs. By evolving so gradually and interactively, biological processes have become endlessly intricate—furthermore, they generally proceed in many small steps that allow ample opportunity for reproductively advantageous positive and negative controls. Even a potentially continuous process such as hormone secretion actually occurs in short pulses with overall output additionally dependent upon the duration of "On" and "Off" intervals.

The Thermodynamics Of Life

In vivo chemical processes that depend upon *external inputs of energy* for production of more complex organic molecules are considered *anabolic*. Those internal reactions that *release trapped solar energy* through the breakdown of complex molecules are known as *catabolic*. The sum of all anabolic and catabolic processes within any organism is referred to as its *metabolism*. And the metabolism of every living thing individually or even of all life on Earth combined as Gaia, remains subject to the very same Laws of Thermodynamics as any steam engine, electric razor or bowling ball. But only the First and Second Laws of Thermodynamics are directly related to life, since the "Zeroth"and Third Laws are concerned with absolute zero temperature and heat conduction respectively.

The *First Law* states that *energy is neither created nor destroyed.* In other words, the energy content of an isolated system cannot change. *Isolated* is the key word here. Quite obviously, no living thing on Earth is ever totally isolated from solar energy inputs. Such inputs include complex organic molecules as well as photosynthetically released atmospheric oxygen and *liquid water* (preserved by biospheric oxygen and kept from freezing by sunshine).

The *Second Law* declares that *all real processes are irreversible.* Restated in other ways, it indicates that perpetual motion is impossible, that heat passes from hotter to cooler objects but not the reverse and that *entropy* (disorder) always increases. In other words, all systems eventually run down unless sufficient energy is added to (at least) overcome heat losses.

The Second Law's statement that things always run down has often been misunderstood. For example, some religious persons claim that life confirms its miraculous origins on a daily basis by not running down in obedience to the Second Law. Indeed, birth and growth obviously rejuvenate and enhance complexity, which is quite the opposite of wearing out or increasing disorder. But once again, while it truly is remarkable, life persists only because far more solar energy reaches Earth than appears on the anabolic energy accounts of all living things combined. The sun is running down plenty fast enough (by converting five million tons of hydrogen nuclei into energy every second) to power much more than all of the increasing order attributable to life on Earth (over 23 billion tons of newly formed carbon compounds per year). So taken as a whole with life included, our entire solar system still is running down and its overall disorder is increasing. Non-scientists often make logical errors of this sort when they pick and choose among scientific discoveries, accepting those that seem compatible with their own beliefs while ignoring or rejecting less favorable findings. But the hard-won reputation of scientists is based upon respect for all evidence, not merely the more palatable. Indeed, any refusal to accept amply confirmed scientific evidence implicitly denies the validity of science—so the selective use (abuse, actually) of scientific evidence invalidates any argument purportedly based upon that evidence (see Prologue).

Life Is A Four Letter Word

Increasing disorder can be viewed as a loss of information and increasing order as a gain in information. The increasing order of life is based upon DNA, and DNA does just one thing: It stores information. Just as computers can express any information with a specific sequence of zeros and ones, all DNA-borne information is expressed in varying sequences of four *nitrogen-containing bases, cytosine, guanine, adenine* and *thymine* (CGAT). There are only four possible two-digit sequences of zero and one (00,01,10,11). But with four different letters like CGAT, there are sixteen possible two-letter sequences (CC, CG, CA, CT, GC, GG, GA, GT, AC, AG, AA, AT, and TC, TG, -TA, TT), or sixty-four different three-letter sequences. And so it came to pass that life's collected wit and wisdom was writ in *triplet* (three-letter) combinations of DNA's four *nucleotides*. Each of those sixty-four possible DNA triplets—or their matching RNA *codons*—carries a specific message. Thus AUG is part of the initiation signal that orders RNA to "start translating here" as well as the codon for methionine, while UAA, UAG and UGA all say "stop translating there". However, most DNA triplets simply call for one or another of the twenty different amino acids that can be utilized for building proteins. With sixty-four possible DNA triplets but only twenty amino acids to choose between, allowing several RNA codons to code for the same amino acid makes it possible for all DNA triplets to carry useful information (with the first two nucleotides being decisive in most cases—and see second paragraph below).

Complementary DNA Strands Fit Together

The semi-isolated central portion or *nucleus* of every complete cell in your body encloses forty-six double-stranded *chromosomes* that display almost all of your unique inherited information (see also Cell). Each chromosome consists of two very long DNA molecules (known as the *sense* and *anti-sense* strands) twisted snugly together. The snug fit of that *double helix* comes about because every C on one long DNA molecule shares three hydrogen bonds with a G on the matching strand and every A is similarly secured by two hydrogen bonds to a T on its neighboring *complementary* strand. Any obvious error that does not fit comfortably, such as C or G opposite A or T, can often be remedied by special enzymes that detect poor fit. The particular nucleotide sequence borne by the sense DNA molecule of any chromosome includes a great many relatively short messages on how to build particular proteins. Each of those short messages is yet another of your *genes*. In order to utilize that genetic information, it must first be *transcribed* from DNA onto a *messenger RNA molecule* which is then *translated* into the specified amino acid sequence of a particular *protein*.

RNA Utilizes U Instead Of T

Each messenger RNA is constructed of nucleotide sequences complementary to the sense strand of the appropriate DNA gene, just as if building another matching (anti-sense) DNA molecule. Being less adherent to DNA, RNA is readily released as it is created by transcription. Like DNA, RNA utilizes C, G, and A—but RNA substitutes Uracil (U) wherever DNA would use Thymine (T). So the RNA sequence complementary to CGAT on DNA would be GCUA (corresponding to GCTA on the anti-sense DNA strand). During translation of each messenger RNA triplet, the designated amino acid is enzymatically bonded to its just inserted amino acid neighbor with a covalent carbon-to-nitrogen *peptide bond*. Any short string of peptide-bonded amino acids is considered a *polypeptide*. Polypeptides containing over 20 to 30 amino acids or coherent combinations of several polypeptides are known as *protein* molecules. So there you have it. DNA stores life's recipes. When and where appropriate, RNA translates that genetic information into specific sequences of amino acids known as polypeptides or proteins (just as a machine tool translates its computerized information into specially shaped pieces of metal). And the entire human DNA *genome* (information bank) contains about 100,000 different genes.

Interestingly, all *mammals* appear to carry nearly identical genes and proteins. What does vary widely between species is the way coherent groups of those genes are distributed among various chromosomes—presumably that distribution influences the onset and duration of gene activation that are responsible for many of the obvious quantitative differences between mammalian species. In any case, were one to subdivide the genes of *any* mammal into about 120 blocks and rearrange those blocks appropriately, one could closely approximate *your* personal gene sequence. Both plants and animals are markedly affected when they end up with an unusual chromosomal distribution or number (due to a chromosomal duplication, for example). Furthermore the need for appropriately *imprinte*d (inactivated) or active maternal and paternal chromosomes to pair up during cell division places another limit on inter-species fertilization among mammals (see Reproduction). But the question is, after dividing the 46 chromosomes bearing your own genes into about 120 *syntenic* sequences (blocks of genes normally carried on the same chromosome) and rearranging them appropriately in 48 chromosome pairs (as in apes), or 40 chromosome pairs (as in mice) or 38 chromosome pairs (as in pigs), are you a man or a mouse? Or a pig? Stay tuned.

Those 100,000 Genes Are The Real You

Obviously your 100,000 genes or their particular translated-to-protein messages are far fewer in number than the many dozen trillion cells that make up your body. Furthermore, each of those cells contains countless millions of molecules including many that are not protein at all (carbohydrates, fats, DNA, RNA) or only partially protein plus some other non-protein component. So rather than representing detailed specifications of all that you are or ever can be, your highly complex proteins might better be viewed as important assembly line tools and structural components not readily obtainable from the environment (meaning your animal and vegetable victims). Evidently humans, elephants, flies, worms and ostriches are all based upon very similar molecules, each creature according to its own kind of genomic arrangement. Certainly you are not what you eat, for all animals convert the same sort of basic building blocks (water, minerals, fats, proteins, carbohydrates, DNA, RNA, and so on) into their particular living breathing selves by processing those ordinary materials through relatively standard intracellular assembly lines in strict accordance with their own unique DNA instructions. Two healthy humans chosen at random may differ in their CGAT sequences by perhaps one in every five hundred nucleotides. And there is about a 1% variation between the amino acids of a typical human protein and its chimpanzee or gorilla equivalent. But information and assembly errors are rare (roughly one per million for DNA and once in a thousand for RNA) so two human parents cannot possibly pass along enough mistakes in their genes to give birth to a chimpanzee or vice versa. Indeed, that many random errors would undoubtedly prove fatal for parents as well as child (and they would never be the right mistakes to produce a happy well-adjusted chimp anyhow).

How The Modular Tools And Processes Of Life Evolve

Evolution is inevitable since inexact replication always produces some individuals who are more suitable or competitive than others. Naturally high mortality rates allow only the most fit under current circumstances to reproduce (thereby amplifying their most-fit genes among numerous descendents). But how can evolution be an effective process for adaptation if almost all random information errors are harmful? There is a three-part answer to that question. First, the massive reproductive potential of all life forms. Second, the modular nature and easy transferability of DNA-based information. And third, the movement of groups of modules or even entire living systems into other life forms (see Cell).

Under suitable conditions, all life forms show excessive reproductive capacity. Most could potentially produce many *orders of magnitude* (powers of 10) more new individuals than the environment would ever support. Yet a great oak tree is considered reproductively successful if it gives rise to just two successfully reproducing oaks in two hundred years. The remaining millions of viable acorns from that first oak simply represent its excess reproductive capacity. The same holds true for dandelions, chickweed, ants, aphids and you. As an average human female, you were born carrying hundreds of thousands of potential eggs. As an adult human male, you may release up to 300 million sperm in each ejaculation. Such a massive overcapacity in sperm, egg and seed production is usual and customary. Those who cannot cover the Earth with their seed after a catastrophic decimation of their own or another competing species are soon overwhelmed and washed away in the protoplasmic flood of the more prolific. But in more ordinary times, the excessive sperm, egg, fruit and seed production of any particular population simply nourishes many others.

When only the most fit survive to reproduce, each succeeding generation ought to be better adapted for conditions of the recent past. However, even the northern side of a clod of dirt can have a very different microclimate from its southern exposure. And such microvariations often determine which seed sprouts first and thereby becomes dominant. Similarly moss or algae may only flourish on the moist northern surfaces of tree trunks standing near a stream. Even birth order plays a big role in the fate of individuals. Indeed, that entire amazingly productive and powerful historical process known as evolution is based primarily upon chance. Furthermore, being fit and surviving under prevalent conditions often has nothing to do with subjective matters such as the improvement or deterioration of a species. Thus the Napoleonic Wars resulted in the entire next generation of Frenchmen being several inches shorter, for Napoleon gave the tallest and strongest young soldiers the most hazardous duty. Under such circumstances, being tall and strong was a significant reproductive disadvantage. Similarly, when all "good" trees have been clearcut, the descendents of remaining "junk" trees and other plants often come to predominate. Apparently reproductive success is value-free, which seems a more acceptable explanation for leprosy, intestinal tape worms and fatal infantile diarrhea than careful design by a just and loving God who shares our sense of right and wrong and beauty.

If each of a million acorns carries different DNA mistakes and only one of these errors might be beneficial, a squirrel will most likely munch that one best hope for future oaks. But given the huge number of oaks, such useful errors are likely to occur repeatedly and eventually appear in a reproductively successful adult. Furthermore, the horizontal mobility of DNA between individuals and even across species barriers by infection, and the way sexual selection may bring together useful traits within a species (see Reproduction) can effectively expand the eventual distribution of such useful errors. Any single acorn or sperm may be up against astronomical odds but an entire acorn or sperm population is huge enough to guarantee the survival of at least some useful errors. Note that our definition of a *useful* trait involves *tautology* (circular reasoning)—if it survives it is useful, if it is useful it survives. But that merely reflects the way Nature chooses without caring.

All things evolve. Nature controls the direction of that change by exterminating the less favorable, the less adaptable, the less fit, while encouraging her current favorites to multiply. Any alteration of inherited information must carry immediate benefit or at least bring no significant harm to its owner if such information and owner are to persist. In particular, there is no selection pressure in support of *potentially* useful genetic information. The same goes for truly new ideas in human history—such ideas are uncommon, frequently premature and rarely rewarded. In contrast, old ideas are profitably rediscovered, reinvented, recombined and recycled as new possiblities arise for their application. So while your greatest ideas will probably never bring you the recognition that you so richly deserve, they may eventually prosper and spread if resurrected when the time is right—which probably holds true for many of the small changes now present in your DNA as well. Thus evolution advances on a broad front with every living thing occupying a high-risk forefront position.

From one point of view, evolutionary progress through maximal reproductive success might seem a tremendous waste of time, effort and information. But such a criticism implies predetermined values and goals, or at least some evidence that certain nucleotide sequences have intrinsic merit or are more deserving than others. Instead, efficiency and effectiveness are the values of evolution, and no goals exist except reproductive success (as opposed to extinction). Evolution only requires that there be some variation between individuals and that good or lucky reproducers do better than the rest. Any changes in a population should then reflect *naturally selected* characteristics. And as the size and territory of any particular population increases, chance and local factors become less important in determining whether an initially uncommon gene or trait carries enough reproductive advantage to warrant passing it along to the next generation.

The second part in this explanation of how evolutionary improvements occur slowly by chance has to do with the modular nature of DNA. Quite clearly, the primary significance of any gene must lie somewhere between the activity or function of its designated amino acid sequence and the conditions under which the gene is translated into that protein (often an enzyme). Usually each gene is built up from a series of smaller components (modules or useful subsegments of information), some of which are the molecular equivalent of a screwdriver, hex nut, hammer or bearing. Interestingly, the more versatile bits of DNA information tend to show up repeatedly, being expressed within a wide variety of different proteins. This suggests that every one of your little DNA subsegments or modules is in constant competition with all of the rest—over time the more useful modular subsegments will come to predominate with a great many copies—they are then more likely to be carried off into other life forms where they may again provide useful molecular suggestions.

Many modules even have a built-in tendency to hop about within their own genome, with or without add-on or hitchhiker modules. For there is always a chance of some better recombination improving the odds of inheritance. But such *transposons* and other modules can only increase in number and hopping rate until their activities become reproductively detrimental to you—your genes may act selfishly but they must benefit you in order to do well. Of course, some nucleotide modules and also entire genes may be replicated excessively by accident when DNA molecules form or rejoin during chromosome replication and sexual reproduction. This is more likely when multiple copies already exist on one chromosome, any of which could pair up with the shorter matching portion on another chromosome during the exchange of comparable chromosome segments that precedes germ cell formation. And once an organism has an adequate supply of glucose grabber or other important molecules in its genetic tool chest, it can afford to develop variations on that theme, for it will continue to grab plenty of glucose even if some of its grabber molecules are damaged or accidentally altered into better fructose grabbers. So modules fight for space within specific genes, every gene is continuously tested against a great many others, each organism competes with the rest and life remains subject to constant attack by parasitic DNA at all of these levels.

Such creative chromosomal combat undoubtedly has contributed to the many bits and pieces of *scrap DNA* interspersed between those modules of *exon* DNA (information scheduled for expression). No doubt some of your *intron* (silent) DNA has important command and control functions, just as a dictionary or phone book can only provide meaning and establish useful access by including far more information than the mere words or numbers that you might seek. Some lengthy introns include a few nucleotide sequences that occupy critically important positions for the successful function of a nearby gene. Even introns that simply serve as spacers to separate exon modules can help those modules recombine and evolve independently in sperm and egg. But regardless of function, once a long messenger RNA has been faithfully transcribed from the equally long portion of a DNA molecule that represents an entire gene, those RNA sequences complementary to intervening DNA introns must be excised by RNA-rich *splicing enzyme* assemblies—often such splicing is made easier because the introns occupy characteristic sites on sequential loops of that messenger RNA. Only after being shortened in this fashion can the final *messenger RNA* of a gene be released through a pore of the nuclear membrane into your cell's cytoplasm for translation into its specified sequence of amino acids.

Bacteria have no cell nucleus and few introns. However, some introns in nucleotide sequences that designate certain enzymes within *mitochondria* (cell organelles that produce ATP) or *chloroplasts* (photosynthetic organelles) are identically located in far different life forms, which implies that these forms all inherited this particular intron from a common bacterial ancestor. Quite possibly, therefore, bacteria have often carried introns but eventually discarded most of them during the mad dash for reproductive advantage through speedier replication. Perhaps the nuclear membrane separating the carefully organized cell nucleus from the rest of the cell only became necessary because eukaryotic cells are so much more complex than a simple bacterium. Certainly a bacterium bearing many introns might risk producing a lot of useless protein from its readily available intron information. But since information means power, that isolation of the nucleus also simplified the coordination of cell components, some of which might otherwise become too independent for the general good. In other words, the cell's nuclear membrane probably helps to maintain intracellular discipline by providing vital information only on a need-to-know basis—thereby keeping those potentially unruly cell organelles "barefoot and pregnant" (see Cell).

100,000 Genes Are Not Enough

Obviously, your 100,000 genes designate way more than 100,000 different body parts, cells and cell parts and functions, and surely you can do more than 100,000 different things. Thus your proteins should be viewed more generally as rules or tools that act in sequence rather than as specific descriptions of what must be built or done under any conceivable circumstance. Indeed, your comparatively few inherited instructions strongly suggest that you really do have great freedom of action—and that your destiny is largely in your own hands (excepting genetic and environmental limitations) rather than being preordained. Fortunately, it is possible to make exceedingly complex structures such as yourself by using very simple rules. For example, a general rule for making a fern plant might be "go out several stem diameters, then branch, then repeat, repeat, repeat". Convincingly real pictures of mountains, clouds and coastlines are readily constructed using the mathematics of *fractals* (simple recursive non-linear equations) that also can model the sort of *chaotic* conditions we encounter daily in our weather. Apparently weather forecasts remain unreliable primarily because of the vast number of relevant and continuously interacting atmospheric variables. Under such circumstances, an infinitesimal difference in starting conditions can turn out to have an overwhelming impact upon outcome—even such a slight disturbance as the flight of a bumblebee could conceivably be amplified repeatedly during endlessly spreading interactions until it significantly influences next month's weather pattern somewhere far away—similarly, a spreading and changing rumor may have a major and unpredictable effect or else simply die out. Our weather, the planetary orbits around the sun and many biological phenomena may be chaotic at one level, yet quite organized (a hurricane perhaps, or an inspiring thought) at another. Certainly there is unlimited room for variable reiterations and interactions among all of your 100,000 genes, even if the average gene only is subject to direct control by perhaps ten others. But regardless of whether an infant's sucking behavior directly reflects the action of a single protein or represents some complex coordination between 17 or even 170 other genes, that reflex will continue to be inherited because it represents an important reproductive advantage.

Further Reading:

Atlas, Ronald M. and Bartha, Richard. 1987. *Microbial Ecology: Fundamentals and Applications.* 2d ed. Menlo Park, Calif.: Benjamin/Cummins. 533pp.

Darnell, James E. et al. 1990. *Molecular Cell Biology.* 2d ed. New York: W.H. Freeman. 1105 pp. (useful for this and subsequent chapters)

Dawkins, Richard. 1986. *The Blind Watchmaker.* New York: Norton. 332 pp. (useful for this and subsequent chapters)

Futuyma, Douglas J. 1986. *Evolutionary Biology.* 2d ed. Sunderland, Mass.: Sinauer. 600 pp. (useful for this and subsequent chapters)

Gleick, James. 1987. *Chaos.* New York: Viking Penguin. 317 pp. (useful for this and subsequent chapters)

Vogel, Steven. 1988. *Life's Devices: The Physical World of Animals and Plants.* Princeton, N.J.: Princeton University Press. 367 pp. (useful for this and subsequent chapters)

Watson, James D. et al. 1987. *Molecular Biology of the Gene.* 4th ed. Menlo Park, Calif.: Benjamin/Cummings. 1163 pp. (useful for this and subsequent chapters)

Wills, Christopher. 1988. *The Wisdom of the Genes: New Pathways in Evolution.* New York: Basic Books. 351 pp.

Tortora, Gerald J., Funke, Bendell R. and Case, Christine L. 1992. *Microbiology, an Introduction.* 4th ed. Redwood City, Calif.: Benjamin/Cummings. 722 pp. (useful for this and subsequent chapters)

CHAPTER 4

ONE HUMAN CELL

Passing Your DNA Onward;... Semipermeable Membranes;... The Corporate Cell;... Cell Components And Assembly: How You Produce Self-assembling, Self-sealing Cell Membranes In Your Watery World—I: In Water, Fatty Acids Clump Up As Micelles;... II: Three Fatty Acids Covalently Bonded To One Glycerol Form Neutral Fat (A Triglyceride);... III: Cell Membranes Are Self-assembling Phospholipid Bilayers;... IV: DNA Determines Membrane Characteristics, Membranes Help to Determine Which Genes Will Be expressed;... V: And There Is More To Life Than That;... DNA And RNA Are Linear Nucleotide Polymers;... By Their Enzymes Shall Ye Know Them;... Chromosomes Are Nucleoproteins;... Many Introns Must Be Excised From Messenger RNA (mRNA);... Upon Release From The Nucleus, Messenger RNA Is Translated Into Amino Acids By Ribosomes;... Proteins Are The Ultimate Molecular Tools Because Of Their Unlimited Variability In Size, Shape, Charge, Solubility And Allosteric Movements (Ability To Change Shape In Response To Changes In Their Environment);... Peptide Bonds Are Rigid Connections;... It Is Tough For A Protein To Get Back Into Shape But DNA Does It All The Time;... As In Politics, New Combinations Are Common, New Ideas Rare;... The Letters Endure: The Patterns Pass;... Immortality Would Be Fatal;... Cell Imports And Exports;... Diffusion, Osmosis And The Second Law Of Thermodynamics;... Photosynthesis, Oxygen And Carbohydrates;... Glucose, Glycogen And Osmosis.

Passing Your DNA Onward

One specific sequence of six billion nucleotide pairs is what makes you unique. Half of that DNA derives from the 23 chromosomes contained in mother's egg—the rest replicates the 23 chromosomes delivered by father's sperm. Exact copies of those 46 inherited chromosomes are found in every nucleus of your many dozen trillion *somatic* (non-germ) cells. You can forward various recombinations of your mother's and father's genes through your own sperm or eggs. Each subsequent generation will recombine and dilute your contribution. With your 100,000 genes only marginally different from those of all other humans, what possible difference can it make whether you reproduce or not? Well if we ignore all non-genetic information passed along through families during the nurturing process, the biological significance of your reproductive success must lie in whether genes being relayed onward through you bring reproductive advantage to your descendents. In other words, if grandma's eyes or father's nose or the family pot belly are associated with reproductive advantage, that trait or some beneficial aspect of it may become more common—otherwise any out-of-ordinary consequences of interactions between specific genes will tend to fade from succeeding generations.

Semipermeable Membranes

All that you are or ever can be is based upon patterns of DNA transmitted forward through time by the reproductive success of your forebears. Their survival against tremendous odds has optimized every cell in your body to face tough challenges of the past. And regardless of how much they may specialize, each of those *somatic* (body) cells serves mainly to stabilize your internal environment. That is their essential contribution to your reproductive success. However, in order to continue functioning, any living bacterium, plant cell or animal cell must also stabilize the composition of its own *intracellular* (internal) fluid within a fatty *outer cell membrane*—an envelope that can separate those watery contents from *extra*cellular (*outside of* cell) influences. Of course, such a cell membrane separating two identical fluids would have no purpose or function—and any *freely permeable* outer cell membrane separating two non-identical fluids would permit the increasing molecular disorder associated with cell death—yet a totally *impermeable* cell membrane would block all of life's essential interactions—so your cells are surrounded by *semi-permeable* membranes and they must work continuously to maintain appropriate intracellular concentrations of ions and other molecules in the face of chemically different surroundings. Some of the *electrochemical gradients* (see Nerve) that result from this ongoing effort represent readily available energy stores (as in a battery) that can initiate or even power various cell processes and functions. And the particular characteristics of your different sorts of cells are in good part determined by the enormous number and variety of proteins and other molecules dissolved into and protruding from their surface membranes.

The Corporate Cell

Every complete cell of your body contains a complex well-organized membrane-enclosed *nucleus* within which is stored a complete and unabridged copy of your inherited DNA information. But only the subset of instructions relevant to that sort of cell is displayed in convenient form for ready reference—the remaining bulk of each cell's information is somehow misfiled in less accessible fashion. Surrounding the cell nucleus is another well-organized living fluid known as *cytoplasm*. At any given moment, thousands of different proteins are under construction within that cytoplasm, each according to its own messenger RNA sequence. Those proteins are then distributed, utilized and recycled as appropriate to their particular role, be it cell tool, component or product. Also scattered throughout each cell's cytoplasm are dozens or hundreds of discrete bacteria-sized centers for *oxidative phosphorylation*. These *mitochondria* convert much of the reducing power (covalent bond energy) of your animal and vegetable victims (your food) into the *high energy phosphate* bonds of ATP that power so many of your important cell activities (see Metabolism). Interestingly, each mitochondrion still retains and utilizes a bit of DNA information that has been replicated and passed down through countless intracellular generations from its remote free-living *purple bacteria* (*proteobacteria*) ancestors. The *centrosome* is another specialized intracellular structure (based upon *microtubules* stabilized by the same *histone* molecules used to coil and control your chromosomes) that plays an essential role in organizing and carrying out the complex process of cell division. The centrosome, too, may have retained a bit of its own DNA from former times. Even those microtubule-filled whip-tail *flagellae* that propel your sperm (or the sperm of your friends) include histone molecules—which suggests that *eucaryotes* (cells with nuclei) have always relied upon microtubule-based mechanisms for achieving an equal distribution of genetic material as well as for flagellar movement—just one example of how a single protein mechanism can find numerous important applications.

It has been suggested that the cell nucleus originated when one *prokaryote* (bacterium) invaded or engulfed another but could not digest it. But whether or not that is the case, every complete cell of your body clearly represents the long-ago-combined talents and interests of an entire community of *symbiotic* (closely interdependent) bacteria. And over billions of years, countless generations of such cooperative intracellular semi-entities—struggling desperately to survive yet another day—have discarded no-longer-needed DNA information on how to live independently. Improved coordination and further gains in efficiency resulted from centralizing (within the cell nucleus) those instructions applicable to all mitochondria. As usual for such an endless progression, the most unlikely but reproductively advantageous modifications eventually became unavoidable, given enough time and the countless number and variety of competing life forms. In essence, the free-living ancestors of today's intracellular organelles exchanged independence for reproductive advantage as they helped their own eucaryotic organization to overwhelm the competition and achieve reproductive success. By now you may have guessed that mitochondrial DNA (which includes certain quite ancient introns) also supports plasmids, and that these plasmids in turn may be parasitized by satellite molecules. Quite a gathering, that one human cell!

Cell Components And Assembly: How You Produce Self-assembling Self-sealing Cell Membranes In Your Watery World

I: In Water, Fatty Acids Clump Up As Micelles

The typical lollipop-shaped *fatty acid* is a linear molecule based upon 12 to 24 carbon atoms. Except for the relatively bulky *carboxyl* (—COOH) end group, fatty acid carbons are *highly reduced* (covalently bound only to each other and to hydrogen atoms). At your normal pH (7.4), the terminal carboxyl group of each fatty acid is mostly ionized (to —COO⁻ plus H⁺). Thus a fatty *acid* is a proton donor, as its name implies. In water, fatty acids self-assemble into *micelles* (spherical clumps) with their polar (thus water-loving) carboxyl heads forming the outer wetted surface while all of their skinny *hydrophobic* (water-hating) hydrocarbon tails remain tucked deep inside where it is nice and dry and greasy.

II: Three fatty acids covalently bonded to one glycerol form neutral fat (a triglyceride)

A *glycerol* molecule (three-carbon alcohol) becomes covalently bonded to a fatty acid by an enzymatically guided *anabolic, endothermic* (constructive and energy absorbing) process known as *dehydration synthesis* in which an H from glycerol's —OH runs off with carboxyl's —OH as water. The corresponding *catabolic* (breakdown) process is known as *hydrolysis*, an *exothermic* (heat producing) reaction in which a water molecule splits a fatty acid from glycerol—that H_2O molecule simultaneously separates into H and OH to cap the raw ends of each fragment. One glycerol molecule covalently bonded to three fatty-acids becomes a *non-polar* (thus *hydrophobic*) *tri*glyceride. The bulk of your strategic intracellular stockpile of excess oxidizable food energy (the *stored reducing power* represented by your triglycerides) accumulates in those nice soft lightweight (hydrophobic and thus dehydrated) *neutral* *fat* deposits. Fat holds and conducts heat very poorly in comparison to water, so fat provides thermal (as well as mechanical) insulation. Being highly reduced (all those C—H bonds), fat releases *9 kilocalories* (9000 calories) per gram when completely oxidized to CO_2 and H_2O while protein or carbohydrate only provide 4 kilocalories/gram. In addition, carbohydrates are too polar (and thus water-loving or *hydrophilic*) for dehydrated storage within your cells. Therefore every gram of carbohydrate stored as *glycogen* (animal starch) within your cells brings along two grams of water—which means your intracellular glycogen delivers only ⅙ as much energy per added gram of body weight as does neutral fat. Prior to long migrations, a hummingbird will rapidly accumulate excess energy until ¾ of its entire *dry* (water-free) body weight is fat. Since that much energy stored as glycogen would weigh six times more, humans and hummingbirds carry surplus food energy as reduced hydrocarbons (neutral fat), retaining just enough in dissolved carbohydrate form to meet their immediate energy needs.

III: Cell Membranes Are Self-Assembling Phospholipid Bilayers

A *phospholipid* molecule is a *di*glyceride with the third glycerol carbon covalently bonded to a polar phosphate-containing "head". That bulky phosphate head (usually bonded to a water-loving alcohol such as *choline*—one of the B Vitamins) is about as wide as both non-polar hydrocarbon tails combined, which encourages phospholipid molecules to gather as *phospholipid bilayers* (flat double layers) in water with their water-loving phosphate heads making up both wetted surfaces and the hydrophobic fatty-acid hydrocarbon tails sandwiched between. Self-assembling phospholipid bilayers naturally create closed shapes with water trapped inside since that form has no free bilayer edges at which hydrocarbon tails could still contact water. Once brought together by hydrophobic forces, fatty acid tails tend to be held more or less side-by-side by their many weak *van der Waals interactions* (minor random fluctuations in electron charge density on one molecule that cause momentary opposite effects within close-fitting portions of adjacent molecules). Even when stabilized in this fashion, phospholipid molecules can still flex and move about freely within their own layer. Furthermore, the thin oily (fluid) hydrocarbon layer formed by all of those mobile fatty acid tails easily admits other non-polar (fat-soluble) molecules such as *cholesterol*—the dominant fatty molecule in the *plasma* (cell surface) membrane of all eucaryotic cells. Unlike the more disordered internal cell membranes (that contain little cholesterol), fatty acids in the plasma membrane remain rather rigidly aligned at right angles to the cell surface by the organizing effect of their adjacent flat-and-stiff cholesterol-type molecules. Such a cholesterol-induced thickening of the plasma membrane bilayer markedly reduces the incidence of temporary holes or cavities between disordered fatty acid tails that might allow passage of smaller *solute* (dissolved) molecules such as glucose.

Molecules having both polar and non-polar portions tend to store their non-polar sections within the hydrophobic interior of a phospholipid bilayer—complex proteins with alternating polar and non-polar segments often achieve an energetically more favorable (so preferred) arrangement by passing in and out of, or even back and forth repeatedly through, the entire bilayer. The function of many specialized membrane *channels*, membrane-based *receptors* and membrane-organizing proteins depends upon such transmembranous positioning—those proteins with short transmembranous (hydrophobic) segments tend to be restricted to the thinner (because low in cholesterol) intracellular membranes. Some partially protruding molecules expose very complex carbohydrate tips that serve important recognition and attachment functions. Indeed, the particular performance of any cell depends quite specifically upon all of those partially fat-soluble intramembranous molecules. This ability of life's neighboring molecules to attract and interact without being connected is critically important, for life would suddenly end if individual adjacent molecules of any living structure were combined by covalent bonds into one huge (and inevitably non-functional) molecule. Comparable, perhaps, to having all humans permanently stick to one another wherever and whenever they happened to touch. Like the humans that depend on them, molecules must roam freely in order to interact productively.

IV: DNA Determines Membrane Characteristics, Membranes Help To Determine Which Genes Will Be Expressed

Every complete cell of your body inherited identical instructions from your originating egg and founder sperm. Yet your first few *undifferentiated* (not yet specialized) cells somehow gave rise to all those trillions of descendent cells in over 250 varieties, ranging from bone and blood cells to brain cells. These remarkably different outcomes were set in motion soon after father's sperm penetrated mother's egg. As post-fertilization cell divisions then continued, the development potential of each subsequent cell generation was progressively restricted until your foot cells no longer could express DNA instructions applicable only to eyeballs. Of course, none of your cells ever utilize more than a few thousand of their 100,000 inherited instructions anyhow. But quite obviously, any local environmental cues that turn genes on or off and otherwise regulate what options remain for each specific cell at its particular stage must act via the plasma membrane (see Growth and Development). In specializing for its eventual role, each of your cells produces specific membrane-bound proteins by the millions—while also making appropriate intracellular arrangements or even direct connections (as with the wiring of a door bell) that permit the detection, amplification and response to relevant and usually faint, intercellular chemical signals. Where signal molecules are few, they tend to activate only the most sensitively tuned cell receptors (see Hormones). However, as increasing signal strength (concentration) saturates the most highly tuned receptors, more and more cells begin to respond in one way or another. This occurs because less responsive cells may have produced just a few receptors of that specificity or because most membrane receptors and channels are modifications of relatively standard modules that tend to cross-react at very high signal intensities—just as your telephone may pick up the strong signals of a nearby radio station (but hopefully not the reverse). Although particular membrane channels and receptors only react to nearby molecules and conditions, their responsiveness also depends upon the availability of appropriate chemical intermediates and energy. Thus membrane structures merely permit but cannot demand or empower membrane function.

V: And there is more to life than that

In any cell at any moment, millions of essential molecular interactions are taking place within and between membrane, cytoplasm and nucleus in response to local chemical processes and signals from near and far. That means any specific cell structure or function being analyzed *in vivo* (in the live state) provides but a minor and transient sample of current molecular events—when all essential ingredients are available, any intracellular chemical reaction can also be replicated outside of the cell in a laboratory. But such *in vitro* reactions are thereby isolated from the endless in vivo interchanges that determine the availability of reagents, as well as from those positive and negative feedbacks that normally affect every metabolic step, not to mention all of the interlocking enzyme sequences and other input, including delayed or secondary effects of reacting and product molecules, plus any controls or feedback to which DNA responds within specific cell limitations. And that self-organizing complexity is only one tiny part of what makes life such a difficult process to unravel, let alone replicate. Even if we understood all possible interactions of your 100,000 genes, the actual responses of any individual cell or cooperating cell group in your tissues and organs might still turn out to be chaotic. As with the weather, where one can never measure *what is* in sufficient detail to forecast exactly *what will be* before it happens, the overall outcome generally remains unpredictable. Although a particular chemical interaction can reliably be anticipated in some very specific local molecular situation, weak chaos (and therefore a lack of predictability) may still be present at higher levels of organization. Nonetheless, the cumulative effect of huge numbers of individualized cell reactions or of locally chaotic feedbacks to certain standard inputs can often be predicted on the output side. Thus we reliably withdraw our hand from a hot stove. And we know that Earth's moving crustal plates must regularly release built-up strain through earthquakes at sites where two or more independently moving plates have locked their solid edges. Similarly, we are not surprised by the onset of monsoon or hurricane season, nor by the sensitivity of deep snow accumulations in mountain passes to springtime avalanche conditions. For the progressive summation of innumerable stresses or exciting variables that naturally approach critical levels on a smaller scale can only find limited modes of expression or release at the next higher scale—where only the timing and magnitude of an event remain cloaked in doubt. That concept of self-organizing criticality helps to account for most physical phenomena and perhaps even life itself. Thus while life is an unstable and therefore delicately balanced and responsive process that follows statistical rules, the future of a specific mosquito, sage bush, herring, hockey game or human romance is indeterminate. As is so often the case, a scientific generalization may be valid and yet not sufficient to eliminate the element of surprise from individual events.

DNA And RNA Are Linear Nucleotide Polymers

Both DNA and RNA spell out life's information through the exact sequence of their nucleotides. Both are very long nucleotide *polymers* (long molecules built up of short more-or-less identical repeating units). Each of those nucleotides consists of a nitrogenous base (C, G, A and T or U) attached via a 5–carbon sugar to one or more phosphate groups. That 5-carbon sugar is *ribose* in RNA and *deoxyribose* (ribose missing one oxygen after an encounter with *ribonucleotide reductase*) in DNA. Each nucleotide attaches readily to its adjacent nucleotides in a long RNA or DNA molecule via those sugars and their single connecting phosphate groups. So any RNA or DNA molecule consists of a long covalently bonded chain in which sugar and phosphate groups alternate while one or another of the five nitrogenous bases projects out sideways at each sugar. The exact sequence in which nitrogenous bases stick out from one DNA molecule specifies the matching sequence of nitrogenous bases that will be formed under enzymatic supervision during fabrication of a complementary molecule—which might be DNA when duplicating inherited information or RNA when that stored information is to be utilized. Recall that C fits G and A fits T (or U in RNA). Fit implies that they match so closely that two or three hydrogen bonds can hold them securely tip-to-tip (the tip being the free end of each nucleotide opposite to where it is sugar-bound). Once such hydrogen bonds form between *two* long complementary molecules of DNA, the self-stabilizing *double* helix is complete. At that point you have two long polymer strands composed of alternating sugar and phosphate units, coiled together side by side, with matching nitrogenous bases projecting toward one another from every sugar on each molecule. If such a double-helix rope could be untwisted, it would resemble a ladder—the individual rungs being formed by two non-identical but matching nitrogenous bases extending inward from opposing uprights composed of alternating sugars and phosphates.

Obviously we hope that each nitrogenous base within our ancient and precious DNA archive is securely held within its traditional (and presumably healthy for us) position along that sugar/phosphate backbone. As is generally true for reversible reactions, the more energy released during the covalent linkage of each additional nucleotide, the more energy must be added to *directly* reverse that insertion process. So quite sensibly the enzyme (a DNA polymerase) that fastens new nucleotides into place on a matching strand utilizes only the *high energy triple-phosphate* form of the adenine, cytosine, guanine or thymine (or uracil in the case of RNA) to be inserted. The polymerization reaction of DNA or RNA nucleotides therefore becomes difficult to reverse, since each *pyrophosphate* group (two phosphates linked by a high-energy bond) is enzymatically split into individual (hence low-energy) phosphates upon being released in the course of nucleotide insertion. Of course, routinely discarded messenger RNA's and ingested or invading foreign DNA's are easily broken down for recycling by other enzymes located elsewhere within the cell. The bonds *between phosphate groups* in ATP, CTP, GTP and UTP (or in dATP, dCTP, dGTP, dTTP—the *deoxy*ribose form used for building DNA) are referred to as *high energy* bonds because it requires a significant application of chemically stored solar energy to bring those negatively charged phosphates together (just as energy is needed to tighten a steel spring). This trapped bond energy is then released as heat during the removal and breakdown of pyrophosphates that accompanies construction of a DNA or RNA molecule. But ordinarily, the various intracellular enzymes known as ATP'ases utilize energy stored in ATP (AMP~P~P) to perform work or drive specific anabolic processes.

By Their Enzymes Shall Ye Know Them

When reference is made to a particular chemical reaction in biology, you can generally assume that one or more specific enzymes are causing it to proceed. Just as an individual piece of wood can be carved into a statue, or used to build a house or a boat, or burned or allowed to rot (thereby feeding fungi, worms, insects and bacteria), so it is with the building materials within each cell. Anything is possible, but little of note happens spontaneously without the application of specific tools (enzymes) constructed and operated in accordance with ancient DNA information. However, enzymes cannot do the impossible. No amount of ancestral advice will allow your cells to violate the rules of chemistry or the Laws of Thermodynamics. For example, enzymes cannot produce high-energy phosphate bonds without an appropriate input of energy. Thus enzymes simply guide and speed potential reagents into one of an almost endless number of possible outcomes. So give credit to your enzymes for turning that luncheon asparagus into such a delightful human rather than a shy rabbit or grumpy grizzly bear. And these highly efficient enzymes that build your sugars, your phospholipids, your bowels and your wonderful nose (rather than that of an elephant) are proteins manufactured in accordance with your personal DNA information. At an appropriate time and place, your DNA is loosened and the correct starting place for copying one of those long DNA nucleotide sequences is identified by certain of your special DNA-binding proteins as well as by RNA polymerase and its assistant protein molecules sliding along that DNA molecule. Only after that specific gene has been transcribed into complementary messenger RNA information can sequential triplets (three nucleotide sets) be translated into the designated chain of specific amino acids. So it takes many steps to convert your unique DNA information into the polypeptides and proteins that express your information. Furthermore, each of your special enzymes controlling transcription and translation is subject to the influence of perhaps a dozen other regulatory genes or their product molecules. And that entire interactive process began anew when your father's sperm successfully triggered the rich and well-organized assortment of messenger RNAs and proteins within your mother's egg.

Chromosomes Are Nucleoproteins

When cell nucleus information is not being transcribed or duplicated for cell division, that DNA generally remains in storage as a double helix wound tightly onto a histone core. Unlike most proteins which are anions, the five basic varieties of *histones* (that also affect gene expression and repression) carry many positively charged amino acid side-chains. And as cell division time approaches, those *nucleoproteins* form the relatively short and flexible *chromosomes* visible through a light microscope (the individual windings of those long nucleotide molecules on their histone protein cores can be seen by transmission electron microscopy). All complete human cells (other than sperm or eggs) carry twenty-two closely matched pairs of chromosomes plus two lovely X chromosomes for the ladies or an X with a much smaller Y for the gentlemen. *Protamine*, the even more positively charged cationic protein present in sperm, draws its surrounding negatively charged DNA phosphate groups into still tighter coils. Of course, sperm have no need for DNA information during their desperate 100 millimeter dash. Their only job is to outswim 300 million siblings (and possibly countless million unrelated competitors), all seeking life's legendary egg at the end of a long dark tunnel. Here tight information packing to reduce sperm size and increase speed is reproductively advantageous even if sperm DNA information then remains mostly inaccessible until delivered (see Reproduction).

Many Introns Must Be Excised From Messenger RNA (mRNA)

A specific segment of DNA must be identified, loosened, enzymatically activated and otherwise released from its histones and other adherent blocking proteins in order to allow construction of an exactly complementary messenger RNA molecule. This original mRNA molecule may be very long. For example, the huge dystrophin gene is 2000 kilobases (two million nucleotides) in length so even at the usual astoundingly fast copy rate of 3 kilobases per minute, an RNA polymerase enzyme takes about eleven hours to transcribe one dystrophin mRNA. Although a single mRNA could include information for making several enzymes that act sequentially, it usually carries instructions for producing just one particular protein. But that mRNA also bears many silent segments or introns that must be excised by the RNA-and-protein-based *spliceosome* mechanism before the final edited version of your mRNA can be extracted through one of the complex multiprotein pores in your nuclear membrane by a waiting *ribosome* (these RNA-plus-protein mechanisms number in the tens of thousands and give your cytoplasm a granular appearance on high-power optical microscopy) for translation into protein. Certain introns occupy the same DNA site in all multicellular descendents of our common unicellular eucaryotic ancestors so they must have been in place for over two billion years. Undoubtedly some introns represent viral attacks of long ago. For example, *humans*, *chimpanzees* and *gorillas* all share a specific *transposon* (jumping gene that may bring along nearby information) at an identical locus on chromosome 22. Presumably this transposon originated as a viral DNA insertion, since it still retains information for producing the viral *reverse transcriptase* that makes multiple DNA copies of viral RNA information. Apparently a random insertion of viral DNA just happened to hit an important germ cell (sperm or egg) from which all humans, chimps and gorillas descend (even a worldwide epidemic caused by the same virus could never have entered and persisted within the ancestral germ cell of all three species by chance—let alone at the exact same chromosomal site). Many other molecular comparisons as well as our undeniable similarity in anatomy and physiology suggest that humans, chimps and gorillas really do belong within the same genus. But there is understandable reluctance to recognize such close kinship between humans and mere wild beasts—"in His image" sounds so much more dignified.

By making it easier for DNA modules to evolve and recombine independently of adjacent exons, introns provide natural selection with more choices that might enhance adaptation (just as plasmids occasionally help bacteria to persist). But with or without introns, neighboring genes are likely to be inherited together (see Reproduction). Even the delay caused by transcribing many long silent intron segments into mRNA can be put to use, for when construction of a particularly long mRNA cannot be completed before the next cell division, that entire transcription/translation process must then start anew. This helps to prevent rapidly dividing founder cells from expressing cell products that only have relevance to later descendent cells—although the production of so much unnecessary nucleotide verbiage suggests a remarkably inefficient design. But, of course, cells were not designed, nor is optimal efficiency the only criterion used by natural selection. Rather each cell is the outcome of endless struggles, compromises and ancestral revisions. One cannot expect repeated adaptive changes to always deliver the best, most efficient or simplest product. Indeed, that which is sub-optimal but effective will tend to persist whenever major changes involving unfavorable intermediate states would be required to reach any (apparently) better state. As usual, "good enough" has practical advantages over "nothing but the best". Most likely introns were there from the start. And they remain because cells (and possibly ancestral bacteria) incorporated and utilized those introns in a great many practical ways. For example, among cells whose best glucose grabbers only grab fructose weakly, it might not take long for descendents of some cell with a better fructose grabber to predominate if all suddenly became dependent upon fructose. Perhaps that sort of thing explains why eucaryotes have found reproductive advantage for over two billion years in not eliminating scrap intron DNA from their functional exon tool chest.

But this "waste not, want not" approach only works because eucaryotes also can choose which DNA information to ignore and which to use. The fact that spliceosomes are powered by ATP and GTP and based upon RNA components (that closely resemble certain introns in their function) is just another indication that the ability to choose exons and eliminate introns probably dates nearly back to when life first organized under the guidance of self-splicing RNA. In those simpler times, with every living thing also a germ cell, it was particularly important to avoid expressing infective or damaged information. Fortunately, many unwanted suggestions might still be ignored by splicing them out. And once an effective splicing mechanism developed out of self-splicing introns that included instructions for their own endonucleases, an ordinary multimodule gene might give rise to a whole series of related and useful proteins by minor variations in the mode of splicing. Such an interlocking of several gene sequences would ensure their inheritance as a functional unit. These splicing skills remain essential for untangling and repairing your own cellular DNA as well as completing all of those crucial recombinations within your sex cells (see Reproduction). Certain introns probably regulate the release of transcribed exon information that controls your cellular organelles (those mitochondria)—others may protect your more specialized cells from irrelevant molecular advice—perhaps by the direct interaction of a newly released intron RNA with critical portions of that not-needed DNA (providing an intron-encoded regulatory RNA). Some fungal and plant mitochondrial and chloroplast introns are capable of acting as mobile genetic elements or transposons that may carry along adjacent exons. By minimizing their information burden, modern bacteria have pretty much eliminated the need for such extra interactions and controls between transcription and translation. Interestingly, your average gene may only be 1000 to 2000 nucleotides in length. So with the 100,000 genes you have inherited from each parent spread among three billion nucleotide pairs, only about 5% of your nucleotides actually specify a particular protein. The rest apparently represent spacers, access codes, genome regulatory controls, past infections and idle chatter.

Upon Release From The Nucleus, Messenger RNA Is Translated Into Amino Acids By Ribosomes

Ribosomes are small two-part multi-component machines for producing proteins. Although located within the cytoplasm, ribosomes have many intriguing similarities to spliceosomes which in turn resemble simpler (and presumably ancestral) versions of self-splicing RNA (e.g. both self-splicing RNA introns and spliceosomes include branched RNA molecules). And the fact that a ribosome minus its protein components can still translate messenger RNA is just another hint that RNA preceded protein along life's trail. The distinctive RNA-and-protein-containing *nucleolus* inside each cell nucleus is where ribosome component molecules are prepared for export via those complex and dynamic nuclear membrane pores. Even though they include more than four dozen different protein molecules and numerous RNA components, ribosomes swiftly self-assemble from solution—apparently those constantly colliding molecules settle most readily into their stable and functional configuration. Fully assembled ribosomes are complex molecular machines with many mobile segments. In addition to being widely distributed throughout the cytoplasm, ribosomes can be found adhering to the outside of the nuclear membrane or attached to the cytoplasmic surface of the *endoplasmic reticulum*—a series of membrane-enclosed cytoplasmic spaces that extend onto the outer of your two nuclear membrane bilayers (those complex nuclear membrane pores actually pass through both nuclear bilayers in the fashion of small rivets).

Ribosomes only stud the endoplasmic reticulum when directed to that location by an initial *signal sequence* of the protein molecule under construction—that initial sequence, along with its protein *chaperone molecule* (see below) must fit a receptor on the endoplasmic reticulum surface. GTP breakdown then provides energy for releasing the *signal recognition particle* that first directed the ribosome to the endoplasmic reticulum. When studded by ribosomes, the normally smooth-appearing endoplasmic reticulum membrane is referred to as *rough* endoplasmic reticulum. Of course, subcellular details of this sort are best seen using the higher magnifications possible with electron microscopy. Having much shorter wave lengths than visible light, electrons can display far smaller structures, just as a floating log can stop or divert water waves of its own size while a small cork has no effect upon such large waves. By definition, an object that has no effect upon passing electromagnetic waves is invisible at those particular wave lengths. Since visible light waves are about half a micron (0.5×10^{-6} meters) across, an ordinary light microscope cannot reveal much sub-micron detail (but *near-field optics* delivering visible light through a subwavelength aperture close by the sample may soon give those electrons stiff competition in viewing the molecules of life).

Protein molecules produced from mRNA by ribosomes at the endoplasmic reticulum surface are extruded into the more oxidizing environment of the endoplasmic reticulum cavity to undergo further enzymatic modification and final shaping in the presence of (and often enclosed by) large ATP-powered chaperone molecules. These chaperone molecules repeatedly release and reacquire any not-yet-fully-formed protein molecules to prevent them from aggregating and perhaps settling out of solution before reaching their final functional shape. Fully formed proteins ordinarily resist such temptations, at least in part because the hydrophobic amino acid side chains that could cause unfolded proteins to stick together (see Sickle Cell Disease) generally end up folded somewhere beneath the polar surface of the completed protein molecule. But since heated proteins tend to unravel, all cells produce *heat shock proteins* (additional chaperone molecules) when under heat stress. Of course, regardless of cytoplasmic location, each linear messenger RNA molecule passing between the large and small subunits of a ribosome is translated by RNA ribozymes (with the aid of protein enzymes) into a similar length strand of amino acids. The advance of a ribosome along its current mRNA is powered by the high energy phosphates of GTP and irreversibly blocked by *diphtheria toxin* (which therefore terminates the cellular synthesis of protein—see Immunity). Many ribosomes may simultaneously adhere to, read and translate from a single mRNA, each stopping at every nucleotide triplet until the matching *transfer RNA* (tRNA) is brought alongside. As you might expect, every tRNA bears an appropriate complementary nucleotide triplet on one side while the amino acid specified by that triplet is held at the opposing surface. Only if that triplet matches perfectly (*A* opposite *U*, *C* opposite *G*) will it remain at the ribosome surface long enough for GTP hydrolysis to bring about conformational change in the *GTP binding protein* that secures the triplet onto the ribosome. That most recently delivered amino acid is then held in position by its tRNA (still hydrogen-bonded to the complementary mRNA) until secured to the adjacent previously placed amino acid by the enzyme-expedited formation of a covalent *peptide* bond. The growing *poly*peptide (*many* peptide) chain releases its tRNA "empties" so they can reload by collision with yet another of the same amino acid. As with most intracellular chemical processes, the rapid reloading of tRNA's reflects the astronomical number of intermolecular collisions per second. In the end, of course, our entire concept of "rapidly" is based upon that same high frequency of molecular collisions. For life's visible activities must always proceed at a snail's pace in comparison to the millions of multistep molecular processes upon which those activities depend.

The same short —N—C—C— amino-acid backbone sequence separates each covalent (C to N) peptide bond from the next, no matter how large a sidechain the just inserted amino acid may bear. While one could easily design hundreds of different amino acids, only twenty sorts (presumably the most stable, useful and likely to have formed spontaneously) have actually become involved in protein manufacture. And *all* of these *amino acids* have the *L configuration*—their mirror-image D forms probably occurred at equal frequency in the prebiotic world but those D forms somehow lost the competition (perhaps they were slightly less stable in the common protein helix or sheet configurations). Similar selective pressures caused *only D configuration sugars* to be made and utilized by life while Earth's L sugars have all disappeared. Furthermore, since *fatty acids* of the *cis* configuration are preferred by life, those *trans* forms produced in various unsaturated vegetable oils by *hydrogenation* (e.g. when stiffening margarine) may have deleterious effects in your body such as increasing your blood cholesterol levels or being deposited at inappropriate sites. Anyhow, because the standard interpeptide distance is less than the width of a nucleotide triplet, the growing polypeptide chain separates easily from its longer mRNA (and tRNA) as each peptide bond is completed. And all of this automatic anabolic activity depends only upon the right molecules being present at the right time, since dissolved molecules zip about incessantly and these construction components only fit together in one way. So given the right assembly line tools and machines (tRNA, enzymes, molecular chaperones and ribosomes), power (ATP, GTP), supplies (amino acids) and instructions (mRNA), specific polypeptide sequences are almost sure to follow.

Proteins Are The Ultimate Molecular Tools Because Of Their Unlimited Variability In Size, Shape, Charge, Solubility And *Allosteric Movements* (*Ability To Change Shape In Response To Changes In Their Environment*)

Proteins have an endless variety of specific and often changeable shapes that allow them to serve equally endless structural and functional applications. As mentioned, individual proteins generally acquire an immature configuration as they are built, taking final shape only with the assistance of specific chaperone proteins that can prevent improper *aggregation* (sticking together) and *precipitation* (settling out of solution) of the immature protein subunits. New proteins are then held in their final functional form by a number of interactions. These include the small momentary charge fluctuations that induce *van der Waals* attractions between adjacent closely fitting molecules, the tendency of non-polar amino acid side-chains to hide deep within the molecule in response to *hydrophobic forces*, the formation of many *hydrogen bonds* (based upon the sharing of a slightly positive hydrogen atom between its covalently bound oxygen or nitrogen partner and some other relatively negative O or N nearby), the *ionic bonds* due to attractions between nearby positively and negatively charged atoms or molecules, and the sturdy covalent *disulphide* linkages (—S—S—) that can help to maintain protein configurations wherever an —SH on one cysteine (an amino acid) lies near the HS— on another. Disulphide linkages are particularly common in *extra*cellular proteins (those located outside of cells) such as *collagen* where stiffness is more likely to be a virtue (see next chapters). Incidentally, the gradual decline in electrical bonding strength as we move from covalent to ionic to hydrogen to van der Waals interactions simply reflects the extent of electron sharing between the atoms involved. And hydrophobic forces occur because non-polar molecules (or sections of molecules) cannot hydrogen-bond to adjacent polar water molecules—naturally those hydrophobic molecules expose the least possible surface to (or are entirely excluded from) water contact. So rather than being entirely separate and distinct phenomena, these interactions simply represent named segments of the entire interatomic bond spectrum.

Peptide Bonds Are Rigid Connections

The rigid covalent carbon-to-nitrogen (—C—N—) *peptide* bonds between adjacent amino acids are critical to the shape of every protein. These peptide linkages are produced by the enzymatically directed anabolic endothermic (energy consuming) process of *dehydration synthesis*. We have seen how dehydration synthesis bonds fatty acids to glycerol and creates carbohydrate polymer molecules (polysaccharides) out of individual sugars. And here again, the catabolic and exothermic process of *hydrolysis* (breaking apart with water) represents the same chemical reaction in reverse. That your water-based cells build and destroy larger molecules by subtracting or inserting water should come as no surprise for evolution favors those that use what they have to produce what they need. Amino acids ionize at body pH and all share the same —N—C—C— backbone with the amino group (—NH_3^+) at one end and the carboxyl group (—COO^-) on the other—so free *amino* acids naturally line up with their plus (amino) end attracted to the minus (carboxyl) end of the next amino *acid*, practically begging to be locked into peptide bondage by an appropriate enzyme ("marriage broker" is the Chinese term for enzyme). A peptide bond becomes rigid as N's hydrogen forms a hydrogen bond to its newly linked C's oxygen. The lone (*alpha*) carbon wedged between the amino and carboxyl groups of each amino acid serves as attachment site for one or another of the 20 specific amino acid *side chains*. Having just a single hydrogen atom as its entire side chain makes *glycine* ($^+H_3N—CH_2—COO^-$) the simplest possible amino acid. Although the *polar* amino acid *backbone* of a polypeptide or protein dissolves easily in water, some of those 20 different amino acid side chains (referred to in chemical shorthand as R_1 to R_{20}) include non-polar hydrocarbons. Consequently, as any growing peptide chain is progressively released from its mRNA into the surrounding watery cytoplasm, that polypeptide tends to fold with its hydrophobic sections inside, shielded from the surrounding water. Other bonds allowed by fit or determined by interactions between constituent atoms or adjacent proteins then come into play to help control how protein shape develops—such interactions often occur between widely separated amino acids along the same polypeptide chain when that chain folds back onto itself, as well as between amino acids on different polypeptide chains as they join to form a larger protein molecule.

It Is Tough For A Protein To Get Back Into Shape But DNA Does It All the Time

As a protein is heated (cooked), its hydrogen bonds are soon ruptured by increasingly powerful thermal vibrations and molecular collisions. That often causes the protein to become *denatured* (unfolded with internal hydrophobic side-chains exposed and then hopelessly aggregated with other proteins as well as entangled about any covalent cross-links) just as an egg stays hard-boiled even after it cools. Admittedly, a hard-boiled egg might appear better organized than a jiggly raw egg but that is simply because its water content has been stiffened by extensive hydrogen-bonding and hydrophobic interactions among interlocking tangles of disorganized proteins that formerly moved separately through the solution. Sometimes a denatured protein can regain its original condition upon cooling in vitro, but erroneous reconstitution is common since the specific functional shape of a complex protein generally reflects the special circumstances of polypeptide creation at the ribosome—such as those chaperone proteins and the special *isomerase* enzymes that must twist some proteins into shape (even with such assistance, certain proteins may take hours or days to achieve their final functional configuration). Heat also melts a DNA double helix by disrupting hydrogen bonds between complementary nucleotides. However, those complementary DNA pairs tend to resume their original comfortably matched double-helical configuration when cooled. DNA molecules reorganize successfully in this fashion because of their uniform diameter, simple hydrophilic linear structure and matching fit. And that is important, for the DNA double-helix must regularly unravel for copying, then rewind to prevent entanglement during distribution to daughter cells (of course, a great many ribozymes, enzymes and other proteins contribute to the success of that complex process).

Misconstructed proteins are usually taken apart by cell *proteases* (various protein-hydrolysing enzymes) isolated within membrane-enclosed *lysosomes* (see below and Immunity) or else inside of *proteasomes* (barrel-shaped ATP-powered digestive organelles found within the cell nucleus and cytoplasm) that accept *ubiquitin*-labeled and other abnormal, damaged, foreign or short-usage regulatory proteins for digestion. Such selective digestion within the proteasome barrel protects nearby functioning cell constituents from proteolytic damage (until cell injury or cell death opens previously hidden proteolytic sites to uninvited proteins and thereby tenderizes your steak). Individual mitochondria also contain proteolytic enzymes that dispose of damaged or no longer needed proteins. However, *oxidised* or *glycosylated* (interconnected by glucose) proteins can cause disposal problems and may form undesirable deposits in various tissues and organs of elderly and diabetic individuals. Certain misshapen proteins known as *prions* appear especially difficult to destroy—indeed, they seem to replicate and cause infectious dementias such as *scrapie* in sheep and the "*mad cow*" *disease* (that has developed in England) as well as *Kuru* (in New Guinea cannibals) and also inherited dementias such as *Creutzfeldt-Jacob* disease. Of course, the entire concept of a protein *alone* (without accompanying nucleotide sequences) being able to transfer sufficient information to cause a progressive chronic disease that can infect others (as demonstrated inadvertently through injections of growth hormone extracted from affected cadavers, or by eating that protein in the case of Kuru or mad cow disease, or even by gene transfer) is difficult to accept and still under intense investigation.

As In Politics, New Combinations Are Common, New Ideas Rare

While hemoglobin and other larger proteins might include several quite similar polypeptide chains, a smaller protein such as insulin can be based upon a single polypeptide. It is good design, economical and therefore reproductively advantageous for you to form larger structures out of smaller preexisting modules rather than reinventing each new protein from scratch, for a great many useful recombinations of short information-bearing DNA segments will occur spontaneously in far less time than an entirely original and functional larger protein could possibly arise by chance. And when standard tools and supplies can build almost any structure, it does not pay to reinvent such items for every new construction project. Since no reasonable investment of skill and care can prevent all mistakes, it is also a lot cheaper to discard or recycle the occasional small module that does not fit (pass inspection) than to junk an entire protein assembly that proves defective upon completion. This particularly applies to larger proteins containing so many parts that error is almost sure to creep in—either during DNA transcription to RNA or when RNA is translated into protein. Thus cellular productivity is maximized through the use of modular construction techniques for polypeptides and by assembling complex proteins from individual polypeptides. The benefits realized through such separation of manufacturing from final assembly and the cost savings and waste reduction that can be achieved by maintaining a minimal "just in time" inventory of manufactured modules have recently been discovered by the auto industry—perhaps consumer product manufacturers will become increasingly competitive through studies of biological systems where efficient production and minimal waste have been matters of life or death for billions of years.

A great many useful or more aggressive DNA modules have replicated themselves in hundreds or even many thousands of copies throughout your inherited genome. In addition, your cells must continually resist, destroy or excise currently invasive DNA modules. Thus there is never any shortage of potential protein tools and parts. Furthermore, amino acids of nearly the same size and water-solubility can often substitute for one another in equivalent or equally useful structures—just as hammers or screwdrivers may vary considerably in material and construction without loss of function. The incredibly huge number of successful sperm and eggs that give rise to Earth's endless variety of living things make it probable that any polypeptide tools and parts that might usefully fit together in some new fashion will eventually meet and self-assemble. And as more protein structures are deciphered (through analysis of protein crystal X-ray diffraction patterns or in nuclear magnetic resonance or neutron scattering experiments) the common modular and multimodule structural motifs are becoming increasingly apparent—often despite markedly different amino acid sequences.

Those many sites at which a relatively steady rate of *neutral* mutations can alter nucleotide and thus amino acid sequences without affecting survival or final protein structure also allow increasingly accurate estimates of the relationship between any two species, including the length of time since both shared a common ancestor who displayed the ancestral protein that gave rise to the proteins now being studied. In any case, eucaryotes seem to have flourished since they first appeared in the fossil record over two billion years ago despite, or more likely because of, their many transposons (jumping genes—possibly of viral origin) and all of those extra modules in their intron DNA scrap pile. For DNA recombinations (whether during the production of germ cells or as transposons move about) often bring together useful molecules or traits in new and advantageous fashion. All of which suggests that sexual reproduction is here to stay, as is the far-less-frequently helpful horizontal transfer of DNA into germ cells by invasion/infection/insertion/ingestion. But almost surely your finest DNA information and potentially best proteins will never see the light of day—such marvels may have to be reinvented a few million more times before they suddenly become the latest fashion. Clearly, you and Nature will always miss many more opportunities than you can grab.

The Letters Endure: The Patterns Pass

When you get right down to it, bacteria, bacteriophages and bacteriologists compete via their DNA sequences. And like letters of the alphabet, those nucleotides (C and G and A and T) really cannot eliminate each other since they could carry little useful information by themselves or as endless one-letter repeats. Only together do C, G, A and T serve life through the information-rich variations of their sequences. So at least at this level, it does not pay to crowd one another entirely out of the picture. But long ago, CGAT effectively outperformed all information-bearing competitor molecules and thereby almost totally monopolized the information end of life. And the ever-escalating frenzied brutal bloody competition for reproductive advantage that has involved all living things since life first stirred in solution can simply be viewed as the natural (easiest) way to separate currently relevant DNA sequences from the less effective. Our world remains so well suited to its living inhabitants because only the most fit (best suited) of each generation have survived to reproduce. In this ermine-eat-rabbit world, there is not the slightest evidence of underlying motives, guiding principles, predetermination or even preferred results. Humans, redwoods and algae simply represent three lengthy DNA sequences out of an entire fairy tale of possibilities. We each happen to have advantages at this time and place. And many others prosper by adapting to the changes and opportunities (often referred to as pollution) that accompany our success and survival. Some copies of DNA persist and some copies do not, often for reasons that are inapparent or as capricious as peacock tail feathers. That is why no individual or group or government persistently outperforms a fair and free market where all individuals have sufficient access to information so that they can act in their own best interests. From a distance, life rather resembles an animal trail winding through the forest and branching frequently. That trail, relatively random to begin with, is easily diverted by soft and muddy spots, fallen trees or landslides. But usually it is easier to follow along for a while than to break an entirely new trail—although most trail branches do seem to peter out sooner or later. Meanwhile, there is always a chance of the unexpected...

If you customarily view yourself as a specific solid object in the here and now, you might wish to recheck your constituent atoms after another five years of constant wear and repair. Few five-year-resident atoms will then remain except perhaps in your teeth (metal or other) or the non-replicating DNA of your nerve cells—and even here, up to 10,000 nucleotides must be replaced each day in the average cell merely to keep up with DNA damage caused by hydrolysis of your *purine bases* (adenine and guanine). So neither you nor those old friends that greet you five years hence will be of the same solid substance (plus the mandatory new wrinkles and pounds). Indeed you might equally well be viewed as a complex turmoil of information propagating itself forward through time by continuously acquiring and shedding atoms, molecules, information and energy (the way a wet dog sheds water, only ongoing and more so). Apparently life always disrupts its environment. And the information that designates people or other ideas need not be useful to persist. As Dawkins points out, "Mary had a little lamb" has successfully infested a great many youthful minds and thereby propagated itself forward through time as a *meme* or non-DNA-based idea. How does that particular meme really differ from any religion or computer virus or the common cold or you? Although we have been discussing proteins and DNA and other supposedly solid stuff, in actuality this is an information world and it really is your information that persists or dies. Furthermore, as mentioned, even your solid atoms are entirely empty space demarcated by a few point-sized packets of energy. Of course, that scientific concept has little bearing upon your own perceptions when you fall on the ice or accidentally intercept a speeding baseball.

Immortality Would Be Fatal

Some of the ideas expressed herein may persist for a while and even prove useful, yet nothing lasts forever. It has been said (with considerable exaggeration) that the only potentially immortal information is that of our bacterial forebears who persistently flood the environment with their almost exact duplicates. For the usual outcome of a bacterial cell division is two identical versions of the original individual, perhaps containing a few unavoidable alterations in that shared ancestral DNA. Your own body sets stricter limits, since even your most prolific cells can only duplicate themselves a few dozen times before they will divide no more. It turns out that their inherited information only allows your cells to survive, grow and replicate *with permission,* since unlimited cell growth or uncontrolled cell replication within your body would be so disastrous. To some extent, your limited life span may be the price that you pay to reduce the risk of cancer—and only cancer cells ever break free of those normal controls to thereby endanger your life (sort of a "Catch-22" situation). Indeed, if you wish to study a cell that reproduces endlessly in tissue culture, you must either work with a cancer cell or else try to "immortalize" the cell of your choice by causing it significant harm or joining it to a cancer cell. That all of your cells bear very specific instructions *not* to go on living without permission sends a very significant message. Consider what might happen to Earth and the human race if some way was discovered to immortalize individual humans. Most likely the decision on who deserved immortalization would fall to politicians (supposedly described by Mark Twain as "America's only natural criminal class")—and immortalized politicians would truly represent cancers on the body politic. Furthermore, consider who even now has the advantage when everyone is competing for the same limited resources. Under our representative form of government, the elderly (who vote) already have several dollars redistributed to them for every dollar that goes into children's programs (kids don't vote). And just how long do you suppose really old immortals would willingly tolerate the young? Indeed, how could someone who might live forever take even such minor risks as eating ordinary food or crossing the street?

Immortality or at least an unusually prolonged life span may or may not have been tried by some life forms in the past but surely it must be the ultimate disaster for any species to have an excess of non-productive oldsters competing with the young for limited resources (and eventually all resources are limited, if only by the effort one is willing or able to make). That is why every healthy cell in your body knows enough to avoid immortality. This strict warning remains in your DNA simply because it is reproductively advantageous. Of course, there is no evolutionary advantage to long life anyway, once reproduction has been assured—indeed, long life would only have become part of your design if reproduction did not diminish with aging. However, many studies suggest that reproductive success in the face of relentless predators and parasites, competition within your own species and environmental change, requires that your limited assets be diverted to reproduction rather than devoted to self-maintenance and durability beyond what is needed for reproductive success. So perhaps you are destined to die simply because you become irrelevant (except as a competing consumer) once your reproductive days have passed. Certainly from the DNA point of view, your body tissues justify only the minimal investment and maintenance that can keep you going until you have achieved reproductive success. Thereafter you can be junked more cheaply than rebuilt (except to their predators and parasites, non-productive old folks or dysfunctional old cars have little value—unless they become scarce or can be maintained or recycled profitably). The above line of reasoning suggests that a complete redesign would be needed before you could even hope to achieve immortality. And any such reworking would require a marked shift away from reproductive success as the guiding principle of life. At least until now, longevity of that sort would have been associated with an impossible-to-overcome reproductive disadvantage.

Cell Imports And Exports

Depending upon circumstances and your current condition and needs, you are able to create some of the fatty acid and amino acid molecules that your body requires by using readily available materials. Other essentials you must tediously capture and extract in their final form from your animal or vegetable victims (or perhaps await their production by your bacterial co-conspirators). Many of those valuable fats, carbohydrates, amino acids, nitrogenous bases, vitamins, minerals and so on, are taken up by your cells through specific membrane channels or at cell surface receptors that are then internalized along with that small portion of cell surface membrane currently enclosing such an important bit of lunch. Once internalized, the now membrane-surrounded *vacuole* can be fused with *lysosomes* (membrane-enclosed containers bearing appropriate digestive chemicals such as enzymes, oxidants, or strong acids) to further modify or break down those captured bacteria, particles or molecules as seems appropriate. The strong digestive stuff within lysosomes only becomes active after its isolation from the surrounding cytoplasm has been ensured. Lysosome membranes do tend to leak after cell death, however, which encourages *autolysis* (self-digestion) of the cell and further tenderizes your aging steak within its unsterile but dry surface (dryness inhibits microbial metabolism). On the other hand, it is best to briefly heat broccoli or mushrooms before storing in order to inactivate enzymes that otherwise would encourage the broccoli to flower and the mushroom to sporulate—which after all are their reproductive goals. We have seen how ubiquitin (a protein found in all eucaryotes) labels protein molecules destined for early recycling by the proteasome, and how membrane-based and ingested molecules are directed toward the lysosome. In addition, prominently displayed amino acid sequences help to define the durability and fate of many different intracellular proteins. Still others are only released close by one or more enzymes that rapidly convert or otherwise recycle their components. Indeed, specific molecular events are often localized within mitochondria or limited to certain other cell sites in order to encourage the interaction of appropriate molecules. While molecular diffusion remains principal distributor for simple ions and molecules within most cells, all eucaryotic cells maintain busy internal pathways along which various motor molecules transport important materials, often in both directions.

The *Golgi apparatus* is an important series of membrane-surrounded intracellular spaces within which proteins and other cell products undergo sequential modification (usually after passing through the endoplasmic reticulum) before being exported in *secretory vacuoles*. A complex series of docking proteins then delivers those secretory vacuoles to the cell surface where they rupture and release their contents while their membrane joins into and expands the cell surface. As mentioned, the cholesterol molecules produced within the endoplasmic reticulum and forwarded to the plasma membrane have a major role in establishing the characteristics of various cell membranes but the mechanism of cholesterol transport is as yet unclear. While specific aspects of cell metabolism are considered later, you should already have a feeling for the highly organized turmoil within each of your many dozen trillion cells—the constant turnover of enzymes, mRNA, carbohydrates and fats—the thousands of self-organizing chemical processes confined to particular areas or pathways—the newly produced molecules that rush hither and yon bearing short molecular signals that ensure their prompt delivery to specific membrane-enclosed or other intracellular destinations (with those markers then being removed and recycled by the addressee's receptors).

Diffusion, Osmosis And The Second Law Of Thermodynamics

Water diffuses relatively freely through momentary intermolecular spaces in most cell membranes and transmembrane channels. Thus water concentrations on both sides soon become equal (a reliable process based upon the way random thermal movements of water molecules inevitably increase molecular disorder—Second Law of Thermodynamics again). So whenever the solute concentration (expressed as total *osmoles* per liter, with an osmole being equivalent to a mole of molecules such as glucose which dissolve intact, or to half a mole of Na^+ Cl^- since that produces twice as many solute particles) inside one of your cells exceeds the solute concentration in surrounding tissue fluids, the higher water concentration outside will cause the cell to swell. Of course, water's tendency to move from areas of low solute concentration to areas of high solute concentration is matched by any solute's tendency to go in the opposite direction (thereby achieving increasing solute disorder, meaning more evenly distributed solute). But your non-polar (hydrophobic) cell membranes are not freely permeable to most solutes since those solutes that dissolve readily in water at ordinary temperatures and pressures tend to be charged ions or else polar molecules that are larger than water molecules. Usually, therefore, it is water that must cross your semipermeable cell membranes in response to transmembrane differences in solute concentration. In actuality, water molecules continuously cross cell membranes in both directions, but any *net* movement of water across a semipermeable membrane (due to differing water concentrations on the two sides) is referred to as the *osmotic* movement of water (and the process is called *osmosis*).

If we place a sugar solution within a sealed semipermeable membrane surrounded by pure water, many more water molecules will cross the membrane toward the sugar solution than away, until *fluid pressure* (think of fluid pressure here as an outward force being exerted by overcrowded fluid molecules) on the *sugar side* increases to a point where equal numbers of water molecules are crossing the semipermeable membrane in both directions. The enhanced fluid pressure on the sugar side reflects the specific *osmotic pressure* caused by that particular volume and concentration of membrane-enclosed sugar solution surrounded by water. In this type of experiment, osmotic pressure is usually demonstrated by an increase in height of the sugar solution surface above the surrounding water level, or by the tight swelling or even rupture of the sealed semipermeable membrane that encloses the sugar solution. The osmotic pressure generated by any sugar solution against pure water is directly proportional to the number of dissolved sugar molecules per liter of that sugar solution. Your own flimsy semipermeable cell membranes are unable to resist much osmotic pressure before swelling and rupturing (so you never administer solute-free water intravenously). On the other hand, bacteria and plant cells usually have rigid outer walls that prevent their distension to the point of osmotic rupture regardless of the *tonicity* (solute concentrations) of the solution around them. The bacterial cell wall also provides a proper framework to encourage a fair division of genes between the two daughter cells. Lacking a rigid external cell wall, eucaryotes had to develop/acquire/accept the dynamic microtubule-dependent intracellular *centrosome* structure in order to ensure an equal distribution of your genes.

Plant cell outer walls of sturdy water-insoluble *cellulose* (a polymer of glucose) reinforced by *lignin* allow plants to enhance intracellular carbohydrate and K^+ concentrations in order to imbibe water osmotically from arid environments without risking phospholipid cell membrane rupture when it rains. The exposed surfaces of long highly polar cellulose fibers adhere strongly to water (attract water molecules). Evaporation of water from its leaves lifts continuous columns of water up through the tree by suction. The numerous tiny water columns found within any tree exert additional lift by *capillary attraction* (comparable to surface tension)—another measure of how strongly polar water molecules cling to each other and to a polar capillary wall. Such molecular interactions help the tallest tree to suck water into its top leaves from barely damp soil. Plants can afford to load up on water in this fashion because they do not have to go anywhere. However, their more mobile seeds often store extra energy in the form of lightweight fats or oils (hydrocarbons) as well as sugars.

Photosynthesis, Oxygen And Carbohydrates

Photosynthesis is accomplished within plant cells by their many *chloroplasts* (these plant *plastids* are unrelated to plasmids). Animal cells are unable to produce carbohydrates from sunlight by photosynthesis unless they successfully acquire, maintain and expose some of those chloroplasts. Chloroplasts still carry DNA inherited from their ancient free-living photosynthetic cyanobacterial forebears. So plants are bacteria communes (chloroplasts, mitochondria, perhaps centrosomes and even the nucleus) just as you are, but don't let that bug you. The overall chemical reaction of photosynthesis goes CO_2 plus H_2O plus solar energy $\rightarrow CH_2O$ (a general formula for carbohydrate) *plus* O_2. It is likely that those two-billion-year-old banded iron deposits reflect the biosphere's first significant taste of "free" atmospheric oxygen. But it could also be that the ferrous iron dissolved in those ancient seas was (at least in part) oxidised to insoluble ferric iron in the course of light-driven bacterial anabolism under anaerobic conditions (rather than all ferric iron deposits being a direct response to free oxygen levels in the atmosphere). But regardless of exactly when atmospheric oxygen levels finally rose toward modern levels, that atmospheric oxygen initially represented a waste product of bacterial photosynthesis (plus the usual minor breakdown of upper atmosphere H_2O by solar radiation). More recently, plant cells exposing their chloroplast organelles to the sunshine have become the main excreters of free atmospheric oxygen. Although that flagrant pollution of the atmosphere with waste O_2 eventually forced many *anaerobic* bacteria away from Earth's sunny surface, their tough luck was your lucky break since waste O_2 allowed your eucaryotic ancestors to develop *aerobically* (become oxygen utilizers).

Now that Earth's atmosphere contains almost 21% oxygen by volume, we aerobic creatures (whose mitochondria so skillfully utilize that gaseous plant waste) can extract up to fourteen times more chemically stored solar energy from the complete oxidation of carbohydrate (O_2 plus $CH_2O \rightarrow CO_2$ plus H_2O plus energy) than early anaerobes ever achieved by its partial fermentation. That oxidative boost in energy release does not violate any Laws of Thermodynamics, it merely demonstrates the eagerness of hydrogen and oxygen to recombine as water. The ratio of C, H and O in *carbohydrates* (as CH_2O, with one oxygen per carbon) explains why glucose provides less *reducing power* (has less oxidative potential) per carbon atom than the more fully reduced carbons in the *hydrocarbons* (CH_2) of your neutral fat—so hydrocarbons carry more chemical bond or food energy (trapped solar power) per gram than carbohydrates. Note that hydrocarbons are carbons bearing *hydrogen* atoms while carbohydrates were still considered a *hydrated* form of carbon (carbon plus intact water) when they were named. *Simple* or *monosaccharide* (non-polymerized) sugars are named according to the number of carbons in each sugar molecule. Thus ribose is a 5–carbon *pentose*, glucose is a 6–carbon *hexose* (most sugars have -ose endings as in cellulose). *Disaccharides* such as *sucrose* (fructose plus glucose) or *lactose* (glucose plus galactose) are created when such simple sugars are bonded together in pairs by good old dehydration synthesis. And *polysaccharides* such as glycogen and cellulose are long (very branched and not branched respectively) *poly*mers of glucose molecules, again assembled through dehydration synthesis.

Glucose, Glycogen And Osmosis

Since glucose is the more immediate source of energy for many cell functions, why bother to store it as glycogen? Why not just let your cells grab and hold as many glucose molecules as possible whenever times are good? Surely, the anabolic effort wasted in making glycogen could be devoted more usefully to reproductive activities? Well it may seem counter-intuitive but one huge molecule of dissolved glycogen has little more *osmotic* effect than one little molecule of dissolved glucose (even though each gram of glucose or glycogen comes already bound to or dissolved in 2 grams of water). So active muscle and liver cells can build their carbohydrate reserves without raising their osmotic pressure by storing excess glucose as glycogen. Unlike the hard-to-dissolve non-polar (hydrophobic) fat globules within each of your fat cells, that polar glycogen then remains easily accessible in solution until its glucose is once again needed. But why should a huge glycogen molecule have only the osmotic effect of a far smaller glucose molecule? Ought not each glycogen molecule exert the same osmotic pressure as all of its glucose parts independently?

Well, like any intracellular solute, dissolved glycogen molecules certainly would diffuse out of their cell in order to equalize inside with outside glycogen concentrations if only they could cross that semipermeable membrane. But with so few of the huge glycogen molecules present, there actually is little difference between your intracellular glycogen concentration (a few) and extracellular glycogen levels (none). To consider the simplest case, no possible rearrangement can increase disorder when only one huge glycogen molecule is present—disorder remains the same whether that glycogen is located within or outside of the cell membrane. On the other hand, having many glucose molecules inside versus far fewer outside represents a sufficient solute gradient to cause marked inward osmotic migration of water.

Enough for now on semipermeable membranes—multicellularity beckons.

Further Reading:

Ahmadjian, Vernon and Paracer, Surinder. 1986. *Symbiosis: An Introduction to Biological Associations.* Hanover, N.H.: University Press of New England. 212 pp.

deDuve, Christian. 1984. *A Guided Tour of the Living Cell.* 2 vols. New York: W.H Freeman. Scientific American Books.

Margulis, Lynn and Fester, René, eds. 1991. *Symbiosis as a Source of Evolutionary Innovation.* Cambridge, Mass.: MIT press. 454 pp.

Dawkins, Richard. 1982. *The Extended Phenotype.* New York: Oxford University Press. 307 pp.

CHAPTER 5

PROCESSING INFORMATION: THE MULTICELLULAR ADVANTAGE

Life Stabilizes The Environment Which Enhances Life;... Life Feeds On Life Which Limits Cooperation;... Multicellular Groups Self-assemble;... When Complexity Causes Information Overload, Specialization Can Restore Productivity;... Employee Specialization Allows An Organization To Process A Lot More Information;... Death Liberates Scarce Resources For More Productive Use;... Your Specialized Cells Utilize A Great Deal Of Information;... Cooperation, Altruism, Ethics And Morality Come About Naturally In A Society Of Closely Related Cells (Or People);... Ethics Are Situational;... All Humans Are Close Relatives.

Life Stabilizes The Environment Which Enhances Life

In the beginning, all was chaos. Then gradually, imperceptibly, over hundreds of millions of years, Earth's self-replicating and ever-interacting energy-rich organic molecules gave rise to and nourished the self-organizing complexity that became life. As emergent self-amplifying life slowly depleted Earth's initial bounty of reduced-carbon compounds, times became more difficult and competitive. Initially those tiny life forms best suited for the collection of solar energy could reach other sunny locations on wind and wave where they might again prosper undisturbed. But every advance in their photosynthetic skills inevitably polluted each new site with organic debris that others could utilize. Before long such detrivores learned to attack still-living photosynthesizers as well as each other with varying degrees of success. So rather than Earth being clothed in peacefully coexisting photosynthetic populations slowly depleting all available resources in idyllic surroundings while smothering in their own wastes, every growing group and its wastes provided new opportunities for the next opportunist. Thus life's expanding diversity of precariously coexisting and interacting individuals progressively enriched and stabilized the biosphere itself. The surprising stability of Earth's growing economy came from the negative feedbacks inherent within its free market as regular minor successes of individual predators, parasites and other recyclers protected the biosphere from more violent oscillations of output (frequent severe population crashes). But only the most efficient producers, predators and parasites of each brief generation could survive long enough to reproduce under such fiercely competitive conditions. Indeed, the ever-growing variety and complexity of life forms devoted to the production or theft of reducing power continually created new information barriers that frustrated all but the most persistent or fortunate predators and parasites, each desperately coveting its neighbor's assets in order to survive yet another day. Many metabolic and reproductive options became increasingly inaccessible except to life forms capable of cooperation.

Life Feeds On Life Which Limits Cooperation

Regardless of population stability, complexity or density, every new sort of competitor gives rise to a wave of disturbance as more established residents encounter new challenges and opportunities. Perhaps an occasional invader has developed a new niche without causing major disruption to a stable ecosystem, but existing systems generally undergo significant rearrangements as incoming populations take full advantage of their new opportunities. Not uncommonly, several (or many) closely related invaders must cooperate to successfully colonize a new locale in the face of natural and biological adversity. For most environments contain plenty of opportunists-in-waiting already probing for new openings or weaknesses. And when the most marginally established local photosynthesizer, predator or parasite (whether bacterium, insect or fig tree) already has reproductive potential far in excess of any possible opportunity, even the most skilled and highly motivated immigrant must arrive in ample numbers and then struggle to access local resources. Thus it takes some "minimum infecting dose" of viruses to establish an infection, given the inevitable losses suffered through failure to reach, enter or successfully replicate within targeted bacteria or cells. Similarly, some minimum number of attacking bacteria, protozoa (single-celled animals) or sperm may be necessary to get the job done, perhaps by cooperatively overcoming host resistance through the release of toxins or digestive enzymes to which they themselves are resistant. And a more effective social structure often helps entering groups of cells, ants, wolves or humans to establish an overwhelming infection or overcome some threatening neighbor. By amplifying individual strengths, such cooperative social interactions create new opportunities for all members.

Life forms are so variable that the entire spectrum of cooperative and hostile behaviors arises naturally within each species. However, most circumstances reward cooperation over kin-destructive interactions. So ants and wolves hunt in groups because that behavior is reproductively advantageous. One might even rate ants ahead of wolves, since some ants have become prosperous ranchers and farmers, milking their aphid livestock or tending weedless and highly productive underground fungus gardens. In comparison, a wolf does well to recover a buried bone. And unlike ants that may bring along the perfect bit of material to wedge open the protective wing covers of a beetle that they propose to attack, wolves are not noted for adapting new weapons to the hunt. But all sorts of close interactions are bound to occur between individuals of widely different species. Sometimes the relationship is brief and one-sided, as between lynx and rabbit. Then again, a *symbiotic* relationship may gradually develop when neither combatant can gain decisive advantage and the relationship drags on long enough for each to adapt—indeed, chronic conflict tends to create codependence as survivors adjust with increasing success to the status quo. But any ongoing relationship (sharing of information) must bring reproductive advantage to all partners if it is to persist—among other things, successful partners usually differ sufficiently in their requirements to avoid direct competition. Thus a photosynthetic alga (itself an assemblage of many chloroplasts, mitochondria, a centrosome and a host nucleus) can often find long-term happiness entangled with a similarly complex non-photosynthetic fungus able to bring in more resources under difficult conditions. And their descendents tend to prosper together as slow-growing *lichen*, a cooperative plant form that thrives where isolated algae or fungi could not long endure. Yet if a lichen is held in the dark for a prolonged period, the fungus eventually devours its currently unproductive algal associate, just as a starving storm-bound trapper finally consumes his hungry sled dogs. Indeed, it is a self-evident truth that some must die so that others may live. Even among relatives, death can enrich the environment by releasing essential nutrients or other assets (a shelter perhaps) to some more productive user. In the end, every individual has more value dead than alive. Life is based upon (and would soon end without) aggressive and opportunistic recycling.

Multicellular Groups Self-assemble

Any incomplete separation of rapidly dividing single-cell life forms soon produces multicellular groups that represent a tempting collection of scarce resources as well as an ongoing source of metabolic wastes for other tiny organisms to utilize.

The predators, parasites and prey drawn to such a feast further enhance nutrients in the neighborhood while themselves attracting additional codependents. Effective cooperation can gain advantage for members of a closely related multicellular group—as when coordinated beating of their flagellae allows the group to move about or bring in nutrient-and-oxygen bearing fluid. But as usual it is not that simple, for when new circumstances suddenly make flagellae redundant, comparable groups lacking flagellae might come to predominate over those still building and beating nicely coordinated flagellae to no avail. And even though a small group of cells may gain access to new sources of food by combining their mechanical or chemical talents, increasing group size also brings new difficulties. For merely doubling a group's diameter increases its volume and cell count eightfold—thus diffusion soon becomes inadequate for delivery of essential supplies to the starving and suffocating center of such a growing group. So without costly extracellular support structures (see Collagen), larger gatherings of *undifferentiated* (identical) cells tend to remain soft, shapeless, immobile, bulky and fragile. Furthermore, the complexities of cooperative living rise rapidly when increasing group size burdens each cell with more information that must be processed in order to determine an appropriate response.

When Complexity Causes Information Overload, Specialization Can Restore Productivity

A solitary cell utilizes only a limited amount of information since it must do everything for itself. As multicellular gatherings enlarge and begin to cooperate, information overload soon reduces the ability of cells within the group to respond— just as you are more likely to aid a drowning swimmer when by yourself than if part of a large group in which you have no established leadership role—after all, leading is an additional risk (and many fear public embarassment as much as drowning). Thus the maximum size of a simple undifferentiated multicellular organism may also relate to how much companion-cell effect each member cell can evaluate. Similar difficulties are encountered by any small and profitable business that wishes to enlarge, for when a responsively run business with limited information processing capability (a few interchangeable employees) starts to grow, the natural tendency is to avoid altering what works, to continue doing everything the same, only more so. But as the total number of loyal hardworking employees expands, the business soon encounters a drop in product quality and output per employee with more and more of the working time of every participating individual devoted to communications (administration) while previously clear responsiblities become diffused.

Furthermore, it seems that there is always too much to do, even though less is being done less well by each employee. When important tasks can no longer be completed in a timely fashion, all but the most recently hired employees are likely to long for the good old days. So it seems that a successfully expanding business or enlarging multicellular organism must undergo regular reorganizations merely to survive its own increasing complexity—otherwise the gradual buildup of potentially useful information progressively interferes with adequate and timely responses. This is particularly a problem in those small undifferentiated companies where everyone does everything.

Employee Specialization Allows An Organization To Process A Lot More Information

As organizations become increasingly complex, the only way to avoid inefficiency from information overload is for workers or cells to *specialize*. Such specialists can then remain productive and responsive no matter how the overall information burden may grow, since they only require information about their own small segment of the task at hand—which makes the timely acquisition and distribution of information a critical administrative function as the organization enlarges. But workers or cells wishing to abandon the complexities of factory life or multicellularity for the simpler days of small business or unicellular life are generally out of luck, even if they still recall how to do it all. For chances are good that formerly open niches in life also have become increasingly competitive and demanding (except for their nostalgia value, blacksmiths and Model T Fords have little to offer along an interstate highway). Although specialization allows productivity to rise more rapidly than administrative costs for a while, organizations of increasing size and complexity tie up disproportionately more resources within their administrative bureaucracy. That is why savvy producers often enhance their own efficiency by subcontracting with smaller firms for more specialized items and services. Nonetheless, as the average growing business (or organism that still produces most of its own vitamins, amino acids and fatty acids) becomes more and more stable and productive (with more dependents and more assets to lose) there is a further inevitable decline in organizational flexibility—a reduced ability to benefit from breakthroughs or recover from mistakes.

Simply put, it takes a greater disturbance to cause organizational discomfort and a greater commitment in people and other assets before larger organizations can coordinate any deviation from the status quo. So while they may initially enrich and enhance the marketplace, such expanding "do it all ourselves" businesses become increasingly unwieldy and conservative (avoid risk and change) even as changing times are causing their entire organization to lose efficiency and relevance. Eventually the resources sequestered within certain specialized parts are far more valuable than the increasingly unresponsive whole (which remains only as effective as its weakest link). And that is why Mother Nature and the more intelligent or lucky "corporate raiders" generally profit from the death of complex and outdated organizations—even though such major disruptions of the status quo always seem wasteful to the many losers and non-beneficiaries at each redistribution.

Death Liberates Scarce Resources For More Productive Use

Unicellular organisms have an exceedingly high mortality rate but at least each daughter cell from any cell division still represents half of its parent—and most unicellular life forms can continue to grow and divide endlessly. In contrast, larger creatures soon become too complex to reproduce simply by dividing in two, especially since so many of their specialized cells only can serve within unstable easily disrupted tissues and organs that have no direct reproductive function. Furthermore, as any multicellular complex enlarges, its reproductive or *germ* cells (*Life's Original Product*) inevitably come to represent an ever-smaller part of the whole. So while many of us may view death as Mother Nature's way of telling us to slow down, it really is just the opposite—for death restores vitality and productivity by removing inefficient organizations burdened with outmoded information. *Rejuvenation* (becoming young and productive again) is possible with some entities or life forms (caterpillar to butterfly, for example), but the major destruction of costly preexisting structures that is needed for such a thorough reorganization generally causes sufficient turmoil so that even pre-planned rejuvenation is unlikely to succeed more than once or twice—as Chairman Mao discovered during his efforts to shake up an increasingly bureaucratic China.

Despite obvious inefficiencies, representative government survives in large part due to the regular death of incumbent administrations consequent to relatively free and honest elections. Similarly the stock market realizes great profit through dismembering over-mature corporations, and young adults become increasingly productive as the older generation unwillingly releases important assets and gets out of the way. On the other hand, under persistent dictatorships—especially in those impoverished third world nations still ruled by and for a few rich families and the church, mosque or other well-established and therefore conservative clique—natural and human resources tend to remain sequestered and underproductive through many generations (a comparable decline in productivity would surely accompany any significant lengthening of the individual human life span).

Unfortunately, it is natural for political leaders to maximize their short-term personal benefit by misleading. Since only long-term leaders can mislead long-term, the inevitable consequence of hereditary misleading is a loss of organizational competitiveness.

Thus the overwhelming misery that always surrounds self-perpetuating long-term rulers and organizations follows from social inertia plus the ongoing graft, corruption and *nepotism* (that unimportant administrative job for your incompetent nephew or mistress) enjoyed by current beneficiaries of such traditional misrule. Furthermore, anyone achieving a central position of power who actually wished to do something useful would soon be overwhelmed and frustrated. For without a major (and destructive) revolution, the information required to revitalize such an antiquated and centralized organization is only available through the same bureaucrats who flourished under the status quo. So while the rich continue to control far more resources than they ever can utilize productively, the persistent poverty that surrounds them merely confirms their sorrowful conclusion that the problems of the (obviously dirty, uneducated and less deserving) poor are beyond remedy—so they might as well enjoy and let others worry.

Yet Nature routinely demonstrates how a competitive free market can prevent the stagnation that comes with an excessive local sequestration of resources. In human terms, of course, that implies a more-or-less healthy ecosystem and not-too-tilted playing field—a regularly removed, relatively honest and representative government, an educated populace allowed open access to information from a free press, and all of those other essential personal and private property rights that return value to individual initiative. Currently these are the minimum requirements for any human organization that hopes to remain or become prosperous and competitive. No other approach will ever alleviate the ongoing poverty of those lands where ruling families and organizations retain power by restricting information, limiting civil rights and regulating private possessions. Indeed, any lesser and often well-intentioned efforts to help by charitable donations may simply prolong and deepen the public misery.

In similar fashion, the productivity of any cell depends upon the freedom of each mitochondrion to make its own educated decision about when to proceed with oxidative phosphorylation based upon locally obvious information such as glucose concentrations (input) and ATP levels (product). Of course, the overall consumption of glucose by all cells must still be coordinated through appropriate hormonal instructions so that organization-wide goals and needs can be met. In the cell as in the nation, an appropriate amount of responsibility must be delegated at every level of the organization. Although excessive centralization of all decision-making functions generally leads to errors and waste, one can also carry such comparisons too far—after all, your cells simply recycle old or unnecessary mito-chondria by digesting them within lysosomes. And a similar fate awaits the entire cell since those that live by the lysosome shall die by the lysosome—which suggests that society is better served when redundant workers and bureaucrats are retrained for more productive pursuits or retired.

Your Specialized Cells Utilize A Great Deal Of Information

You are a complex multicellular organization. The many dozen trillion cells that make up your being must cooperate in an enormous variety of tasks, all essential to your survival and reproductive success. But those cells also represent a lush concentration of scarce resources. So as long as you live and for quite some time thereafter, you will be subject to vicious attacks by all manner of creatures, great and small. Rest assured, however, that it is nothing personal. It is simply that the ongoing survival and reproduction of those viruses, bacteria, insects, worms, vultures, bears and tigers depends upon their ability to access your resources or those of your neighbors. Not surprisingly, life's incessant "do or die" competitive pressures to optimize efficiency and enhance productivity have forced your individual cells to forsake many skills including even their basic ability to reproduce—as far as sex is concerned, your non-germ cells are now restricted to the home team cheering section. Of course, your mere existence demonstrates that efficiencies were gained at every step along the way as your multicellular ancestors diverted precious resources into specialized non-reproductive cells. But just as a complex computer depends upon software, your personal multicellularity would swiftly grind to a halt without ongoing access to all of your inherited DNA information. Nor could you utilize that valuable information if your cells were not so specialized. And that huge information content brings reproductive advantage only because it helps you to stabilize your internal environment and respond more effectively to the complex world around you.

Cooperation, Altruism, Ethics And Morality Come About Naturally In A Society Of Closely Related Cells (Or People)

Every organization has rules of conduct that regulate relationships between its members, for the burdens and responsibilities of specialization only bring reproductive advantage when co-dependents cooperate. While any sacrifice by one cell to benefit another might seem foolish, such altruism actually pays long-term rewards if it leads to production of more (and more closely related) descendents. Indeed, harsh selective pressures often force closely related individual cells, animals or humans to cooperate and even sacrifice for one another. We tend to admire such cooperation and sacrifice—in contrast, antisocial behavior may pay short-term dividends for the bank robber or cancer cell but it generally fails to bring lasting advantage or enhanced reproductive success.

An interesting computer challenge has clarified some aspects of how cooperation (reciprocal altruism) can benefit unrelated individuals (even though at first glance natural selection might appear to favor purely selfish behavior in social encounters). Our simplified version includes a buyer, a seller and an exchange of packages. One package allegedly contains certain goods while the other package allegedly contains an agreed upon payment, with packages to be opened later. The challenge was to devise a strategy that maximized the accumulation of wealth by any player as buyer or seller. Many complex programs were entered in the contest but as the number of transactions increased, only a simple program called *Tit for Tat* remained consistently among top winners. That program merely directed its player to start honestly and thereafter respond in kind (either honestly or dishonestly) according to the previous move of its opponent. Because Tit for Tat does not initially seek to exploit, it suffers early setbacks when competing with a large population of selfish participants. But a turnaround occurs as the population of suckers diminishes, since exploiters are unable to do business with each other once their reputations are established. Reciprocators then flourish on the benefits of cooperation. Interestingly, if the contest continues and occasional errors permit these competing programs to evolve, eventually a generous and still more successful version of Tit for Tat develops that will forgive the occasional lapse in order to retain the benefits that come from cooperation. A more recent entry seems to have an even more effective strategy known as "win-stay, lose-shift" that can prevent all trading from coming to a halt (as when Tit for Tat stops trading because only cheaters remain).[1] So even amongst computer programs, a reputation for rectitude, cooperation and reciprocity appears to be important for successful long-term relationships. Not surprisingly, society too may benefit by forgiving the occasional transgression rather than always insisting on getting even. For sometimes it is just not practical to be perfectly honest or completely fair, or even to drive within the speed limit.

---

[1] Nowak, Martin and Sigmund, Karl. 1993. *"A Strategy of Win-Stay, Lose-Shift That Outperforms Tit-for-Tat In The Prisoner's Dilemma Game". Nature* 364. pp 56–58

Ethics Are Situational

In order to survive, you too have learned to respond in Tit for Tat fashion (and its more generous version) as well as with win-stay, lose-shift behavior. Of course, cooperative behavior begins naturally at home, since most of your early dealings are with relatives. Except for such basic rules, however, different groups tend to view morals and ethical conduct quite differently, depending upon whether they are headhunters or Quakers, lobbyists for a tobacco firm or Amish farmers, members of a large corporate heirarchy or Iranian Mullahs, radical environmentalists or anti-abortionists (as is inherently obvious in ordinary conversation "He has the ethics of a cannibal!" or "She has the morals of a snake"). For ethical conduct eventually encompasses any behavior that is rewarded by social approval because it brings reproductive or other advantage to a greater number of related individuals or interests than other competing behaviors that were allowed a fair trial. This is why tribalism has been such a basic and tough habit to break, although tribal behavior does decline during good times when those primitive and often unkind support systems seem less essential for survival—which suggests that you personally will be better off when most others are better off as well.

So you definitely are not just a blob of undifferentiated cells. Rather you are a cooperative and internally consistent (therefore moral) multicellular entity, clearly separated from the environment by an attractive skin wrap, able to move about gracefully on strong bones and muscles, tied firmly together by collagen fibers and continuously guided by chemical and electrical messages transmitted via your blood vessels and nerves. While only a tiny percentage of your individual body cells can participate directly in the production of a similar being, your overall reproductive potential (sperm or egg count) still exceeds any conceivable openings/opportunities/requests for your DNA information by many orders of magnitude. And that entire ongoing complexity is easily supported by an occasional animal or vegetable sacrifice. Your victims are a lot like you, of course, since all living things are at least distantly related. So their chemical constituents closely resemble yours and their DNA information can and often does work within you and vice versa. Although each group eventually evolves its own reproductively advantageous rules of behavior, all of life's eucaryotic participants are coated with bacteria and ultimately even composed of germs. Most remarkable.

All Humans Are Close Relatives

The chapters that follow explore some of the roles served by specialized cells in various of your tissues. We will see how those tissues combine into functional organs and how such organs cooperate within their various systems to maintain such a marvelous multicellular creature as yourself. For you are an integral and quite typical part of this relatively tiny living world that spins and circles so unceasingly about its distant fusion source within our limitless Universe. And if our story line ever seems trite, keep in mind the subatomic chaos, the atomic turmoil, the seeming molecular madness within each of your many dozen trillion cells—all harmonizing in their duties while simultaneously regulating immense numbers of semi-enslaved organelles. Given this instability, improbability and interactive complexity at every level, quite obviously we can only examine a few of the concepts that seem most basic to your powerful, durable and beautifully functional design. But even though individual humans differ in many details of their appearance and chemistry, we still resemble each other for one very good reason. Through our common bacterial, then eucaryotic, multicellular, fish-like, amphibian and eventually therapsid reptilian ancestors as well as via endless intervening and many more recent relationships, all humans are very closely related. Considering the fact that modern humans (homo sapiens) have been around just a few hundred thousand years, you've come a long way baby, even if you are far from perfect.

After all, Earth is over 4.5 billion years old. Earth's oldest known rocks solidified about 4.2 billion years ago. Fossil evidence of life may extend back 3.8 billion years. Multicellularity has persisted in animals for over 750 million years. The earliest mammal and dinosaur fossils are more than 225 million years old and the last dinosaurs died tragically just 65 million years ago. Indeed, some evidence suggests that the big break for dinosaurs may have come about 205 million years ago when previously established life forms were mostly annihilated as Earth suffered severe impacts from large meteorites. And copious evidence certainly confirms an enormous meteorite impact on the Yucatan shore of the Gulf of Mexico about 65 million years ago that directly or indirectly eradicated almost all life—indeed it killed every single member of most species then on Earth. But that lamentable termination without notice of those dominant and well-adapted dinosaurs once again leveled the playing field for all surviving contestants. And this time, when the dust finally settled, mammals had become Earth's dominant players. With plenty of meteorites still orbiting out there, who will be next?

Further reading:

The Economist, London—A weekly business and news magazine featuring excellent writing and interesting analyses of current events.

Business Week—McGraw Hill, Publishers—Another interesting business weekly.

Withers, Philip C. 1992. *Comparative Animal Physiology.* New York: Harcourt Brace. 949 pp.

CHAPTER 6

SKIN

Skin Is Taken For Granted;... Your Skin Touches The World;... Hair;...
Epidermis;... Skin Color;... Dermis;... Heat And Fluid Losses Through
Normal And Damaged Skin;... Pimples;... Your Sweaty Hands And
Feet;... Modified Sweat Glands;... Local Heat Loss;... Humidity;...
Temperature And Reproduction;... Burns;... Skin Grafts.

Skin Is Taken For Granted

We relate to the world through our skin. We expect it to locate hot stoves or drafty doors in the dark without contact, and to detect tiny insects in time for a preemptive strike. We assume that our water-resistant covering will neither leak body fluids nor allow osmotic swelling in the hot tub. We are not surprised when our attractive comfortable seamless skin unfailingly makes alterations to suit our changing size and shape over a lifetime, or that it warms us in winter and cools us in summer. Nor are we impressed when it darkens in response to excess light or becomes increasingly translucent under northern skies to ensure ongoing Vitamin D formation. While we know that our skin will stretch without binding to allow our most violent movements, we still expect it to adhere tightly so we won't slip about inside it as we run or lose our grip on a tree branch or baseball bat.

We gripe about how long a cut bleeds or a scrape takes to become invisible. We complain endlessly about pimples and freckles, our hair's color and location, loss or gain. We are annoyed that our skin flushes when we are embarrassed or menopausal. We hate the wrinkles, smelly perspiration drives us up the wall, and just one little skin infection can make us completely forget the many trillion times our skin has skillfully and silently defended us from bacterial attack. Furthermore, we simply cannot leave well enough alone. So we paint myriads of strange patterns in rainbow colors on our exposed surfaces. And any skin that protrudes is in constant danger of circumcision, costly reshaping, or being poked full of holes from which to hang ornaments. Even when our skin smells nice, we smear and spray it with offensive chemicals. Even on hot days we cover our beautiful skin with uncomfortable, unattractive clothes. We should be thankful that skin is not transparent. Can you imagine what high fashion might do if our kidneys or brains were visible?

Your Skin Touches The World

When you reach out and touch someone, you touch skin to skin. If something is as irritating as a splinter, you have let it get under your skin. Not only do we manipulate the world through contact with our skin, even the way we move or display our skin can drive men, women and mosquitoes wild. About two square meters of skin provides you with a surface that is very sensitive to your surroundings. It can detect tiny temperature changes or the lightest touch. Through pain signals, it faithfully identifies all manner of hazardous conditions. And persistent pain draws your attention to matters of above average importance such as having your finger caught in the car door. While ordinary sensations of pressure and temperature may not reach your attention, you can always inquire of any skin surface by concentrating on that part. Tongue, lips and fingertips are particularly sensitive areas due to their closely spaced receptors, but even the thick skin of your back has sufficient sensitivity to warn of potential problems, if not evaluate them in detail. Except for pain receptors which continue to complain, your skin usually reports only changes rather than absolute values, for little is gained from messages of nothing new to report. When two individuals at normal body temperature maintain steady contact, the lack of reports from their opposing skin surfaces can have a soothing effect. Young pigs, people of all ages, and many other creatures appreciate and seek out this sensation of a disappearing surface between individuals whenever an opportunity arises to cuddle.

Hair

Our African and Southeast Asian ancestors did not need much hair for insulation—indeed, being mostly hairless seems to reduce fluid requirements in hot and sunny climates for those that maintain an upright two-legged stance. Thus human populations entering colder climates have generally appropriated the fur and feathers of their animal victims. Yet your hair still stands on end when you are cold, frightened or feeling a surge of patriotism. And when a dog barks right behind you, those goose bumps represent individual *hair follicle muscles* contracting. By increasing an animal's apparent size, such a hair-raising experience could frighten some opponent—even your widely scattered body hairs tend to trap additional motionless air to better insulate your skin surfaces from heat loss. The greasy (non-polar and thus non-wetting) secretions of *sebaceous glands* that empty upon the base of each hair within its hair follicle help to waterproof both hair and nearby skin. That makes it easier for water on skin surfaces to form into droplets which then fall from your generally downward-pointing hairs. So unshaven hairy legs stay warmer and drier when walking through wet grass. Leg and body hair also help you to feel your way along at night or when you are not looking where you are going. For every hair is a delicate lever that detects and amplifies air movements caused by approaching animals, or reports contact before insects or stationary objects can touch the skin. A sensitive nerve ending at each hair root picks up such incoming information and refers it centrally for evaluation.

Your hair, outermost skin layers and fingernails are primarily composed of a water-resistant *piezoelectric* (demonstrating mechanically induced, electrical effects) fibrous protein called *keratin* (this family of about 30 related proteins is found mostly as internal strengthening filaments within epidermal cells). It is unclear how important those piezoelectric effects are to skin sensation. The ten different keratin proteins of your hair are strongly cross-linked and polymerized to form that tough fibrous final product. Keratin is also the principal constituent of hooves, feathers, claws and beaks, fish scales, silk, spider webs and rhinocerous horns. Note the economy in finding different uses for the same versatile keratin gene rather than developing entirely new genes for every possible purpose. Dead cells make up all but the growing root of each hair. These cells are layered like shingles in an outward direction, so hair only combs easily away from your body. As you age, your hair roots lose their ability to produce and incorporate pigmented melanocytes. And your hair length is limited even if you do not cut it, for hair follicles intermittently stop work and degenerate for 4-6 months after 2-5 years of growth. The discontinued hair falls out during that non-functional time or when production of a new hair resumes from below.

Progressive hair loss based upon diminished production and thinning of individual hairs as well as more shallow hair follicles, can be due to your genes and to hormonal effects as you age. Hair loss may also be caused by fungus infections, chemical exposures and other changes in the body including normal or *pathological* (abnormal) *autoimmune* processes (in which you make antibodies against yourself) as well as various medical treatments. Most young adults have about 100,000 scalp hairs. Those of Asians are round in cross-section (therefore Asian hairs are straight) while African hairs are flat (curling easily) and Caucasian hairs are elliptical. Many forms of chronic poisoning (arsenic, for example) can be detected by chemical analysis of your hair. Even a single hair shed at the scene of a crime usually carries sufficient DNA to allow identification of the criminal if that former owner provides a blood, tissue or hair sample for comparison with the unknown. Tiny DNA samples of this sort are enzymatically amplified (by DNA polymerase) into a great many copies before being compared.

Epidermis

Rub your dry skin near a beam of sunlight and airborne skin cells will create tiny visible dust motes. Those flat dead keratinized skin surface cells constantly being shed by your entire surface make up a significant proportion of house dust. There are even microscopic insects that subsist entirely on discarded skin flakes. Perhaps the occasional tiny bite or itch that occurs when there are no flies or mosquitoes to swat can be blamed on those little bugs, too impatient to wait, prying loose still-adherent skin cells for lunch. If you continue to scratch and remove skin flakes, the progressive exposure and stimulation of fine nerve endings and receptors causes increasing discomfort at that site. Persist in rubbing and soon clear fluid seeps from the denuded skin surface—ordinarily this tissue fluid from between deeper still-living *epidermal* (outermost layer of skin) cells cannot escape through your keratinized and greasy skin surface. Indeed, a pocket of fluid may accumulate beneath your superficial epidermal cells when any *shear* (friction) injury separates those epidermal layers. Some of that blister fluid undoubtedly represents tissue fluid but tight bulging blisters are probably pumped up by sweat gland secretions that no longer can reach the skin surface through their sheared-off sweat ducts.

The bottom layer of your *epidermis* is where *keratinocytes* (skin cells) continuously grow and divide into daughter cells, and it is from here that surplus cells are gradually pushed to reach the the skin surface in three or four weeks. As these cells move away from their nourishing blood circulation, they suffer various chemical changes including loss of cell nucleus, dehydration and death—eventually reaching the surface as flattened water-resistant shingle-like keratinized *corneocytes* (skin cell remnants). There they protect underlying tissues from abrasion until in turn rubbed off. A thin non-cellular *basement membrane* layer, reinforced and also anchored down by fine collagen fibers, underlies the *germinal* layer that produces all of your epidermal cells. The wavy irregular contour of this basement membrane gives it a much larger area than the actual outer skin surface. One advantage of such a wrinkled basement layer is the oversized surface made available for producing skin cells that soon must shrivel and keratinize. That expanded basal surface also allows more solutes and gases to diffuse across from *capillaries* (tiny nutrient blood vessels) in the *dermis* (deeper supportive layer of the skin) to nourish the overlying epidermis. And the increased contact area provided by the irregular epidermal/dermal junction enhances adherence between epidermis and dermis.

Epidermis contains no blood vessels and will soon die if it loses contact with nutrients diffusing from underlying dermis, so it is very important that skin surface friction not cleave epidermis from dermis. While *intracellular* (inside each skin cell) keratin fibers markedly strengthen the epidermal cells themselves, those cells also interconnect at sturdy reinforced "spot weld" sites known as *desmosomes*. Furthermore, each epidermal cell is embedded in a fibrous protein that in turn is surrounded by fatty layers—an arrangement that makes skin very resistant to penetration by water or fat. Given this sturdy but layered construction, it is hardly surprising that superficial blisters occur far more readily than deeper dermal/epidermal separations. Even when a blister is deep enough to injure blood capillaries within some dermal projections (thereby becoming a blood blister containing bloody fluid), such deep separations are unlikely to follow the uneven collagen-fiber-reinforced basement membrane. A new blister is likely to contain sterile fluid. Removing the epidermal cover of that intact blister exposes sensitive nerve endings and allows deeper still-living skin cells to dry and crack apart. Not surprisingly, your deep epidermal cells repair damaged skin surfaces more quickly and comfortably if shielded by an uninfected blister.

Skin Color

Ordinary variations in skin color between individual humans are mostly based upon the amount of a dark brown pigment *(melanin)* present within epidermal cells. Special pigment cells called *melanocytes* alter their production of melanin in response to sunlight or hormone signals from the pituitary gland. Various human races differ in the amount of melanin ordinarily present within their epidermal cells. In general, the optimum (and thus prevalent) skin color of long-established populations at any latitude is determined primarily by how much melanin is needed to protect their underlying germinal layer epithelial cells from excessive exposure to ultraviolet light. At the same time, enough light must penetrate that melanin layer to alter deeply located skin chemicals that give rise to the Vitamin D necessary for bone growth (see later chapters). Those with naturally darker skin generally suffer less long-term *actinic* (sunlight) damage—similar wrinkles and abnormalities of skin cell pigmentation or growth may result from chemical exposures or cigarette smoking. Asians often have a yellowish skin color attributed to the dominant presence of *carotene* (another light-absorbing pigment related to Vitamin A). Caucasians deposit less melanin so the pink color of their underlying dermal capillaries often shows through. Occasional individuals of any race (and other mammalian species as well) may carry no functioning genes for the production of melanin pigment in skin, hair or eyes. Such *albino* individuals appear particularly pale white or pink.

Melanocytes actively transfer their melanin out through long *protoplasmic* (living cell substance) extensions. This melanin is then engulfed by live skin cells nearby. Such an incremental method of skin coloration differs from the rapid alterations of skin color brought about by the many long-armed *chromatophores* (small pigment-containing sacs) of varying colors found in squid, octopus, flounders, chameleons and many other creatures. These individuals may rapidly enhance their color by contracting skin muscles that spread their pigment, or fade out by withdrawing that pigment under one silvery reflective area. Often this is done in a patchy fashion that closely mimics background coloration. Squid expel a mucoid glob of ink as a decoy when threatened, while themselves fading to transparency in order to flee. The complex patterns and pulses of color that pass over a dying squid on the beach are due to the uncoordinated relaxation and contraction of a great many chromatophores. Would that we might die so beautifully!

Dermis

While keratin-rich *epi*dermal (*outside* of dermis) cells adhere to each other at internally reinforced sites where their membranes come into contact, the sturdy underlying *dermis* is extensively reinforced by strong *extra*cellular (*outside of* cell) collagen fibers. Animal skin dermis becomes *leather* when all of those extracellular tissue fibers are *tanned* (cross-linked) into a sturdy sheet. Tanning of leather is unrelated to a suntan. The granular surface of a football demonstrates the normal irregularity of cattle dermal/epidermal boundaries—those little pores were once hair follicles. Both hair follicles and sweat glands are deeply rooted within the dermis—hair follicles require deep dermal support for their muscle attachment and nerve supply—sweat glands must be near larger blood vessels in order to produce copious quantities of perspiration. Of course, all blood vessels are located as deeply (and well-protected from injury) as possible. That is why your outermost dermis contains only those tiny tufts of permeable capillaries to nourish the epidermis. Other than blood vessels, lymph vessels, fat cells and specialized nerve endings, the dermis mostly contains fiber-forming cells (*fibroblasts*) and various mobile defensive cells that are always ready to attack bacteria and other invaders. These include many bone-marrow-derived *Langerhans* cells that station themselves within the epidermis so that they can rush samples of new invaders to your lymph nodes for identification (a bad idea in the case of AIDS virus—see Immunity).

Fibroblasts secrete long extracellular protein fibers (mostly *collagen*, some *elastin*) that form strong microscopic-size bundles oriented in all directions within a bed of *glycoproteins* (see next chapter). But eventually growth and gravity bring most collagen bundles into alignment with the predominant tensions in each skin area. Because we all grow and swell in similar fashion, there is little difference between individuals in the preferential direction that skin will split when punctured with a sharply pointed cone-shaped object. Surgical incisions or accidental skin tears that parallel these skin *cleavage lines* will heal with the least collagen production. In contrast, a skin incision across the prevailing lines of skin tension has a tendency to pull apart, which encourages additional fibrous tissue deposition and a more prominent scar. To see why collagen bundles line up as they do, you need only struggle into Norwegian Net long underwear that is several sizes too small—a look in the mirror will show you how the mesh openings elongate horizontally and sag under the weight of your more prominent fatty (or even muscular) areas.

When a surgical incision must cross skin cleavage lines to gain better exposure of underlying tissues, it is best kept as short as possible since even a vigorously overstretched short incision usually produces less scar (due to having more support during the healing process). Longitudinal incisions or full thickness skin injuries on the neck, arms or legs, especially in front of or behind a moveable joint, are subjected to a lot of pull while they heal. Those tensions and consequent healing delays (along with ongoing normal wound contraction) often stimulate sufficient scar production to restrict movement. Therefore surgeons may prefer to access such joints through a longer (perhaps "Z" shaped) incision or even two parallel incisions, or to approach via a longitudinal skin incision on the side of the extremity rather than cutting lengthwise down the front or back. A surgical incision on a fetus or newborn infant heals very rapidly, often with little or no scar formation. The healing of surgical or other wounds is a complex matter that involves multiple hormones and matrix-bound locally active hormone-like growth factors, various contractile and other molecules released by fibroblasts, cell migration, intercellular signals, conversion of one cell type to another, blood clotting and the dissolution of blood clot, collagen fiber breakdown and deposition, production of new ground substance, removal of debris by specialized cells, growth of new vessels and the production of antibodies. Furthermore there may be extensive shifts of extracellular fluids in response to inflammation, movement, position and many other factors. Wound healing also requires adequate nutrition (including substances that the body ordinarily can produce for itself) and an appropriate intake of Vitamin C. These matters are discussed throughout this book.

Heat And Fluid Losses Through Normal And Damaged Skin

Although individual water molecules change velocity with every intermolecular collision, water temperature is defined by their average speed. Any water molecule bounced out of solution by unusually vigorous collisions has acquired boiling or steam temperature. Since that evaporated water molecule carries off far more than its share of heat energy or molecular momentum, those left behind will be cooled (just as the departure of an honor student reduces average test-scores in that class). It turns out that the evaporation of a cubic centimeter (cc)—also known as a milliliter (ml)—of water will cool 500 cc (500 grams or just over a pound) of water or blood or other body tissue by more than 1°C. Incidentally, water has a higher *heat capacity* (meaning it absorbs more heat energy with less change in temperature) than other common materials except for rock. Yet persons living near large bodies of water enjoy a more moderate climate than their friends living among rocky mountains because heat penetrates rock so slowly that deep rock temperatures actually reflect past climatic conditions—in comparison, convective turnover allows the uppermost fifty meters of sea water to exchange heat with the atmosphere during a single season. Frugal rural mothers often provided hot baked potatoes for each coat pocket in the winter—to warm the child's hands on the way to school and then serve for lunch. Like those potatoes, you are mostly water, so your body temperature only changes slowly upon exposure to hot or cold air (water has over 4000 times the heat capacity of air). When your skin must warm a great deal of passing air, the additional *wind chill* can markedly enhance cooling. In contrast, the warmed air trapped next to your body by whiskers or a fur coat will insulate you from (reduce your heat loss to) a cold environment.

Evaporation of water can keep you comfortable on a hot day so you sweat, run through the sprinkler or take a cold shower. Your body heat also escapes into cold water without evaporation, since objects at different temperatures that are in close contact soon reach the same temperature unless additionally heated or cooled. The outer surfaces of non-tropical marine mammals are usually cold and wet. Most therefore insulate their warm heat-producing inner selves from that cold skin surface with a thick layer of *subcutaneous blubber* (fat beneath the skin). Because it both conducts and stores heat so poorly, the fat layer just under your skin can remain very cold while deeper fat layers on muscle and bone stay warm. Fat is a good thermal insulator because of its minimal water content, its intracellular (thus discontinuous) mode of distribution, and its light weight. Furthermore, those long and mobile non-polar molecules transfer thermal molecular movements to adjacent molecules very poorly (akin to poking at them with a tiny, limp spaghetti).

The fatty acids of cold-water-tolerant sea creatures tend to be *unsaturated* (include rigid double-bonds between their carbon atoms). These sites of molecular angulation impair the usual close fit between linear hydrocarbon molecules. And that is important, for while the attractive or repulsive force between ions falls to ¼ of its former strength when those ions are twice as far apart, van der Waal's attractions between close-fitting molecules drop to ¹⁄₁₂₈ of their former strength as the molecular separation doubles. Not surprisingly, therefore, *mono-* and *poly*unsaturated fatty acids (bearing single or multiple double-bonds) keep single-cell high-latitude marine-life phospholipid bilayers loose and pliable at low temperatures (those same unsaturated fats then help to maintain flexible cell membranes and soft fat deposits all the way up the food chain). In contrast, tropical oils are almost fully *saturated* (completely hydrogenated—without double bonds) as otherwise they would be too runny at tropical temperatures.

At rest you put out about as much heat as an 80 watt light bulb, barely enough to warm a bed with several blankets in October. Yet just a few fermenting herbivores can nicely warm an entire dairy barn in January. Physical activity markedly increases your heat output, of course, for your animal and vegetable victims are inefficiently converted into movement—a lot of their chemically trapped solar energy is lost by you as heat rather than in performing work. Fortunately, the insulation value of your subcutaneous fat is easily adjusted by increasing or decreasing blood flow across that fat barrier to your dermis. The smooth muscle cells that control your blood vessel diameters are regulated through nerve, hormone and other chemical signals to those vessel walls. So your subcutaneous and dermal vessels dilate when you exercise, allowing your flushed and warm skin to transfer or radiate (as infra-red electromagnetic waves) excess heat energy from your hot blood directly into adjacent air, water or solid surfaces that are at lower temperatures. When moving slowly or standing quietly, the constant updraft of heated air next to your warm body delivers odors from near and far into your downward pointing nose.

Your deep dermal blood vessels dilate to filter more water into the ducts of your sweat glands as you warm up. Sweat then pours onto your skin surfaces so evaporation can cool you down. Sweat also minimizes the insulating layer of still air next to your body by flattening and clumping your wet hairs. A small quantity of salt and other solutes is left behind as sweat water evaporates. Eventually these salts may form a visible whitish deposit on your skin surface or hat band. The salt that you lose in sweat is primarily NaCl so you prefer salty food on hot days and after heavy physical exertion. Replacing both salt and fluid losses can reduce symptoms of fatigue and muscle cramps that may follow strenuous summertime activities. In earlier days, table salt (NaCl) was often an expensive delicacy delivered by camel caravans across far deserts, for salt production generally occurs in warmer latitudes where shallow bodies of water evaporate easily. Sodium chloride is the main salt of those extracellular fluids that you must carry about as replacement for the salty seas within which your ancient single-celled ancestors first thrived—you can easily taste that salt in your own sweat or the blood dripping from roasting meat. When you are salt-deprived, your sweat glands reduce their salt output and alter other sweat solutes in cooperation with your renal (kidney) regulation of *urine* water and solutes. Sweat and sebaceous gland secretions (*sebum*) exert a moderate antibacterial effect that helps to limit your skin surface bacterial populations. Such bacteria compete viciously with each other and with fungi for living space on your more moist and nutritious skin areas. Fortunately, the crowding out and chemical suppression of harmful bacteria by *saprophytic* (ordinarily harmless) bacteria is common.

Pimples

Although bacteria generally cannot get under (penetrate the outer keratinized layers of) your skin, certain bacteria and even tiny insects may enjoy sumptuous living inside your hair follicles and their connecting sebaceous glands (some of these little insects can be acquired when you share cosmetics). Sometimes the walls surrounding your blocked sebaceous glands become *infected* (invaded by microbes) and therefore *inflamed* (warm, red, swollen and tender). The platoons of white blood cells attracted to that inflamed *pimple* from adjacent dilated capillaries then help to corral bacteria that have eaten into the surrounding dermis. You can assist your body's efforts to control such invasions by not manipulating or squeezing the infected area, for any increase in local tissue pressure spreads bacteria-laden fluids and may lead to an even larger infection (now a *boil*) full of whitish purulent fluid (*pus*—see Immunity). Your inner and outer surfaces support more individual bacteria than you have cells. Indeed, you personally provide for a bacterial population many thousand times greater in number than the entire human population of our crowded planet—and without a doubt you mean the world to them. Which also makes you a never-ending (while you live) battlefield, inside and out. Not only are you supplied and built of as well as attacked and defended by bacteria from birth to death, but your accumulated organic resources will be recycled by bacteria for quite some time thereafter.

Your Sweaty Hands And Feet

Sweat glands continuously moisten the palms of your hands and the soles of your feet. Despite bible-based old wives' tales about the adverse consequences of "spilling your seed" during intercourse (by penile withdrawal before ejaculation) and hence the presumed equally grave risks of *masturbation* (sexual self-stimulation), hair will never sprout from your palms or soles. Nor is there any evidence that those who masturbate regularly will thereby increase their risk of other infirmities such as going blind or even requiring glasses (see Reproduction). Furthermore, because hair and sebaceous glands are so closely associated, your palms and soles lack sebaceous glands as well. And that is reproductively advantageous, for hairy or greasy palms and soles would easily lose grip or traction. But why must your hands get so embarrassingly wet when you are nervous?

That also has to do with grip. For the countless thousands of sweat ducts that open onto the fine ridges of your palms and soles greatly improve grip. And a good grip was particularly important for your ancestors when they were frightened, so they could more safely swing off through the tree branches, or grip a rock or club or sword more securely. The tough guys in old movies used to spit into their own hands and rub them together before engaging in a bar-room brawl. The subliminal message there was, "My hands are real dry 'cause you don't scare me a bit". Baseball pitchers find that a moist hand improves control and delivery while persons with abnormally dry skin often drop dry objects such as clean silverware because their continually loosening surface epithelial cells then act as surface roller bearings. So sweat helps all of those tiny ridges and furrows on your hands and feet maintain your grip and traction, just as moistening a suction cup improves its adherence to a smooth dry surface.

Modified Sweat Glands

Certain *apocrine* sweat glands (in your armpits and around the sex organs in the groin) secrete increasingly complex chemicals after puberty under the influence of your male and female hormones. Armpits and groins can become quite smelly as those complex secretions are broken down by always-present skin bacteria. Thus the onset of adolescence often has friends recommending anti-perspirants to dry such secretions or deodorants to suppress the odor-causing bacteria. Of course, when lesser remedies fail, you might try a shower or bath. Armpits sweat more and may smell worse when you are nervous, for your sweat glands modify their output in response to nerve and hormone signals. If you worry that your smell is noticed by others, you probably are correct in a way, for sweat smells include certain poorly understood odors that serve as *pheromones* (unconscious sexual signals or chemical messages having hormonal effects) that may disturb the physiology and rational thought processes of others nearby without them even being aware of the cause (see Hormones). Thus we land animals, whose remote fish and amphibian ancestors had naturally wet surfaces from living in the sea or along moist shores, have developed kidney-like sweat tubules that can moisten our skin surfaces for gripping and cooling even in the desert. In addition, we have economically modified those sweat glands to broadcast messages and export specific molecules as necessary, all without changing the basic sweat gland motif.

Even our finest conversion of the basic sweat gland design remains subject to nervous system and hormonal controls, and its function dependent upon the ability of specialized sweat glands to markedly modify the nature of their secretions, from more or less salty water to a much more complex fluid. Certain cloud formations and the entire class of *vertebrates* (animals with backbones) to which humans belong have been named in honor of those highly admired, modified sweat glands— long glorified in art, poetry and song and variously known as mammary glands, udders, teats or breasts. For it is they (as well as your hair and specialized inner-ear bones) that define your class of vertebrates as *mammals*—those who suckle their young. So the next time you pour modified sweat onto your breakfast cereal, just think of how it all evolved. By the way, those lovely mammatocumulus clouds with the hanging teat-like formations hold only water vapor. We discuss mammary glands in more tedious detail later (Reproduction, Growth and Development).

Local Heat Loss

The rich sub-surface blood supply of your face, scalp and nasal lining helps to cool your hard-working brain. This same copious circulation explains why a hat so effectively preserves body heat on cold days (cold nights too—flannel nightcaps were regularly worn to bed in winter before central heat). You can lose large volumes of sweat from the face and scalp, and your nose or mouth unavoidably undergo evaporative cooling as they humidify each breath en route to your lungs. Prominent irregularities within your moist nasal cavity further enhance its evaporative surface area and ability to cool adjacent blood and brain. And blood entering your brain may lose a little additional heat by *counter-current heat extraction* as it passes close by surface-cooled venous blood returning to your heart (see Circulation, Respiration and Kidney).

Humidity

As temperatures rise, water molecules travel at correspondingly higher average speeds and are less likely to remain together as droplets. Therefore warm air can hold more invisible water vapor than cold air. Conversely, when moist air cools, its water excess is revealed as clouds or fog (suspended water droplets) or in dew deposits upon cool surfaces. Clouds, fog and dew simply show that the local humidity has exceeded 100%. Although a cubic meter of air at room temperature can hold more than 30 milliliters (an ounce) of water as an invisible gas, evaporation ceases once air becomes water-saturated. Thus you drip sweat on hot and humid days without enjoying significant evaporative cooling—yet an equal output of perspiration on hot days with low humidity leaves you comfortably cool and dry.

Temperature And Reproduction

Those familiar with testicles know that they must be held several degrees below body temperature in order to function properly. The scrotal sac cleverly varies its length in accordance with body and external temperatures. Thus warm testicles hang loose to optimize cooling while scrotal skin muscles contract and the scrotum shrivels upward when the testicles become cold, as well as at other times (to be discussed). Interestingly, the temperatures at which reptile eggs are incubated often determine whether the outcome will be male or female. It is even possible that sudden warming or cooling of the Earth could have caused all dinosaur hatchlings to emerge as individuals of the same sex—a frustrating way to end The Great Age of Dinosaurs.

Burns

While your *epithelium* (skin) tolerates a lot of mechanical, chemical and electromagnetic abuse, too much will result in immediate or delayed damage. Examples include road rash abrasions from falling off your bicycle, acid burns from spilling your car battery, alkali burns from cleaning solutions, ultraviolet wave injuries (sunburns) or infrared wave (heat) injuries from contact with a hot stove. And widespread skin damage often results from contact with rapidly oxidizing materials (burning house, clothes on fire, contact with exploding hydrocarbon fuels and so on). The significance of skin burns is determined by their location, extent and severity. And location can be critical, for even a minor airway burn may cause death by interfering with adequate gas exchange. Burn extent commonly is described in percentage of body surface area damaged, burn severity as *first, second* or *third degree.* Although the latter classification resembles one generally used to categorize murders, it is reversed in severity since a *first degree burn* only causes redness, swelling and pain like an ordinary sunburn while a *second degree burn* implies painful partial thickness skin destruction with blistering or exposure of raw-looking deeper layers and a *third degree burn* means full thickness epithelial loss. Initially a third degree burn is insensitive except at its less severely burned margins—and unassisted healing will depend upon ingrowth of new skin from nearby live skin edges. Of course, regardless of degree, a murder always causes death. Thus the different legal classifications of murder refer only to mitigating circumstances with regard to culpability.

The many difficulties that arise following thermal injury to skin simply reemphasize the important roles of healthy skin. For example, any skin burn can lead to *swelling* as damaged capillaries leak fluid into the burned area. And burns of increasing depth and severity allow tissue fluids to leak out through damaged body surfaces. The major circulatory fluid losses consequent to a severe body burn encourage evaporative cooling that can cause critical *hypothermia* (low body temperature) within hours. Modern acute-burn therapy often requires intravenous replacement of up to several gallons per day of extracellular fluid losses. While skilled supportive care in a hot high-humidity hospital room can frequently bring a badly burned *hypovolemic* (low blood volume) individual through the time of most severe fluid losses, all sorts of opportunistic bacterial invaders then become a major threat to survival. Initially, these bacteria attack locally wherever skin protection has been lost. Later they may invade bowel wall, lungs and blood stream as the greatly weakened immune defenses of the badly injured victim collapse in the face of devious bacterial toxins and overwhelming bacterial attacks. Adequate nutrition remains a serious problem for the burn patient who must maintain a normal or even feverish body temperatures despite continuously wet surfaces and damaged superficial blood vessels that no longer can constrict to reduce heat losses. Necessary anabolic repairs, control of infection and an elevated metabolic rate all enhance the need for an adequate intake of high-energy nutrients as well as supplements of vitamins, minerals and so on.

When full-thickness skin loss occurs over a significant area (again, the definition of *significant* differs depending upon whether the burn involves eyelid or buttock), one hopes to achieve replacement skin coverage as soon as possible since the healing/scarring process continues for as long as the wound remains open. Even though some scar may be tolerable on a large flat surface, scar thickening rapidly interferes with function at sites where mobility is required. Thus early skin coverage is usually best although we occasionally delay wound closure to control infection, or very rarely even to increase wound strength by allowing additional scar formation. Severe skin burns present many more problems than we have alluded to here, and their treatment has become an important medical subspecialty.

Skin Grafts

The classical surgical method for covering areas denuded of skin involves live-skin transplants from elsewhere on the same individual. These transfers often consist of epidermis plus some dermis—a *split thickness skin graft* (about 0.3mm thick in the average adult) that leaves epidermal cells behind within trimmed-off sweat glands and hair follicles so that the *donor* site can heal. A *full thickness* skin transfer often improves final appearance and function (the less important donor area can then be closed by pulling it together or by application of a split thickness skin graft). A graft of skin plus some underlying tissue (e.g. skin plus fat plus muscle) can also be shifted with blood supply intact (or transplanted and blood vessels reconnected at the recipient site) when necessary for the repair of more complex injuries or in order to close chronically infected areas.

Skin substitutes, epidermal-growth-stimulating factors and tissue culture techniques have markedly enhanced burn treatment options. Under most circumstances, grafts of new skin can only be placed after all visibly dead tissue has been *debrided* (removed) and local infection controlled. Without a healthy skin cover, damaged tissue cannot be protected from hungry microbes or even *maggots* (sometimes those cute baby flies are a big help in the debridement process). Split thickness skin grafts necessarily include only a thin layer of fibrous dermis since by definition they cannot include deeper structures such as hair follicles, sebaceous glands and sweat glands. Even after healing, split thickness graft recipient sites are more fragile than normal skin, so ongoing graft shrinkage and scar formation may lead to persistent problems. Treatment of a severely burned person rarely restores normal appearance and function—quite clearly, burns are better prevented than treated.

Your *stratified* (layered) *squamous* (flat) epithelial skin cells provide a wonderfully functional, constantly renewed *outer* surface. But any build-up of skin debris within closed body cavities (as may occur when stratified squamous epithelium grows through a perforated eardrum into the middle ear) causes ongoing infection and increasing difficulties. So the linings of your closed body spaces are constructed quite differently, as we shall see.

CHAPTER 7

MULTICELLULAR MORALITY BEGAN WITH COLLAGEN

Cooperation Created Fiercer Competitors;... Collagen Caused Kindness Through Specialization Of Cells;... Collagen;... Good Proteins Come In Small Modules;... You Are A Monkey's Aunt (Or Uncle);... After Life Comes Glue;... Extracellular Space Can Be A Busy Place;... How Your Tissues Hold Their Water;... Proteoglycan Is The Answer.

Cooperation Created Fiercer Competitors

In the beginning, all were adversaries. While solar-powered photosynthetic life competed to the death for light, space and nutrients, all others preyed upon photosynthetic life and each other. Laboriously gained solar energy was the prize ever again ripped from the still living body of its most recent owner. Nothing was sacred or secure. The battle for life could be won repeatedly but lost only once. Those displaying particular nucleotide sequences might come and go like mirages but DNA itself would abide. At its most basic level, life's turmoil and frenzy represented unending unorganized elimination bouts between randomly revised wraith-like momentarily embodied information packets. And those combatants who survived long enough to pass along a slightly revised DNA sword and shield thereby armed their descendents for further mindless and ever more complex battles. While this might seem an incredibly wasteful way to proceed, advances came from everywhere, everyone played, everything served. With no goal in sight, all changes had equal significance. Though the latest innovation rarely endured, that passing blur was life in its usual fast-forward setting.

However, all was not just "kill or be killed". At times, unplanned cooperation suddenly became the only pathway toward survival. Not uncommonly this occurred when one bacterium or cell penetrated or engulfed another, yet neither could dominate. Under such chaotic circumstance, cooperation sometimes brought unexpected dividends. And if certain of these interlocked former combatants came to control critical information or metabolic processes, the improvement in cell capability and organization often brought significant reproductive advantages to all participants. So after one or two billion years of life's fumbling and bumbling about, some descendents of formerly free-living individual bacteria found themselves in new roles as chloroplasts and mitochondria that came with full eucaryotic benefits. But despite their release from the burdens of solitary existence, these symbiotic subcontractors remained under intense pressure to enhance efficiency—otherwise they and their cell mates could not prosper. Increasingly it paid to be helpful to other subcellular organelles, for the success of that home cell determined the survival of all. Yet just as in politics, long-range planning did not exist at any level, the public interest was irrelevant and even though progress was defined as any outcome of competition by any means, relentless changes continuously added ever-growing complexities to the overwhelming information burden.

Despite many constraints, innumerable simple partnerships also developed between individuals or groups of free-living bacteria or cells, with utterly selfish participants inextricably entangled in some mutually sustaining reproductively favorable relationship. Larger gatherings of closely related cells also occurred. However, the organization of any multicellular cooperative still depended primarily upon adherence among soft and easily torn cell membranes. And when such a group of cells was ripped apart, any surviving cell had to fend for itself or it left no descendents. So the individual cell partners within most multicellular arrangements remained very similar in their abilities and needs, since few would gain reproductive advantage by irreversible specialization within what might be just a temporary group. The inevitable competition between such similar cells placed severe constraints on group size and information processing abilities in the pre-collagen era.

Collagen Caused Kindness Through Specialization Of Cells

Animal evolution never really took off until sturdy extracellular collagen frameworks became available to stabilize multicellular systems and thereby guarantee individual cells lifetime employment. As in subcellular communes, closely related member cells remained under desperate pressure to enhance the group's efficiency in the face of ever-growing challenges by competing organisms. Although surpluses generated were immediately reinvested, each evolving design soon succumbed to newer, more fortunate or effective organizations. However, by making it possible for animal life to take this next huge step up the organizational ladder, collagen brought the first measure of kindness to intercellular relationships. Before collagen, every cell had to fight for its own reproductive opportunities. After collagen, it took cell specialization to keep a multicellular animal in the running. And under that relentless pressure to specialize, most member cells gave up their own reproductive potential so that the group might survive—placing their reliance upon those pampered sex cells to move the group's shared DNA into the next generation. Thus your brains, brawn, beauty and metabolic skills reflect the successful reproductive efforts of your ever more efficient, specialized and talented multicellular ancestors as a myriad of less well-endowed creatures (our subjective judgment) fell by the genetic wayside. Of course, having such a huge number of specialist non-sex cells in charge of the production and distribution of your eggs or sperm explains why being female or male has played such a major role in your growth and development. And why you have so much trouble getting your mind off sex. It is all quite delicately balanced. If you devoted any more time and attention to sex it could become a serious reproductive disadvantage, for you might starve or otherwise neglect your daily needs or safety. On the other hand, any less time and effort devoted to reproductive concerns might leave the best opportunities to those more eager.

Certain conclusions seem inevitable. One reason children usually survive childhood is that they have become tough through natural selection—after all, not one of your direct ancestors ever died in childhood. Furthermore, you are here only because your parents took a lot of extra time and went to a great deal of trouble to engage in frequent sexual intercourse rather than indulge in other less strenuous or repetitious activities then available to them. This despite the fact that those other activities might have left them healthier, wealthier and wiser, or at least more rested and better fed. And fortunately for our fully populated Earth with all of its nearly identical people, it has again become possible for humans to separate the enjoyment of sexual information transfers from the duties and pleasures of reproduction, just as their bacterial forebears did at the start. Not surprisingly, when given the opportunity and despite all religious teachings to the contrary (for many religions thrive on subverted sex drives and guilt), most humans tend to favor other currently beneficial activities over maximizing their personal reproductive success. Indeed, where decent health care and reliable *contraceptives* (means for the avoidance of pregnancy other than giving up sexual intercourse) are readily available, the birth rate automatically settles toward population sustaining levels. Thus human experience confirms what you already knew *instinctively* (via inherited information)—that there are *two separate fundamental drives—sexual desire and the urge to reproduce*. Ordinarily, the more frequent desire to enjoy heterosexual intercourse obscures the more basic urge to have children. That sexual activities are usually so pleasurable simply emphasizes their importance in encouraging reproductively advantageous behavior.

Thus it really shouldn't matter if sexual intercourse originated as deviant bacterial behavior caused by a perverted attack of mutant plasmid information. It only matters that such DNA exchanges proved reproductively advantageous for both the plasmid and its bearers who, once they learned how, lived sexually ever after (and apparently enjoyed it). Only when sexual intercourse becomes difficult or distasteful, or where contraception reliably prevents pregnancy, does reproduction depend primarily upon the wish to bear children. Your strong primary sex drive certainly makes evolutionary sense, for it often takes repeated sexual intercourse to bring about a successful pregnancy. But it also helps to have another even more basic urge to fall back upon if sexual interactions somehow fail to please. So effective and safe contraception remains a blessing from any reasonable point of view, for it helps man and woman to bond in a warm mental/physical/sexual relationship according to their needs and desires, while reproduction becomes a preplanned, eagerly anticipated event rather than an unexpected, often dreaded or even intolerable burden (see Reproduction).

Collagen

A glance from the window of an average American home suggests that Earth is owned and operated by humans—yet we are but the tiniest sample of all the wild and wonderful creatures that have roamed through sea, land and sky. This may be your big moment in life's endless succession but collagen is what allowed the entire competitive escalation of multicellular complexity and information-processing capability to begin—for only collagen could ensure that altruistic sacrifices made by closely related cells for their peers would not be in vain. Small wonder that collagen is the most common animal protein, one quarter of your dry weight.

Life struggled ceaselessly for three billion years before collagen was invented. Undoubtedly, an endless number of other equally essential and even more immediate concerns also held up evolutionary progress. But just think how far you might have evolved and what sort of place Earth might now be if this sturdy extracellular protein had entered common usage one or two billion years earlier. So why did that insoluble extracellular fibrous framework take so long to develop? Might it have been a mere conceptual impasse rather than a complex biochemical problem? That a cell cannot afford to specialize in the absence of collagen fibers, yet a cell must specialize in order to produce collagen fibers? Perhaps some of what is known about collagen can help you to contemplate that question.

While it comes in many varieties, the fundamental collagen polypeptide is a helical (spiral-shaped) molecule about 1000 amino acids in length. Three such polypeptide chains—tightly wound together and hydrogen bonded—create a triple-stranded rod-like protein molecule 3000 A° long and 15 A° wide (to keep that in scale, remember that one *Angstrom* =10^{-10} meters, so 10 billion Angstroms = 1 meter. More importantly, individual cells often measure about 10 microns across and 10 microns = 10^5 A° or one hundred thousand Angstroms—which provides some sense of the difference in size between an ordinary cell and a very large extracellular molecule). Each long and rope-like *procollagen triple helix* is initially secreted as part of a loose fiber bundle by elongated cells known as *fibroblasts*. Once delivered into the extracellular space, the procollagen molecule has its protective amino acid end-tufts trimmed off by extracellular enzymes known as procollagen peptidases. Those freshly trimmed procollagens, now referred to as *tropocollagens*, rapidly self-assemble into orderly layers with 400 A° gaps between adjacent tropocollagens in the same line—those 400 A° gaps are staggered between rows so that only every 5th row matches. The stiffness of the resulting collagen fibers and sheets depends primarily upon the extent of covalent cross-linkages among tropocollagens—particularly between lysine and hydroxylysine—amino acids that are especially common in collagen.

So there you have the simple basic structure that possibly delayed your arrival for billions of years—a regularly repeated sequence of three amino acids (commonly glycine-proline-hydroxyproline) in each of three long tightly twisted amino acid chains that together form a straight fiber capable of resisting great tension. Once in the extracellular space and trimmed, those fibers rapidly self-aggregate into the sturdy cross-linked water-insoluble extracellular fibrous meshwork that holds you together while also resisting and transferring pressures and tensions due to muscle contractions, gravity and other accelerating forces.

Good Proteins Come In Small Modules

Interestingly, every collagen gene includes about 50 short exons and half of these exons are just 54 nucleotide pairs in length—which allows them to specify a helical polypeptide 18 amino acids long that makes exactly 6 full turns. This arrangement positions a glycine centrally within the triple helix during each complete three-amino-acid turn. And that is essential, for were a larger amino acid to take glycine's internal position, the resulting unsightly bulge would prevent hydrogen bonding between the $-C=O$ and $H-N-$ of adjacent polypeptide chains, thereby allowing easy unraveling and a disastrous loss of tensile strength (just as a localized separation anywhere along its length will weaken a fully closed zipper).

Collagen is certainly an unusual protein (if one can call something that common "unusual") but other strong extracellular protein fibers such as silk and elastin are similarly based upon regularly repeated amino acid sequences. In addition to resisting tension, elastin can stretch to several times its original length and still spring back to the starting condition. That elasticity may result in part from the way covalent bonds crosslink a great many elastin molecules into a meshwork pattern with easy extensibility in all directions. Furthermore, the many hydrophobic amino acid side chains of each 700-amino-acid-long elastin molecule tend to form tight water-excluding hydrogen-bonded spirals in between those sturdy covalent crosslinks—the way such hydrogen bonds separate and reform apparently gives the entire elastin structure a spring-like response to being stretched.

Elastin is a major component of elastic tissue, along with *fibrillin*, a connective tissue protein that is particularly abundant in elastic tissues (and also found in the periosteum of your bones and the suspensory ligaments of your eye lens). Elastin has minor importance in most collagenous connective tissues except where its ability to store tension can reduce work, as in certain modified leg tendons that give more bounce to horses and kangaroos, or in the neck tendon of grazing animals that must frequently raise their heads to detect approaching carnivores, or in your own stretchy aorta that stores your arterial blood at elevated pressures (see Circulation). But the ongoing formation of covalent crosslinks causes elastin to gradually stiffen with advancing age.

You Are A Monkey's Aunt (Or Uncle)

Certain modifications of the primary collagen molecule allow it to become more heavily cross-linked in locations such as your Achilles tendon where stiffness is a virtue. Other variations include carbohydrate to encourage formation of collagen sheets rather than tendons. But successful collagen formation always requires an adequate supply of *Vitamin C* (ascorbic acid), for this strong reducing agent maintains *prolylhydroxylase* (an essential enzyme) in its reduced ferrous (Fe^{++}) *active* state. The least cross-linked (hence lowest-melting-point) collagen is found in soft-fleshed cold water fish, while heavy hydroxylation of prolines is necessary to stabilize the more chewy collagen of warm-blooded creatures. Most animals produce their own Vitamin C, but all *primates* (you and your ugly ape and monkey relatives) and guinea pigs as well, have lost that ability. Presumably that need for dietary Vitamin C arose by separate but similar mutations in an ancestral primate and an ancestral rodent since no other mammals have that problem—or does God perhaps require Vitamin C, just like man in His image? Although hunter/scavenger/gatherer diets generally consist of fresh (living or recently deceased) foods containing plenty of Vitamin C, the inadequate preservation of your animal and vegetable victims for future use could lead to a diet deficient in Vitamin C. And if you subsist exclusively on such an inadequate (because long-dead) diet for several months, your continuously reconstructed collagen may then become inadequately hydroxylated—such poorly bonded collagen fibers will melt (separate) at ordinary warm-blooded body temperatures.

On the other hand (for reasons that remain unclear), the sturdy collagen of deep sea hydrothermal vent worms includes little hydroxyproline, yet it remains stable at far higher than boiling temperatures (maybe even up to 250°C) under pressures that exceed 260 atmospheres. Perhaps those pressures squeeze such collagen fibers together sufficiently to enhance their van der Waal's interactions (just as high undersea pressures reduce cell membrane fluidity)—in which case, these worms should pull apart easily if rewarmed (nearly boiled) at the sea surface. Signs of *scurvy* (Vitamin C deficiency disease) include loosening of teeth, spontaneous skin bruising, poor healing of wounds, weakening of blood vessels and a great many other distressing complications of weak collagen up to and including death. Vitamin C has additional important functions unrelated to collagen—for example, in the metabolism of certain amino acids and the construction of your adrenal steroids. Ordinary amounts of Vitamin C appear to enhance your resistance to infection—the effects of larger doses are still under investigation (e.g. as antioxidants to minimize intracellular accumulations of those oxidised and therefore dysfunctional proteins that are associated with aging).

After Life Comes Glue

So how did this family of sturdy, insoluble, fibrous proteins ever get a name like *collagen* (meaning "to produce glue" in Greek)? What possible relationship is there between glue and such strong stringy stuff? Well, not so long ago, as automobiles rapidly replaced horses, the perplexed driver of an obviously disabled vehicle at the roadside often received merry but unwanted advice such as "Get a horse!" from passers-by whose own transport still happened to be working. Large numbers of dead and disabled vehicles soon littered the countryside, providing a marked contrast to the quiet patiently plodding hayburners of yesteryear that only rarely were encountered in a disabled or dead condition. In good part, of course, the lack of end-stage horses lying about was explainable by their large stores of readily recyclable organic materials. Unlike a dead car that might remain more or less intact for recycling at anyone's convenience, dead horses rapidly became very offensive when undergoing spontaneous decay with the aid of bacteria, worms, insects, birds and other reducing-power-attracted opportunists such as your freshly washed dog. On the other hand, if delivered under its own power in a timely fashion to a glue factory (rendering plant), that faithful worked-out old dobbin could return to service as soap, explosives, cat food, leather belts, cowboy boots and fertilizer. Any collagenous left-overs (bones, tendons, ligaments, noses and hooves) were readily boiled down to create *animal glue* or produce those fine flavorless packets of cooking *gelatin*. So how does boiling convert a naturally tough stringy insoluble protein into a soluble denatured protein that firmly fastens unfinished wood surfaces together or tastefully stiffens a hot liquid into a jiggling solid as it cools?

You may recall that a great many *hydrogen bonds* (which occur when the relatively positive hydrogen already in covalent bondage to one O or N is irresistibly attracted toward an adjacent relatively negative O or N) stabilize each triple-stranded tropocollagen into a very strong protein string. But when such protein strings are heated to boiling, the added thermal energy causes such a flailing of individual polypeptide strands that the hydrogen bonds between them are progressively disrupted. As they then whip about ever more freely, these long twisted polypeptides become inextricably entangled about each other and their covalent crosslinks. Thus in contrast to linear DNA (which lacks covalent crosslinks and is most stable when complementary double-strands match perfectly), collagen becomes so entangled during complete denaturation that it can never regain its linear triple spiral form. Of course, all of those disrupted hydrogen bonds make denatured collagen very attractive to adjacent polar water molecules, which encourages denatured collagen fibers to dissolve in hot water and then produce a stable jelly as those cooling gelatin molecules whip about less vigorously. In other words, gelatin stiffens water by forming an interlocking water-attracting tangle or lattice structure throughout the solution. That allows boiled turkey bones and other dead animal parts to become nicely jellied soups as they cool. And gelatin glue naturally adheres to any nearby polar (thus wettable) surface as its multitude of hydrogen bonds become available through evaporation or absorption of water molecules into the wood. Despite that useful role as glue, gelatin lacks an organized fibrous structure, so it would not resist your tissue tension even as well as the most defective collagen produced without Vitamin C. Or as they say, ask a simple question…

Extracellular Space Can Be A Busy Place

Collagen is laid down in large parallel bundles at locations where it must transfer a great deal of tension between two bones (as a *ligament*) or between a muscle and a bone (as a *tendon*). But the collagen fibers of most *connective tissues* are deposited in relatively random directions throughout the *extracellular* space (between cells). Many of your mobile cells are able to travel along collagen fibers and other fibrous connective tissue proteins such as *fibronectin*. For example, the *integrins* are a family of cell membrane receptors that specifically bind to and release fibronectin. Growth and development, healing of wounds and control of infection all depend upon efficient cell migration through the extracellular matrix along such pathways. And the way various matrix constituents retain or concentrate tissue growth and repair factors at the site of injury helps to direct repairs and keep wound healing from becoming a systemic response for a local problem. Yet under the microscope, many connective tissues appear mostly empty except for fibroblasts, fat cells and a scattering of collagen, elastin, fibrillin and fibronectin fibers. Functionally, this *a*cellularity (*absence* of cells) in your connective tissues makes sense, for your body requires flexibility between its more solid tissues to allow the limited changes in your shape that occur with different positions and movements—for example, when tying your shoes. The stresses of accommodating to such body deformations are best taken up by those soft, fatty, fibrous and not very cellular connective tissues.

How Your Tissues Hold Their Water

So what fills all the space between your connective tissue cells and those many fibers? Clearly *extracellular* (tissue) *fluid* must be present because connective tissues that are low in fat have nearly the same density as water. Furthermore, the extracellular water, ions, gases and small nutrient molecules being transported by your blood stream can readily enter and diffuse through those connective tissue fluids. And as is evident from your occasionally sunken eyes, parched lips and dry wrinkly tongue, tissue fluids return readily to your circulating blood when you are *dehydrated* (water depleted and thirsty). This dynamic distribution of your extracellular fluid makes it even more remarkable that tissue fluids do not quickly settle to the lowest portion of your body. Admittedly, your ankles may swell a bit with prolonged standing, but what stops all of your extracellular fluid from flowing to your buttocks when you sit or sloshing down around your ankles when you stand? What keeps just the right amount of tissue fluid in the right place so you don't puff down there as you shrivel up here? How does your extracellular fluid know where to go and how much of it is required?

Proteoglycan Is The Answer

It appears that those rude individuals who whistle and say, "It must be jelly 'cause jam don't shake like that!" are correct. For the connective tissue matrix within which your collagen fibers are deposited and through which water, gases and other solutes must diffuse, actually is jellied or stabilized by very large soluble extracellular *proteoglycans*—polyanionic (bearing multiple negative charges) molecules that are about 5% protein and 95% carbohydrate. Their polar carbohydrate side-branches and multiple negative charges allow proteoglycan molecules to dissolve readily and attract many cations (especially Na^+, the dominant cation of your extracellular fluid). So between all of the water that it takes to dissolve such highly charged and polar proteoglycan molecules and all of the additional water osmotically attracted by those many electrically held cations, it is not so surprising that your extracellular fluids and solutes tend to stay where each enormous proteoglycan molecule happens to be produced and deposited. And since proteoglycan molecules (which resemble microscopic ferns or feathers with many interacting fronds) can be several microns in length, there is little likelihood that those proteoglycans will ooze and slither down to your low spots. Thus your tissues do jiggle 'cause it's jelly and that jiggling becomes most evident when your tissues are not already stretched by gravity—in a swimming pool, for example, or in outer space.

Several decades ago, the reliable delivery of intravenous fluids through reusable and resharpened hollow steel needles often represented a difficult technical challenge, especially in persons with "poor veins". So essential fluids were sometimes delivered via a sturdy *trocar point* (saw-toothed) needle drilled into a bone marrow cavity (tibia in children) or as a *clysis* delivered through an ordinary intravenous-type needle into the fatty subcutaneous tissues of the thigh or abdomen. *Hyaluronidase* could be added to speed the egress of clysis fluids from swollen subcutaneous tissues into the circulation. The action of this enzyme was to depolymerize *hyaluronic acid*, a prominent connective tissue proteoglycan. In similar fashion, your naturally high concentration of corneal hyaluronidase expedites the local removal of extracellular fluid from the corneas of your eyes so they will remain transparent. But there is far more to multicellularity than just collagen, proteoglycans or even the very localized expression of those common hyaluronidase genes...

CHAPTER 8

COLLAGEN PULLS, CALCIUM LIFTS, CARTILAGE SUCKS

A Skeleton Made Of Sugar Sucks;... A Flexible Water Skeleton That Can Suck Or Blow;... Some Seek Shape And Shelter Within Their Own Metabolic Wastes;... Your Dry Bones;... Shock Absorbers Reduce Structural Stress;... Collagen Pulls While Cartilage Sucks;... You Were Patterned In Cartilage;... Perichondrium;... Bone Formation;... Your Inner Rocks;... Bones Rarely Break;... Bone Remodeling Reduces Risk Of Fracture While Controlling The Calcium Concentration In Your Blood;... Avoid Osteoporosis, Rhubarb And Spinach.

Strong as it is, collagen alone could never hold you up. To stand tall, or even short, you also need a rigid incompressible framework that can resist gravity and other accelerations. So just as tent poles bring structure to a floppy cloth tent, your bony skeleton provides functional form to your otherwise soft and shapeless substance.

A Skeleton Made Of Sugar Sucks

Support alone does not guarantee mobility, of course, for otherwise trees would wander in and out of the shade as you do. Instead trees are immobile strongly rooted porous laminations of polar (thus water-loving) cellulose fibers bonded by lignin. Those sturdy and wet cellulose fibers enclose glucose-filled osmotically active plant cells that draw great masses of polar water out of barely moist soils. Entrapped water then unfurls and supports many new branches and leaves in order to capture more solar energy in reduced carbon (glucose) form. Their water burden makes larger trees enormously heavy—about as heavy as the same volume of water. That is why freshly cut live-tree logs other than balsa do not float very high. Lumber shrinks as it dries, losing both weight and flexibility.

A Flexible Water Skeleton That Can Suck Or Blow

A squid actively traps water for support within its tough cone-shaped mantle-muscle-surrounded body cavity. The weight of this trapped water is no deterrent to streamlined travel through the equally dense surrounding seas. Indeed the inertia of its internal water column ordinarily helps the squid move slowly astern with minimal effort while also stiffening it into proper squid shape. But for an urgent departure, the squid forcefully ejects much of that skeletal water through a jet nozzle that can swivel in almost any direction. Its reduced size and inertia then allow the just-contracted squid to accelerate more rapidly. Caterpillars rely upon internally generated hydraulic pressures for their support and locomotion since they too lack rigid internal or external support structures. As a result, their locomotion on land is slow and energetically costly. However, one soft-bodied legless fruit fly *larva* (maggot) is able to proceed more efficiently by jumping 12 cm (5 inches) at a time. It achieves that mode of muscle-assisted legless hopping by curling tightly and then utilizing the elastic recoil of sudden release to help it evade predation by ants.

Some Seek Shape And Shelter Within Their Own Metabolic Wastes

For over 600 million years, single and multicelled organisms have sought support and shelter from predators within sturdy external deposits composed of their own waste CO_2 co-precipitated with spare calcium as calcium carbonate ($CaCO_3$). Even giant clams find such shells economical, functional and reproductively advantageous, but a similarly proportioned structurally sufficient limestone *exo*skeleton (*external* skeleton) for you would weigh many hundred pounds. To achieve easy mobility, animals must excrete or abandon all non-essential weight.

Your Dry Bones

By weight you are one-half to three-fourths water. That water fills out and even inflates various of your tissues and organs. But your bony skeleton is a rather dry composite of strong collagen fibers interlocked with metabolically essential (and potentially toxic) calcium phosphate salts. Being quite insoluble and more dense than water, bones sink to the bottom of your soup. Within their hollow cores, your bones either contain fat cells (*yellow bone marrow*) or blood-forming cells (*red marrow*). Most humans enclose enough body fat plus gases in lungs and bowels to barely float their bony burden in fresh water or more easily in heavier salt water. Thus live people are nearly of water density (which is about 1 gram per ml.

or 2.2 pounds per liter or a bit over a pound to the pint). Bird-bones enclose air but live seabirds float high or low depending upon how much air is trapped within their lightly oiled and therefore water repellent plumage. Unfortunately those non-polar insulating feathers readily absorb spilled hydrocarbons—heavily oiled birds that do not become *hypothermic* (cold) or drown are soon poisoned by oil ingested during self-grooming. Your relatively lightweight but rigid *endo*skeleton (*internal* skeleton) provides sturdy support for your tissues. To allow mobility, that incompressible skeleton is broken into hundreds of independent metabolically active bones. In turn, each of those appropriately sized collagen-reinforced shatter-resistant skeletal struts and levers is strengthened, aligned and securely positioned to transfer all reasonable forces safely and effectively—which allows you to range widely and avoid predation as you seek water, steal solar energy from plants and devour other *herbivores* (animals that prey upon plants) or *carnivores* (such as fish).

Shock Absorbers Reduce Structural Stress

Heavy collagenous restraining straps and hinges at each of your *joints* (where two or more bones meet) place strict limits on the movements of your bony *struts* with respect to each other, for an excessively mobile bone could easily injure adjacent tissues or allow catastrophic collapse due to loss of leverage. Furthermore, the strong muscles (and their tendons) that surround and *spring* those joints, markedly reduce the effect of sudden strains or impacts by resisting or giving way slowly and elastically. But only your soft water-filled cartilages can provide essential hydraulic cushioning to absorb *shocks* where rock-hard bones contact and move across one another. Clearly you are held in shape by an endless array of interwoven collagen fibers, threads, strings, ropes and cables ranging from the fine meshwork of your loose connective tissues and skin to the taut fibers that transfer the pull of each muscle fiber into the thick collagen bundles of your tendons. In turn, those tendons penetrate the fibrous *periosteal* wrapping of your bones to meld with the organized collagen bundles of your skeleton. Obviously there is much more to life than just collagen, but if you could somehow dissolve away their other molecules while keeping each collagen fiber intact and in place, your best friends would remain readily recognizable. They might appear a bit pale, ghostly and worse for the wear and tear, but their entire structure, even teeth and eyes, would remain fully formed.

Collagen Pulls While Cartilage Sucks

In common with other connective tissues, cartilage consists mostly of collagen (and numerous other fibrous proteins), various proteoglycans and jellied (trapped) extracellular water containing the usual solutes. Cartilage maintains a higher osmotic pressure than surrounding tissues through its increased concentrations of highly anionic and hydrophilic proteoglycans and the many Na^+ cations that these proteoglycans attract. So the extracellular matrix of cartilage tends to imbibe water until it becomes tightly distended—that is, until its collagen fibers are fully stretched out and tense. Therefore the firmness and compressibility of your cartilages reflects their high osmotic pressure. Where joints require only limited flexibility (between your vertebrae, for example), both bones can be securely bonded to a single intervening *fibrous* (collagen-rich) cartilage. But the more mobile *synovial joints* of your extremities depend upon the separate bearing surfaces of each bone end being individually cushioned by a smooth *hyaline* (more watery/less fibrous) *cartilage* cap. The slippery gas-and-nutrient-rich *synovial fluid* secreted by surrounding *synovial membranes* lubricates your hyaline joint spaces. Hyaline cartilages require far less collagenous reinforcement than fibrous cartilage because their opposing surfaces ordinarily slide so smoothly that hyaline cartilage encounters only compressive (and not shear or friction) forces. However, fully compressed hyaline cartilage resists sliding so a hyaline cartilage forcefully pinned into position can be torn by severe shear stress—the meniscus cartilage of the knee is commonly injured in this fashion when football players are hit while their foot is firmly planted. And badly torn cartilage cannot heal properly, for *chondrocytes* (the cells that first deposit and then live on within your cartilage) generally reside too far from their own blood supply to engage in vigorous healing. Indeed, chondrocytes are sustained by the rather slow diffusive exchange of gases and solutes with adjacent tissue capillaries (often via the synovial fluid).

That diffusion of gases and nutrients is greatly speeded by the repeated cartilage compressions of ordinary use. So cartilage tends to *atrophy* (shrink) with disuse or *hypertrophy* (flourish and thicken) as a result of ongoing vigorous but not excessive joint activity. However, the persistent and extreme stresses applied to various joints by gymnasts and distance runners can eventually disrupt cartilage, as can ordinary wear and tear (by the time you reach old age) or any local bacterial infections or ongoing damage consequent to *autoimmune* (misdirected at self) antibodies. The repair of joint damage generally involves inflammation, enhanced circulation and deposition of scar tissue by fibroblasts—all three may interfere with smooth and painless joint function, especially since any ingrowth of blood vessels encourages cartilage conversion to bone. A multitude of generally painful joint dysfunctions are descriptively grouped as various forms of *arthritis*. Although cartilage rebounds from an impact (brief compression), it tends to *creep* (deform progressively) under ongoing pressure. This accounts for your half centimeter (about ¼ inch) loss of height during an ordinary day's upright activities. You regain that height at night when those decompressed cartilages are again free to imbibe fluid (weightless space-persons may gain a temporary inch in height by that same mechanism). It seems that all vertebrates must walk (swim, fly) on trapped water in this fashion, for without cartilage there would be nothing to prevent your bones from cracking and grinding together until eventually they became solidly *fused* (connected through their surfaces by scar tissue and calcium phosphate deposits), leaving you an immobile pushover.

You Were Patterned In Cartilage

Every surface cartilage of your many fitted synovial joints seems perfectly sculpted for its setting. Detailed design specifications need not clutter your DNA, however, for cartilage tends to develop and assume proper shape in response to conditions at sites where it is needed. Even a *fractured* (broken) bone that fails to mend properly may develop a *psuedarthrosis* (false joint) at the break site. Such a psuedarthrosis is commonly associated with pain and malfunction since any weight placed upon the false joint tends to cause buckling with forceful displacement of broken bone ends into surrounding soft tissues. Interestingly, both bony ends of a chronically non-healed fracture tend to develop smooth cartilaginous caps of the sort one might find in an ordinary synovial joint. Similarly, when a congenitally displaced hip (without a normal hip socket) is properly positioned soon after birth, it causes a fully functional well-shaped hip joint to develop—complete with cartilage and ligaments—at the site of persistent contact with pelvic bone. Easily modified flexible cartilage skeletons allow sharks and rays of all sizes to prosper—however, adult fish and land-dwelling animals gain reproductive advantage by replacing the cartilaginous fetal skeletal pattern with a more rigid structure. Most larger land-based herbivores must soon be ready to flee for their lives so they are born with quite advanced bone formation. In contrast, the primarily cartilaginous construction of human infants reduces risk of injury to mother and child while still providing sufficient support for effective screaming, eating, excreting and smiling.

Perichondrium

An external collagen-fiber-reinforced envelope surrounds cartilage except at its bony or free (synovial joint) surfaces. This firmly attached *peri*chondrial (*around* cartilage) layer carries nutrient blood vessels that serve the underlying *avascular* (lacking in blood vessels) cartilage—perichondrium also produces new cartilage. Forty or fifty years ago, wrestlers competing without ear protectors often suffered repeated mat friction (shear) injuries that separated a patch of ear perichondrium from its underlying cartilage. Repeated episodes of new cartilage formation along the inner surfaces of such displaced perichondrium (over blister-like collections of blood) eventually led to unsightly "cauliflower ears". Similarly, when rib cartilage is surgically removed but some of its surrounding perichondrium remains in place, new and irregular cartilage or even bone will soon regenerate from those disrupted perichondrial surfaces. But do not be misled by how readily the tasty perichondrium or *peri*osteum (*around* bones) comes away from those shiny barbecued cartilage or rib surfaces at a picnic—that only occurs because the heat and moisture of your skillful barbecuing has converted their sturdy collagen to gelatin.

Bone Formation

During normal growth and development, the human fetal skeleton *ossifies* (turns to bone) under the influence of transforming growth factors as blood vessels invade the cartilage. Because this process follows a standard schedule, the current stage of skeletal maturation is referred to as the *skeletal* or *bone age* of the fetus, infant or child. With each defined in terms of the other, bone age and chronological age are normally about the same. So a skeletal X-ray examination should reveal the approximate age of any youngster. Chondrocytes give way to *osteoblasts* (bone-producing cells) under the influence of a variety of related bone-growth proteins (that may originate in the kidneys) and other signal or support molecules. These *osteogenic proteins* or *morphogens* also encourage other tissues to become bone. While size, shape, location and use all affect their final structure, bones generally develop a hollowed-out (often cross-braced) interior surrounded by a sturdy dense-bone *cortex*. Ordinary tensile and compressive forces act mainly upon outer bone edges so this more-or-less hollow design reduces weight without much loss of strength. Dense cortical bone is mostly laid down in tiny side-by-side cylinders—each surrounding a central nutrient blood vessel derived from periosteal vessels. The thin outermost bone layers that surround all of these concentrically layered cylindrical *Haversian systems* are deposited just under the periosteum. Once fully enclosed within bone by their own activity, osteoblasts become known as *osteocytes* (a bit like changing your name after painting yourself into a corner). Individual osteocytes (*osteo* means bone, *blast* means producer, *cyte* means cell) transfer nutrients, wastes and gases to neighboring cells via thin protoplasmic extensions that pass through tiny *canaliculi* (interconnecting bone passages).

Following ossification of the original skeletal cartilage, bone growth continues circumferentially from periosteum and lengthwise (in long bones) from transversely positioned cartilaginous growth centers. These *epiphyseal growth plates* are located where each end of the *diaphysis* (bone shaft) widens out to become the proximal or distal *epiphysis*. By widening near each joint surface, long bones are better able to resist any oblique compressive strains transferred across the adjacent synovial joint—those larger joints are more easily stabilized and cushioned. The expanded epiphyses sufficiently reduce pressures applied to joint surfaces so that bony epiphyses can be built of lighter *cancellous* (spongy or porous) bone. In contrast, the compact bone of which the narrow diaphyseal shaft is made permits some slight twisting and bending that can improve resistance to temporary overloads (a hollow cylindrical bone-substitute of uniform diameter and equivalent strength would be totally inflexible and far heavier). Some flat bones of the skull form directly from periosteum through *intramembranous* bone formation rather than via *endochondral bone formation* (by conversion of a cartilage model into bone).

Where flexibility has advantages over rigidity, cartilage may persist in the final structure. So cartilage gives form to your ears and the tip of your nose as well as those *anterior* (front) rib cartilages—although the latter do ossify slowly as you age. The growth rate of your skeleton (and soft tissues) is subject to the influence of various growth and steroid hormones. Too much *growth hormone* secreted by a pituitary gland tumor during childhood or adolescence (before those epiphyseal plates mature into solid bone and finally terminate longitudinal growth) can lead to excessive height and size. *Anabolic steroids* and *testosterone* (male sex hormone) have similar effects on bone and muscle growth, causing an epiphyseal-plate-based growth spurt in children and adolescents. Unfortunately, anabolic steroids are frequently utilized (abused) by athletes seeking improved performance or greater muscle mass. But steroid-induced skeletal-growth-spurts markedly accelerate bone maturation (just as your own adolescent growth spurt soon tapered off) so the net result is *premature closure* (maturation into bone) of all epiphyseal plates and cessation of bone growth (at least in length). Thus the use of anabolic steroids can bring about a shorter adult stature than would otherwise have resulted. More important adverse effects of anabolic steroids on brain, liver, testicles and immune system are discussed in later chapters.

Your Inner Rocks

Hydroxyapatite or $Ca_{10}(PO_4)_6(OH)_2$ is the crystalline mineral that stiffens your bones and teeth. Like most rocks, hydroxyapatite is rather insoluble in plain water. The flexibility of a bone varies with its mix of collagen fiber and calcium phosphate. A high concentration of collagen allows deer antler bone to absorb repeated severe impacts with low likelihood of antler fracture (or of severe headache that might delay mating). On the other hand, whale ear bones are mostly rigid mineral in order to transmit sound vibrations with minimum loss and high fidelity (so they are easily broken by nearby underwater explosions). In all cases, collagen provides an essential site where bone deposition can begin, with the first tiny hydroxyapatite seed crystals being laid down within those 400 A° gaps between the tips of neighboring tropocollagen fibers. By thus interlocking strong collagen fibers with sturdy calcium phosphate crystals, the advanced composite material known as bone has provided life-saving strength and resilience for several hundred million years. But tiny changes can occasionally cause big trouble—as in *Osteogenesis Imperfecta* ("brittle bone disease")—this well-known, generally lethal bone condition is so strongly selected against that its prevalence depends entirely upon the rate of new genetic mutations. The fact that this disease is associated with frequent bone fractures simply demonstrates the crucial importance of adequate collagen fiber reinforcement to bone strength. One common cause of this disease is a single nucleotide error (T for G) that leads to cysteine (a larger amino acid) replacing a glycine near one end of each collagen polypeptide. A number of other substitutions and even exon deletions can produce a similar partial unfolding of the collagen triple helix—leading to marked weakness of those abnormal collagen fibers (see Collagen).

Bones Rarely Break

To split firewood, you must initiate and then advance a crack that parallels the majority of structural fibers. By forcefully prying apart the two sides, the entering wedge or axe head delivers much of its force at the point of the crack. But a split may not progress if too much of the force is dissipated by excessive flexibility of the sides—that is why a harder strike is far more effective, and frozen or dried wood so much easier to split. A knot (old branch crossing deep inside the wood) can similarly stop a split as tension at the crack tip then spreads out along (and is resisted by) those knotty cross-fibers. This explains why your axe is now stuck in the log, although it may not help you to extract it. And that short crack in your automobile windshield may soon lengthen as stresses or movement between the two sides exert leverage at the tip of the crack—the longer that crack (lever arm), the more rapidly it will advance. Sometimes a glass crack can be stopped at an early stage by drilling a small hole through which to evacuate air and inject a strong plastic glue of appropriate optical density into the split (equivalent to placing strong cross fibers or knots in a log). Like trees, your tissues generally have a grain (predominant alignment of collagen fibers) along which separation or splitting is more readily achieved. Of course, all natural *cleavage planes* are generously cross-braced by fibers to resist such overloads. But small cracks in bones and minor tears in other tissues are an inevitable accompaniment to life's activities, so the fibrous (and other) proteins of your body are regularly and routinely resorbed, redeposited and remodeled in order to reduce the risk of catastrophic structural failure—perhaps 5% of your bone mass is under construction at any moment.

Reproductive advantage goes with speed and maneuverability as well as sturdy construction. Being unbreakable is no advantage if that makes you too heavy to catch lunch or escape. So your structure has been optimized by your long-suffering ancestors through an unending series of actual life-and-death field trials. As a result, the collagen content, size and mineralization of your skeleton is probably sufficient to resist the occasional 50% overload (above your usual maximum). That safety margin may not exist or suffice in individual cases, so fractures (strut failures) and *dislocations* (joint disruptions) do occur in the face of repeated as well as excessive stresses. Although your bones often crack (that persistently sore spot you refer to as a *bone bruise*) they infrequently break and then only due to stresses well above their structural limits. A *pathological* fracture occurs in a bone already weakened by osteoporosis, cancer or other disease. But even normal bones can be overstressed with repeated flexion when there is insufficient rest for reparative processes. Such *stress fractures* may occur in a foot bone of the military recruit who carries a heavy pack on an excessively long march. In other words, unreasonable stress not subject to ordinary commonsense limitations (by soreness and fatigue) can certainly cause a bone crack to progress into a fracture, or lead to inflammation with painful swelling of the affected soft tissues (as seen with tennis elbow and many occupation-related *repetitive motion injuries*). Bones and other support tissues generally respond to less extreme stresses by thickening and becoming stronger, while disuse (when immobilized in a cast, for example) promotes calcium removal with weakening of the bone structure. The prolonged weightlessness of space flight is associated with bone demineralization and reduced osteoblast activity. All of which suggests that you are a responsive and adaptable creature built of responsive and adaptable cells, tissues and organs—up to a point.

Bone Remodeling Reduces Risk Of Fracture While Controlling The Calcium Concentration In Your Blood

Even if not broken or cracked, your bones undergo continuous remodeling, especially while you still grow. An integral part of that lifelong remodeling process is removal of bone. *Osteoclasts* are very large *multinucleated* cells (each cell containing many nuclei) that share ancestry with certain blood cells and specialize in bone removal. The rate at which they perform that duty is determined by local intercellular signals and also *systemic* (body-wide) conditions. For example, the proper function of your cells requires an appropriate and stable level of free calcium ions in your extracellular fluids. So whenever blood Ca^{++} concentrations decline, your blood *parathyroid hormone* levels rise and stimulate osteoclastic destruction of bone until blood levels of calcium and phosphate ions return to normal—any excess of blood $PO_4^=$ is taken up by cells or excreted in the urine to prevent *ectopic calcification* of non-bony tissues. On the other hand, high blood Ca^{++} levels reduce parathyroid hormone secretion, allowing osteoblasts to produce more bone than is being destroyed by osteoclasts so that elevated blood calcium levels then return to normal.

Calcitonin, a hormone from the *thyroid gland*, stimulates osteoblasts to produce more bone when body fluid (extracellular) Ca^{++} levels are high. And *Vitamin D*, the fat-soluble "sunshine" vitamin, is involved in your Ca^{++} metabolism as well. But first its precursor molecule must be altered by ultraviolet light while in the dermis of your skin—that Vitamin D molecule then undergoes additional chemical modifications in your liver before finally being activated in your kidney. The fully effective form—Vitamin D_3 or *calcitriol*—increases calcium absorption from the gut and reduce calcium losses in the urine, thereby helping to maintain adequate blood levels of calcium and phosphate for bone mineralization. Vitamin D has further direct and indirect effects upon osteoclasts and osteoblasts, as well as other body-wide actions (see Hormones). With Vitamin D supplements routinely provided in bread and milk, we no longer see *rickets* (simple Vitamin D deficiency disease) as a cause of inadequate skeletal mineralization in growing children (who must absorb a lot of calcium). Not surprisingly, children with rickets often became bow-legged or knock-kneed due to excessive flexibility of their weight-bearing long bones.

Rather than noisily drilling through your hard bones by mechanical means, your osteoclasts silently secrete strong hydrochloric acid solution on their advancing side (in essence creating an external lysosome). Trailing along behind each large osteoclast in its newly created tunnel are many smaller osteoblasts that deposit bone in concentric layers until the new central nutrient blood vessel is tightly enclosed within its reconstructed Haversian system (complete with canaliculi). As a result of ongoing osteoclastic activity, the Haversian canals of older bones are no longer all arranged in parallel fashion. Indeed, the variable orientation of new osteoclast-bored Haversian systems strengthens your bones against splitting as knots do a tree—by stopping most bony cracks before they can turn into a fracture. That lack of uniformity may help to explain how geese can migrate thousands of miles without regularly suffering wing-bone stress fractures (engineering calculations predict such fractures after about 100,000 wing strokes).

Large quadrupeds reduce stresses on their muscles, tendons, bones and joints in various ways such as regularly switching lead limbs during galloping in order to redistribute peak impacts. With a horse's feet hitting the ground 7000 times per hour of trotting, even its normal safety factor of three times the usual loading in a single event may become inadequate, especially with those repetitive impacts taking place on uneven surfaces. The trot/gallop transition of large four-legged mammals is designed to minimize peak forces rather than reduce energy costs—however, migrating herds travel within narrow speed ranges that minimize energy output for the gait being utilized.

Many local and systemically controlled (body-wide) repair processes come into play at the site of each new bone crack or fracture. These include invasion of the blood clot by *primitive* (undifferentiated) bone cells as well as *macrophages* (see Immunity) and the deposition of bridging collagen fibers between bone ends by fibroblasts while osteoclasts and osteoblasts remodel damaged bone nearby. Growth hormone, transforming growth factors and naturally produced anabolic steroids also have a role, as do the electrical effects of stresses exerted upon the crystalline piezoelectric bone matrix. Appropriate levels of muscular activity and adequate blood albumin (protein) and red blood cell concentrations are among other important factors in the healing process, which in many ways resembles a localized reversion to an embryonic stage of growth and development. Any fracture involving a long bone epiphyseal growth plate during childhood or adolescence may limit further lengthening of that bone. To avoid having longer leg bones on one side than the other, it formerly was common practice to restrict growth of the opposite (normal) cartilaginous epiphyseal plate. More recently, bone lengthening by very gradual stretching of the abnormally short bone has become an effective alternate treatment. With this technique, continuous traction is applied after dense cortical bone has been divided (while leaving periosteum and the centrally located, metabolically more active cancellous medullary bone intact). Provided such progressive stretching does not exceed normal bone remodeling rates, it can be continued for as long as necessary. Orthodontists have long used comparable tension devices to move teeth about in the jaw (those dental braces).

Younger individuals generally heal appropriately treated fractures, *sprains* (torn ligaments) and wounds quite rapidly—their active bone growth and remodeling may also encourage harmful calcium deposits on transplanted heart valves of biologic origin (e.g. "pig valves"). Tissue injury, tumor and elevated blood levels of calcium and phosphate occasionally induce new bone formation as well. Ectopic bone may even arise in an ordinary midline abdominal surgical scar, especially if the incision has extended onto the cartilaginous *xyphoid* (lower tip of sternum). As with cartilage production, bone formation and remodeling are responsive to local conditions while still subject to specific instructions and general rules carried within your DNA.

Avoid Osteoporosis, Rhubarb And Spinach

While human males generally have heavier bones and muscles, it is human females who must transfer their bone calcium and phosphate reserves directly into the growing fetus and into breast milk. Osteoclast activity rises when normal estrogen levels decline so menopausal females tend to develop *osteoporosis* (excessive bone demineralization) regardless of whether they have had children or supplied milk by breast feeding. Instituting female hormone replacement therapy at the menopause, along with a sufficient intake of calcium and Vitamin D_3 (among the individuals who become osteoporotic are some with defective Vitamin D receptor genes) seems to reduce the likelihood of postmenopausal osteoporosis. Rapid menopausal demineralization makes an older woman susceptible to low stress fractures of vertebral bodies (backbone) and extremities. Osteoporotic bone fractures tend to occur in metabolically active cancellous bone (within vertebrae and at the ends of long bones) due to the greater calcium depletion of cancellous (versus compact) bone. Slender young women who exercise so strenuously that they often miss menstrual periods may increase their risk of osteoporosis in later life. Women with frequent non-ovulatory cycles and thus reduced progesterone production also reach menopause with a less-well-mineralized skeleton. Men whose puberty was delayed past the age of 13.5 years (puberty is the time of maximal bone mineralization) tend to have a reduced skeletal calcium content and higher likelihood (than other men) of developing osteoporosis in old age. Elderly women gradually lose half of the calcium in their cancellous bone and a third of the Ca^{++} in cortical bone while men average 30% and 20% bone Ca^{++} losses respectively.

Menopausal women may rapidly mobilize bone lead (Pb^{++}) accumulated over a lifetime from leaded gasoline, peeling paint, soldered water pipes, soldered food cans, etc. If hazardous levels of lead are detected on blood tests (high lead levels are more likely in symptomatic menopausal females who have never had children or lactated), that blood-borne lead can be *chelated* (tied up chemically) within penicillamine or EDTA molecules for excretion in the urine. Pb^{++} exerts its nasty effects on physical and mental functions because it often enters into the same reactions as Ca^{++} but then does not let go and leave when calcium would. Interestingly, people who rapidly lose weight may similarly poison themselves as their shrinking fat deposits release fat-soluble pesticides built up during a lifetime of minor exposures. "Everything in moderation" (even weight loss) seems appropriate advice for those seeking to preserve their health. In fact, repeated bouts of unsuccessful dieting pose a greater health risk than remaining plump. And despite Popeye or your mother's advice, children ought not eat much spinach or rhubarb since both contain high levels of oxalate (which binds calcium within the gut, thereby reducing the percentage of dietary calcium that can be absorbed to serve normal needs).

Further Reading:

Hildebrand, Milton et al. 1985. *Functional Vertebrate Morphology.* Cambridge, Mass.: Harvard University Press. 430 pp. (This and subsequent chapters)

CHAPTER 9

YOUR HUMAN SKELETON

Life Depends Upon Instability;... Life Is Unstable At Every Level;... Your Bones Are Unstable Too;... Using Leverage;... Your Skeleton Is Not Very Graceful;... Location Is Everything;... Would Three Legs Be Better?;... Reproduction Matters;... Symmetry;... Weight And Balance.

A rock may contain a great deal of information but a rock does not do much. Your skeleton is far more than a rock. Indeed, your skeleton is a responsively re-modeled fiber-reinforced system of struts and levers that also serves as a readily available storehouse of essential minerals. However, the incessant physical and metabolic activities of a skeleton terminate abruptly at death. Thereafter the bones of the deceased are quite durable, often outlasting surrounding soft tissues and any wooden containers for centuries or even millennia. On rare occasions, buried bones can become heavily infiltrated by minerals and thus attractively fossilized. Only then is that skeleton just another pretty rock.

Life Depends Upon Instability

Every living creature requires an uninterrupted flow of energy to maintain its integrity. That is why, under authority granted by the Second Law of Thermodynamics ("everything runs down"), you have been sentenced to steal reducing power from other life forms for the rest of your days. Strange, that something as important as life must rely upon instability for its very existence. Rocks don't have that problem. Indeed, life resembles our legal system in being unplanned, costly, wasteful and far more concerned with sustaining its own cumbersome *process* than with the generally rotten result. While subject to slow incremental evolution in many of their parts, both life and the law rely heavily upon precedent and current inertia to guarantee a fat future. Nonetheless, you generally cannot acquire more of life's process (the essence of life) than you already possess. Furthermore, no matter how often your life may be saved or how highly it is valued, your life cannot be conserved in other than its usual temporary unstable form—and because taking more than you need can be as harmful as not getting enough, moderation remains un-surpassed as a long-term strategy. Perhaps life is best viewed as an exceedingly complex balancing act at many levels. In this unequal contest between instability and death, there are no reliable rules nor dependable guarantees. But since each defines the other, it takes both life and death to bring meaning into existence.

Life Is Unstable At Every Level

At the subatomic scale we encountered high energy levels with the strong, electromagnetic and weak interactions controlling tiny particles (localized packets of energy, actually) circling each other at very specific, relatively enormous distances and incredibly high speeds. In this emptiness, indefinite things changed simply because they were measured—the very nature of subatomic particles prevented us from simultaneously determining their position and momentum (Heisenberg's *Uncertainty Principle*). So the more you know about where such a subatomic particle is, the less certain you will be about where it is going and vice versa (similar rules may apply to adolescents). Although based upon wispy but energetic electron clouds jealously surrounding distant star-struck nuclei, atoms appeared reassuringly substantial in comparison as they interacted electromagneti-cally (chemically) in ways that seemed easier to understand. Brokered by those relatively huge enzymes, life's atoms could often be lured into long-term covalent commitments in order to complete their sparkling electron rings. Yet here again, selectivity and instability were essential since your life surely would end most abruptly if your atoms suddenly bonded together into one huge stable molecule, no matter how cleverly constructed. So most relationships among your component atoms remained impermanent—a quick bump and change in the heat of the moment brought about by that perfect fit or the temporarily irresistible charms of one with the opposite electrical persuasion. As for all of those frantically writhing molecules, their average shape and disposition depended not only upon the exact sequences of atoms they had secured in covalent bondage but also upon the water solubility of their various parts—how they attracted, repulsed and were folded amongst the ceaselessly crashing polar molecules of your internal salty seas. At this scale, life merely seemed agitated, crowded and confused.

It was a relief, therefore, to leave molecules behind and focus on relationships within those living assemblages of *endosymbiotic* (cooperating internal inhabitant) bacteria known as cells—even if slightly disconcerting to see yourself in times mir-ror as an animated bacterial cooperative. However, the excitement of finally deciphering the first few mystical words and even phrases of that ancient four-letter DNA language, faithfully recopied and passionately passed forward over billions of years, soon made us forget that these runes also confirmed your less than exalted origins. As is so often the case when ancient records finally are translated, it turned out that your early chromosomal literature mostly concerned mundane matters—the keeping of trade accounts, matters of ownership, deliveries and obligations. Indeed, it was surprising to see how your primitive ancestors using unsophisticated tools evolved such ingenious methods for monitoring and regulating the endless trillions of simultaneous transactions that make up your own internal economy and external relationships. While uncovering the 100,000 genes that send out those 100,000 different protein commandments bearing your "Thou shalts" and "Thou shalt nots", modern translators of DNA have also unearthed endlessly jumbled and shattered fragments of apparently meaningless words and phrases. As they now try to fit those shards together and retrace linguistic relationships between different forms of life, these researchers enhance our understanding of how evolving molecular species contribute to efficiencies, affect the flow of energy, alter the balance of trade and improve control over profits and losses, oxidations and reductions, thereby ever again contributing to that bottom line—your living metabolism.

Those most concerned with intracellular and multicellular relationships have long recognized that one can only understand cellular politics by tracing the flow of energy dollars. As always, some transactions are readily visible, others tend to be obscure and based more upon a wink or a nod and who knows how to get to whom. Generally, however, your energy dollars have been laundered through numerous accounts after being seized from the photosynthetic peasants who so laboriously trap solar energy from first light to dusk—only to lose most of their glucose product to unsympathetic herbivorous tax collectors—who in turn, rarely relinquish their growing hoards of reducing power to the carnivorous authorities without a struggle. And though you may currently eat high on the food chain, all living things eventually tumble back to the bottom for recycling (even cremation cannot alter that fate). In the meanwhile, you had better distribute your spoils of glucose and other reduced compounds in optimal fashion among the many cells that are you so that each of those cells can bribe, cajole or coerce their own endosymbiotic colleagues and slaves to produce the high energy phosphates and electrochemical gradients that will keep your entire thieving organization together until the end of your days.

So far, so good, even if sometimes less than inspiring. But as we now approach the entire organism, we find that there is a lot more to the whole than its parts. For somewhere along the line, all of this incredible complexity has been subtly transformed into the real you, and clearly you are far more than just a rich collection of point-like subatomic particles, cooperating microorganisms and stolen nutrients. Furthermore, your atoms, molecules and cells have been subject to constant repair and revision ever since the fateful day when sperm met egg—indeed, almost every part of you has been replaced or rebuilt over the past few hours, days, weeks or years. So when you get back together with that adult friend after five to seven years, both of you will have been just about totally exchanged. So who or what do you encounter just half a decade later?

It is only a pattern, a complex design, a unique arrangement of information being continuously impressed upon new material. The material part is essential, of course, for the design of your friend could not persist through any intervening discontinuities or dematerializations. Just as some computers lose all track of what they were doing if the power fails, the special coherence of your wonderful information will be lost to us the moment your superbly run organization loses its ongoing flow of energy. Perhaps the experts might one day recover much of that information but still, dead is dead. Apparently the real you is an uninterruptible sequence of operations that requires a fail-safe power supply. And while you actually exist only in the present (and that present is most accurately defined as the shortest possible instant we can identify—which could well be 10^{-43} seconds according to the mathematics of Big Bang theory), it also is clear that you have been and will continue to be a stream of information and energy imposed upon matter rather than a complex molecular collage that somehow walks and talks. Nevertheless, we can learn a lot about any creature from its bones, whether fossilized, scattered about or still in use.

Your Bones Are Unstable Too

Both the squirrel skeleton freshly deprived of supporting soft tissues and the magnificent dinosaur skeleton just released from its rock of the ages will immediately collapse into a disorganized heap of unattached bones under the influence of gravity—even though countering the influence of that gravity was the main reason for their existence. However, unlike the average jigsaw puzzle, either of those sets of bones could be reassembled in many different ways or even artfully combined with bones from some unrelated species without giving offense to the untrained eye. But regardless of how correctly or incorrectly any skeleton is displayed, no bone in that arrangement will even remain in place (let alone exert leverage upon the outside world) unless firmly wired to its neighbor—and such ill-fitting bones support your own sturdy and smoothly mobile structure as well. Once again, on this even larger scale, you seem to have successfully integrated thousands of individually unstable, dynamically interacting bones, cartilages, ligaments, muscles and tendons into a functional living whole. Yet despite their numbers and evident redundancy, you would not work nearly as well if even a few of those many parts had been left out or removed in order to simplify matters for the average anatomy student. Apparently instability and complexity both bring major advantages when sufficient processing capacity ensures timely detection and coordination of important information, appropriate output and adequate maintenance (a hot topic among designers of ever-smarter military aircraft).

Using Leverage

Muscle contraction performs work. Muscle relaxation is passive. Many muscles contract with great power but no voluntary skeletal muscle can actively shorten beyond two thirds of its resting length—even a muscle the length of your thigh can only pull in a few inches. So how are such short and powerful muscle contractions converted into the agile "fight or flight" movements that kept your ancestors alive and reproducing against all probability? Sometimes the best answer to a question is another question. So we ask, "How does a seesaw work?," and "How can an ordinary person exert sufficient muscular force on screw jacks to lift an entire house?" An obvious partial answer to the last question must be "very slowly and carefully," for we all recognize the reciprocal relationship between speed and power, between the gearing of a sports car and that of a tractor. Similarly, we anticipate obvious differences between the body build of a champion high jumper and that of a successful sumo wrestler. And long ago we learned that a seesaw would balance if the lighter person sat farther out from the *fulcrum* (pivot point)—which meant the lighter one traveled a greater distance at higher speeds during each up-and-down cycle. Hence one could say that the lighter one used the *mechanical advantage* provided by a longer lever arm to lift the other, or else that the heavier chap used the shorter lever arm at a *mechanical disadvantage* to propel the lighter blighter at greater speed over a longer distance.

By analogy, how muscle attaches to bone with respect to a neighboring joint (fulcrum) ought to determine whether that muscle sacrifices power to gain range of motion and speed or sacrifices speed to gain power—just as you willingly give up speed to enhance your power when you slowly turn the screw jack with a long handle in order to lift a house. The enormous mechanical advantage provided by the screw jack is determined by the circumference of the circle made by your working hand on the jack handle (pi times the diameter [d] of that circle—the diameter being about two times the length of the jack handle) compared to the gain in height (h) achieved during that one circle (namely, moving the house up the width of one thread on the screw jack). In other words, pi times d/h. Or consider the seemingly complex, multiple pulley systems often utilized for lifting heavy loads. To determine your mechanical advantage with such a system, simply divide the length of rope you have pulled in while raising that load by the distance the load has been lifted. Indeed, if that inverse relationship between power and speed did not exist, some clever person would have invented perpetual motion a long time ago, thereby liberating all humans from the cruel dictatorship of the Second Law of Thermodynamics. Since that has not occurred, you must continue to swing a long handle over a wide arc in order to ratchet your car up a single notch on that bumper jack. And watch that the handle does not snap back at very high speed if the jack somehow fails to catch and is pushed back down one notch. For then the automobile would be using its mechanical disadvantage to exchange power (its great weight moving over a short distance) for speed (the rapid and wide swing of that jack handle).

Your Skeleton Is Not Very Graceful

When compared to the animated glucose-grabbing version of a skeleton regularly encountered at Halloween, the actual human skeleton is rather ungainly. Indeed, your adult skeleton becomes noticeably thicker and heavier from top to bottom since lower sections must support everything above. The soft skull of a newborn is obviously molded and reshaped by its passage through the birth canal. Usually such gradual and temporary deformations are harmless. But even after birth, it is important that enlarging skull bones around the brain not *fuse* (grow firmly together) until rapid brain growth has been completed. In the meanwhile, a soft *anterior fontanelle* (midline skull gap) remains where *frontal* and *temporal* skull bones meet. That site may bulge noticeably when the infant screams and strains or when there is increased pressure within the skull from other cause. Eventually the skull plates thicken and fuse at those long interlocking *sutures* in order to provide life-long protection for the brain. Shallow *sinuses* (air-filled cavities) at the front of the skull provide a crumple zone to soften any major impact on the front of your solid brain case. Sinuses also reduce skull weight, increase the resonance and carry of your voice, contribute secretions to your nose and, along with the *zygomatic arches* (cheek bones), your frontal and maxillary sinuses protect and support your eyes. Your upper teeth are rooted in the stationary *maxilla* and thus firmly attached to the skull while your lower teeth are equally firmly planted in the mobile *mandible* (lower jaw bone). Unlike your fishy ancestors, you no longer have to open wide in order to ingest your dinner in one struggling piece—so you have flat cutting teeth *(incisors)* in front of those grabbing/slicing *canines* while your thick grinding *molars* provide powerful backup (see Digestion).

The seven *cervical vertebrae* (backbones of your neck that support your head) must allow back and forth as well as rotating movements, for it is reproductively advantageous to be able to tilt or turn your head rapidly without also having to tilt or turn your body. Thus the weight-bearing surfaces of vertebral bodies in your neck must be narrow front-to-back so that the spinal canal and its enclosed spinal cord remain undisturbed at the center of rotation. Indeed the first cervical vertebra is little more than a bony ring that supports the skull and pivots on an upward extension *(odontoid)* from the second cervical vertebra. Your *thoracic* and *lumbar* vertebrae show increasingly heavy construction and marked anterior-to-posterior elongation since little rotation and far greater strength is required here. The soft-centered *fibrocartilaginous discs* between each of your vertebral bodies readily absorb all reasonable compressive forces while still permitting adequate movement. These discs gradually stiffen and indent adjacent vertebral bodies as those light cancellous bones become osteoporotic, so older people lose height and have to walk or jump more carefully. At times, soft disc content will *herniate* (bulge out) through the more fibrous disc capsule, perhaps compressing a nerve root or even the spinal cord. Such a "slipped" disc is most common in the strong and flexible lower back but also frequent in the less strong, more flexible neck. Although urgent surgical disc removal may be essential in order to preserve nerve or spinal cord function, less severe situations often stabilize and improve slowly with appropriate rest and exercise. Aging and aching vertebrae that remain in direct contact can eventually form a bony union—this reduces the likelihood of further intervening-disc-material extrusion but any spontaneous or surgical *fusion* of adjacent vertebrae also increases leverage and hence the stress exerted upon nearby disc spaces that still remain mobile.

Location Is Everything

At this point one could reasonably ask a few general questions about how you are put together. For example, would there be any advantage in placing your central nervous system or sensors more centrally? Well, such a location might speed messages and keep your ears warmer alongside your abdomen but it certainly would interfere with your wide visual fields, unobstructed hearing with minimal echoes and smelling of things other than self. Furthermore, you barely float in water as it is and there are many streams to cross in your endless quest for food and romance. So it seems advantageous to bunch important sensors and your air intake where all will usually remain above water. Since your eyes are in front, it must be more important for you to see where you are going and what you are pursuing than to look where you have been or evaluate what is chasing you. Halibut and flounder derive benefit from moving their bottom eye around to the top side where there is more in view than just mud. On the other hand, herbivores generally look sideways, exchanging the advantages of binocular vision for wider visual fields that allow earlier detection of carnivores and keep them in view during the escape. Certainly one cannot flee or fly safely while continuously turning the head to look back. And it makes sense to bunch your more delicate functions and sensors up front where they can best serve and be protected from cold and injury—the other (back) side can then be tougher and stronger.

Would Three Legs Be Better?

Why walk on two legs when three would improve stability? To a considerable extent, mobility and stability are opposites. You do not stay alive because of how steadily you stand. Rather you may survive because of how rapidly you can move in any direction from a standing start. A nervous moose puts all four feet very close together until it decides which way to go. Small mammals such as chipmunks routinely run on bent legs despite the obvious inefficiencies and reduced leverage of such a gait since that prepares them for an instant leap. Although a single leg would be even more unstable, you would then fall over easily and one-legged hopping is too much work anyhow. So what advantage does rear-limb drive have over making your forward limbs stronger? Only those mammals that must dig or fly for their reducing power need stronger front limbs than rear. Bats (and birds) use powerful chest wall muscles to flap those wings but diggers and runners must place more muscle bulk out on the limb itself to stabilize the joints of that structure against heavy and variable loads. Rear limb drive allows you to raise your sensors above tall grass or low scrub (related weight and balance considerations make it preferable to locate your light lungs ahead of your heavier guts—see Respiration). When four legs are on the ground, more weight is transferred to the rear on starting and to the front during braking so rear limb drive emphasizes starting power over braking while also providing sturdy support so those weaker forelimbs can reach, grab and fight under direct sensor supervision. As your front limbs move forward in four-legged running, you simultaneously *extend* the vertebrae (arch the back) to increase that reach. Your front limbs can then pull the ground farther back by *flexing* (hunching) your back vertebrae before the next big push by the rear limbs. For the galloping horse, back extension is a routine part of the powerful leap while back flexion enhances the follow-through. Furthermore, longer legs and longer strides reduce the energy costs of transportation since each step carries a similar per-pound-transported expense for any size mammal moving at comfortable speeds. Among running animals, back and buttock (rear leg extensor) muscles normally are stronger than flexors. And those easily displaced soft tissues and organs in front of the lower vertebral column offer little resistance to marked flexion of the torso.

Reproduction Matters

Why is your pelvis so wide, with the greater trochanters of the femur coming off sideways? Why not place the pelvis directly over the legs? Here again, there is value in pelvic width and instability. Bringing the hips more *medially* (toward the middle) could increase stability but also reduce muscle leverage (see olecranon below) as well as make a running start more difficult. For in order to run, you must first fall away from dead center. Having very laterally placed hip bones allows you to initiate a quick start in any direction simply by reducing support on that side. It also allows you to have thicker, more muscular legs without them interfering with (rubbing against) each other. The female pelvis has a relatively larger circumference than the pelvis of a similar-size male in order to allow passage of the infant head at birth. Not surprisingly, that relative increase in female bony girth (and therefore weight) reduces maximal athletic performance in many sports, especially since fertile females also carry extra energy as neutral fat in order to successfully complete a pregnancy and lactation. Why don't pelvic bones eventually fuse into a solid ring as skull bones do after brain growth has been achieved? Since the female pelvis is barely large enough to allow passage of an infant head anyhow, it clearly is advantageous to loosen those pelvic joints before giving birth. An even larger pelvis that might make such loosening unnecessary would be heavier still—just as a water pitcher needs thicker walls than a water glass. And when catching lunch, escaping from carnivores or leaving lost battles, it would be a persistent reproductive disadvantage to run more slowly on account of additional pelvic weight. A lighter and less muscular upper torso also reduces stress on the female pelvis. As in automobile design, weight saved in one place allows further reductions in the strength and weight of other components, which then improves performance or reduces power requirements.

Symmetry

Why are animals symmetrical between their left and right sides? Except under unusual environmental conditions (e.g., both halibut eyes above) symmetry seems advantageous since right and left turns are equally important. So unless special circumstances reward particular modifications (e.g., barn owls with one ear slightly higher than the other to improve localization of sounds in the dark), mobility is generally enhanced by left/right symmetry (although the enlarging shells of slow-moving snails may curve one way or the other [sometimes in a temperature dependent fashion] to minimize materials invested). For example, alpine cows with shorter legs on one side would have to circle an entire mountain-top to return for each milking. And there is some evidence that animals instinctively seek a more symmetrical mate (presumably because normal growth and development is one mark of "good genes"). So birds seek mates with symmetrical tails (and even symmetrical leg bands) while humans often judge a computer-combined female face more attractive than any of the less symmetrical individual faces averaged into that composite. As for design and construction, symmetrical animals are the natural outcome of releasing diffusible growth and development substances along the midline. The symmetrical decline in promoter molecule concentrations away from the midline (along with counter-diffusion gradients of peripherally released or more localized inducer substances) then provides cells in local tissues with critical information on where they are and what they should do or become during growth and development (elbow cells versus finger cells, for example, or muscle fibers instead of fibroblasts). Even your egg's original radial symmetry soon gave way to left/right symmetry after that plucky sperm won entry (see Growth and Development).

Weight And Balance

Your bulky limb muscles are placed *proximally* (as close to your body as possible) to reduce unnecessary weight farther out on the lever (limb). For it is easier to flap your wings, fling a rock or punch and kick rapidly when tendons transfer forces out onto lighter limbs from large trunk muscles. And most synovial joints place one rather concave surface in contact with another that is convex in order to stabilize the joint throughout its permitted arc of motion. Though it might seem to make little difference whether the convex surface occurs on the proximal or distal side of a joint, in actuality the design of each joint is almost inevitable. This is most obvious at the hip where a convex pelvic bone that had to fit onto a concave femoral head would require far heavier construction on both sides—especially to avoid breaking the lips off such a hypothetical hollowed-cup femoral head. Furthermore,

more muscle mass would be necessary to maintain joint stability after switching convexities in this fashion. Anyhow, inserting a large flat bone *into* a rod-like long bone violates *common sense* (lessons learned from past mistakes). And not surprisingly, the depth of the *acetabulum* (hip socket in the pelvic bone) increases along with muscle power available for leaping in various vertebrates.

On the other hand, enclosing your knee in a deep socket would interfere with the proper cushioning of impacts by cartilage and muscle and tempt injury by setting rigid limits on joint movement. So you rely upon strong *medial* and *lateral collateral* ligaments to limit side-to-side motion while your sturdy centrally located *anterior* and *posterior cruciate ligaments* control anterior and posterior slippage. Your knee joint is additionally cushioned by the somewhat mobile *meniscus cartilages* that intervene between tibial and femoral bearing-surface cartilages. And finally, the entire knee apparatus is enclosed, stabilized and sprung by your powerful *quadriceps muscle* and its mobile *patella* (knee-cap bone) anteriorly as well as by numerous other tendons and muscles—all of which encourage your convex distal *femur* to remain centered upon that hardly hollow (thus no thin breakable edges) proximal *tibial* plateau. Moving the quadriceps line of action forward through the patella improves its leverage, but quadriceps contraction remains at a serious mechanical disadvantage for straightening the knee because its muscle pull so closely parallels the rigid femur and tibia rather than coming in at an angle to the tibia. So your forward kick is fast but relatively weak despite the great strength of those springy knee-stabilizing extensor muscles. Given their major mechanical disadvantage, it is not surprising that your powerful quads soon tire when you walk down a steep slope since that requires them to both oppose and transfer all of the potential energy represented by you upon a hill as it is converted into kinetic energy.

A heavy truck starting down a mountain pass has an even more significant problem. It has utilized the solar energy trapped in many gallons of *fossil* fuel (hydrocarbon reducing power derived from ancestral life) to get to the top, and a significant fraction of that energy will soon be recovered as heat (or else out-of-control acceleration) during the long descent. Obviously it would not take the heat energy from very much fuel to set the brakes on fire so truck brakes serve only as the back-up system—most of that gravitational potential energy (heavy truck on high hill) being converted into kinetic or heat energy during descent must therefore be directed back through the same heat-resistant engine, exhaust stack and radiator system that so successfully dealt with the fuel energy released en route to the top. That is why highway signs prior to long descents order truckers to shift into a lower gear. Furthermore, by altering the timing of engine valve closure, the so-called Jake Brake temporarily changes a truck engine into a noisy compressor (BAP! BAP! BAP!) that more effectively converts the truck's kinetic energy into hot compressed air for release through the engine's exhaust system. If fuel prices rise sufficiently, perhaps regenerative braking equipment will allow some of that potential energy to be salvaged as hydrogen fuel, or maybe some other way can be devised to couple ascending trucks with the energy released by descending trucks (perhaps electrically, rather than mechanically as between two cable cars) in order to reduce overall fuel consumption on particularly steep and heavily traveled terrain.

As with left/right symmetry, the head to tail development of animal embryos is partially determined by differences in the concentration of diffusible regulator substances—which in turn induce or reflect activation of the *homeobox* positional gene sequences common to most multicellular life forms. Many simpler organisms such as worms and insects have an obviously segmented design in which structures simply repeat on most segments in a symmetrical fashion. Human ribs and their associated vertebrae, vessels and nerves bear silent witness to an ancient segmental design, as do the multiple gill slits and aortic arches of every vertebrate embryo.

Your massive and powerful legs are stabilized and their range of movement limited by your pelvis and its associated musculature. In contrast, your arms would be far less effective if attached to a fixed upper pelvis—such a pelvis would unduly restrict your breathing as well. So each of your arms is suspended from a separate mobile muscle-enclosed *scapula* that can shift and rotate around your rib cage to extend your reach, power your pull, "crack the whip" on your throw (making baseball pitchers look strange when photographed in mid fling) and improve your air exchange as you swing off through the branches (see Respiration). Note that there is no direct bony connection between your vertebrae and your scapulae—flimsy and mobile *clavicles* provide the only bony links between those scapulae at your shoulders and the rest of your skeleton at the sternum. The human clavicle is a functionally unimportant bone that serves mostly to steady the arc of shoulder movement and protect underlying nerves and vessels to the arm. Horses lack clavicles entirely and derive all of their powerful scapular support from very strong muscles attached to the elongated *spinous processes* of nearby vertebrae—those muscles and their *dorsal* (posterior) vertebral spines create the hump in front of your saddle. By eliminating clavicles, cats have maximized scapular mobility and running efficiency for the short high-speed dashes that allow them to separate fleet-footed herbivores from their once stolen reducing power.

Your modern sedentary life-style involves a lot of sitting, which means those two pelvic *ischial tuberosities* must bear your weight (ordinary upright sitting in a chair puts little pressure upon your *coccyx*). A very narrow bicycle seat places your weight on the soft tissues between those ischial bones (including the base of the penis). While excessive weight bearing on the base of the penis could possibly harm sexual function, a narrow bicycle seat would be even more hazardous if human males had an *oosik* (penile bone, subject to bruise or fracture) as so many other male mammals do. Incidentally, the walrus oosik is similar in size, ivory-like appearance and retail value to walrus tusks. But whether female or male, *your* only single, isolated and symmetrical midline bone is that U-shaped *hyoid* located transversely and deep to the front fold where chin becomes neck. The hyoid bone may be fractured when strangling someone with a rope or by hand, but otherwise it is rarely injured.

There is close correspondence between the bones of your upper and lower extremities except for the patella in each quadriceps tendon—and your patella serves a leverage function comparable to that of the *olecranon* (elbow extension) on each *ulna*. But a mobile patella is less likely to be injured (and presumably better suited to falling or kneeling) than some rigid olecranon-like extension of the tibia would be at the exposed front of the knee. The lower extremity emphasis on strength and stability over maneuverability is evident from the way your tibia serves as dominant weight-bearing leg-bone between knee and ankle joints. Compare this to the forearm where your *ulna* dominates the elbow joint while your *radius* provides the main bearing surface at the wrist. Although far less sturdy, that alternate arrangement of arm bones markedly enhances your dexterity and agility by permitting *pronation* and *supination* (palms down and up) of the hand. Note also that while the radius contributes slightly to the elbow joint, the fibula has no role at all in the knee. Incidentally, the relatively small and round radial head of a child is easily dislocated from its circular fibrous cuff at the elbow joint if an adult jerks the child about vigorously by the hand.

Your wrist bones permit a wide range of hand movements even though analogous bones form the relatively immobile but springy plantar arch of your foot. And while your nimble fingers contribute almost as much area to the palmar surface as the palm itself, your rather awkward toes provide barely enough additional surface to pick up your dirty laundry and help your weight-bearing foot balance and shove off. The relatively long flat feet of bears and people allow their strong calf muscles to work at a marked mechanical disadvantage through the *achilles tendon* for a fast and moderately strong, running takeoff. A *bunion* (prominent painful head of the first metatarsal) is the result of persistent lateral displacement of the great toe, often consequent to wearing pointy-toed shoes and high heels. Permanent discomfort eventually results from the abnormal concentration of weight on the head of that first metatarsal. Presumably this inane clothing fashion has persisted (despite causing innumerable sore feet) because the attractive balancing movements that high heels impart to the female pelvis also bring reproductive advantage.

So there you have it. Reproductive advantage comes to those who are better at utilizing available information. Bats, whales, horses and humans have each found markedly different uses for their finger bones. And every reproductively advantageous application has encouraged further appropriate changes until structural limits were reached or a local optimum achieved. Evolutionary progress depends upon death removing less well-adapted players from the field before they can reproduce—nonetheless, changing times have a way of randomly moving the goal posts and reproductively advantageous traits are often surprisingly inapparent, counterintuitive or situational. So while the currently familiar outcomes of evolutionary selection seem both natural and obvious, it could have turned out quite differently had initial conditions been even slightly altered. In view of the many other reproductive opportunities available to your undoubtedly charming parents and the trillions of sperm your father produced over his lifetime—any one of which might have met any other of your mother's million original eggs to produce an entirely different you—your personal existence must rank among the most improbable of events. And if we expand our view to include life's long history of unending variations and chaotic interactions, it becomes quite apparent that the most exacting rerun of life's evolutionary processes could never again bring you into being. Truly you are unique.

Further Reading:

Any well-illustrated textbook of anatomy

CHAPTER 10

WHY MUSCLE?

Information Is;... Marketing Your Message;... Move It Or Lose It.

Information Is

Any arrangement of atoms within a molecule carries information. So does any ongoing relationship between molecules. Indeed, organization on any scale both requires and displays information. At times even disorder may inform. Thus information was growing long before life came along. Not only were matter and energy already organized according to the general laws of physics, but many complex information-bearing molecules guided and energized First Life. Since then, by its competitive solar-powered more or less accurate replication of information-bearing molecules and larger structures, life has progressively enhanced the complexity of our world. And every advance in complexity has provided new opportunities for the further expression, expansion and analysis of previously irrelevant information.

Useful information brings reproductive advantage. Of course, particular bits of information may be helpful, harmful, irrelevant or any combination of the above. This holds true whether that information is carried by a molecule or on the financial pages of your newspaper. Information that harms you may be faulty or true. Similarly you may profit from sending out or receiving incorrect information. And sometimes the only good use for the financial page is to start a fire. Apparently the value of any information lies in its effect rather than its accuracy. Furthermore that effect may remain unknown until circumstance and user have been specified, and perhaps even then. The information swirling around you often seems irrelevant because you cannot collect and analyse it in a meaningful manner. So as your ability to store and manipulate data evolves, you increasingly utilize all sorts of previously irrelevant (because unavailable, overwhelming or otherwise incomprehensible) information.

As conditions change, new and useful information can bring marked reproductive advantage. Bacteria have always led in the development and application of life's information. Their leadership is based upon enormous populations, extreme variability, frequent horizontal transfers of information and high mortality rates—all of which encourage evolution by giving free rein to natural selection. With other life forms already constructed out of bacteria as well as inhabited by them, there are endless active, passive, selective, random and passionate interchanges of DNA and RNA information between all living things—ranging from accidental nucleotide entry and attacks by small viral packets to the acquisition of entire bacterial systems with their information intact and the transfers of uniquely recombined chromosome sets during the endlessly varied, long-ago-choreographed interactions leading up to sexual intercourse.

Even the tiniest information upgrade may have small or large effects, depending upon a recipient's status and stability. Certainly the acquisition of new endosymbiont bacteria (e.g. chloroplasts or mitochondria) causes major and lasting changes (see Metabolism). Interestingly, evolution tends to proceed in fits and starts (according to the fossil record) with long periods of apparent stability separated by times of remarkable change. Some of those sudden transitions that seem to affect almost all life forms over wide areas or even around the world may result from catastrophic meteorite impacts and/or periods of extreme volcanic activity and/or other major environmental changes. The many more limited extinctions or ordinary destabilizations of individual populations (species generally persist just a few million years anyhow) could simply reflect a failure to match up to a new competitor (who might even be a relative bearing new genes or entire endosymbiont systems) or the decimation of an entire species consequent to particularly virulent viral or bacterial epidemics (although parasite production generally rises and falls along with host population).

The current AIDS epidemic again demonstrates how larger life forms remain susceptible to attack by still-evolving information in our environment. Furthermore, in altering that environment beyond recognition, we provide endless new opportunities (as well as the threat of extinction) for previously latent, localized and often infectious information. But when attacked, we can now fight back by seeking more information, changing our behavior and responding in other ways, rather than simply waiting for all susceptibles to suffer and die off as our ancestors had to do. Thus humanity need no longer rely upon the few survivors of the latest harm to replenish an entire population through their excessive reproductive capacities. So while the natural selection of your remote and recent ancestors has made you what you are today, the ongoing evolution of humans by natural selection has almost ground to a halt—assuming we can avoid various disasters consequent to overpopulation and our ever-increasing abilities to destroy one an-other. In comparison, those rapidly replicating insects, garden slugs and rats continue to evolve in real time, matching any of the purposeful or inadvertent changes we have managed to bring about in their environment.

Life's information has always been revised, replicated and recycled. But mod-ern humans have so enhanced and accelerated their non-inherited computer-based information acquisition and handling skills that you can neither isolate yourself from that flood of new data nor hope to keep up with it. So the escalating informa-tion burden of modern life increasingly forces individual humans to specialize and subspecialize in order to remain useful and competent in their work. While most specialists already rely upon information-handling machines, the further develop-ment of these machines may eventually free the educated public from its dependence upon many of those same specialists (who must then retrain since they know so very much about so little that has suddenly become irrelevant—but retraining has become essential for every occupation anyhow). Still there is too much informa-tion. So is it chance alone that delivers information? Indeed, how does life's information compete?

Marketing Your Message

Great effort goes into the preparation of any book for publication. Once labo-riously arranged and endlessly rearranged, that information is readily replicated. Furthermore, a few extra books printed during a press run only increase overall expenses by the minor cost of that extra paper—the leftover blank paper on that roll might have been wasted anyhow. However, printing information in a book actually lowers the value of its otherwise blank pages unless that book then reaches someone who can use that information. So almost inevitably our greatest, most organized and costly efforts go into marketing of our information. Quite clearly, nothing can stop an outstanding idea whose time has come. Once released, a useful idea usually escapes the control of its originator, who thereafter tends to become irrelevant to the success or effect of that idea. And since we all interpret any new information in the light of previously acquired and recombined knowledge, the same idea may affect a great many people in markedly different ways. But every ordinary book enters a marketplace already saturated with hundreds of thousands of other titles, not to mention other competing forms of *communication* (informa-tion exchange). In the end, whether a book reproduces its ideas successfully in many minds or simply disappears without a trace may relate more to who you know and how well you advertise than what your book has on offer.

Life's information follows rather similar patterns, beginning with those un-countable trillions of your more remote ancestors whose lives were undoubtedly nasty, brutish and short (a great name for a law firm). The primary goal of each ancestor's particularly desperate efforts to survive and reproduce against incredible odds was to forward their invaluable DNA endowment to you. So now you repeat-edly replicate those once-rare-and-remarkable ancestral contributions at negligible cost to yourself, going through about four hundred complex ovulations and men-struations in a lifetime or preparing up to 300 million mobile information-bearing potentially life-giving sperm for release in every ejaculation. Despite their minimal cost in time, energy and materials, each new egg or sperm bears a uniquely recom-bined version of the particular DNA information that you inherited from your father and mother. Yet every tiny packet of your invaluable information is worth less than the individual nucleotides used to write it unless that information packet can be marketed. From an evolutionary point of view, sperm and eggs are destiny. The rest of your body is just hype. So the relatively huge metabolic investment in your brains, guts and muscles (compared to your minor outlay on sperm or egg production) simply demonstrates how far the competition for reproductive advan-tage among multicellular creatures has moved away from the basics.

Interestingly, that evolutionary arms race has now reached a point where one of its consequences—our combined intellects—has inadvertently created a sepa-rate reality. For the first time in life's endless struggle, our success is no longer measured solely by how effectively we pass inherited information along to the direct descendents that we tend to view as our crowning achievements (even when they are young and have only future value, if any). Instead we are at a point where most who have the option will choose education, health, comfort and greater riches over the endless sharing, other-directed investment and even impoverishment that comes with excessive reproductive success. Of course, both education and reproduction depend upon the replication and distribution of slightly modified information. And our growing communication skills increasingly drive us to share useful ideas by verbal intercourse. Therefore, like Dr. Seuss, you may eventually have millions of intellectual descendents without ever having faced the challenges and rewards of child rearing.

With humans directly or indirectly consuming perhaps one-third of the bioproduct of this Earth, any reduced emphasis on reproductive success is surely no bad thing. For there can only be negative survival value in humanity continuing its unearthly rate of reproduction when we no longer face the very high mortality rates that formerly weeded out all but the most fit or fortunate. Human evolution cannot possibly keep pace with this rapidly changing information world anyhow—thus the way our non-DNA-based information is handled will determine humanity's future (rather than further natural variations in any individual's DNA or that of their relatives and descendents). Even so, adult humans still activate several eggs or produce billions of sperm each month simply because the associated labor, material and energy costs are so low that it does not pay to stop and restart those assembly lines. Furthermore, you just never know when a newly matured egg or two or a fresh batch of sperm might just hit the spot or even represent your only remaining opportunity for reproductive success. As with any manufacturing process, the more production exceeds demand, the less significant an occasional flawed product or unexpected disruption in one of many parallel assembly lines within your two *gonads* (where those eggs or sperm originate).

From the point of view of any mathematically inclined egg, sperm, pollen or seed, this is an incredibly huge and unbelievably dangerous world. Indeed, there is but a vanishingly small chance that any of the sex cells currently graduating from your ovaries or testes will be fortunate enough to contribute their unique information to a reproductively successful adult. And all eggs, sperm, pollen and seeds not utilized in accordance with their enclosed instructions are simply discarded to nourish other life forms—just as the vast majority of all living things die and are eaten long before they are able to reproduce. That is as true for alfalfa sprouts, mosquito larvae and cuddly little bunny rabbits as it is for fish, clams, calves and piglets. Only the occasional fertilized frog, mouse or salmon egg survives long enough to replenish the species. Small wonder then that those unaware of her personal preference for a scratch behind the ear, who have never smelled her sweet breath or seen her affectionately nuzzle her calf, can find it so easy to view a cow as hamburger on the hoof or an intermediate stage between grass and milk.

Despite the fact that natural selection has lost much of its impact on humans residing in first world nations, you continue to act as if the particular information borne by your own eggs or sperm has overwhelming value. Evidently the sincere and even passionate marketing of that obvious surplus was a more successful reproductive strategy than any other available to your ancestors. Although it remains self-evident that your eggs or sperm are almost entirely unwanted, your enormous overproduction of sex cells represents a never interrupted, repeatedly modified process that dates back to very ancient times. For without penis and vagina, your earlier ancestors had to rely upon wind or wave to distribute their tiny information packets, just as many viruses, bacteria, pollen, sperm, eggs, dandelion seeds, larvae and embryos, coconuts and other small plant and animal forms still do. Under such circumstances, the efficient mass-production of those information packets clearly was advantageous. Innumerable other interactions also played a role, but the present excessive production of human sex cells probably harks back to when DNA information was competitively broadcast to any open receiver. Of course, this apparently wasteful although actually inexpensive process requires no justification for it works.

Move It Or Lose It

It is all very well to market your information widely in hopes that a few copies will land where they can flourish and themselves reproduce but "almost" or "close" or "nearly" are not necessarily winning reproductive strategies. So while innumerable small packets bearing life's ancient information are passively distributed over great distances on land, sea and air, many are capable of some purposeful movement that can improve their options once they have nearly arrived. For a sperm capable of swimming those final few centimeters is far more likely to succeed than one unable to propel itself and a bacterium may gain significant advantage by moving just a few millimeters or even microns. Certain bacteria of spiral shape simply flex and twist ahead through the water, which to them is a very dense medium. Others incorporate rotating flagellae to propel themselves forward or back. And many single-celled protozoans row about with cilia or scull along with flagellae (perhaps acquired along with some bacterial endosymbiont) while amoeboid forms travel efficiently but tediously along surfaces by extending their leading edge while retracting the trailing portion of their phospholipid-bilayer-surrounded protoplasm. To achieve such purposeful movements (and also redistribute chromosomes and other essentials within the cell) requires the production of special proteins that can convert chemical energy into movement—the same necessities have driven multicellular animals to invest heavily in specialized *muscle* cells (often referred to collectively as *meat*) that can easily lift many times their own weight.

However, your muscles only justify their cost when they help you to support and then deliver those sperm or eggs in style. Fish gotta swim, birds gotta fly, and you gotta be in the right place at the right time if you want to catch them. As in real estate, the three most important things are location, location and location. And muscle is what allows larger animals to relocate for refreshments and reproduction once they have exhausted local possibilities. For as animal size and cost increase (and their populations correspondingly dwindle in accordance with what the environment will support), passive transport upon wind and wave no longer counts among winning strategies—the losses are too high and the chance of success too low. Furthermore, if you represent a lot of reducing power and you cannot out-number or outsmart 'em, you had best be able to outrun or outfight 'em. But regardless of scale, there are only a few basic ways to travel through any environ-ment in a purposeful fashion. One can push ahead, pull ahead, row ahead or rotate (screw) one's way to success. By using many little cilia as oars in a coordinated fashion, small groups of cells move fluids past their combined surfaces. This changes their position in the solution, or the solution at their position if they are anchored. However, for larger animals the basic choice was between pushing and pulling. And it really was no contest, for pushing requires multiple rigid internal pushrods that are difficult to direct. On the other hand, pulling creates tension and tension results in alignment and alignment allows cooperating forces to be applied to the straight line between puller and pullee. Since something pushed is likely to go elsewhere than where you intended (thereby reducing the efficiency or even useful-ness of that push), it is a classical sign of poor judgment to put the cart before the horse—and even a specialized push-device such as the wheelbarrow remains slow and difficult to handle.

So while single cells often advance by pushing their cell membrane forward in front as they retract their rears, larger groups of muscle cells find it reproductively advantageous to all pull together. Thus *muscles never push directly* while doing work or, to put it more accurately (see octopus below), muscles *contract* (shorten) *actively* and *lengthen passively*. In other words, muscle cells only perform work during con-traction. That means your skeleton is both stabilized and operated by groups of muscles pulling at each joint. Some of those muscles are therefore located so that their contraction *extends* (straightens out) a joint while other muscles actively *flex* (fold) a joint by their contraction—and because of the way they cross one or more joints, a few muscles also twist. In fact, whether the tendons of some muscles that cross several joints flex or extend a particular joint by their contraction may depend upon current skeletal position and the contractile state of other muscles. The strength of various muscle groups adapts rather quickly to how much power is usually re-quired, which generally reflects the customary workload and how much *leverage* (mechanical advantage or more commonly, mechanical disadvantage) is involved. Furthermore, muscles generally manifest some *tone* (contractile activity) while work is under way, regardless of whether that work is primarily powered by themselves or by their opposing muscle groups. For it takes a smoothly yielding counterforce to provide positive (moment to moment) control and thereby prevent unopposed flinging sorts of movements (just as an effective opposition and free press can re-duce erratic behavior by the party in power). Your interacting *skeletal muscles* (also known as your *voluntary* muscles) can contract in a great variety of combinations and sequences. In addition, every voluntary muscle contains large numbers of separate *motor units* (groups of *muscle fibers* whose coordinated contractions are controlled by a single *motor nerve cell*). Thus there are endless ways to adjust the output of each individual muscle. However, even maximal stimulation can only shorten a muscle to about two-thirds of its resting length.

Of course, where the load is excessive, a muscle may not shorten at all despite your best effort—indeed, it may even lengthen. When muscle activation leads to definite muscle contraction but no noticeable muscle shortening, that effort is referred to as an *isometric* contraction. And the isometric contractions of a great many opposing muscle groups continuously stabilize all of your weight-bearing bones and joints as you stand or walk—otherwise you would collapse in a heap. At any given moment within any particular voluntary muscle, some motor units are likely to be relaxing while others contract—for certainly you would not want all motor units of an entire muscle to relax or contract simultaneously during normal physical activities (unless you are a frog making that big jump). So how is muscle contraction accomplished? What mechanism is it that shortens an individual muscle cell or skeletal muscle *fiber* (a long row of muscle cells fused into one unit)? Must an entire fiber contract in order to exert tension on the collagen meshwork that binds it? And what are some implications of such cell or fiber shortening?

Well, an entire muscle fiber certainly shortens and pulls on its surrounding network of collagen fibers in unison with the other muscle fibers of its motor unit. And since the principal ingredient of all cells is incompressible water, as those muscle cells or muscle fibers shorten they also must widen (since their internal volume remains unchanged)—which allows the simultaneous contraction of many individual muscle cells lying transversely across an octopus arm or lizard tongue to suddenly make these structures thinner and therefore longer. Thus the contraction of longitudinally oriented muscle cells or muscle fibers shortens a structure while the shortening of transversely oriented muscle cells can constrict your blood vessels or forcefully and rapidly drive that octopus arm or lizard tongue forward. Octopus arms need no bones to stiffen them underwater since they too are of water density. And the lizard only propels its tongue through the air after first reefing or gathering it accordion-style on an internal forward-pointing extension of the hyoid bone. It has even been suggested that the very last heart muscle cells to shorten during an ordinary *systolic contraction* (heart beat) may be so arranged that their simultaneous widening smoothly initiates the subsequent cardiac reexpansion.

Any stationary display of muscles in a beauty contest or for intimidation is made more impressive when forceful contraction causes bulging of opposed muscle groups. But muscles that have been enhanced primarily through prolonged and strenuous isometric exercises may fatigue more readily. Thus regardless of their formidable appearance, such heavily muscled individuals often have little endurance (so hopefully you can outrun them). It seems that many such trade-offs are made among the three principal sorts of muscle fibers, which therefore are quite variably distributed throughout the different muscles of different individuals.

Isometric or weight-lifter's exercises tend to enhance your *fast-twitch white* muscle fibers. Such muscle is glycogen-rich but relatively poorly supplied with the iron-containing mitochondria that produce high energy phosphates for muscle contraction. Fast-twitch white fibers also include very little iron-containing *myoglobin* (which stores extra oxygen within muscle tissues) and have comparatively few *capillaries* bringing in fresh blood. Fast-twitch white muscle fibers therefore tend to predominate in upper extremity muscles that must occasionally or intermittently produce powerful, quick and possibly lifesaving contractions. On the other hand, postural muscles need to contract more slowly and evenly. Often they must remain partially contracted over long periods of time while you are active. Not surprisingly, such postural muscles contain high concentrations of mitochondria, myoglobin and capillaries, and consist predominantly of *slow-twitch red* muscle fibers. And as you might also expect, the percentage of those slow-twitch red fibers tends to decline (by disuse) during a prolonged space flight.

Finally, *fast-twitch red* fibers are found in highest concentrations where both speed and endurance are essential, as in the running muscles of the lower extremities. Although fast-twitch red fibers might seem ideal for every purpose, it turns out that slow-twitch red are far more efficient for body support functions since each muscle fiber *twitch* (brief contraction) carries a similar energy cost. And fast-twitch white muscles bring advantage because they can be called upon for short but major efforts without diverting your limited blood supply away from your essential red muscles at their moment of greatest need (having arms that rarely tire does not help if your legs then give out). Although any muscle can be made stronger or more resistant to fatigue by appropriate exercises, there are marked differences between individuals in what can be achieved. Age, nutrition and health are obvious limiting factors but quite surprising gains can occur in appropriately exercised muscles. For example, an ordinary *latissimus dorsi* (chest wall muscle) with a lot of fast-twitch white fibers can be converted to a predominately slow-twitch endurance muscle by ongoing appropriately timed electrical stimulation of its nerve. Thereafter that trained muscle can be transferred into the chest and wrapped around a damaged heart in hopes of improving its pumping capacity.

Further Reading:

The New England Journal of Medicine, is a prestigious weekly publication devoted to normal and abnormal human anatomy and physiology (especially as relevant to the practice of medicine and surgery).

CHAPTER 11

MUSCLE RULES

The Muscle Arms Race Eventually Depends Upon Death;... On The Importance Of Muscle For Coveting;... What Was The Question?... Better Muscle Contraction Through Chemistry;... The Actual Machinery Of Muscle Contraction;... Actin Holds;... Myosin Pulls;... Ca++ Controls;... ATP Pays;... Overview Of The Muscle Contraction Process;... Deathly Disorder;... Panting Repays Oxygen Debt To Muscle (Via The Blood);... Getting Your Muscles To Do What You Want;... Cardiac (Heart) Muscle;... Smooth Muscle.

The Muscle Arms Race Eventually Depends Upon Death

Our earliest ancestors were totally dedicated to the process of replication. At an appropriate time they simply split in two. All parental resources transferred automatically to both daughter cells. Nothing was left over. Fierce competition permitted no waste. In marked contrast, your entire body now represents one of the most advanced, cooperative, state-of-the-art systems available on Earth for the care, feeding and delivery of sex cells. And that tremendous investment in marketing your information is a natural outcome of the ever-escalating biological arms and legs race that began with First Life. As with all arms races, you now get far less bang for the buck (many fewer descendents per unit of stolen reducing power) than did your simpler multicellular ocean-dwelling ancestors—some of whom could produce millions of eggs per year. Despite the far greater metabolic investment in you as an individual, you will leave comparatively few descendents no matter how hard you may try. Furthermore, like all departments of defense, you have become very adept at sequestering resources. Thus, for the sake of those who follow, you are bound to die so that your assets can be recycled after your reproductive days have passed.

In the good old single-cell days, death came often but not to all. And newly arrived single-celled competitors often started out more or less even with others of their sort—quite unlike your own descendents whose survival, sustenance and education requires uninterrupted access to the resources that you struggle so hard to acquire. But the more successful rarely recognize when they have enough—and if humans were to live on endlessly and persist in their mindless accumulation of resources, the old ones would likely dominate, compete with and impoverish the next generation rather than becoming their inevitable, if often reluctant, benefactors. Quite clearly, the possibility of a markedly lengthened life span with sustained vitality represents an unprecedented danger for human generations yet to come. For soon enough, some newly evolved disease or other major challenge would then eliminate the remaining aged, decreasingly adaptable and ever less self-sufficient human race.

Those most desperately seeking immortality or at least a greatly prolonged and healthy existence could in fairness point out that the rules of evolution have only just changed—with natural selection and reproductive advantage no longer driving human progress, it is our *civilization* (the way we process and pass along non-genetic information) that will seal the common fate. And since older humans are our storehouse of traditional information, they also remain an essential resource to be prolonged and cherished. Right? Well, not necessarily. For the significant advances in information production, utilization, storage and retrieval upon which our future depends comes from younger individuals. Indeed, most oldsters are more or less computer illiterate. As for the old ceremonies and traditional wisdom that brought continuity and meaning to the daily lives of Australian aborigines and the Inuit, or to Balinese and Chinese farmers, or even the mysteries and myths that still lend authority to aged leaders of antiquated religions, somehow all of those persistent admixtures of historical fact and dream-time beliefs just seem increasingly outdated and irrelevant to the young. Thus change generally threatens authority and is resisted by the old who quite correctly say "Over my dead body!"

On The Importance Of Muscle For Coveting

To determine how it all evolved, this emphasis on muscle-based speed and power, we must extend our imaginations back to the good old days of early multi-cellular life. There we encounter all manner of weird and wonderful life forms displaying various contrivances for getting even or getting ahead, some of which were far more effective than others. Among the most vicious and violent participants in that passing parade, through the desperate utilization of their first primitive muscle cells, your newly specializing ancestors somehow managed to survive by seizing essential reducing power from others less muscular or less fortunate. Of course, every improvement in those still rudimentary but increasingly powerful muscle cells resulted in a greater concentration of nutrients which made the more muscular even more highly coveted by their hungry neighbors—forcing them to fight ever more fiercely for their next red meat dinner or run away yet faster to avoid becoming one. Nowadays, most humans are too fat, skinny, sick, old, weak, young, inept or otherwise clueless or crippled to successfully forage or even defend themselves in the traditional cold and cruel world. So except for the most con-firmed survivalists among us—a significant subset of whom are angry religious fundamentalists eagerly awaiting *Armageddon* (worldwide disasters anticipating the final battle between good and evil as supposedly predicted in the bible—according to the regularly disproved but ever best-selling self-designated experts on biblical interpretation) to afflict *others* as "their just desserts!!" (for doing what?)—we should be grateful that science and technology finally have relegated human muscles to a minor supporting role. Now it is our ability to process great quantities of non-DNA-based information that prevents various microorganisms or even the remaining large carnivores from culling current human populations down to more healthy and environmentally sustainable levels—given half a chance, those carnivores would gladly consume the old, weak and diseased human majority in order to restore "The Balance of Nature"—whatever that might momentarily mean in this dy-namically unstable biosphere (presumably more of them and fewer of us). But humans presently rule the world with their construction skills and advanced weap-onry, ranging from antibiotics and refrigerators, houses and window screens to clubs, spears, guns, fire and explosives. Those powerful defenses have replaced our comparatively puny muscles as the most effective way to ward off the predators and parasites (including other humans) that lust endlessly after our assets.

Furthermore, to limit the development of unfavorable attitudes, we make it a point to prevent large carnivores from achieving reproductive success after they enjoy a single meal of live human flesh—any large creature demonstrating such propensity is promptly and selectively blown away by the local populace, out of physical shape as they may be. Indeed, this has become so routine that we feel no surprise when some pot-bellied weak-limbed attorney with a few extra dollars is able to best the mightiest bear. Thus as long as your muscles can pump enough air and blood to get you in and out of the car and grocery store, those muscles will probably help you to achieve reproductive success as well—which allows you to relax confidently in your easy chair despite being a complex collage of bacterial origin, still totally dependent upon the other bacteria-based life forms that you enslave or slaughter. For it is their repeatedly stolen solar energy, so laboriously trapped within covalent chemical bonds, that must power your every muscle contraction until the last.

What Was The Question?

In the never-ending quest for a more advantageous way to distribute sperm and eggs, multicellularity has come up with some amazingly complex *Rube Goldberg* (an early twentieth century cartoonist renowned for his descriptions of more complex and indirect ways to complete simple tasks) mechanisms, including you. As one might expect of designs by a Whole Earth Committee, with innumerable change orders and regulatory interjections by endless biological bureaucracies over hundreds of millions of years, you have almost entirely lost sight of the original goal. Often your only purpose seems to be the immediate gratification of some newly recognized want or need. But still you remain susceptible to the old signals. So you suffer uncomfortable diversions and highly emotional dislocations of your self-centered subjectivity when exposed to even the most superficial reproductive opportunity. And challenges to your reproductive power or its social equivalent (your money, your weapons, your tribe or your religious beliefs) tend to provoke the most violent (sometimes even muscle-dominated) reactions.

Better Muscle Contractions Through Chemistry

So how on Earth can sunshine trapped within covalent bonds of organic molecules (your solar-based reducing power) be converted into those still-important muscle contractions? How could electron-sharing between atoms possibly help you to impress men with your grace or women with your strength and skill as you drive that ball into the basket? What sort of micro-engine fits so neatly within the smallest of your muscle cells and operates so quietly and efficiently at your normal body temperature? Well first of all, the forceful shortening of any cell is likely to involve traction upon fine filaments within that cell. And only an equal pull in both directions will allow an elongated muscle fiber to contract symmetrically and in place. Although muscle shortening could proceed by actively trimming (or else winding-in) individual filaments, it seems far simpler and more controllable for muscles to contract by ratcheting groups of filaments past each other—and that is how muscles work. You can power that ratchet with a remarkable range of fuels, from roasted ants and pan-fried armadillo steaks to dried sunflower seeds, boiled zucchini and salted herring roe on seaweed. But before any of that stolen reducing power can be utilized by your muscles, it must first be extracted from those unwilling donors by digestion outside of your cells and then further refined through intracellular processing before finally being converted into ATP—for it is adenosine triphosphate that powers the heavy lifting and transportation jobs within your muscles and other cells.

Quite obviously there must be a long sequence of metabolic and micromechanical events between enjoying such gastronomic delights and the actual shortening of muscle. And we since we already know that everything runs down, there will be energy losses associated with each of these intermediate steps—indeed, most of the covalent bond energy in free fatty acids or glucose is lost as waste heat prior to or during your muscle contraction rather than being expended in the performance of work—so it is hardly surprising that physical exertion warms you up. During mild exercise, your muscles preferentially "burn" free fatty acids rather than glucose. But glucose (readily available from intracellular glycogen stores) releases its stored energy more rapidly and efficiently in vivo than fatty acids do—that is why you boost the proportion of carbohydrate in your fuel mixture during heavy exertion, and why carbohydrate-depleted muscles must work at the slower rate determined by fatty acid mobilization and oxidation. Muscle glycogen stores can apparently be enhanced before an athletic event by exercising to exhaustion and then carbohydrate loading for several days. Glucose (but not fatty acids) also allows muscles to perform some work under low-oxygen (anaerobic) conditions (see Metabolism).

The Actual Machinery Of Muscle Contraction

Early muscle cells were ordinary cells that utilized their constituents in a slightly different and more effective fashion. Chief among those constituents was *actin* which represents at least 10% of total cell protein in a wide variety of cells. Easily assembled and disassembled actin filaments are an important part of the fibrous skeleton that helps cells to transfer intracellular organelles and other molecular essentials as well as stay in shape and move about. Thus actin was destined for an important role during the gradual upgrade of already actin-dependent cell lengthening and shortening processes into the more organized contractions of true muscle. Of course, easily assembled and disassembled *microtubules* (made of the protein *tubulin*) provided another option and some of the five major varieties of *intermediate filaments* (keratin, for example) may also have been available. In addition to these fibrous components of the cell skeleton, the average cell included an assortment of ATP-powered motor molecules *(myosin,* several sorts of *dynein,* multiple types of *kinesin)* that could travel along or exert pull on actin filaments or microtubules. As it turned out, your early ancestors gained reproductive advantage by depending upon actin and myosin to enhance their mobility.

Actin Holds

Actin is a *globular* (more-or-less spherical) protein molecule. When a number of actin molecules are present in solution, they tend to line up and self-assemble in helically twisted double-stranded filaments. That tendency to form filaments of up to 10 microns in length can create a problem when many cells die suddenly and release their contents into the blood stream—indeed, a major injury or severe illness can overwhelm the circulating proteins of the *actin scavenger system* (these proteins include *gelsolin* and ordinarily prevent any accidentally released actin from forming filaments that could interfere with blood flow and initiate clotting). Gelsolin and its associated proteins also regulate the usual formation of long thin actin filaments inside of your cells, controlling their assembly, disassembly and even overall length. Each molecule in the sturdy actin filament provides attachment sites to which other proteins easily connect and disconnect. Some of those proteins interlock with actin as internal or just-under-the-cell-membrane support structures. Others such as myosin may rapidly connect and disconnect while pulling themselves along the actin filament—although apparently only in one direction, due to the polarity of the actin filament.

Myosin Pulls

Myosin is the other preexisting intracellular fibrous protein that became directly involved in muscle contraction (*myo-* refers to muscle and *muscle* derives from the Latin term for "little mouse"). Myosin molecules include an active ratcheting mechanism for moving along the actin filament. Various modifications of myosin and other motor molecules can be found hauling goods about within most sorts of cells. Myosin controls the separation of daughter cells during cell division and it moves the pigment granules peripherally in your melanin-producing cells. Kinesin sorts out chromosomes at cell division time, hauls mitochondria about and also distributes intracellular *vesicles* (membrane-enclosed sacs) containing wastes or useful chemicals such as neurotransmitters—moving its loads at a speed of about one micron per second. Dynein (like kinesin) transports essential materials along the microtubule tracks that organize intracellular transport generally—such microtubule tracks also extend throughout the long thin nerve cell extensions known as axons. Every muscle fiber contains billions of myosin filaments that readily self-assemble from numerous smaller precursor protein modules. When completely assembled, a myosin *thick* (in comparison to actin) *myofilament* in a skeletal or cardiac muscle is a symmetrically double-ended molecule. It rather resembles two golf bags joined end to end at their bottoms and jammed full with firmly stuck clubs of variable length—each of those club heads being directed outward. These myosin heads are *allosteric* molecules (able to change shape in response to local conditions) that can *flex forcefully* and repeatedly under appropriate circumstances, with each active flexion powered by chemical energy stolen from a tightly bound *ATP* molecule.

A free myosin head rapidly adheres to any adjacent *actin thin filament* unless blocked—that attachment encourages the myosin head to flex and release its P*i* (the third and no longer covalently bound so now *inorganic* phosphate) before discarding the rest of that depleted ATP as adenosine diphosphate or ADP. The flexed myosin head thereafter remains stuck to its actin filament unless fresh ATP is available. In that case, since myosin prefers fresh ATP to actin, it releases the actin and straightens out to grab that new ATP. But myosin is an ATP'ase (an enzyme that always hurts the one it loves) so the just-attached-to-myosin-and-much-beloved ATP is swiftly split to ADP plus Pi, with myosin temporarily left holding the energy that had been locked within the last interphosphate bond of ATP. At that point myosin would rather return to actin than live with such a broken-down old ATP (just as adolescents complain about the family car after they trash it). So myosin again locks onto actin and the myosin head again flexes forcefully and again releases its Pi and then ADP. In this fashion, the myosin heads festooning each end of the double-ended thick filament in *striated* (skeletal or heart) muscle can independently ratchet along the nearest of their six surrounding actin filaments. By pulling these thin filaments centrally from both directions, myosin draws the far ends of the voluntary muscle fiber or cardiac muscle cell together. Thus muscle shortening results from the repeated nodding of myosin heads with that movement being powered by the breakdown of ATP. Note that neither actin nor myosin filaments change in length, they simply slide past one another.

Ca^{++} Controls

To be useful, muscle contractions must include ON and OFF controls that act in a responsive and timely fashion. In a way, the timely part is solved because both predator and prey depend upon the same basic muscle mechanisms. In other words, whether your fastest muscles are contracting speedily or slowly may simply depend upon your perspective or leverage—who is being chased and which has the longer stride. Your ON controls depend upon yet another self-assembling set of allosteric fibrous proteins known as the *troponin-tropomyosin complex*—that complex normally protects actin from nearby myosin heads by lying snugly over the helical actin chain. However, any rise in the intracellular Ca^{++} ion concentration causes troponin-tropomyosin to change shape and move out of the way, allowing myosin unobstructed access to actin. Then, as long as intracellular Ca^{++} ions and fresh ATP remain available, each myosin head protruding from the heavy myosin filament can continue to ratchet itself along an adjacent actin light filament at a rate of about 30 strokes per second, moving 10 *nanometers* (billionths of a meter) per stroke, until both ends of the myosin filament eventually bump into those transverse protein supports (the *Z lines*) that hold actin filaments in place (kinesin similarly advances along its microtubule track in 8 nanometer strides that correspond to the spacing of tubulin dimers in that microtubule). Given the need for actin and myosin filaments to overlap from the outset, it is not so surprising that a complete or full muscle contraction only shortens an entire muscle by about 30%. Yet even if the multiple myosin heads in a striated muscle fiber could advance somewhat more rapidly than the measured 300 to 450 nanometers per second, that still is not a very useful rate of muscle contraction—for a muscle 10 centimeters (4 inches) long would then take nearly nearly two weeks for a complete contraction. But, of course, *each striation* in a striated muscle fiber represents *another* of the two-micron-long *sarcomere* units positioned end-to-end along that entire fiber, and it is all of those sarcomeres shortening simultaneously that allows an ordinary muscle *myofibril* (with 5000 such striations per centimeter) to achieve maximal contraction within a couple of seconds. Anyhow, as soon as intracellular Ca^{++} ions are again expelled, the troponin-tropomyosin complex swiftly swings back up out of the helical actin grooves to cover those actin attachment sites—provided enough ATP remains to tempt all myosin heads into letting go of their actin. The muscle fiber then passively returns to its resting length (aided by contractions of its opponent muscle groups) as actin and myosin filaments separated by troponin-tropomyosin again slide easily past one another.

ATP Pays

The rapid ATP-powered removal of Ca^{++} ions from intracellular fluid is a necessary talent of all cells and involves a number of different proteins and transport systems. This fundamental need to eject calcium derives from the crucial role played by phosphate ions in life's intracellular energy exchanges (in ATP, GTP and so forth). Since intracellular phosphate levels must always be high, intracellular Ca^{++} levels have to be kept correspondingly low (by constant out-pumping) for otherwise calcium and phosphate ions would *precipitate* (come out of solution together)—forming the relatively insoluble equivalent of intracellular bone. So Ca^{++} ions in high concentration are poisonous to living cytoplasm and every cell must always be ready to expel calcium. That allows you to utilize brief purposeful Ca^{++} incursions as economical signals for specialized cell functions—frequent small Ca^{++} incursions also help to keep important Ca^{++} exporting mechanisms in good shape—sort of like combining mandatory fire drills with other useful functions (e.g. group exercises, an attendance check or a communication opportunity). While your extracellular fluids contain sufficient free Ca^{++} to stimulate such cells, your muscle fibers have accelerated their responses by storing additional Ca^{++} within the widely distributed membrane-enclosed *sarco*plasmic reticulum (as the specialized endoplasmic reticulum of a muscle fiber is known)—and those intracellular sarcoplasmic reticulum stores release their Ca^{++} into the muscle fiber *sarco*plasm (*muscle* cytoplasm) in response to the *electrical depolarization* and Ca^{++} entry that initiate each muscle fiber contraction.

Overview of the Muscle Contraction Process

The next chapters provide more details on muscle and nerve cell-membrane depolarizations. But in overview, a *motor nerve cell* stimulates its *motor unit* (every muscle fiber under its control) by sending out an electrical nerve impulse that releases *acetylcholine* molecules into every *neuromuscular synapse* (special signaling site where a nerve branch tip lies adjacent to its muscle fiber). Those acetylcholine molecules activate specific muscle fiber membrane acetylcholine receptors, causing electrical depolarization of the *sarcolemma* (surface or *plasma* membrane of the muscle fiber) along with its transverse *T tubule* (inwardly directed) extensions. The depolarized muscle fiber plasma membrane allows Ca^{++} entry which encourages the sarcoplasmic reticulum to dump additional Ca^{++} into the intracellular fluid. As long as Ca^{++} then remains free within the sarcoplasm, troponin-tropomyosin stays bent out of the way and myosin can ratchet itself along the actin filaments. This ratcheting (the basic process of muscle contraction) ordinarily ceases when cytoplasmic Ca^{++} has again been expelled into surrounding fluids (or else restored to the sarcoplasmic reticulum) so that actin attachment sites are once more covered by troponin-tropomyosin and the muscle fiber can relax. The ratcheting process also comes to a sudden halt when the supply of ATP runs out—for Ca^{++} cannot be pumped back into the sarcoplasmic reticulum without ATP, nor will myosin ever be tempted to let go of actin. This accounts for *rigor mortis* (a several-hour-long period of persistent postmortem muscle rigidity due to the termination of mito-chondrial ATP production brought on by a lack of oxygen and nutrients).

Cytoplasmic Ca^{++} levels may rise sufficiently in slowly dying tissues to coprecipitate with the phosphate ions usually found near mitochondrial surfaces (see Metabolism).

Deathly Disorder

Another way of describing life is "the right stuff in the right place at the right time". Death results when inadequate energy inputs (whether by illness, starvation or circulatory arrest) allow disorder to increase in accordance with the Second Law of Thermodynamics. Clearly, life is far more than a simple mixture of exactly the right ions and molecules, for that could equally well describe death (or soup). Soon after death, the activation of proteosomes and release of enzymes from intracellular lysosomes brings about a breakdown of actin and myosin, and later of collagen as well. That is why fresh (recently deceased) meat is tougher than properly aged (partially decomposed) meat—assuming both are served rare (rather than boiled for a long time as when converting the tough collagen of a pot-roast into gelatin).

Panting Repays Oxygen Debt To Muscle (Via The Blood)

Muscles capable of continuous heavy work require a copious capillary circulation as well as many mitochondria for producing new ATP. Trained athletes display an enhanced utilization of fatty acids that delays depletion of their muscle-fiber glycogen stores. But even healthy muscle with a good circulation is easily overworked. So sprinters breathe heavily for some minutes after completing a race in order to pay off the oxygen debt incurred by their maximally exerted muscles—in the meanwhile, such poorly oxygenated cells cannot complete the combustion of glucose to CO_2—so temporarily over-exerted muscle fibers release *lactic acid* into local tissue fluids for reprocessing by the liver. The local build-up of potassium (K^+) and hydrogen (H^+) ions and near exhaustion of muscle fiber ATP supplies probably contribute to symptoms of acute muscle fatigue as well (see Nerve). The several days of muscle fatigue and soreness that can follow excessive exercise may represent some minor, diffuse, more slowly reversed injury to your muscle fibers or muscle capillary walls. Ordinarily there is a rapid turnover of all muscle fiber components—your actin filaments are regularly broken down and rebuilt within days. Chronically overworked muscles in healthy young adults commonly *hypertrophy* (enlarge) to meet any reasonable demands. Heavy exercise is associated with significant heat production so copious perspiration is required to achieve evaporative cooling. Under such circumstances, the adequate replacement of salt (NaCl) and fluid losses reduces the likelihood of muscle cramps and aches.

Getting Your Muscles to Do What You Want

Your *voluntary* or *skeletal* muscle fibers contract in coordinated local groups (motor units) under the control of individual motor nerve cells. Presumably the motor unit of a single irritated (overactive) motor nerve cell accounts for that occasionally bothersome twitchy eyelid or lip that others can hardly see—the few dozen fibers of such a tiny motor unit must have minimal impact on the function of an average muscle containing a great many motor units. So your usual easily seen or felt muscle contractions represent the combined efforts of a great many motor units. And the endless number of possible sequences and combinations of individual motor units that can be activated in your various muscles explains why most motor skills require a lot of practice (experience, trial and error) to learn. But while an entire muscle may contract slightly, briefly, persistently, slowly, quickly, a great deal and so on, muscle contraction is an all-or-none affair for individual motor units. Of course, an individual motor unit may receive only a single nerve stimulus or it may get several or even a whole series of stimuli in a row. Because a motor nerve recovers (returns to its resting state) far more rapidly after a single electrical discharge than muscle fibers can possibly complete their depolarization, contraction and relaxation, any rapid sequence of motor nerve cell discharges can induce sustained motor-unit contractions by maintaining persistently high sarcoplasmic Ca^{++} levels in those muscle fibers. Not surprisingly, skeletal muscles fatigue rapidly when their persistent contractions (as in isometric exercises) compress local capillaries sufficiently to block blood flow—and it costs less energy for you to shorten a slowly contracting and slowly relaxing muscle fiber one time than to repeatedly depolarize a more rapidly contracting fiber, even though the resulting prolonged contractions might seem quite similar. So it is economical and therefore reproductively advantageous for your muscles to include appropriate proportions of both slow-twitch and fast-twitch fibers. We have noted that a dead person needs no energy at all to sustain that last muscle contraction brought about by persistently high intracellular levels of Ca^{++} upon exhaustion of cellular ATP stores. The large ants whose mandibles are placed to approximate the edges of a wound before their bodies are broken away, then remain clenched in position by a similar mechanism (Mother Nature's own skin clips). Incidentally, the pre-tensioned strike of a trap-jaw ant takes only a third of a millisecond—making it the most rapidly responding natural mechanism known.

Cardiac (Heart) Muscle

Your *smooth muscle cells* and *striated heart muscle cells* are both considered *involuntary* since (in contrast to voluntary skeletal muscle *fibers*) they are not under your direct conscious control. We have seen how the actin and myosin fibers in skeletal and cardiac muscles are arranged in regularly repeated units or *sarcomeres*, with many end-to-end sarcomere units forming each *myofibril* and many long bundles of myofibrils extending the full length of each skeletal muscle fiber or cardiac muscle cell. However, cardiac muscle cells generally have but a single nucleus while each of the much longer voluntary striated muscle fibers includes a great many cell nuclei—those many nuclei can provide their important information without complicating cell division because muscle fibers are mature forms that no longer divide. The skeletal muscle hypertrophy that results from regular vigorous exercise usually reflects enlargement of individual muscle fibers but new muscle fibers also form under some circumstances from locally available muscle precursor cells.

Cardiac muscle cells contract and relax quite slowly although both processes speed up or slow down along with the heart rate. Your gradual cardiac contractions minimize the work necessary to squeeze incompressible blood out of those heart chambers. Cardiac muscle cells exert tension on each other at the reinforced, very irregular (therefore surface-expanded) *intercalated discs* where cell ends meet, as well as through the fine meshwork of connective tissue fibers that surrounds each cell. Your heart needs no tendons or ligaments since cardiac chambers contract more or less concentrically rather than exerting their combined efforts at a distance—that allows those short heart muscle cells to pump blood more efficiently than any hypothetical arrangement of longer muscle cells or fibers. Unlike voluntary muscle fibers, cardiac muscle cells are not directly driven by motor nerve cells. Instead, all heart muscle cells tend to depolarize spontaneously. An electrical impulse from whichever heart muscle cell is first to depolarize then spreads swiftly via small adjustable intercellular openings called *gap junctions* (direct intercytoplasmic connections between adjacent cells). Your two upper-heart chambers (*atria*) contract together as a unit, as do your two *ventricles*. Being in less of a rush to contract and relax, cardiac muscle cells obtain more of the Ca^{++} needed for each heartbeat from surrounding tissue fluids—each influx of Ca^{++} through the depolarized cell membrane and T Tubules also triggers Ca^{++} release from the cardiac muscle cell's own limited intracellular (sarcoplasmic reticulum) stores. Unlike voluntary muscle fibers, cardiac muscle cells will not respond to a new electrical impulse until sufficient time has elapsed to allow relaxation from the previous contraction. Such a built-in *refractory period* protects cardiac muscle cells from harmful sustained contractions that could arrest your circulation.

Smooth Muscle

An individual *smooth* (versus striated) muscle cell contains a single nucleus and derives most of the Ca^{++} needed for each slow contraction from its surrounding tissue fluids. Instead of having actin and myosin arranged in numerous parallel bundles, your smooth muscle cells have an extensive internal meshwork of actin and myosin that allows them to stretch to several times their fully contracted length and contract effectively from almost any length. This makes them ideal for service within your intestinal wall, urine bladder and other expansile internal passageways (blood vessels, bile ducts and so on). In contrast, striated muscle contraction becomes impossible once your cardiac or skeletal muscle is overstretched to the point where the actin and myosin filaments of each sarcomere no longer overlap. Your smooth muscle cells contract in response to *autonomic* (non-voluntary) nerve signals or various local physical or chemical changes, or as a result of electrical depolarizations transmitted from adjacent smooth muscle cells via protoplasmic bridges. As with cardiac cells, some smooth muscle cells have *internal rhythmicity* (a tendency to depolarize spontaneously). Because they contract and relax very slowly, smooth muscles consume relatively little energy when they must remain partially contracted (e.g. while raising your hair in the course of some hair-raising experience, or when regulating your blood vessel diameters). Thus you rely upon smooth muscle cells to bring about the many slow, involuntary, internal adjustments not requiring (hence not subject to) your conscious control (see Autonomic Nervous System).

All muscle cells transfer their intracellular contractile power to collagen fibers outside of the cell through regularly spaced adhesion plaques that attach to and transfer forces across the cell membrane. Severe diabetes mellitus or starvation can lead to muscle cell or fiber disassembly as myofibrillar proteins are catabolically degraded. The production of new myofilaments begins just under the cell membrane in close association with the adhesion plaques and then proceeds centrally into the cell. So the striated muscle cell (or muscle fiber) membrane plays an important role in organizing and orienting all of those continuously reconstructed myofilaments and myofibrils.

CHAPTER 12

NERVE RULES

Life Depends Upon A Separation Of Charges;... Transmembrane Ion Gradients Mean Life: Life Means Work Plus Risk;... Nerve And Muscle Cell Function Depend Upon Electrochemical Gradients;... Overview;... Cycle I: There Are Several Steps;... Cycle II: How Does A Nerve Cell Decide When Or Whether To Fire (Depolarize Or Discharge)?... Why Does A Nerve Cell Transfer Its Message Chemically?... Cycle III: The Electrochemical Events That Accompany Membrane Depolarization;... Na^+ Leakage Is So Exciting;... K^+ And Cl^- Leakage Inhibit Neuron Depolarization;... Calcium Channels;... Myelination And Saltatory Conduction Helped Your Ancestors To Prevail.

Life Depends Upon A Separation Of Charges

A living cell is an enormous number of ceaselessly interacting molecules. If each molecule of that huge number could somehow be grouped with others of its own kind, every cell would yield more than a thousand distinct groups. Yet no matter how skillfully you might recombine such groups within an appropriate droplet of salt solution, the resulting creation would be soup. For life requires both substance and order. Order implies the right number of the right molecules in the right place at the right time. And life's uninterrupted order extends back to the time when all began. Ever since then, each living thing has either responded appropriately to its constantly changing environment or it has suffered increasing disorder unto death (as in soup). Life's exceedingly complex and precariously balanced system clearly requires the continued expenditure of chemically trapped solar energy to sustain it. About one-third of your total energy expenditure at rest simply maintains different ion concentrations on opposite sides of your several thousand square meters of cell membrane—such a major energy investment by all living cells surely identifies transmembrane ion gradients as fundamentally important to life. We have seen how transmembrane differences in the concentration of atoms and molecules must hark back to the first phospholipid bilayer that separated something alive from the adjacent salty sea, for First Life undoubtedly enclosed many more complex and interactive molecules than were contained in equal volumes of nearby sea water. And some of those organic molecules must have carried electrical charges that attracted certain ions and repelled others.

Of the four smallest cations with a single positive charge (H^+, Li^+, Na^+, K^+), Na^+ has always been the most plentiful in oceans with K^+ but a distant second. For similar reasons (availability and solubility) Cl^- has dominated oceanic anions. Although it contained many molecules at higher concentrations than found in surrounding waters, First Life still had to stay in osmotic balance with those waters. That made transmembrane ionic gradients inevitable from the start—for with internal salt concentrations already at oceanic levels, the additional water drawn in by life's extra solutes would have caused First Life to swell and rupture. Thus the ejection of Na^+Cl^- (Earth's dominant solute) became key to life's survival. Quite obviously, life's origin and prosperity required effective use of locally available materials. And the endlessly varied interactions between life's essential molecules clearly demonstrate how adaptation to inevitable consequences has always been associated with reproductive advantage—so with transmembrane concentration gradients of common ions being unavoidable, life has cleverly utilized those gradients to facilitate as many essential cell functions as possible. But how could the innate tendency of transmembrane ion gradients to run down (thereby increasing disorder in accordance with the Second Law of Thermodynamics) be utilized most advantageously? Indeed, how are such gradients produced and maintained in the first place?

Transmembrane Ion Gradients Mean Life: Life Means Work Plus Risk

In order for transmembrane ion concentration gradients to be useful, a cell must be able to drain and restore them. It turns out that an early step in photosynthesis involves the solar powered build-up of electrically charged ions (protons in this case) on one side of a semipermeable membrane. Furthermore, each of your many quadrillion (10^{15}) mitochondria utilizes photosynthetically stored solar energy in order to recreate such transmembrane proton gradients and thereby drive the ATP production that pays for establishing the essential transmembrane ion gradients by which you sense, process and respond to information about your next victim or reproductive opportunity. Mitochondrial ATP also supports a great many intracellular anabolic processes and functions. Life performs work and increases order whenever it pumps Na^+ ions out of a cell (where Na^+ is present in low concentration) into the surrounding salty sea. And when Na^+ ions are allowed back into that low-in-Na^+ cell interior, some of that stored energy becomes available to drive essential cellular reactions. This method of power storage is analogous to keeping your car battery charged, or to an electrical utility using spare (off-hours) generating capacity to pump water into an elevated lake for later release through hydroelectric generators at times of peak electrical demand. Intracellular proteins are usually anions (with a net negative charge) so Cl^- is easily expelled from the negatively charged cell interior. In contrast, the expulsion of Na^+ from a cell soon creates an impossible-to-overcome electrical gradient unless some other cation is simultaneously imported. Their single charge and ready availability make H^+ and K^+ the natural cations for exchange with intracellular Na^+. K^+ is included in most fertilizers (as *potash* or potassium carbonate, formerly recycled from wood ashes) because fungus and plant cells sustain their osmotic pressure and healthy turgor with far higher intracellular K^+ concentrations than animal cells require. Bacteria, fungi and plants have come to rely upon proton pumps (that expel always-available H^+ in exchange for K^+) since they often inhabit sodium-depleted waters or soils. On the other hand, animal cells exchange Na^+ for K^+ since variable H^+ levels within the cell could harm performance by disturbing many of life's ionizations and molecular interactions—animal cells generally acquire enough Na^+ from their victims anyhow (see Digestion).

Whether you view transmembrane ion gradients as savings or expenditures, they clearly mediate a great many transactions. And all transactions involve transfers of information. Thus life expends energy to maintain transmembrane ion gradients so it can respond to information and capture sufficient energy to maintain such gradients. While the complex instability of life makes it totally dependent upon outside energy sources, life's acquisition and utilization of such energy makes it increasingly stable. But the more successful and stable any life form becomes, the more its neighbors (large and small) will covet that enhanced reducing power, which again raises risk and increases instability. Apparently there are no rewards without risk in this competitive world, so life remains a gamble and the thin edge between order and chaos is where we seek reproductive advantage and find death.

Nerve And Muscle Cell Function Depend Upon Electrochemical Gradients

Nerve and muscle cells typically utilize transmembrane ion gradients in rather unique fashion for their own purposes. With many essential details, this story is best related in several cycles of increasing complexity.

Overview

An *electrical wave of depolarization* sweeping down a *motor nerve cell's* long *axon* somehow brings about contraction of every voluntary muscle fiber in its motor unit.

Cycle I: There Are Several Steps

Every electrical depolarization of a motor nerve cell axon releases thousands of *acetylcholine* molecules at each of its neuromuscular *synapses* (those sites where an axon-branch-tip nestles within 50 nanometers or 50 billionths of a meter of its muscle fiber—a gap several hundred hydrogen atoms wide). When acetylcholine activates specific receptors embedded in the adjacent muscle fiber *synaptic* surface, an electrical wave of depolarization or muscle *action potential* is initiated that quickly spreads from the synapse over the entire *sarcolemma* (phospholipid bilayer surrounding the muscle-fiber) including its T Tubule sarcolemmal extensions (that dive into the muscle fiber). In this way an electrical impulse traveling down a motor nerve cell axon depolarizes the membranes of all its associated muscle fibers—which in turn permits Ca^{++} to enter and initiate contraction of those particular muscle fibers.

Cycle II: How Does A Nerve Cell Decide When Or
Whether To Fire (Depolarize Or Discharge)?

At any moment, a typical nerve cell must decide whether or not to initiate a depolarization. Basically that is all it can do. To help it make that decision, every nerve cell receives a great deal of trans-synaptic input directly from other nerve cells as well as indirect chemical advisories (released into tissue fluids) that activate other specialized membrane receptors. Some of this incoming information *excites* the nerve cell, some has an *inhibitory* effect. If sufficiently excited, a nerve cell will discharge. If sufficiently inhibited, it will not discharge. But when nerve cell discharges do occur, each is the same as every other—hence depolarization is an *all-or-none* phenomenon. Of course, a very excited nerve cell is likely to discharge periodically or far more frequently than a similar cell that is less excited. And with experience a nerve cell can alter the value that it sets upon any specific exciting or inhibitory input. In contrast, muscle fibers have neither opinions nor options since any motor nerve cell depolarization is sufficient to bring about contraction of all relaxed muscle fibers (or enhance and prolong any contraction already underway) within that motor unit.

Why Does A Nerve Cell Transfer Its Message Chemically?

A direct cytoplasmic connection between cells (such as a gap junction) could carry instructions electrically from a nerve axon to its muscle fibers. That would eliminate the short delay caused as acetylcholine is released and diffuses across the narrow *synaptic cleft*. But speed isn't everything—indeed, increasing speed without simultaneously improving control can quickly kill the speeder. As usual, there are a number of factors to consider. First of all, the acetylcholine effect is rapid and brief since diffusion across a tiny space is swift and acetylcholine molecules are quickly broken down by *acetylcholinesterase* into acetate and choline (which are promptly returned to the nerve tip for reuse). So each new jolt of acetylcholine is eliminated from the synapse before the nerve cell recovers sufficiently to undergo its next depolarization. That brief delay in output provides time for some feedback to keep you posted on how (and what) you are doing. Furthermore, acetylcholine is but one of many *neurotransmitters* released by a great variety of nerve cells. Of course, every one of those signal molecules only affects the specific parts of potential re-cipient cells that bear appropriate receptors. Acetylcholine happens to be especially useful at voluntary neuromuscular junctions where its very brief duration of action enhances fine motor control. And the way acetylcholine is delivered into and also broken down within the synapse markedly decreases the risk of either too wide a distribution or insufficient local acetylcholine effect.

The many (over 50) other neurotransmitters that you use vary widely in their mode and duration of action before being diluted to insignificance, destroyed by enzymes or else returned to their source by special transporter molecules. Using neurotransmitters of appropriate onset, duration and distribution at properly se-lected sites minimizes the possibility that the wrong nerve cell might be stimulated or inhibited by overhearing a chemical message intended for some other cell within your complex and crowded *central nervous system* (brain and spinal cord). And the specific interactions of neurotransmitters with their own particular receptors al-lows intercellular fluids to simultaneously distribute a number of messages that can readjust various nerve cell activities over differing periods of time with minimal interference. In comparison, a nervous system based entirely upon direct electrical connections might be more rigidly organized and difficult to retune as a cooperat-ing whole during times of rest or maximal stress. It might also be more difficult for such a nervous system to *learn* (modify its responses to incoming information). Furthermore, the fact that these chemical messages only affect specific recipient cells bearing exactly appropriate membrane receptors prevents any prolonged muscle fiber depolarization from secondarily reverse-depolarizing its more rapidly recovering (so again electrically responsive) motor nerve cell. That sort of accidental backfire from slave to master cell might easily occur if nerve axon and muscle fiber were electrically joined via a gap junction. And even the rare mix-up between input and output of any *neuron* (nerve cell) during a fight or flight could represent a serious reproductive disadvantage.

Cycle III: The Electrochemical Events That Accompany Membrane Depolarization

A tiny *electrode* (electrical conductor such as a wire) placed within the high-in-K^+ interior of a resting nerve or muscle cell and another in the (K^+ depleted) extracellular fluid surrounding that cell will show an electrical gradient of about 0.07 volts (70 millivolts) across the cell's non-polar plasma membrane. That inside-negative potential of -70 mV results from the net outward diffusion of positively charged K^+ ions through a few persistently open K^+ channels. But with K^+ the only ion crossing that phospholipid bilayer, each additional K^+ ion that escapes will be leaving an increasingly minus cytoplasm for an ever-more-plus outside. Consequently a growing electrical pull is exerted upon all outward-bound K^+ ions until a balance is reached between *inward* moving K^+ ions (being repulsed by the positive charge outside or attracted by the negative charge inside) and *outward* moving K^+ ions (diffusing away from the far higher concentration of K^+ inside). This balance between an electrical pull in one direction and a concentration gradient shove in the other is referred to as an *electrochemical gradient*. And the electrochemical gradient established by a slow outward leak of K^+ ions is responsible for maintaining the -70 mV potential of a resting nerve cell with respect to its surrounding fluids.

Na+ Leakage Is So Exciting

Some neurotransmitters cause nerve cell membrane (or muscle sarcolemma) Na^+ channels to open briefly. The consequent rush of extracellular Na^+ ions into the negatively charged and low-in-Na^+ cytoplasm rapidly reduces the usual 70 millivolt gradient across the postsynaptic cell membrane. When several such stimulating signals from one or more sources reach the same portion of nerve cell membrane at almost the same time they are likely to drive the local transmembrane electrical potential below 60 millivolts (each motor neuron discharge has that same critical effect on all muscle fibers of its motor unit). Such a voltage drop is enough to open adjacent *voltage-sensitive Na+ channels* which further boosts Na^+ ion leakage across the membrane and thereby decreases local transmembrane voltages even more, thus triggering additional nearby voltage-sensitive Na⁺ channels to let still more Na⁺ enter. Once successfully initiated, such an electrochemical disturbance expands outward as a progressive wave of rapidly opening voltage-sensitive Na⁺ channels. Such a *wave of depolarization* usually originates at the *axon hillock* (where a motor nerve cell body joins its axon) and passes on down the nerve axon from there. Or in the case of a voluntary muscle fiber, the wave of depolarization expands from the neuromuscular synapse to sweep over the entire muscle fiber membrane. Yet nothing actually accompanies that measurable electric change as it races over the axon or muscle membrane. Rather, as each voltage-sensitive Na⁺ channel opens, a few Na⁺ ions simply dart a short distance inward across the membrane—comparable to the way many individuals sequentially waving their placard can create "the wave" seen at ball games. As in a water wave or a long series of toppling dominoes, it is the disturbance that travels—the people (ions, electrons, water molecules, dominoes) stay put.

Anyhow, that sudden inrush of positively charged Na⁺ ions boosts the initial -60 mV inside voltage to +35 mV or so. Then all voltage-sensitive Na⁺ channels gradually close—their initial swift opening and subsequent slow closure are both triggered by the same voltage change acting upon different portions of the sodium channel. Simultaneously being turned ON quickly and OFF slowly can usefully achieve a brief and self-limited effect (just as those annoying faucets in airport lavatories manage to rinse off a little of that smelly soap before shutting themselves down to reduce risk of flooding—see also G proteins). The triggering of voltage-sensitive K⁺ channels allows a brisk outflow of K⁺ ions that rapidly restores the usual inside-negative voltage after an initial overshoot to -85 mV or so. Thereafter the nerve (or muscle) transmembrane electrical potential drifts back to the normal resting voltage as open K⁺ channels mostly close and invading Na⁺ is pumped out of the cell in exchange for K⁺ (every 3 Na⁺ out and 2 K⁺ in costs one ATP hydrolysed). The original slow leakage of K⁺ then maintains the -70 mV resting transmembrane potential. The large number and variety of K⁺ channels in nerve, muscle and secretory cell membranes allows great variability in cell responsiveness, repolarization rates and patterns of repeated depolarizations. Interestingly, a cell cannot again be depolarized by any amount of stimulation until full K⁺ outflow has reset all of those voltage-sensitive Na⁺ channels—such a temporarily non-responsive cell is in its *absolute refractory period*. And the *relative refractory period* that follows represents a time of reduced responsiveness due to membrane *hyperpolarization* (to -85 mV) when ordinary stimuli are unlikely to open enough voltage-sensitive Na⁺ channels to reach the 60 millivolt gradient necessary for a new depolarization.

In summary, neurons depolarize whenever exciting stimuli sufficiently out-weigh inhibitory inputs to push local transmembrane potentials below a critical level of about 60 mV. Rapid opening of nearby voltage-sensitive sodium channels then allows an inrush of positively charged Na⁺ ions with progressive triggering of adjacent voltage-sensitive Na⁺ channels on down the axon. That wave of depolarization traveling the length of the axon (or over the entire muscle fiber surface after synaptic stimulation) is followed by a wave of repolarization as rapidly opened voltage-sensitive Na⁺ channels gradually close and voltage-sensitive K⁺ channels open widely. Eventually the customary slow leak of K⁺ ions reestablishes resting conditions.

K⁺ and Cl⁻ Leakage Inhibit Neuron Depolarization

So nerve or muscle cell *stimulation* opens Na⁺ channels locally, while excessive K⁺ leak is inhibitory to further excitation. But with K⁺ channels so heavily involved in repolarization, there should be some separate way to *inhibit* nerve or muscle cells—and, in fact, inhibitory impulses usually open one or another sort of Cl⁻ channel through which negatively charged chloride ions can rush into the muscle or nerve cell interior—although those concentration-driven incoming Cl⁻ ions are soon increasingly repulsed by the ever more negative cell interior as local trans-membrane voltages rise to around -80mV. Thus Cl⁻ entry is as inhibitory to cell depolarization as K⁺ outflow, since either can hyperpolarize the cell. Indeed, muscle-fiber Cl⁻ channels turn out to be more important than K⁺ channels in preventing muscle-fiber *hyperexcitability* (spontaneous *action potentials* or depolarizations). So chloride entry helps to maintain the usual relative refractory state in which your voluntary muscle fibers quietly await their next motor neuron depolarization, and abnormal Cl⁻ channels underlie the muscle stiffness and impaired relaxation seen in congenital *myotonia* as well as various medical problems encountered with *cystic fibrosis*. Since just a few ions crossing a membrane can bring about significant voltage changes, a neuron may depolarize hundreds of times without greatly altering intracellular ion concentrations. In any case, an ongoing exchange of intracellular Na⁺ for extracellular K⁺ maintains high cytoplasmic K⁺ levels and minimizes cytoplasmic Na⁺ levels. By moving Na⁺ and K⁺ simultaneously in opposite directions, the Na⁺/K⁺ exchange pumping mechanism reduces its work against electrical gradients. And as usual, your various ion pumping proteins are coupled to ATP breakdown in a fashion that makes more energy available than is required for moving K⁺ in or Cl⁻ and Na⁺ out against their concentration gradients.

Calcium Channels

Cell membrane Ca^{++} channels also come in multiple varieties. Muscle membrane Ca^{++} channels admit Ca^{++} ions from extracellular fluids or sarcoplasmic reticulum stores. And there is a momentary inflow of Ca^{++} into tiny nerve endings during the brief period of positive inside voltage just after nerve cell membrane depolarization—such presynaptic Ca^{++} inflows trigger the release of tiny vesicles filled with acetylcholine molecules into the synapse. At other times and places, Ca^{++} inflows open certain K^+ channels that link membrane potential with intracellular regulatory enzymes. In addition, Ca^{++} entry is necessary for strengthening a synapse—an aspect of memory known as *long-term potentiation*. Ca^{++} inflow also stabilizes muscle-membrane acetylcholine receptors at what will become a permanent neuromuscular synapse (a *post-synaptic* effect). As previously mentioned, intracellular Ca^{++} concentrations are held low to avoid co-precipitation of Ca^{++} with intracellular phosphate. Thus a minor inflow of Ca^{++} can serve as an urgent signal that (like a fire drill) also maintains your calcium-expelling and *cell membrane repair* (using vesicle membrane as a patch) mechanisms in a state of readiness by giving them a regular workout. An ATP-powered expulsion of calcium from the nerve cell begins immediately and is associated with a slightly excessive Na^+ inflow that electrically boosts Ca^{++} egress (the less dangerous intracellular Na^+ is then handled in the usual, more leisurely fashion). Similar Na^+/Ca^{++} exchangers are active in muscle fiber and sarcoplasmic reticulum membranes.

Inherited defects in sarcoplasmic reticulum calcium-release channels can lead to *malignant hyperthermia* in susceptible humans undergoing anesthesia. Here the standard inhalation agent *Halothane* along with *succinylcholine* muscle relaxation (see next chapters) can trigger massive intracellular Ca^{++} release leading to muscle rigidity, hypermetabolism, high fever, electrolyte disturbance, severe organ damage and death if not promptly detected. *Porcine stress syndrome* is a comparable condition of swine that can result in devalued meat products if triggered by the stresses involved in going to market. Nonetheless, the mild myopathy of porcine stress syndrome has remained a reproductive advantage in the view of swine breeders since the associated spontaneous muscle contractions lead to a leaner, more heavily muscled carcass (the result of unintentional isometric exercises by those lazy swine).

Myelination And Saltatory Conduction Helped Your Ancestors To Prevail

The speed of nerve conduction and muscle contraction have been upgraded constantly during the never-ending race for survival and reproductive advantage.

One way to make a nerve cell conduct messages more swiftly is to increase the diameter of its axon (squid mantle neurons have the largest axons). A more practical way to improve performance within a very compact nervous system is to electrically isolate those tiny, closely packed axons by applying many short segments of insulation—a process known as *myelination*. So the nerve cells that control your voluntary muscles are myelinated—but then again, so are the motor neurons of your predators and most of your animal victims (it is hard to keep up, let alone get ahead of the competition). Anyhow, a myelinated axon resembles a long string of sausages with each segment bulked-up by a separate spiral wrapping of phospholipid cell membrane (including stabilizing proteins and carbohydrates) around the axon membrane. Every (electrically insulating) myelin segment remains in continuity with the separate support cell that contributed it. Furthermore, the short length of exposed axon membrane between myelinated segments bears unusually high concentrations of Na^+ and K^+ channels in clusters—which allows depolarization to proceed by rapid jumps since enough Na^+ enters at one uncovered gap to initiate opening of voltage-sensitive Na^+ channels in the next gap. Through such *saltatory* (jumping) *conduction* of nerve impulses, your ancestors speeded their nerve membrane depolarizations in order to fight or run away more swiftly so that their reproductive success could eventually result in you.

CHAPTER 13

YOUR 100 BILLION NEURONS

There Is No Such Thing As Smart Enough;… Conflict, Competition, Cooperation And Codependence;… Truth Is More Novel Than Fiction;… Survival Of The Fittest Microcircuit?… Information Tends To Spread, (And Also To Reduce Disorder—Take That, Second Law!);… Competition Increases Wealth;… The Input Is Overwhelming;… So What Is Behavior?… The Incredible Non-shrinking Of Your Personal Program;… What Can We Learn From Sensory Deprivation?… How About Anesthesia?… Blocked Motor Output—Muscle Relaxants;… Mind And Brain;… Your Central, Peripheral And Vegetative Nervous Systems Are One;… Your Internal Display Is Subjective And Defective;… So Your Behavior Is Often Weird;… And Possibly Ambivalent.

There Is No Such Thing As Smart Enough

Ceaseless interactions of matter and energy had saturated Earth with information long before First Life even stirred. And it has steadily been getting worse ever since. For willingly or no, all living things inherit, create, amplify and distribute their own unique information. Each must sense and respond to important environmental changes, incoming predators, potential prey. Every new interaction or relationship enhances complexity. Opportunities and dangers lurk. Nothing is simple. All decisions are final. Your many trillion ancestors had to resist parasites, evade predators and fight or love one another in order to reproduce. Any slight advantage might delay death—the most primitive chemical signals between neighboring related cells could carry great risk or bring big rewards. And as soon as specialization of cells within multicellular life forms proved advantageous, strong selective pressures favored the better organized—if a few cells of some multicellular gathering became especially adept at sensing and signaling, that further improved chances for a more effective group response to the ever-changing environment. Endless elimination bouts benefited those best able to sense and respond. Even the limited information processing skills of early neurons would have seemed miraculous had their tiny multicellular competitors been smart enough to be impressed. Only the most appropriate responders to environmental cues remained to reproduce. With competition fierce, relentless, unending, there was no such thing as smart enough. Inevitably the number of nerve cells increased. Regularly they developed more effective interconnections that clarified which would sense, which command and which obey. But still there was too much information, no matter how efficient an individual organism might become. So cooperation became increasingly vital.

Conflict, Competition, Cooperation And Codependence

The road to reproductive success is paved with good reducing power. But no life form can travel that trail by itself. Even photosynthetic bacteria depend upon others to fix their nitrogen in an accessible form or to break down residual organic materials. And all other life forms rely upon photosynthetic bacteria (either free-living or residing as chloroplasts within plant cells) to trap solar power and maintain the biosphere. Furthermore, all cells must cooperate with their mitochondrial endosymbionts if both cells and mitochondria are to persist—just as any multicellular creature must coordinate its member cells in a more effective fashion than the competition in order to survive. Individual multicellular plants fight to the death for water, sunlight, minerals and CO_2. The roots of many plants form massive free-enterprise zones with fungi and bacteria. In most such close relationships, the plant exchanges glucose for improved access to water, minerals and nitrogen. Birds and bees have always transported plant seeds and pollen in exchange for fruit and nectar (that sugar again). Ants herd and protect aphids in order to secure their sticky sweet secretions. Certain ants even protect particular thorn trees from encroachment by insects or vegetation—again the payment for such out-of-species services is mostly in sugar. So who is working for whom? Well, ordinarily, we view the employer as boss. She who pays the workers and other suppliers calls the tune. But relationships in the biosphere or between businesspersons, politicians and bankers really are far more complex, and these alliances only endure while they enhance the success of all parties.

Although politics and politicians are best understood in terms of the flow and redistribution of dollars, the key to biological relationships is an understanding of the flow and redistribution of glucose units. Of course, politicians and their lobbyist keepers only flourish as long as they benefit their less visible co-conspirators, the *special interests* (whose goals inevitably conflict with the public interest since it is primarily by misuse of public funds that the politically connected achieve their wealth, power and prestige—the human equivalent of great reproductive success). But note again the instability of politics, based as it is upon competition and conflicts-of-interest at all levels. See how often it results in cooperation as well as codependence—and how any regulators and regulatees caught with their hand in the till (e.g. in those savings and loan scandals) either claim to be public benefactors or victims themselves.

Similarly, while humans speak of having successfully domesticated rice, beans, corn, wheat, potatoes and sunflowers, one might equally accuse those special interest photosynthesizers of having bribed farmers all around the world for thousands of years to clean out competing species, collect and plant their seeds, fertilize, weed, protect and harvest—all in exchange for a few surplus glucose units (carbohydrates). Domesticated dogs, pigs, sheep, chickens and camels also seem to have gained reproductive advantage over often-smarter still-wild competitors, predators and parasites by forging a close alliance with humans. Indeed, the benefits of domestication have been so great that it would be hard to say which side is more dependent upon the other (just as recipients of large bank loans may have far greater influence at the bank than small borrowers—even in cases where those large loans didn't go to relatives or friends of bank officials). So is the hard-working shepherd in charge when he provides food, shelter and protection so his flock can enjoy a life of leisure? Or did his sheep make the better bargain when they exchanged waste body hair and their freedom to roam for better health, a diminished mortality rate and (often) guaranteed annual reproductive success? And where does the sheep dog fit in?

It is hardly surprising that plants and animals have cooperated so readily in their own domestication. Of course, other life forms such as garden slugs, coyotes, malaria organisms, nettles, dandelions and deer have done well through their association with humans without entering into formal codependency arrangements. But in all such cases, the reproductive success of the less mobile or less intelligent has grown along with their reliance upon humans or human activities. For while humans may well have the superior intellect, their extensive control over resources cannot help but attract ever more effective and specialized predators, parasites, competitors and codependents. Which in turn increases the information burden for all sides, thereby encouraging still more cooperation, codependence and conflict. Only by pooling neurons with those of relatives and other allies can ants and fish, geese and humans handle and respond appropriately to far more information. And that is the very least it takes merely to remain in the running since only the most effective social groups (those schools, tribes, anthills and flocks with good luck, good members with particular talents and good leadership) are able to help their members survive.

Thus it came to pass that socialization and specialization inevitably begat civilization out of information overload, while the socially empowered (those with the resources to reward allies) gained increasing advantage over the solitary—no matter how strong their armor, sharp their claws, rapid their reflexes, sturdy their skeletons or clever their responses. Even today, competitive human cultures rise or fall on how appropriately they obtain, process and respond to information. Those social structures that most nearly guarantee each individual an effective education with free access to information and private property rights inevitably maximize creativity and wealth. And their victory is displayed on well-stocked grocery shelves rather than in heavy armor rolling past during May Day parades. In contrast, totally repressive societies isolate themselves from so much information that they cannot compete while partially repressive societies eventually disrupt as their citizens become aware of how they have been impoverished through the *propaganda* (lies and inadequate information) distributed by self-serving leaders. So repressive societies are now in decline, particularly since information can reach any radio or television set from satellites stationed in outer space—radio and TV may carry lots of propaganda but eventually they confirm any wide disparity between word and deed.

Regardless of social structure or type of organization, fierce competition continues among members of all groups. You cannot long escape sibling rivalry, teenage rebellion, ambition or the urge to get ahead—even while out for a relaxing Sunday drive. In this ongoing battle, greater intelligence usually brings reproductive advantage. So when any single nervous system successfully coordinates larger numbers of more versatile nerve cells, the pressure grows upon others to upgrade their performance as well. Yet no matter how dominant the social organization, how brilliant its members or how easily they can access relevant information, every group is continuously and inevitably overwhelmed by the sheer volume of incoming data. That is why all organizations are blind to important input, stupid in their processing and incompetent in their behavior. And why the distant leaders of larger organizations tend to be most out of touch with local realities. Thus *crisis* remains the usual state of affairs in every organization. Indeed, larger groups are generally unable to redirect their collective attention toward the next most threatening input except by declaring a crisis—and sometimes not even then.

In view of all this, it is hardly surprising that your 100 billion neurons (a current estimate) barely suffice to keep you out of the rain. You are lucky to get by, let alone deal appropriately with any tiny sample of that uncontrollable information deluge. So you detect very little of what goes on around you, you ignore most of what you detect, you forget most of what you do not ignore, you incorrectly evaluate most of what you do not forget, and you usually end up wishing you had responded differently. Welcome to the real world, or rather the insignificant bit of it that you even attempt to display internally in your own inadequate illogical hormone-ridden fashion.

Truth Is More Novel Than Fiction

You should not feel too badly about your own information processing limitations, however, for no one else seems to know why that last rain dance did not work or why your team lost although you wore those lucky shorts, or why your head aches and aspirin won't help. At least you can usually figure out whether to fight or run, you generally recognize which food is rotten and you probably will reproduce successfully—although you can never become the perfect parent you once vowed to be during those embattled adolescent years—for conflict, competition and cooperation are fundamental to all relationships and the family was your (hopefully gentle) introduction to life's realities and burdens, victories and defeats. Furthermore, you may be about as smart as humans are likely to get—there simply is no biological prospect that ever more intelligent descendents will evolve to solve all problems on Earth. Indeed, it is far more likely that your own descendents will continue to cause many more problems than they even recognize, just as you have.

Your remote multicellular ancestors often got the jump on their competition by including a few more neurons. Since each additional nerve cell could interact with so many others in so many different and useful combinations, those starting with relatively few neurons often found a modest expansion the easiest way to upgrade their information processing capabilities. Currently, however, we may be approaching the maximum useful number of neurons per human individual. Adding a few generalized neurons to your present hundred billion or so is unlikely to noticeably improve your intelligence. Achieving more than such a minor increase within current design limitations (support, circulation, cooling and birth constraints) appears fraught with difficulties.

Even now the infant head is painfully large for pelvic delivery. Often enough, it gets stuck and a *caesarian* (surgical) delivery must be arranged through the abdominal wall. Yet the female pelvis has widened about as much as useful mobility and strength will allow. And to pass the infant head, that female pelvis still must be loosened at every joint by hormones and enzymes, often to the point of unpleasant instability. Already the newborn's brain continues to grow and develop for many years after birth, and certainly the infant remains totally helpless (as an extrauterine fetus, if you will) and the child completely dependent for about as long as most parents can endure and support. Indeed, merely to reach the stage of brain development displayed by other primates at birth, you would have had to remain inside the uterus for 21 rather than nine months—an obvious impossibility. So does this apparent size limit on human brain development permit any useful predictions?

Survival Of The Fittest Microcircuit?

Interestingly, this biological limit has become increasingly irrelevant to human progress as modern civilizations reinvent the ways they produce, detect, process and respond to information (using satellites, radar, lasers, fiberoptics, electronics, computers, automation and so forth). But even if a dramatic expansion of individual human mental capacities remains out of reach, we have not nearly achieved effective use of those neurons already in possession. For the manner in which life's information is formally presented for processing often takes little advantage of natural human interests, or how and when in life humans learn most easily. So it seems likely that some of the new information now accumulating at an exponential rate will help us to improve the manner in which we process the rest. The practical limits of this newest and most profound interaction of humans and their thinking machines are quite distant and unclear, as are the social systems and entire civilizations that might result. Nonetheless, fierce competition will undoubtedly continue to force technological changes upon us at an ever-increasing rate as the same old biological wants and needs drive us across this immensely rich, uncharted and increasingly abstract wilderness—which means we had better try to understand ourselves as well as possible.

Information Tends To Spread
(And Also To Reduce Disorder—Take That, Second Law!)

Inherently formless and progressively unrestrictable flows of electronic information will finally make it possible for all humans to share in the Commonwealth of Information. The term Commonwealth here takes on new meaning for it cannot be used up. Indeed, the more widely and rapidly any information is *disseminated* (as in sowing seed), the more wealth and useful new information it will produce as it reaches those best positioned to benefit. And the more obvious the inevitable burdens of war, environmental degradation, poverty and overpopulation, the less acceptable such forms of information abuse will become, particularly as more appropriate, effective and affordable systems of rewards and punishments evolve. Free access to information means never having to reinvent the wheel. And far more brain power can be tied up puzzling over incomprehensible things than in understanding a proper explanation by someone who is able to make the subject seem simple. Thus our children obtain a more useful knowledge of physics (while presumably tying up far less brain power) in their beginning high school classes than we ever received from our college professors, since the confusing evidence of that day had to be memorized as received and then regurgitated in the same awkward, disorganized fashion.

Competition Increases Wealth

Preparation for war still provides great profit to many, but war itself has been outmoded by more profitable forms of business competition. Military power based upon new technologies may still defeat brute force attacks from the outside but the technological edge in warfare is increasingly transient, costly and even counterproductive as information disseminates ever more widely. Similarly, the benefits for all humanity in eradicating pestilence and poverty are increasingly apparent. Although even Mother Theresa must compete ferociously to achieve her goals, humanity finally is at a point where the harder we fight for knowledge, the more we tend to benefit others as well as ourselves. And the better off everyone is, the more the birth rate goes down as couples seek control of their own lives. All of which should allow us to deal more responsibly with Earth's biosphere (our environment). But in the meanwhile, what you do and become as an individual depends largely upon how well you utilize those 100 billion neurons. For what you sense, how you process it and the ways that you respond are still vital to your individual survival and reproduction—which brings us to your senses.

The Input Is Overwhelming

Day and night, wherever you may go, you are immersed in electromagnetic waves of all possible frequencies and energies. Some come to you directly from the sun and other stars. Others reflect from the moon in June or your dog in heat. Still other electromagnetic waves are emitted by radon, molten lava, your warm spouse and the street light at midnight. Furthermore, this is a disturbed world of earthquakes and quaking aspens, water waves and sound waves, screams, growls, creaking doors, guns, firecrackers and children singing. And it is a chemical world of new mown hay and lilacs, burnt toast and sewage, fresh-squeezed lemons, bacon frying and coffee percolating. You cannot possibly keep track of it all. Indeed, if you picked up any more information you probably could not even function. Yet your ancestors all functioned for quite a while. And you have inherited many of their abilities to selectively filter important signals out from that overwhelming input. So information critical to your survival or reproductive functions is likely to arrive priority-labeled for your immediate attention by its attendant discomfort, pain, pleasure or other strong emotion. Indeed, the reproductive advantage in having strong emotions and painful miseries comes from the insistent way they draw your attention to hunger, thirst, predators, tissue damage and sexual urges—matters that must be dealt with in a timely fashion. Thus it is neither miracle nor accident that your normally fine judgment and keen eyesight become blurry when you are faced with a reproductive opportunity. Or that you tend to swat biting flies with emotion or feel warm and defensive about your children despite their "terrible two's" and teenage rebellion. Similarly, your neighbors and close relatives stir up greater interest and concern than far finer folk living elsewhere. And even by law, your children or distant cousins get first draw upon any treasures (although they are not saddled with any debts) that you may have accumulated—once lawyers and accountants have taken more than their fair share. It seems that much of the behavior and judgment "hard wired" into your genes is also reflected in the social systems handed down by previous generations.

So What Is Behavior?

As yet, comparatively little is known about how your nerve cells integrate their discharges to bring about that appearance of evaluating and reacting to input often referred to as behavior. At a time when no one really knows whether your skull encloses 50 or 100 billion nerve cells, it is hardly surprising that we lack detailed descriptions of the ways those neurons interact. We can, however, categorize your nerve cells as *sensory* (input) *neurons, interneurons* (processing) and *motor* (output) *neurons.* Furthermore, we know that interneurons make up the vast majority of nerve cells in your brain and spinal cord, and that activity-dependent alterations in the impact of interneural synaptic connections are basic to the development of neural networks, memory and learning. Also that your brain is protected inside the solid brain case portion of your skull while its major extension, your spinal cord, passes downward from the brain within the bony security of your vertebral arches to terminate at about the level of the second lumbar vertebra, close to your waist. An average interneuron interacts directly with perhaps ten thousand others. Some synaptic connections are incoming, often to tiny extensions of nerve cell *dendrites.* Some are outgoing, often from tiny branches of the nerve cell *axon.*

Certain nerve cells tend to discharge hundreds of times per second while others might depolarize just once in that time, but all of those discharges either stimulate or inhibit their target cells. Thus any single interneuron can alter the output of many thousand others in an additive or subtractive, organized and exceedingly complex fashion. So your behavior is brought about by a basic inherited pattern of neuronal interconnections that has been modified endlessly in response to random organizational events and individual life experiences. Your inheritance is unique. So are your experiences. And so is your behavior, within limitations set by those ancestral factors and chance events.

That does not mean your behavior is necessarily unpredictable, however, or that it cannot be modified. Your behavior is unique just as a finger print or artist's sketch is unique, but each is an ordinary output based upon certain rules and tools. Even an artist's sketch repeated on the same topic can be done differently every time—it may be affected by further input and additional experience, new interactions and materials, or different movements, values or moods. But within limitations set by your inherited (in)ability to draw, your sketch expresses your current feelings and skills. On the other hand, your fingerprints can only be altered by damage. Their hard-wired design differs from the fingerprint designs of your friends, but those differences are mostly in rather minor details. Humans are very similar in their behavioral limits and responses as well. To a considerable degree, each of us shares not only the potential for a great many skills and abilities but also the potential for criminal or destructive behavior. *Nature* (your DNA), *nurture* (your life experience) and *chance* combine to determine how you will turn out. Within those limits, your normal behavior depends upon normal sensory input, normal processing and normal motor functions.

Of course, the processing part may accompany or even follow a behavior or output, especially in the case of certain *involuntary* reflexes where sensory input directly activates motor output while also sending signals centrally for processing. Thus your eye blinks at an approaching insect or baby's finger before you can even consider swatting the fly or praising baby's obvious intelligence and curiosity, or vice versa. Similarly you swiftly withdraw your hand from a hot stove prior to any review of the potential for tissue damage by directly conducted electromagnetic energy in the infra-red wave band. But it is less easy to explain why the conscious urge to make a movement may occur one third of a second *later* than the initial neuronal interactions that prepare you for that movement—almost as if the conscious urge represents an attempt to explain what already is underway rather than instigate it.

The Incredible Non-Shrinking Of Your Personal Program

A psychiatrist may give you medication that somehow helps you to process information more readily or control your depression but otherwise the usual job of a psychiatrist or psychologist is to *listen* in order to help you clarify emotional matters that are interfering with your function. Such emotional matters might relate to the disparity between what you were taught and what you expect versus what you want to do or what you get. Often such verbalization allows you to grasp the reason for your unhappiness/inefficiency. Then your uncomfortable internal program can be revised, but only by yourself (or rarely by others under exceptionally stressful circumstances such as hostage/torture). Your problems are not unique. Many people, including mental health professionals, have similar problems, frequently worse. So if your problems begin to interfere with your reasonable function or happiness, they can often be dealt with quite readily by consulting such expert assistants. Of course, when others see things differently—or pretend to do so because that brings reproductive benefit—psychiatric attention may be devoted to the wrong people (in our opinion). For example, psychiatric hospitalization was one (mis)treatment for those who did not appreciate Communism in the Soviet Union. Furthermore, "normal" (obviously a culture-based definition) individuals growing up in families that are quite abnormal may come to feel that something is wrong with themselves rather than the rest. For just as it is reproductively advantageous to adjust to the world rather than set strict standards that the world must meet, so everyone has an innate tendency to conform and readjust internal perceptions until they more closely match those of the family, pack, flock, herd, church group or lynch mob. But the modern dissemination of information that makes us increasingly aware of marked differences between cultures can also confuse matters. Does a Moslem adolescent find more happiness within an arranged marriage plus strong family support or are the freedoms that allow you to expand your own life/neuronal-connections/experiences as an individual more important? Is there more to life than happiness? Like what? Such questions are considered in later chapters.

What Can We Learn From Sensory Deprivation?

Your biased and unstable internal display of reality differs from that of others in part because past interactions, relationships and experiences contribute so much to your current interpretations of events. Thus each of us continuously creates and responds to a relatively unique internal representation of the world outside. Defective though this process may be, it still represents a great evolutionary advance. For it allows humans and other higher animals to experiment internally with attacks, deceptions (that bird dragging its "broken" wing to lure you away from its nest) and other strategies until a more appropriate/less dangerous response can be devised to what seems to be going on. Clearly such a dynamic and ongoing process requires uninterrupted input to confirm and coordinate your endless balancing act between past and future, between real and unreal. Yet you can voluntarily enter a state of minimal sensory input in the awake state by floating totally relaxed in a dense salt solution at body temperature within a dark sound-proof box. At such times your *mind* (whatever that is) tends to lose its steady interpretive control over your ordinary *stream of consciousness* (flow of ideas and perceptions) so you may hallucinate. Presumably this intentional reversal of your normal signal-to-noise ratio leads to defective sampling of apparent sensory inputs and inappropriate pattern recognition as your random or low-level nerve cell discharges are no longer drowned out by the usual overwhelming sensory inflow. The mind is therefore left treating accidentally generated internal noise as a signal, trying to piece together the current reality from such random and irrelevant inputs. An occasional near-death experience or startlingly real vision seen during illness or other abnormal mental state (perhaps while fasting or in relation to epileptic seizures) may depend upon similar alterations of nerve cell function induced by the same sort of stressful (low-oxygen or low-glucose) state or abnormal nerve cell discharge pattern that has given rise to so many religions. Extreme fatigue and many chemicals also can alter sensory inputs and perceptions to the point of hallucination.

How About Anesthesia?

This term refers to an induced state of "not feeling", usually associated with a relatively complete interruption of processing, conscious memory and voluntary responses. Many people apparently enjoy entering this state or at least its early stages, judging from the number of potent anesthetic agents (nervous system blockers or depressants) that currently are abused (e.g. nitrous oxide, ether, heroin, alcohol, various solvents). Sometimes, as too often seen with alcohol and illicit drugs, toxic effects such as death can occur soon after or even before full anesthetic levels are reached. Many anesthetics induce altered sensations and perceptions. All are accompanied by changes in motor neuron function and abnormal behavior, especially during induction and recovery. Common fat-soluble anesthetic agents may affect certain cell membrane components such as the proteins that make up ion channels. For example, isoflurane is an effective anesthetic that seems to open nerve-membrane K^+ channels, while barbiturates and ethanol open Cl^- channels. Either effect would inhibit depolarization of the affected nerve cells (some anesthetic molecules have additional effects while being metabolized or eliminated). Local anesthetics prevent membrane depolarization by blocking voltage-sensitive Na^+ channels. On the other hand, heroin tends to overstimulate receptors that ordinarily would respond to one or another sort of *endorphin* (pain-suppressing neurotransmitters that help you to deal with stress and promote a sense of well-being—as in the "jogger's high"). And among numerous other effects, caffeine alters adenosine receptors, Valium acts upon gamma-aminobutyric acid receptors, marijuana on anadamide receptors, nicotine stimulates dopamine release centrally while cocaine interferes with the dopamine re-uptake transporter that returns dopamine to its nerve cell for reuse, amphetamines pry the re-uptake system open so dopamine leaks back into the synapse—thus nicotine, cocaine and amphetamines provide a similar dopamine high (although nicotine has five to ten times more effect at the same dose). Of course, the net effects of these drugs can be quite unpredictable for they depend upon alterations induced throughout the nervous system—such derangements often have widely separated and even antagonistic effects that vary markedly between individuals and under different circumstances.

Blocked Motor Output—Muscle Relaxants

Tetrodotoxin from puffer or blowfish is the apparent active ingredient in Haitian *zombie*-producing mixtures. In much smaller doses, it brings high prices in Japanese *Fugu* restaurants. Like *saxitoxin*, the closely related and equally dangerous chemical produced by the dinoflagellates of toxic *red tides* (the usual cause of clam and mussel poisoning), tetrodotoxin blocks Na^+ channels. Small doses allegedly give a pleasant buzz or tingle (an acquired taste, perhaps). Larger doses can lead to near metabolic arrest (zombie state) or death. Witch doctors undoubtedly bury more zombies than they are able to resuscitate, but those who can be resurrected under such difficult social circumstances often require little convincing that they are now controlled by others. The effect of tetrodotoxin in blocking the *synchronized rhythmicity* of cerebral cortical cells presumably contributes to brain dysfunction (confusion) as well.

Some South American Indians still use curare-dipped darts to induce motor nerve blockade in their animal victims. Biologists studying tigers and grizzly bears often depend upon similar muscle relaxants to expedite capture for the processing of needed information. Too much muscle relaxant kills, as the animal also stops breathing for so long that it suffocates (fortunately diaphragm muscle is quite resistant to such blockade and recovers first). Not enough muscle relaxant endangers the biologist, who then must rapidly remove his or her reducing power from the vicinity. Moderation in all things—avoid excesses in life. Curare-related molecules allow anesthesiologists to bring about prolonged (½ hour or longer) skeletal muscle relaxation in the operating room. Curare inhibits muscle depolarization/contraction at the synapse by competitively displacing acetylcholine from acetylcholine receptors. By activating those acetylcholine receptors, *succinylcholine* provides minute-to-minute control of muscle relaxation through temporary muscle depolarization (while often leaving the same lingering soreness one gets from excessive exertions—see also porcine stress syndrome in preceding chapter). Of course, anesthesiologists must support breathing and other body functions as long as their patients remain relaxed by such drugs—and the use of voluntary-muscle relaxants at surgery also requires administration of an anesthetic to control discomfort since the patient remains unable to move or otherwise complain of pain.

In *Myasthenia Gravis*, the individual's own antibodies cause muscle weakness by attacking those acetylcholine receptors. Inhibition of acetylcholinesterase by chemicals such as *physostigmine* may therefore help to reduce symptoms (see Immunity). Acetylcholinesterase is permanently deactivated (within stable chemical complexes that no longer can split acetylcholine) by *nerve gases* and similar *pesticides*. Not surprisingly, accidental pesticide poisoning often kills people (e.g. through respiratory muscle paralysis due to overstimulation by high levels of acetycholine) as well as wildlife and beneficial insects—in addition to endangering the health of many. *Polio* virus damages and destroys motor neurons, thereby leading to varying degrees of temporary or permanent muscle paralysis. The release of *tetanus toxin* into body fluids causes *spastic paralysis* (continuous muscle contraction) by blocking neurotransmitter release within the central nervous system. The classical muscle spasms of *tetanus* (also known as *lockjaw*) produce facial muscle contraction (grimaces) and painful hyperextension of the trunk since back muscles ordinarily must be stronger than abdominal muscles (despite the mechanical disadvantage that comes from being located so near to your spine) in order to power your leap. Like tetanus toxin, *botulinus toxin* blocks neurotransmitter release by cleavage of the *synaptobrevin* (synaptic vesicle membrane protein) molecule, but *botulinum-B toxin* (one of several sorts produced by the responsible bacterium) irreversibly blocks acetylcholine release at the neuromuscular junction so *botulism* causes *flaccid paralysis*. Tiny doses of this toxin can be injected to weaken spastic or other muscles (for example, to restore balance between muscles that direct the eyeballs and thereby help bring about conjugate eye movements—or to weaken the adductor muscle spasm that can contribute to stuttering and difficult speech). *Black widow spider venom* activates a presynaptic receptor, causing excessive acetylcholine release and then neuromuscular blockade.

Mind And Brain

With the *real functional you* so dependent upon current input, processing and output, you are easily disrupted. Perhaps a useful definition of mind would include intact sensory and motor systems as well as brain, for changes anywhere along the line can significantly alter behavior. It is true that some extraordinary individuals are able to rise above (function effectively despite) severe disabilities, but it is also apparent that injuries which seriously alter input, processing or output will significantly change a person. Those who claim detailed knowledge of an afterlife, as well as writers of other fictions about brain transplants, should keep such limitations in mind when discussing the out-of-body transferability of an intact persona—regardless of whether they devoutly hope to ensure endless torment for others and eternal rewards for themselves (promoting the former might seem to make one ineligible for the latter) or merely wish to maintain the unchanged and reliable function of an old brain in a new body (or bucket of sterile salt solution). Furthermore, as Sperry has shown, consciousness cannot be separated from the readiness to respond—apparently the uniqueness of your being does not derive from ordinary (hence widely shared) sensory inputs and common perceptions, rather it is established by your own individualized output or behavior.

Your Central, Peripheral And Vegetative Nervous Systems Are One

Your entire body is surfaced and permeated by exquisitely sensitive, continuously interacting sensors tuned to important electromagnetic frequencies (heat, light), mechanical stimuli (sound and other vibrations, pressure, movement, position, muscle and tendon tension) and specific molecules (taste, smell, pain and all manner of other specialized chemoreceptors). Most of these sensors respond to specific signals by changing their cell membrane polarization—specific reports are then forwarded via particular neurochemicals released at one or more synapses. Almost all important changes that take place in your internal or nearby external environment are at least potentially detectable. Where detection only comes about as a result of delayed tissue damage—e.g. after exposure to high energy X-rays or gamma radiation—we can assume that such inputs remain immediately undetectable because they were irrelevant to the survival of your ancestors—if widely dispersed radioactive wastes ever do become a problem, your descendents might find reproductive advantage in carrying or even implanting radiation detectors. But for now, your *brain* receives all of its sensory input (information about the outside world) through your twelve paired *cranial* (within the skull) nerves and your *spinal cord.* Those incoming action potentials are then displayed internally in very specific and complex fashion by the neurons of your *central nervous system* (brain and spinal cord) for your ongoing amusement, interpretation and response.

Your Internal Display Is Subjective And Defective

So is our internal display of your internal display, but both seem truly amazing (at least that is how we seem to interpret our subjective internal displays). And your internal display differs considerably from that of others. In the first place, what was noticed and not noticed and incorrectly noticed will depend upon the condition of your sensors, your overall state of awareness and your preconceptions, all of which vary over time. How you then process that input for display and interpret that display are subject to further great variation. Not surprisingly, your behavior (what you do about all of your input) is even more variable. A bit of this variability can be demonstrated in the differing descriptions provided by each of several witnesses to a crime, or by the disagreements between jurors at any trial, or in the way that various individuals might view the same familiar town, building, meatloaf or fruitcake—and even your simplest observations tend to be influenced by context or subjective introductory terms such as large, small, sweet, fast, ugly, risky, good, right, bad, smart, dumb, rich and so on.

So Your Behavior Is Often Weird

Mental states can greatly affect perceptions. Depressed and angry individuals and those who have been subjected to violence tend to misinterpret and react more violently to challenges or perceived threats that others might be willing to ignore or discuss. Higher testosterone levels in males can increase aggressive behavior, as may the intake of alcohol and many other drugs. Michener describes how adult males in Afghanistan and other Muslim cultures are often accompanied by young boys who serve to satisfy their sexual needs until marriage. Although those young victims of sexual abuse must feel violated, helpless and angry, such abusive behavior tends to be repeated by them upon the next generation. Are sexually abused males more likely to become fierce and angry warriors? Can similarly abusive behavior by the dominant males in prison populations or other primate social settings cast light upon why such behavior develops and persists (presumably because it formerly brought reproductive advantage), and how it can be eliminated?

In any case, your sensory input is broken down and distributed to various sites for analysis in many different ways. Somehow it is then coordinated between the separate sites that consider *what* is occurring and *where* to produce a relevant reconstruction of pre-existing modular memories (including past responses) so that you can deal with the present reality as it is displayed. Interestingly, no single nerve cell or group of cells is in charge of integrating such a group/school/herd/flock type of neuronal interaction—nor could one be, given the endless combinations of neurons that represent your unlimited experiences, thoughts and responses. While a great deal of input is routinely deleted at every level of integration, much of that passing information still can be accessed voluntarily in *real time* (at the moment) if attention is directed to it (what your right index finger is feeling just now, for example). This sort of hierarchical organization makes it possible for smaller modules of nerve cells in your brain to eventually collate and respond more or less sensibly to enormous inflows of information from the huge numbers of sensory nerve cells located at and within your body surface.

And Possibly Ambivalent

The two largest parts of your brain are the right and left cerebral hemispheres. If you are right handed, it is usual for the left cerebral hemisphere to dominate. That left hemisphere apparently analyses the world more or less by sequential processing, while the right seems rather a parallel-processing "Big picture, forget the details" side. Fortunately your left hemisphere dominates your language functions, vocabulary, speech and manual control, leaving the right to rule over spatial reasoning, face recognition and perception of emotions (tone of voice, gestures and attitudes). The right hemisphere also dominates your left facial expressions of emotions, although the left will often cut in if the right brain is surprised and your face too truthfully expresses distaste for someone or something (that brief, especially left-sided flicker of distaste before the hearty greeting). Having the left side of the face dominant in expressing emotions leads some individuals (such as ex-President Bush) to unbalanced emotional expressions which suggest that the right hemisphere feels one way while the left thinks and says the opposite (a subconscious public recognition of this lack of commitment might account for wide but shallow electoral support—with many preferring the message but not the man).

At such times of internal conflict, your speechless right brain may still express its subconscious mind through the musical talent department if your left brain is willing to listen. Imagine that you suddenly encounter your boss who is not really such a fine person, although perhaps a legend in her or his own mind. You decide it would be reproductively advantageous to smile and say "Hello" and pretend to be very busy and enjoying your work. So far you have been successfully guided and controlled by your analytical left brain. But if you are interested in a second opinion from the right brain, listen to the catchy tune you happen to be humming as boss and retinue march off. For that particularly persistent melody may hum along quite relentlessly but unnoticed until you suddenly find yourself mouthing such relevant words as, "You can take this job and shove it!" Apparently, speaking your right mind could be dangerous, so the right is properly speechless, which helps you to act and speak decisively as if of one mind. But the ongoing urge to sing your otherwise silent and often frustrated right mind may explain the public frenzy and huge incomes achieved by those minimal musical talents who happen to tap into (speak to or for) your right brain.

Perhaps that intensely satisfying religious babble known as speaking in tongues is an indirect way for the non-verbal right brain to complain that your analytical left brain is not handling big problems in a right-satisfying (or righteous?) fashion. And when you solve a problem by sleeping on it, is it because right brain input has finally reached your attention through dreams after being ignored all day? Might certain varieties of stuttering, some aspects of consciousness and even self-awareness itself (whatever that is) somehow result from chronic cooperation and competition between the two sides? And is the supreme relief that follows really successful communication simply the mute right brain saying "Well! It's about time that was said". Does listening to music while studying simply distract the right brain so the left brain can learn more effectively? Although some of the rare individuals born with only one hemisphere may seem quite normal, one might anticipate that they would be unusually decisive. Just as a one-armed economist is less likely to say, "Well, on the other hand…".

In any case, your two large *cerebral hemispheres* are interconnected by many great bundles of nerve axons passing through the *corpus callosum* so that they can remain in close touch. And your complex human behavior is finally determined in this highest part of your brain—the so-called *neocortex*. Ultimately it is the incomparable information processing powers of those two cerebral hemispheres that ensures your access to the stored reducing power of most other life forms rather than vice versa. For human civilizations no longer depend upon your abilities to run, jump or hit—important as these may be. Rather, civilizations are based upon your ability to play freely with information.

The next chapter views your nervous system in more detail.

Further Reading:

Trevarthen, Colwyn, ed. 1990. *Brain Circuits and Functions of the Mind. Essays in Honor of Roger Sperry.* Cambridge, England: Cambridge University Press.

CHAPTER 14

YOUR NERVOUS SYSTEM

Complex Interconnections Increase Neuron Versatility;... Fatigue;... Overview Of Your Nervous System;... When Little Is Known, We Much Prefer To Express It In Latin;... Your Autonomic (Self-governing) Nervous System;... Cerebrospinal Fluid And Meninges;... Your Blood Brain Barrier;... Memories Are Made Of This?... A Few Hints;... Just Think About It;... All Creatures Provide Evidence;... You Think, Therefore You Are (So You Think)—A Very Brief Anatomical Review;... The Thalamus And Hypothalamus (Under Thalamus).

Complex Interconnections Increase Neuron Versatility

Brain size increases more slowly than body size between species of vertebrates since the total number of useful neuronal interconnections establishes information processing capacity and these interconnects increase far more rapidly than the neuron count. Of course, you couldn't process anything at all when your tiny new nervous system first appeared as a midline infolding of your embryonal *ectoderm* (surface layer). Those few pioneer nerve cells had little chance to get lost as they moved or grew along underlying surfaces or followed faint chemical trails (an ability inherited from their independent unicellular ancestors). Many of these neurons then elongated with your growing nervous system, providing signals and pathways for other nerve cells to follow. Early neurons organized and oriented themselves through contacts with specific cell surface molecules as well as by attractive or repulsive gradients of cell products or more widely diffusible organizer and growth factors. As usual, each sort of nerve cell population was strictly limited in its numbers and site of occurrence—perhaps a particular number of cell divisions exhausted some essential supplies or even wore down certain important structures such as the *telomeres* (nucleotide repeats that serve as chromosomal end-caps and appear to protect/preserve important genes located near the chromosome ends). Enlarging neurons routinely sent out many thousands of exploring *axonal* (nerve cell outflow) and *dendritic* (input side of nerve cell) branches and twigs along and across various tissue planes. Neural, glandular, muscular and surface targets drew their innervation from appropriately directed and chemically attracted cell processes.

Your *cerebral* nerve cells also traveled outward from their sites of origin on *lateral ventricle* surfaces. Soon they formed distinct columns as well as layers reminiscent of their epithelial origins, with each layer committed to different axonal and dendritic patterns by growth factors, neuropeptides, hormones, neurotransmitters from ingrowing afferents and other chemical instructions. The newest cerebral neurons migrating outward had to pass previously positioned nerve cells to reach their currently outer-brain-cortex position. En route, each gave rise to and also intercepted thousands of potential interconnections, many of which were then dragged along to produce the preliminary version of your incredibly complex and uniquely individualized cerebral wiring diagram. Most of your pioneer neurons played only temporary scaffolding and guide roles before being replaced in the course of further growth and development. Although your fetal cortical cells persisted, a great many of their less appropriate connections atrophied and disappeared in response to disuse, competition or suppression. For a time, your new cerebral neurons grew and developed within individual electrically interconnected (by gap junctions) processing units of a few dozen neurons where they blindly activated ancient instructions and sorted out new relationships through repeated group depolarizations. Gradually each neuron became more isolated within its unit and better focused upon regularly required, coordinated and chemically appropriate connections and duties—which included concurrent communication with a great many more neurons outside of those modules in parallel fashion.

With the status and responses of each interneuron modulated by its millions of membrane receptors and ion channels as well as through thousands of *afferent* (incoming) and *efferent* (outgoing) synapses, interneuronal communications are far more complex than those relatively straightforward depolarizations at neuromuscular synapses (which themselves are not nearly as simple or straightforward as preceding chapters would suggest). Furthermore, a single interneuron may use many different chemical outputs for signaling, and bear receptors for dozens of different neurotransmitters such as acetylcholine, simple or modified amino acids, peptides and hormones—all of which allow the individual axonal or dendritic branches of that neuron to participate a great many different ways in a great many different circuits (after having attracted, diverted, directed and repulsed exploratory nerve cell branches from all over). Some neurochemicals and their receptors play entirely different roles in non-nerve cells elsewhere in your body. That is properly frugal.

There is no need to design a new doorbell mechanism for every house, store or office. Nor does one expect the same response when ringing up a flower shop as a prison. There are many possible mechanisms by which such receptors might activate different varieties of K^+, Na^+, Ca^{++} or Cl^- channels—some of these activations are direct while others proceed via various second messengers (see Hormones) of differing strength and duration. For example, your brain includes separate neuronal systems sensitive to three closely related derivatives of the amino acid *tyrosine* (*dopamine, norepinephrine* and *epinephrine*). And certain other amino acids such as *gamma-aminobutyrate* (GABA), *glycine*, glutamate and aspartate are so heavily utilized as neurotransmitters that they are present within many nerve cells in large amounts. GABA is a transmitter of inhibitory neurons so it serves to depress neuronal activity in the brain. Glycine has a similar role in the spinal cord—binding to chloride ion-channel receptors.

Those afflicted with "*Familial Startle Disease*" (also referred to as the "Jumping Frenchmen of Maine" disease—and perhaps identical with the "Goosey" disease encountered in the Southern United States) display exaggerated startle reflexes to loud sounds or other unexpected stimuli such as being tapped on the nose, shocked electrically or goosed (causing a marked jump, repetitive and often foul language and a reflex carrying-out of commands). Apparently the jumping results from a single amino acid alteration in inhibitory glycine receptors and the condition can be treated by medication that enhances GABA neurotransmitters. (Although jumping when startled and using foul language when goosed could be considered normal adaptive behavior, the reflexive speech and disinhibited carrying out of harmful commands clearly qualify this condition as a disease requiring treatment—as does the high infant mortality rate associated with sustained post-startle muscle rigidity in the first year of life). *Glutamate* and *aspartate* (the two amino acids with acidic side branches) and apparently also *ATP* have powerful exciting effects on the brain and may be the most common signaling molecules at central nervous system synapses. *Serotonin* (5-Hydroxytryptamine), a derivative of the amino acid *tryptophan*, usually acts as an inhibitory neurotransmitter that opens K^+ channels—it can also be found in the gut wall and within blood platelets. When released from platelets, serotonin exerts a localized blood-vessel-constricting effect.

So you utilize a great many signal and receptor mechanisms. And a lot of those are somewhat modified in various locations to have different impacts or to play different roles. All in all, very functional and economical—but hardly strong evidence for a preplanned and coherent design. Of course, we still do not understand much of the experimental evidence relevant to single nerve cells. And there is little doubt that we lack the brainpower to ever comprehend all possible interactions of those 100 billion intracranial cells, let alone figure out much of what is occurring in the world outside. Furthermore, life's trials and tribulations have a way of changing a person. You are not the same as you once were. Every day leaves its mark and your ongoing activities cannot help but change your brain as its ceaselessly modified neuronal connections continue to reflect life's information burdens—even nerve cell deaths are a normal response to disuse as well as injuries. Certain connections also become more durable and dominant as you grow older, for particular neurons grow, develop, myelinate and retain their structural plasticity during different phases of your development. That is why some lessons and skills are best acquired at certain ages. Old dogs often resist learning new tricks—perhaps it no longer is worth the effort. But experience suggests that a one year old child usually cannot be toilet trained while three-year olds tend to toilet train themselves. It also appears that language sounds and then meanings are easier to grasp at some ages than others. But in general, you can continue to learn new information and skills as long as your neuronal interconnections are still capable of change.

A neuron may last a lifetime if it continues to receive and transmit a lot of messages. If ignored, it eventually undergoes apoptosis. The continued loss of unnecessary interconnections and even unused neurons may correspond with the increasingly focused, often narrow-minded behavior that you display as you continue to improve your skills and acquire new information. Although your options may decrease with each trimming of redundant or unused circuits, your skills and abilities will simultaneously improve through routinization of your responses and the progressive myelination of your neural fibers. Useless or depleted individual brain cells die and are disposed of with little fuss. As is true elsewhere in your body, this active and orderly process of *apoptosis* requires the doomed cell to activate new genes and quietly destroy itself in order to minimally inconvenience neighboring cells. In contrast, when functioning neurons are severely damaged they may suddenly release large stores of neuroactive amino acids and peptides. The resulting overwhelming stimulation may damage or destroy nearby cells—especially those neurons most tuned to specific excitatory molecular signals such as glutamate. The cumulative effect of repeated small blood vessel and nerve cell injuries in professional boxers is often enhanced by the area-wide or mass depolarizations associated with each *concussion* (knock-out). Many boxers eventually develop *dementia pugilistica*, a condition that somewhat resembles *Alzheimer's Disease* with shrunken brain tissues showing extensive neuronal damage and widespread deposits of *beta amyloid*. But even though boxers are encouraged to exchange their brain cells for money, any sale of their other body parts would surely be viewed as a dubious ethical proposition.

Fatigue

No biological mechanism is designed for continuous output. All biological systems fatigue and require rest. Even your most essential functions such as breathing, pumping blood and depolarization of nerve cells, rely upon intermittent activations with intervening rest periods for recovery. Fatigue is not well understood. Muscle fatigue probably reflects a number of factors including local exhaustion of nutrients and oxygen, ion shifts, a buildup of waste products and perhaps some sort of structural fatigue. Although muscles tire and even malfunction if they are exercised beyond the limits of their circulation and other resources, physical rest ordinarily allows full recovery of normal function. All higher animals also require mental rest or sleep. While you might not view nightmares as very restful, sleeping and dreaming clearly serve numerous important functions. Any student knows that one can find plenty of energy for a movie or dance despite being totally exhausted by one's studies of biology. For once applicable circuits of your brain are full, a new interest can still be dealt with but the part devoted to the old topic requires rest to restore effective function. Then, during a nap, the newest information somehow becomes better integrated (or lost). That allows a return to the task with your capacity renewed. Indeed, "Let's sleep on it" is a traditional ploy for coming up with an improved approach to a puzzling problem. And common experience confirms that unusual problems or difficult tasks are often more easily solved or completed on the day following careful evaluation and repeated mental rehearsal as well as a good night's rest.

Sleep appears to serve numerous maintenance and restorative functions. Undoubtedly it originated in response to the regular light and dark cycles that result from rotation of the Earth. Once it proved reproductively advantageous and therefore likely to persist, sleep was integrated into many of your most essential functions and routines (just as dogs have converted their need to urinate into an elaborate message system that includes information on identity, diet, sexual condition, time, territory and undoubtedly much more). So regardless of its original purpose, sleep now allows you to run through matters of recent emotional or other impact for collation and disposal. Those vivid and disconnected dreams probably represent momentary extrapolations of images flashing by as you examine and clear your circuits with the recording function turned off. Sleep also interrupts a train of thought, thereby reducing the risk of ongoing obsession and perseveration in particular circuits—comparable, perhaps, to the way a computer or VCR avoids holding one display for long enough to permanently mark the screen. Sleep disorients, therefore you require regular reorientation to surroundings that might otherwise be increasingly misperceived as they are more and more taken for granted. And just like night and day, sleep is associated with an ebb and flow of your own body cycles and functions (your *circa*dian or *about*-one-day rhythm—see Hormones) that helps you to minimize stress while rebuilding your cellular reserves.

Sleep provides regular opportunities for a general testing of circuits at a time when these can safely be disconnected from most motor functions through hyperpolarization of particular interneurons—although that particular disconnect seems defective in *sleep walkers*. On the other hand, *narcolepsy* is a sleep disturbance that includes unanticipated daytime attacks of sleepiness and other symptoms such as *cataplexy* (a sudden loss of motor tone associated with excitement or emotion—presumably due to activation of the same inhibitory interneurons that keep you from injuring yourself or others during your nightmares). And sleep allows you a quiet time for routine drills that maintain the integrity of important but less-utilized neuronal connections and body-wide reflexes. Thus your potentially life-saving fight-or-flight responses and life-giving reproductive functions are checked and rechecked during those several periods of rapid eye movement (REM) sleep each night. Indeed, penile and clitoral erections must be run far more regularly than fire drills to ensure that your circuits and tissues are able as well as willing when an urgent reproductive opportunity arises. Active REM sleep occupies about eight hours per day of an infant's much longer sleep time. Apparently the developing infant brain needs a lot more circuit testing and electrical stimulation than an established adult brain. Cetaceans (whales and dolphins) sleep one cerebral hemisphere at a time to reduce risk of drowning.

Overview Of Your Nervous System

Your delicate symmetrically shaped 1500cc (on average) brain with its spinal cord extension is weightlessly suspended (actually it weighs about 50 grams when submerged) within nourishing cerebrospinal fluid. Besides being stabilized and supported by three concentric fluid-filled membrane wrappings of increasing strength, your brain is firmly crated inside your skull while your spinal cord is insulated by overlying fatty tissues and sheltered under your bony vertebral arches. From that warm dark quiet least-disturbed location of your entire body, your central nervous system sends out its many symmetrically paired extensions. Those *nerves* bring *sensory* (incoming) information and issue *motor* (outgoing) orders.

Specific, often consciously processed instructions move out through voluntary motor nerves to your striated skeletal muscles. Routine moment-to-moment housekeeping advice is usually delivered via *autonomic* nerves. The more *exciting* of those autonomic messages pass through *sympathetic nerves*—those dealing with *restorative* functions pursue *parasympathetic* pathways. While we customarily speak of your central, peripheral, voluntary, motor, sensory, autonomic, sympathetic or parasympathetic nervous systems, you really have just one closely integrated nervous system, inseparable from the living body that provides all of its input and expresses each of its responses.

When Little Is Known, We Much Prefer To Express It In Latin

Twelve pairs of nerves emerge from your brain within the skull. Three of these *cranial* nerve pairs are purely sensory (smell, sight, hearing and balance). The rest carry messages in both directions. By custom, these nerve pairs are numbered 1 to 12 from top down (front to back). Thus cranial nerves #1 would be the two olfactory nerves reporting on smells; #2 the two optic nerves delivering visual information; pairs #3, 4 and 6 control your eye muscles while #5 and #7 innervate scalp, face and mouth; pair #8 report hearing and balance information; pairs #9, 10, 11 and 12 innervate lower face, mouth and neck. Pair 10, the long *vagus* ("wanderer" in Latin) nerves also carry sensory information from and deliver parasympathetic orders to your thoracic and abdominal viscera.

Your lengthy spinal cord gives rise to 31 pairs of *spinal nerves* and is somewhat expanded where additional neurons and axons are needed locally at the levels sending nerve branches to your arms and legs. The cord then tapers down and ends near the second lumbar vertebra (it once filled your fetal vertebral canal but thereafter the spinal cord elongated more slowly than your body). Below that L2 level, your vertebral canal still encloses numerous *nerve roots* that descend to emerge between lower vertebrae. Those descending nerve roots are collectively referred to as your *cauda equina* ("tail of the horse" in Latin). Each spinal nerve contains both sensory and motor axons that travel together after emerging from the spinal cord via separate *sensory* and *motor nerve roots*. Sensory nerve roots enter the spinal cord *posterolaterally* (behind and to the side) while motor nerve roots emerge *anterolaterally* (at the sides of the front surface). To reduce non-essential clutter within the cord, your *sensory nerve cell bodies* are situated within *dorsal root ganglia* (posterior nerve root enlargements) outside of the spinal cord. Being located off to one side of its long incoming dendrite-axon allows each sensory nerve cell to nourish and maintain that extension without affecting the passing inflow of messages. In contrast, motor neurons both receive and respond to information so their cell bodies are conveniently placed within the forward portion of your *centrally located* ("H"-shaped) spinal cord *grey matter* (grey matter refers to tissue composed primarily of neurons—*white matter* is mostly made up of myelinated axons).

After individual spinal nerves pass laterally out of the vertebral canal between your vertebral arches, their branches often travel jointly with branches of other spinal nerves over varying distances. The meshwork pattern that results from different spinal nerves branching and joining (often en route to a limb) is known as a *nerve plexus*. Plexus patterns develop because growing axons preferentially follow preexisting axons that seem headed in the proper direction. Then, sooner or later, stronger or more localized chemical signals draw individual axons or dendrites off toward the particular target cells they are destined to innervate. But even after being so diverted, axons often encounter and again grow along other axon bundles during another segment of their trip. Thus the sympathetic, sensory and motor *fibers* (axons and dendrites) making up a particular *peripheral* nerve differ markedly in source and type from the nerve fibers initially traveling together within any specific *spinal* nerve. And the way various body structures draw their innervation (and also blood supply) from high and low tends to reflect how those structures migrated and were reorganized during your embryonic and fetal days—or perhaps the level at which comparable tissues originated in your remote ancestors.

We have seen how the immutable laws of physics and chemistry were established at the time of the Big Bang. Those laws cannot change—once discovered and understood, they allow reliable deductions and predictions to be made about past and future events and conditions. In marked contrast, human anatomy and physiology cannot be figured out according to any general rules less complex than the processes and structures that they attempt to describe—since each structure or function could equally well have come about in countless other ways. For over the past billions of years, life has undergone endless modifications and revisions—each justified entirely by *reproductive advantage at that time* (the otherwise incomprehensible layout of plumbing and electrical circuits in many older, frequently remodeled business or professional buildings is similarly explained by historical contingencies). As a result, it is usually far simpler to memorize anatomical and physiological facts than to figure them out. Of course, neither life nor anatomy need justification as long as they work (remain competitive). The way things evolved under specific long-forgotten circumstances therefore leaves biology with few hard-and-fast rules—and those rules have many exceptions.

Eight pairs of *cervical* spinal nerves (*C*1 to *C*8) emerge above and below the seven cervical vertebrae of your neck. Nerves C4–8 and the first thoracic spinal nerve (*T1*) all contribute to the *brachial plexus* of nerves. That plexus in turn gives rise to the three major nerves of each arm and hand (*radial, median* and *ulnar* nerves). You are probably most familiar with the ulnar nerve at the site where it passes behind the inner aspect of your elbow. Bumping that "funny bone" can send quite a jolt down the *medial* (little finger and also ulna bone) side of your forearm and hand. An infant's relatively inelastic brachial plexus may suffer permanent damage if sufficiently overstretched (or even torn) during a difficult delivery. The resulting *brachial palsy* (a birth injury) often leads to deformity and diminished growth of that weak or paralysed arm. At any age, a very strong and low blow to the side-of-neck/top-of-shoulder zone can cause similar damage. The lowest spinal nerve branches to the brachial plexus are least likely to be injured (overstretched) by the impact of a downward blow since they angle upward from their site of emergence (between vertebrae T1 and T2 within the chest) in order to pass out *over* the first ribs and *under* the clavicles en route to the arms—however, those lower brachial branches are susceptible to pressure damage from being stretched across an abnormal first or cervical rib. Your important *phrenic nerves* (made up of branches from spinal nerves C3, 4 and 5) descend through the chest (along each side of the heart) to innervate their respective halves of the flat *diaphragm* muscle (that migrated downward from your neck region during embryonic life and now separates your heart and lungs from your abdominal viscera).

Twelve pairs of thoracic spinal nerves emerge bilaterally from below the twelve thoracic vertebrae of your chest to innervate symmetrical *segments* (more-or-less horizontal and repetitive layers) of that thorax and upper abdomen. Except for autonomic nerve branches to your mobile guts, each side of your body is innervated independently—those paired sensory and motor nerves never extend across the midline in front or in back. Presumably this rule is rigidly enforced because any confusion (about which side is being bitten or which side should respond) could prove fatal—furthermore, muscles on opposite sides of the body often oppose each other or alternate in their contractions, as with fish swimming or alligators walking. So following sensory nerve injuries, a denervated (numb) area may become smaller or even disappear by ingrowth of adjacent sensory nerve fibers, but such fibers always derive from the same (damaged) side.

The *lumbar plexus* and *sacral plexus* innervate your lower body and legs and are well protected from injury within the pelvis. The two principal leg nerves are the *femoral* nerve which enters your anterior upper leg at the groin and the *sciatic* nerve which leaves mid-buttock posteriorly headed for your lower leg. The sciatic nerve may be injured by an undesirably low and posterior hypodermic needle in-jection into buttock muscle (this can cause foot drop as well as litigation). The sciatic nerve can also become symptomatic from chronic sitting upon an overfilled wallet (a visit to the doctor can cure that problem).

Your Autonomic (Self-governing) Nervous System

The *sympathetic* portion of the autonomic nervous system is often referred to as your *fight or flight* (or stress response) system. Sympathetic fibers (your *thoracolumbar autonomic nerve* axons) emerge from the spinal cord along with your *thoracic* and *lumbar* anterior motor nerve roots—those sympathetic nerves soon separate from their spinal nerves in order to distribute housekeeping instructions to a long chain of sympathetic *ganglia* (interconnected groups of sympathetic nerve cells) located along each side of your vertebral column. Additional sympathetic ganglia (and extended sympathetic nerve plexuses) are found on the front surface of your large abdominal aorta (artery) and in nearby retroperitoneal (behind the guts) locations. Thus your *pre*ganglionic sympathetic fibers (axons) usually extend only a short distance from their sympathetic nerve cell of origin (which is located laterally within the central grey matter of your spinal cord) *to* reach their target sympathetic ganglion cells. Many relatively long *post*ganglionic sympathetic fibers (axons) then extend out *from* those ganglion cells to amplify and coordinate the delivery of sympathetic messages to all relevant tissues—so your sympathetic sys-tem broadcasts its emergency type messages quite widely. For example, postganglionic sympathetic fibers travel in the outermost layer of your many blood vessels to regulate their smooth muscle *tone* (state of contraction). Injured postgan-glionic sympathetic nerve fibers tend to regenerate rapidly, and temporarily denervated smooth muscles and glands still maintain some spontaneous activity. However, skeletal muscle fibers remain paralysed once their motor nerve axon has been cut—indeed, such *denervated* skeletal muscles atrophy quite rapidly unless artificially stimulated (perhaps a locally effective Cl^- channel blocker or Ca^{++} chan-nel activator might encourage low-grade spontaneous muscle-fiber contractions and thereby maintain denervated skeletal muscle in a more useful state pending nerve repair/regrowth).

The soft central core of hormone-producing tissue within each fragile adrenal gland is known as the *adrenal medulla*. Embryonal ganglion cells turn into adrenal medulla cells under the influence of *cortisol* (adrenal cortical hormone). Otherwise such ganglion cells become sympathetic neurons under the influence of fibroblast growth factor and nerve growth factor. Therefore your adrenal medulla cells secrete *epinephrine* and *norepinephrine* (adrenaline and noradrenaline) into the blood circulation as *hormones* when stimulated by their preganglionic sympathetic fibers—just as ordinary *post*ganglionic sympathetic nerve cells generally release norepinephrine at their target organ synapse. However, *acetylcholine* remains the *pre*ganglionic sympathetic fiber neurotransmitter-of-choice—for if norepinephrine were to deliver preganglionic orders to the adrenal medulla, your responsive adrenal release of more norepinephrine would act as an even stronger signal to release yet more norepinephrine (an unopposed positive feedback). Acetylcholine also serves as postganglionic sympathetic neurotransmitter in the few locations where norepinephrine's direct small-vessel-constricting action would be inappropriate— as when enhancing your sweat gland output so you can break out in the cold sweat that makes your body more slippery or dilating skeletal muscle blood vessels for your next fight or flight. But even acetylcholine can only dilate blood vessels having an intact *endothelial* (inner-vessel-surface) lining since acetylcholine causes endothelial cells to release a short-lived *endothelial-derived relaxing factor* (which is *NO* or nitric oxide). Local sympathetic constriction of blood vessels is similarly dependent upon your intact endothelium releasing *endothelin* (a peptide) and possibly also *superoxide* (O_2^-).

Parasympathetic nerve axons (your *craniosacral* autonomic nerves) emerge from the central nervous system in close association with your *cranial* nerves and *sacral* spinal nerves before branching off to their separate and often distant visceral targets. Parasympathetic ganglia are small and located at the target organ, be it heart, bowel wall or gland. Both preganglionic and postganglionic parasympathetic fibers utilize acetylcholine as their synaptic neurotransmitter so parasympathetic effects tend to be brief and localized. In contrast, it may take 20 minutes for a full sympathetic flight-or-fight reaction to wear off—you calm down gradually as the epinephrine and norepinephrine are slowly cleared from your circulation or taken up by many local nerve endings at synaptic sites of release. Except during dire emergencies, your parasympathetic and sympathetic functions tend to remain in dynamic balance. But while it is important for your fight-or-flight reactions to activate all sympathetic reflexes together, you would not want to call upon all of your parasympathetic functions simultaneously—so the more specific postganglionic parasympathetic distribution allows your tissues and organs to be activated individually. Thus some parasympathetic fibers enhance your intestinal circulation to promote food absorption while others empty your bladder or rectum, or reduce your heart rate, or bring about penile and clitoral erection (ejaculation is a sympathetic reflex) and generally help you to relax, recover and enjoy the good times—as will be discussed in more detail when we review other systems.

Cerebrospinal Fluid And Meninges

Your brain and spinal cord (together known as your *central nervous system*) float within a layer of protective *cerebrospinal fluid.* Almost all of that nutrient fluid is produced by secretory *choroid plexus* structures (tufts of capillaries covered with secretory cells) within each of the four fluid-filled *ventricles* (cavities) of your brain. You have two *lateral ventricles,* one deep within each cerebral hemisphere. Both lateral ventricles drain centrally into a *third ventricle.* In turn, that third ventricle connects to your *fourth ventricle* (located within the brain stem) from which cerebrospinal fluid escapes by one *dorsal* (back) opening and two lateral openings to enter the fluid-filled space surrounding your brain and spinal cord. A thin central fluid-filled canal runs inside your spinal cord grey matter and communicates with the fourth ventricle. Your brain and spinal cord are covered by 3 concentric membranes or *meninges.* The innermost is the *pia mater* ("delicate mother" in Latin). This flimsy layer carries nutrient vessels and moors the spinal cord via lateral and terminal pial extensions. The middle membrane is the *arachnoid* ("spidery" in Latin) *layer* which encloses your *cerebrospinal fluid. Subarachnoid bleeding* from nutrient blood vessels on or near the brain surface is usually associated with damage to brain tissue (as in a *stroke*). The outermost of the three *meninges* surrounding your brain and spinal cord is the fibrous layer known as *dura mater* ("tough mother" in Latin). Where dura contacts your skull, it also includes a sturdy *periosteal* (bone-depositing) layer. Several large venous blood *sinuses* (channels) flow between stretched layers of dura mater, an arrangement that provides external tension to hold those venous channels open (see below).

A *lateral* (on the side) skull fracture may tear the *middle meningeal artery* just outside of the dura mater. If that artery continues to bleed, the injured person may first awaken from the concussion, then fall into *coma* due to increasing pressure exerted upon the brain by the expanding *epidural* (outside of dura mater) blood collection. This acute surgical emergency requires immediate drainage of the space-occupying blood clot and closure of the torn artery—otherwise death soon follows. Before death the patient's condition may be desperate but good recovery remains possible if this condition is recognized and treated promptly. But the situation is often confusing, since a hard bump on the head also can cause *subdural bleeding* (between dura and arachnoid). Interestingly, the principal accumulation of subdural blood often occurs on the side opposite to the bump, consequent to small veins being torn as brain jerks away from the suddenly accelerated skull. If a subdural *hematoma* (collection of blood) rapidly reaches significant size it too can cause a relatively *acute* (sudden) brain compression, perhaps again after an intervening *lucid interval.* Unfortunately, such acute collections of subdural blood are frequently associated with injury to underlying brain as well as brain swelling. The condition of a patient with an acute subdural hematoma may deteriorate more slowly than that of someone with a typical epidural hemorrhage, but an acute subdural hematoma is more likely to result in death or permanent damage despite prompt treatment. On the other hand, a minor subdural accumulation of blood may cause no symptoms at the time of some apparently insignificant bump on the head. But as that small subdural clot becomes walled off, its surrounding membrane of scar tissue permits water entry while still confining most of the progressively smaller solutes derived from broken-down clot—such a situation encourages the gradual osmotic enlargement of the (by now) *chronic subdural hematoma.* Thus an initially small subdural clot can eventually cause the insidious onset of *symptoms* (complaints) or *signs* (abnormalities detectable by others) over subsequent weeks or months. When such a chronic subdural hematoma finally is evacuated surgically, the patient often makes a very good recovery.

There are two sturdy infoldings of the dura mater that help to support and restrain the soft brain within the skull. One, the *falx cerebri,* forms a flexible incomplete partition between the two cerebral hemispheres. This curved falx ("sickle" in Latin) contributes to the *three-way traction* holding open your important *sagittal* (meaning *in the midline plane,* front to back) *venous sinus* (see below). The other dural partition, the *tentorium cerebelli,* roofs over your *cerebellum* in its posterior *fossa* (Latin for "ditch" or "pit") of the skull. Thus the tentorium underlies the *occipital* lobes of your two cerebral hemispheres. Any major pressure increase within the skull (whether caused by tumor, blood clot, brain swelling or some other space-occupying process) can force the bottom part of the *brain stem* down into the *foramen magnum* ("big hole" in Latin) of the skull—that *herniation* compresses the *medulla oblongata* ("long soft thing") of the lower brain stem along with lower corners of the cerebellum (the cerebellar tonsils) within the foramen magnum, which swiftly leads to unconsciousness, loss of control over blood pressure and pulse, and cessation of breathing—death soon follows.

Inside of your vertebral canal, several small but important nutrient blood vessels pass through the anterior *epidural* fat layer to reach the spinal cord. By flexing his or her body well forward with knees to chest, a person can separate the spinous processes of adjacent vertebrae. It is then possible to insert a long hollow needle through skin and underlying tissue layers in order to puncture the dura posteriorly for a *spinal tap*. Of course, any needle injury to the spinal cord itself can cause serious damage, both by the direct impact of a sharp needle on delicate spinal cord neurons and by the harmful effects of any bleeding within the soft spinal cord. On the other hand, individual nerves of the cauda equina consist of axons that have far more collagen and myelin support than the cord—such nerves are generally pushed aside without harm by the long spinal needle. That means a spinal tap is more safely performed somewhat below the second lumbar vertebra (where the spinal cord ends after dividing into the numerous branches of the cauda equina).

Mid-to-low spinal cord damage by any cause may lead to *paresis* (weakness) or paraplegia (paralysis of the lower *extremities*). Severe upper (cervical) cord injuries can cause *quadriplegia* (paralysis of all four extremities). High spinal cord injuries often stop respiration if they involve the phrenic (C3, 4, 5) level or higher. Recent evidence suggests improved recovery from potentially disastrous blunt spinal cord injuries at any level if early therapy includes high doses of intravenous adrenal steroids—the goal is to reduce the spreading, progressive nerve cell damage that is at least partially consequent to the formation of toxic superoxide radicals and cell membrane breakdown (see Metabolism). Surprisingly, it seems that adequate lower body activity can be maintained if at least 10% of cord axons remain functional following a diffuse cord injury. That suggests a significantly greater margin of safety in neural tissue than in the skeletal system, for example. The initial recovery from severe nerve cell damage due to polio virus gives similar evidence. Difficulties that sometimes arise many years after such an attack may reflect reactivation of the intracellular polio virus.

*Epi*dural (*outside* of dura mater) placement of a hollow needle allows one to deposit local anesthetic where it can block spinal nerves emerging from the dura without affecting the cord itself. On the other hand, a *sub*dural spinal tap allows determination of spinal fluid composition and pressure (pressure may be elevated by infection, subarachnoid bleeding and other causes) as well as induction of *spinal anesthesia*. A concentrated glucose solution mixed with the *local anesthetic* (local anesthetics block voltage-activated Na^+ channels and thereby prevent neuron depolarization) makes the spinal anesthetic solution much heavier than adjacent spinal fluid. That allows one to readjust the spinal anesthesia levels upward or downward by tilting the patient. If a large-diameter needle is used for such a spinal tap, there is increased likelihood of prolonged post-spinal-tap headache due to persistent escape of cerebrospinal fluid through that not-yet-sealed needle hole into the epidural space (from whence fluid readily returns to the blood stream). While headaches are not always so easily explained, it clearly is more comfortable to have one's brain entirely submerged in cerebrospinal fluid.

Under normal circumstances, you produce new cerebrospinal fluid at about 30 to 60 ml. (one to two ounces) per hour so a similar amount must leave your rigid skull via *arachnoid villi* located within the *sagittal venous sinus* of your midline falx cerebri. Ordinarily the venous blood within that sinus remains at subatmospheric pressure due to being in arrested free-fall toward your heart—and the same mutual love of polar water molecules for each other that holds your sagittal sinus venous blood together as it hangs above your heart (thereby improving your cerebral venous return) also siphons your cerebrospinal fluid into the sagittal sinus whenever there is sufficient fluid to inflate those thin-walled arachnoid villi above your brain. Of course, any interference with the free flow of cerebrospinal fluid can lead to increased cerebrospinal fluid pressure and *hydrocephalus* ("water on the brain"). If such a blockage occurs within or between the ventricles (perhaps due to a developmental problem, tumor, injury or infection), the obstructed ventricles enlarge and press upon surrounding brain tissues from within. A blockage farther downstream (as with unusually thick-walled arachnoid villi consequent to infection or perhaps even a clotted sagittal venous sinus) raises cerebrospinal fluid pressures more uniformly. Hydrocephalus can often be corrected surgically by installing a drainage device that transfers surplus cerebrospinal fluid to some other sterile location within the body (into a vein perhaps, or another body space where extra fluid does no harm until it can reenter the circulation).

Your Blood-Brain Barrier

The smooth inner surfaces of all your *blood vessels* (the hollow tubes that carry your circulating blood) are lined by a single continuous layer of flattened endothelial cells. A vast majority of those blood vessels are tiny *capillaries* whose channels barely allow your disk-shaped red blood cells (which measure 7 microns or 7 millionths of a meter in diameter) to squeeze through in single file. Those capillaries are where your blood and nearby tissue fluids exchange water, solutes and gases. Although fat-soluble molecules cross intact endothelial surfaces quite readily, polar molecules do not. Therefore some endothelial cells have small openings between them and others provide pores passing through. Furthermore, many endothelial cells actively engulf and transfer fluid (along with any circulating molecules that become attached to specific endothelial-cell membrane receptors) across into the tissues. So there is a relatively free exchange of smaller solutes (but not of proteins or blood cells) between blood fluids and most tissue fluids *except* in the brain where capillary endothelial cells are tightly joined without gaps or pores—

those endothelial cells do not drink freely and they also bear peptidases that rapidly break down many polypeptides. In addition, all brain capillaries are tightly enclosed by a great many *glial* (supporting) cells that determine what and how much will cross the capillary wall in either direction. *Astrocytes* are multibranched glial cells that nourish specific neurons and also form a live scaffolding to support their axons. *Oligodendroglia* myelinate axons within the central nervous system (*Schwann* cells provide the myelin wrap for peripheral nerve axons). *Ependymal* glial cells line inner-brain (ventricle) surfaces where they help to secrete and circulate cerebrospinal fluid. *Microglia* are brain phagocytes that clear out dead cells and bacteria. Among its many other roles, your *blood-brain* barrier helps to exclude bacteria, viruses and other parasites that might gain entry to your blood. Normally this barrier also interferes with white blood cell (and antibiotic) access, although brain capillary permeability is enhanced by inflammation.

With many glial cells packed tightly about each neuron, there is little space in the central nervous system for extracellular fluid—especially in comparison to other soft tissues. Your many sorts of astrocytes are as yet poorly understood—often they display some of the same surface receptors as your hundreds of varieties of neurons—it has even been suggested that astrocytes may depolarize in a similar fashion to neurons (some also seem to communicate directly with their neuron via gap junctions) and thereby possibly contributing to your thinking. Whether or not that is true, astrocytes definitely transport solutes, wastes, gases and nutrients between capillaries and individual neurons. And those half dozen layers of nerve cells in the 2–4 mm rim of grey matter surfacing your brain are additionally nourished by cerebrospinal fluid flowing past. The relative absence of intercellular (tissue) fluid in the brain suggests that astrocytes also serve as extracellular fluid and ion reservoirs, absorbing excess K^+ ions or contributing Na^+ ions as necessary to maintain optimal conditions for neuron depolarization. And since your brain size must be tightly controlled whether you are drinking gallons of beer or extremely dehydrated, your astrocytes may prevent dangerous brain swelling or shrinkage by continuously altering their own internal osmotic pressure—which would allow them to exchange appropriate amounts of water with blood under almost all conditions.

However, very young (your blood brain barrier was poorly developed during infancy) or old persons who have become severely dehydrated should not be allowed sudden free access to all the water they might wish to drink. For such rapid dilution of their previously dehydrated (thus markedly hypertonic) circulating blood by large volumes of water (less desirable in any case than providing solute-laden soups, juices, milk and so on) could allow water to enter brain tissues more rapidly than astrocyte osmotic controls were able to readjust. Severe *brain swelling* might then cause coma or even death. The brain swelling that sometimes follows significant or recurrent head injuries could reflect a comparable disruption of astrocyte osmotic controls by massive releases of neuroactive chemicals from many injured neurons. Certainly the extreme close-packing of your neurons and glial cells would be impossibly precise if your brain were to swell and shrink in accordance with your state of hydration as other tissues do—without the blood-brain barrier, a firmly enclosed central nervous system of the same size might safely hold far fewer neurons (even if adequate cooling and separation of circuits were assured). That close packing of your neurons has important implications for the efficiency of your brain, just as every computer designer lusts for the improved performance that comes with successfully packing more processing units onto a smaller chip.

Many dozens of different neurotransmitters and neuromodulators have critical signaling duties in the brain. The majority of such signaling molecules are simple amino acids, modified amino acids and small peptides that often are plentiful in the other life forms upon which you prey—and elsewhere in your own body those same neuroactive chemicals may serve quite different signaling functions. So by erecting an effective blood-brain barrier and placing astrocytes in charge of fluid and solute intake, your ancestors markedly enhanced the versatility and reliability of their brains (and yours). For that barrier allowed them to utilize preexisting cell signals and receptors in a variety of new ways rather than waiting while new signal molecules and matching receptors evolved for every possible brain cell function. Thus your blood-brain barrier prevents your thoughts from being scrambled by the eggs that you eat or the hormones with which you regulate your gall bladder and pancreas.

Memories Are Made Of This?

Your brain selectively acquires, stores, integrates and responds to information. That is all it does. That is all it can do. Furthermore, the cognitive activity of your brain is what defines you as a living human. Remarkable as they are, human sperm and eggs simply transmit ordinary human information. Incapable of living on their own, the huge numbers of sperm and eggs scattered about by all adult humans are thus the moral equivalent of a virus. Indeed, every *diploid* somatic cell that you eventually shed into the environment (totaling hundreds of pounds of hair and surface cells) bears twice as much of your invaluable human information as any haploid egg or sperm (see Reproduction). Nonetheless, the particular information combined by mother's egg and father's sperm was critically important as your embryo developed and grew in strict accordance with ancient rules embedded in the DNA program of all animals. Even when electrical activity began to flicker and then pulse through your fetal neurons, helping them to grow, develop and organize, you still followed those instructions mindlessly. Although empty of unique content at birth, your mind had prepared itself to acquire information from that new and ever-challenging outside environment—and you have been submerged in a flood of information ever since. As an infant, your limited circuits dealt mainly with immediate needs. Slowly you learned to interact, to listen for meaning as well as sound, to express yourself, trying always to help others to serve you more efficiently. You experimented, investigated, cooperated, defied and tested. Gradually you acquired skills. Progressively you sought to impose your will. Inevitably you were socialized.

Eventually you defied traditional authority, declared your independence and entered relationships outside of the family, only to fall prey once again to your inherited instructions and those hormones. And since reproducing you have come under the increasing control of your spouse and children, your employer and customers. So rather than being ruler of all you survey, you now compete madly just to stay even—and that is probably the best possible outcome for you as well as the world you intended to redesign. For radical redesigns rarely improve upon what has evolved in a more or less free market. Just as human anatomy could be much simpler, so could the human world. Neither was meant to be this way, they just happened—and those endless stabilizing and destabilizing interactions at every level inevitably make social systems more difficult to understand, untangle or improve than anticipated. But incremental changes are always underway in such unstable systems with their many problems. So if you find society less than satisfactory and are willing to try, you just might be the person to help bring about important improvements.

You gaze absent-mindedly at those bundled-up people hurrying past in the downpour. Suddenly, "You aren't listening to me! You never do!" Without the slightest hesitation, you reply, "Of course I was, Dear!", and as proof, you repeat back verbatim those last less-than-memorable sentences while simultaneously searching them for content to see if a reply was indicated. Similarly, you can dial a phone number without first writing it down, provided whoever gave it to you will just shut up while that number still echoes in your mind. And it would be easy to walk barefoot through your home at night without turning on lights if only the children picked up their toys. Quite often you recall an irrelevant fact or event from years ago, yet you regularly lose your train of thought when interrupted or distracted.

Indeed, you cannot concentrate clearly on the simplest idea or picture without having it fade away after just a few minutes—presumably that helps you to avoid persistent after-images and perseverating neuronal circuits (which also suggests that concentrating upon particularly annoying or distasteful ideas might be a good way to banish them).

The present is a moving knife edge that divides past from future. The significance of any moment lies buried in the past. The outcome of that moment depends upon processes already in motion. Without memory you could not learn from experience. You rely upon your visual-spatial memory to find your bedroom or a handkerchief in the dark, or to bring your car to a stop on a winding country road when the headlights fail. And your separate short-term memory systems for visual and auditory events remain essential for the acquisition of vocabulary, for reading, writing and verbal language skills, and for understanding what was just heard as well as checking back to see whether you said what you meant to say while those recent sentences still echo in your mind. The warmest memories of your most unforgettable moments and the ongoing function of your brain both depend upon processes that are simultaneously very complex and very unstable. In some fashion the input from a chaotic world is superimposed upon the steadier spontaneous rhythms of your basic circuits to modify the excitability and maintain the appropriateness of specific axonal and dendritic connections. Of course, your uniquely encoded version of *what was* has to be constantly compared with your currently displayed version of *what is* so that you can promptly attend to novel inputs—which requires you to maintain some record of familiar inputs at meaningful (above spontaneous nerve discharge) levels for comparison.

But why should a glance or smile from your classmate be so memorable and those new hit tunes so easy to recall when the subjects taught in those classes have so totally escaped from your memory? Perhaps one day it will be possible to enhance learning chemically (see hints below). In the meanwhile, there surely is a valuable lesson in the fact that carrot cake is more enjoyable than Socrates. Perhaps carrot cake interests you far more than some annoying old Greek because information is more readily acquired and reinforced in your memory when it has relevance to your preexisting world view or if it brings direct or indirect reproductive advantage. In any case, we easily understand our relationship to cake and never forget it, while school lessons often vanish despite our most valiant (although perhaps last-minute) efforts. Yet if we persist there comes a point (perhaps with an outstanding teacher) where we begin to develop a more abstract value and reward system. Or more likely, we rediscover the inborn curiosity that somehow was stifled by boredom, irrelevancies, authoritarian instructors and adolescent peer pressures to be cool. As we then learn how this interesting and new (to us) information can indirectly bring competitive and reproductive advantage, we find pleasure and reward enhancing our ability to learn—until eventually a new fact or idea becomes more welcome than that extra-rich cream cheese frosting. But facts or ideas must fit in. Just as onions have their place but not on carrot cake, information that is appropriate and rewarding is easier to learn than information based upon foreign concepts or beliefs. And punishment rarely teaches the lesson intended—it is far more likely to instill an ongoing distaste for learning as well as for those in charge. But just imagine how easy it would be if one could unforgettably imprint all of mathematics or all of economics or all of art upon anyone's neuronal circuits as they slept.

Surprisingly, that might be no great favor to anyone who values her or his own individuality. For surely it must be the ultimate rape to have the always questionable values, belief systems, concepts and approaches of another inserted along with the tortuous information gathering and reward system that made it all possible. Far better to keep that material nearby in a book or machine where it can be consulted when desired. And if at first the effort required for learning seems all out of proportion to the gain, just keep looking for the big picture and remember that (for better or worse) you are building the only mental house you will ever inhabit. Indeed, it makes no difference whether you are rich or poor, female or male or part of any minority (all persons belong to some minority), in the end your own happiness and worth to others will depend upon the care, love and effort that you put into your own mind. So let others influence you with qualities or facts that you respect and wish to share but don't let them into your head unless you would march to their drummer. Of the several billion people now on Earth, no two are even nearly alike (barring identical twins, triplets, etc.) That variety allows you to choose, compete, enjoy alternate sources of information and discover fresh and useful ideas as multiple viewpoints all fight for their moment in the sun. To take away variety and competition would guarantee stagnation and decline. But good things never come easily, especially when the effort is an inextricable part of that good. So while new subjects may seem especially difficult at the start, later supplements bring new insights and other treats to enjoy—after all, even your monkey relatives may work harder to satisfy their curiosity than for treats. And there are plenty of hints about how you learn and remember (but interesting information rarely surfaces in a comprehensive format, as will become apparent if you choose to glance over the somewhat scattered and technical details in the next subsection).

A Few Hints

Most neurons send out many more branches than eventually persist. Even at the neuromuscular junction, several motor axons may initially innervate each new muscle fiber. But through timing of impulses, repetitive stimulation, utilization and coordination among adjacent muscle fibers, the synapses of a single neuron soon become dominant. Branches from other nerve cells then disappear, for neuronal input fades once a synapse loses functional significance. Obviously this process of *potentiation* affects both sides of the synapse since the neuron that becomes dominant must also be the one to whose synapses the muscle fiber responds. Neither presynaptic or postsynaptic adjustments are yet well understood but there are some clues. For example, postsynaptic (muscle fiber) acetycholine receptors are stabilized (but only at an active synapse) by the Ca^{++} inflow associated with repeated muscle depolarization. Furthermore, that strengthening of a neuron-to-neuron synapse (as opposed to its stable function or weakening by disuse or lack of effect or molecular mismatch) known as *long term potentiation* (an activity dependent synaptic enhancement that appears important for memory) also seems to affect both sides of the synapse.

Of course, the outcomes of individual excitatory or inhibitory synaptic events between nerve cells are far less predictable than the overwhelming likelihood that a single motor neuron discharge at an active neuromuscular synapse will produce a muscle fiber action potential. For one thing, the effect of a particular neurotransmitter on its targeted interneuron may persist for a considerable time. So when a large number of different neurotransmitters are released (on either the input or the axonal side of that neuron) by different axons for different purposes to different receptors, all sorts of interactions and gradations of cumulative effects can be anticipated. On the recipient (usually dendritic but sometimes cell body) side, weak presynaptic inputs tend to enhance Ca^{++} entry into tiny branches while stronger signals affect larger dendrites as well. After repeated stimulations, the effect of such Ca^{++} priming may last for minutes. Elevated postsynaptic Ca^{++} levels may lead to long-term presynaptic and postsynaptic changes as well. On the postsynaptic side, some cation channels can be modulated to reduce their ion flow. That effect may be brought about by associated intracellular regulatory complexes involving *phosphorylation* by *protein kinase C* or *dephosphorylation* by various *phosphatases*. Modifications of this sort can change both the speed and extent of synaptic response. And calcium-activated K^+ channels also link membrane potential with second messenger effects that additionally alter the target cell.

Furthermore, glutamate excitation of a synapse may lead to ongoing post-excitatory effects as slow-opening channels are repetitively reopened (due to the slow rate at which glutamate unbinds). Glutamate is especially important in early development, synaptic modulation and memory formation. It is also involved in the nerve cell damage that can follow seizures, neuronal degeneration and ischemic releases of stored neurotransmitters. There are two types of excitatory receptors, fast ones that open cation-selective (Na^+ or Ca^{++}) channels and slower ones that function via intracellular G proteins (see Hormones). One can experimentally block damage from excessive nerve cell excitation by raising Mg^{++} levels to reduce the release of neurotransmitters. GABA-*ergic* (GABA-*like* amino acid neurotransmitting) agents also inhibit acetycholine release (GABA-ergic stimulation of neurons opens Cl^- channels which is inhibitory) or block certain acetylcholine receptors much as scopolamine/atropine do (see Eye).

Excessive Ca^{++} entry may be harmful directly or via its effect on the Ca^{++} regulatory protein *calmodulin* which turns on *nitric oxide* (NO) *synthase*. NO is actively produced by endothelial cells (to dilate blood vessels by relaxing smooth muscle), by GI tract lining cells (to regulate gut contractions), by immune cells (to activate bacterial killing) and by nerve cells (as a post-synaptic to pre-synaptic neurotransmitter). The NO receptor is an iron atom (on enzymes such as guanyl cyclase, which produces cyclic GMP, an important cell regulator). But too much glutamate and excessive Ca^{++} entry releases too much NO, causing cell damage or death. Apparently NO synthase has structural similarity to cytochrome p450 reductase (an important liver enzyme involved in drug metabolism—just another example of how much easier it is to modify a functioning protein for a new purpose than to start over).

In vivo, NO and *carbon monoxide* (CO) are short-lived gases that seem to serve as neurotransmitters. NO travels in solution from the postsynaptic to the presynaptic neuron to affect presynaptic G proteins which then enhance presynaptic neurotransmitter release. Ordinarily NO acts only at a few local, simultaneously discharging synaptic sites before being destroyed (½ life of 4–6 sec). Intravenous *sodium nitroprusside* is often utilized to reduce elevated post-operative blood pressures—nitroprusside releases NO and thereby also strengthens synapses in similar fashion to long-term potentiation. On the other hand, free hemoglobin absorbs both NO and CO, so free hemoglobin in brain tissues might block any extracellular retrograde messenger effect of NO and CO and thereby make it more difficult to learn (see Blood). Interestingly, aged or damaged cells tend to release destructive oxygen free-radicals. There is some evidence that older gerbils regain their ability to learn a maze when treated with anti-oxidants. So perhaps some sort of anti-oxidant such as Vitamins C and E (water-soluble and fat-soluble anti-oxidants respectively) may one day help you to preserve your aging intellect and also reduce the risk of arteriosclerosis (see Circulation) or even the accelerated loss of neurons often seen in the aged or those with Amyotrophic Lateral Sclerosis (Lou Gehrig's Disease) or possibly Alzheimer's Disease—even simple anti-inflammatory agents are among many substances being evaluated in efforts to reduce progression and affect various facets of the latter condition (whose underlying cause remains to be defined). The above selections are just a few of the more interesting hints currently under intensive investigation—many *controlled studies* are underway comparing outcomes in experimentally treated individuals with the results in (a control group of) similar persons not (or otherwise) treated. Some of these studies are *double-blinded* with investigators only learning later which patients received the particular treatment under evaluation and which received an alternate treatment or merely a *placebo* (sugar pill).

Just Think About It

Their neuron-like membrane receptors raises the question of whether some of your uncounted trillions of glial cells are directly involved in neural function and thinking (or only indirectly—by their close coordination of your glial support functions). If your astrocytes do have a role in information processing, that would tremendously expand the already impressive estimates of your total brain capacity. But even without that potential glial boost, you probably have about 10^{11} (100 billion) neurons that may average 10^4 (ten thousand) connections each. If we conservatively disregard the fact that some connections stimulate and others inhibit, we still have 10^4 connections raised to 10^{11} power as your potential number of possible brain states. This excessively large number already exceeds the likely number of atoms in the Universe. And if neuron-like glial cells also contribute, the computing potential of your brain becomes even less comprehensible to your limited brain. So how is all that brainpower tied up?

Well, what did you have for lunch last Tuesday? Were you sleepy after lunch? Was the traffic noisy? Dogs barking? What program was on T.V. when you stopped by your aunt's house last time? How many steps on her porch? About how many boards? What color was your first bicycle? How was the weather on your last vacation? The food and the company? The season? The flowers? Was there a lamp on the table? What did you discuss? How many songs do you know? How many roads and trails can you remember? How many faces do you recognize? What was the name of that cute girl in first grade? Where is the tallest tree you have ever seen? The biggest tree stump? The most spectacular sunset? The most boring class? How many books have you read? Where did you leave those insurance papers? When? How many ball games/movies/television programs can you recall? How many street lights between here and there? How often have you worried about your children or yelled at them? Does Grandmother have a little bump on her nose?

Apparently no single neuron has sole responsibility for recognizing grandmother. Rather your grandmother is taken apart and analysed in separate areas according to color, form, movement, voice and her many other delightful qualities. These various circuits then recombine in some fashion to create an active internal display in *real time* (while it is still happening). Of course, all of that information is not needed in order to recognize either grandmother or Beethoven's Fifth Symphony. Just a few lines on a clever sketch, that familiar voice on the phone, a few notes of a well-loved theme and your previously imprinted memory circuits fire up to fill in the blanks. A great many such modular processes may go on simultaneously, often involving some of the same or mostly different neurons. That is how you come to your often incorrect conclusions. That is what accounts for your not-always-perfect behavior. For you are so submerged in ordinary day-to-day information that even your superb sensors and processors cannot hope to handle it all. And despite the fact that you regularly lose thousands of nerve cells and forget a great many details, almost surely you will still recognize Grandma. Similarly, one can spend days painting even the semblance of a lovely landscape. An exquisitely detailed photograph of the same scene takes but a moment, yet it also contains far too much information for instant perusal, analysis and internal replication. Our brains continually cut corners, extrapolate, guess, consolidate—so deer hunters are sometimes mistaken for deer and a jury of his peers acquitted the man who shot his mother-in-law after mistaking her for a raccoon.

All Creatures Provide Evidence

Our understanding of how various portions of your brain function and interact is relatively primitive but growing rapidly. Until quite recently, the situation was somewhat analogous to someone using a rifle to investigate how an automobile works by shooting into various sectors and observing what went wrong. As recently as 1966, Russian scientists operated upon a cat and entirely divided its nervous system at the midbrain level. Despite no remaining connections between the upper and lower parts of its central nervous system, they discovered that regular electrical stimulation of the cut surface of the midbrain allowed the cat to run on a treadmill, admittedly in an absent-minded fashion. But even with such running on a flat surface at a steady rate, an endless variety of positions and tensions had to be computed in order to detect and correct errors and smooth out instabilities of movement—while locally acting learning rules continuously reported back and strengthened best efforts through trial and error. All of which allowed that cat in the neurosurgical hat (undoubtedly an experienced runner) to maintain balance and coordination using *proprioceptive* (position, muscle and tendon tension, and other internally generated) *feedbacks* from lower nerve centers that could only go as high as the cut midbrain.

Many subsequent studies of (similar-to-human) animal brains have made it clear that higher and lower brain centers interact extensively—however, lower non-conscious modalities handle routine body affairs without detailed instructions from higher centers—which is hardly surprising since higher centers evolved long after such matters already were efficiently coordinated. So you think and drive and chew gum and scratch, all at the same time. And even the simple neural networks of worms and jellyfish provide insights on nerve cell development and function, just as the giant axons controlling squid mantle muscle played a crucial role in early investigations of nerve cell depolarization during the 1950's and 1960's. Further-more, ongoing evaluations of humans who (by accident or disease) have suffered all sorts of damage to various parts of their central nervous system continue to generate much useful information, although the relationship between an injury and some of the resulting functional deficits can be quite indirect.

You Think, Therefore You Are
(So You Think)—A Very Brief Anatomical Review

Your two cerebral hemispheres account for ⅞ of your entire brain by weight. Their active cellular surfaces are expanded by *gyri* (bulges) between *sulci* (grooves) and each cerebral hemisphere is customarily viewed as having four subdivisions—the *frontal lobe* anteriorly, the *occipital lobe* posteriorly, the *parietal lobe* in between and the *temporal lobe* on the side. Your *basal ganglia* are paired subcortical collec-tions of nerve cell bodies that appear to be essential for the planning, generation and automatic execution of voluntary movements as well as the development and retention of motor skills (such as playing basketball, brushing your teeth, driving an auto and so on). Nonetheless, the bulk of your cerebral nerve cells are in the *grey matter* (outer layer of your cerebral cortex—which greatly simplifies heat transfer into the copious circulation of your skull, scalp, face and nose) while your *white matter* axons are conveniently placed inside. The reverse holds true for your spinal cord where great quantities of incoming and outgoing information make it more practical to position locally required nerve cell bodies centrally to avoid interfering with the mostly myelinated ascending and descending axons passing by en route to your brain or spinal nerves.

The deep central sulcus between your frontal and parietal lobes separates the *sensory* gyrus (postcentral gyrus) where body sensations are displayed from the corresponding precentral primary *motor* gyrus in the frontal lobe where you decide what to do about it. But as usual, things are not nearly that simple. For many nearby and distant areas subtly send, receive and interact with numerous side branches of neurons en route to their principal functional area. Such a preexisting structure of ancillary connections may be unmasked and strengthened if the primary area suffers damage, which could account for much of the recovery sometimes achieved through diligent efforts after localized losses of neurons (especially in human females, who seem to have less dedicated and specialized circuitry).

The *right* cerebral hemisphere receives information primarily from and also controls the major musculature of the *right* side of your head and *left* side of your body. The *left* cerebral hemisphere is correspondingly in charge of *left* head, *right* body. This pattern results from large bundles of axons crossing the midline in the brain stem. The reproductive advantage in having right brain control the left arm and leg while left brain controls right arm and leg is unclear but possibly such an arrangement helps the two cerebral hemispheres to better coordinate their activities. Or perhaps this is just another left-over from the days when your fishy ancestors barely escaped from some threatening sensory input by first contracting trunk muscles on the opposite side in order to pull away more quickly.

In any case, a *stroke* (dead brain tissue—usually due to arterial blockage or arterial rupture) that causes right facial muscle weakness and left arm and left leg muscle *weakness* (left hemi*paresis*) or *paralysis* (left hemi*plegia*) involves the right cerebral hemisphere, while comparable left-sided brain damage can produce left facial with right-sided trunk and extremity disability. Such well-known complexities of brain and spinal cord axonal circuits often allow a *neurologist* (a physician specializing in the nervous system) to specify the site of brain or spinal cord injury or evaluate progression of nervous system disease by eliciting abnormalities in peripheral sensory and motor function. But X-ray and nuclear magnetic resonance studies generally determine the site and extent of brain damage more accurately.

A great many axons carry messages from your cerebrum to lower centers via the *internal capsule*. Similar white matter fibers crossing between cerebral hemispheres in the *corpus callosum* help your two hemispheres to stay in touch. The *midbrain* is about an inch in length. Eye and ear input and reflex eye movements are coordinated at this level. Loss of cells in the midbrain structure known as the *substantia nigra* ("dark substance") can lead to the progressive tremor and other difficulties of movement seen with *Parkinson's disease.*

The Thalamus And Hypothalamus (Under Thalamus)

The thalamic area is a reflex center for all sensory input except smell (which goes directly into olfactory bulbs and cerebrum). Some sensory input is interpreted in the thalamus, including pain, temperature, light touch and pressure. And some basic emotions seem to spring from this level. That is hardly surprising, since most mammals have quite similar lower brains and basic emotions. Major differences between mammalian brains often reflect the extent to which each depends upon certain inputs such as smell (dog, wolf or bear) or sound (bat or dolphin). Your memories form at this thalamic level and in medial temporal and nearby nuclei such as the hippocampus, as well as in higher centers.

The hypothalamus is in charge of *homeostasis* (keeping your body in stable healthy condition). It controls your *autonomic* nervous system, integrating those vegetative motor and sensory functions with your *endocrine* (hormone) system via its regulation of your *pituitary* (master) gland. In the hypothalamus we find some aspects of aggression (as well as the uncontrollable rage of rabies), also your temperature controls, the regulation of your carbohydrate and fat metabolism, blood pressure, hunger and thirst, excretion, sleep, sexual reflexes and biological rhythms— the basic functions that you share with all other mammals. While some of these modalities have highly localized trigger areas, the overall function tends to be more widespread (involving feedbacks with gut peptides, blood hormone and solute levels, neighboring midbrain centers, etc.). Thus there is body-wide activation of sympathetically innervated structures during rage or fright—your heart rate and blood pressure rise, red cells are squeezed from the spleen into the circulation, skin and gut arteries and veins constrict to boost your skeletal muscle blood supply and your skeletal muscle fiber contraction is enhanced, as is your muscle uptake of K^+ (many of the above are direct effects of epinephrine, which also raises blood glucose and enhances various brain and sensory functions).

Your *cerebellum* coordinates complex movements. Note that as we move from higher (cerebral hemispheres) to lower (cerebellum and brain stem) brain functions, these functions become less voluntary or conscious, more automatic or primitive. Below the cerebrum your brain functions in about the same fashion as the brain of any other primate. Indeed, even when using your cerebrum, you have been known to make a monkey of yourself. And that is not so surprising, for your close resemblance to the great apes and monkeys comes through common ancestry rather than God having made them *almost* in His image. But while your proteins differ from those of chimpanzees by less than 1%, your behavior almost surely differs from theirs by much more than that—which again shows that you do not need a new gene to account for each new behavior. For example, it is easy to imagine how some very slight modification of just one gene could have allowed your ancestor's cerebral cells to divide one extra time. That might have increased the ancestral cerebral neuron count by ⅙ and their possible neuronal interconnections by very much more. Yet such a major anatomical change would have derived from an apparently minor genetic cause. Of course, every change usually requires a great many compensatory alterations elsewhere before its full benefit is realized. But any sudden increase in total neuron count might have been integrated quite easily, since it is the input of incoming axons that seems to organize the cortex into functional units, and there are lots of spare axon branches that then atrophy from disuse or inappropriate connection.

However, there is far more to brain function than any description of the complex cellular mechanism within your skull could even imply. Furthermore, your own peculiar behavior depends directly upon your chemical state as well as that enormous number of neuronal interconnections. Consider, for example, how epinephrine and similar chemicals increase the sensitivity of your cells to signals (increase your signal to noise ratio) up to a maximum and highly stressful level. Or how a chemical that simply *decreased* your sensory signal-to-noise ratio might tend to make your behavior repetitive and unsure. So it is not surprising that certain chemicals or antidepressants may increase or reduce *obsessive* thoughts (persistent ideas) or *compulsive* (repetitive) behavior. Or that your hormones play other tricks on your mind, as we shall see a few chapters hence.

Further Reading:

Gilbert, Scott F. 1988. *Developmental Biology.* 2d ed. Sunderland, Mass: Sinauer Associates. 843 pp. (useful for this and other chapters)

CHAPTER 15

SPECIAL SENSES

Just What Is Going On?... We Are All In The Soup Together;... How You Taste;... How You Smell;... What You Smell;... We Seek Pleasure;... Tastes Vary;... Some Inputs Are More Important Than Others;... It Has Always Been Harder To Listen Than To Talk;... Your Hearing Machine;... Inner Ear;... Balance;... Hearing;... Cells Surrounded By High-Potassium Solutions Must Accumulate Sodium Ions.

Just What Is Going On?

By its existence, life-style and metabolism, every living thing disrupts and pollutes the environment. Directly or indirectly, all of these disturbances represent dangers and opportunities that other life forms must sense in order to survive. To observe how life's disturbances can spread and be detected, we will now explode a series of large identical imaginary firecrackers outdoors in the dark, so brace yourself. At a distance of 200 meters, the light generated by each explosion will pass before the bang arrives. This is no surprise, for you already know that light travels a million times faster than sound. You also remember that light easily crosses a vacuum which is impenetrable to sound, while sound readily passes through murky waters or solids that would absorb or reflect light. But different as they are, both light and sound energy move outward at their usual speeds in all directions from the exploding firecracker until constrained by the surroundings. Ideally, therefore, such bursts of light and sound will expand as spherical fronts. And even after passing your position, those spreading light-and-then-sound frontal disturbances can still startle fine folk farther away while you relish the return of dark and quiet. Of course, since those initially concentrated surface disturbances enlarge evenly in all directions, their perceived energy intensity must decline with your distance from the source squared (for the spherical surface area of any light or sound front is proportional to its radius squared). So doubling your distance from the firecracker should make its sound and light just one-fourth as intense—and if you continue to move away, eventually both light and sound energy from those continuing explosions will become too dilute for detection by your unaided eyes or ears. At those points, you no longer can separate that light or sound (the signal or information) from other lights or sounds (the background noise). However, if a brisk breeze shifts the expanding sphere of noise toward you and thereby exposes you to a smaller younger sphere with higher surface energy, you may once again hear that firecracker sound despite being at a normally inaudible distance. Of course, the only way a breeze can alter the intensity of a light signal is by stirring up dust.

There are endless ways in which light or sound waves can *interfere* (reinforce or cancel each other) or be *blocked* (creating a shadow), *diffracted* (with white light this gives a rainbow), *absorbed* (warming the target), *reflected, refracted, diverted, directed, diffused, focused* or *funneled*—the resulting alterations in energy intensity can often be put to practical use. For example, modern digital signal processors make it possible to generate antinoise or antivibrations that actively produce silence or reduced vibration at the source. Hopefully it will not be long before one can match any sound source with waves of equal intensity and frequency but opposite phase. Allowing both sounds to cancel each other produces a pleasant artificial silence in which one can converse, listen to music or even think. And imagine its frustration when the noisy neighborhood dog finally receives that anonymously donated anti-bark-generating collar. By incorporating such active noise control units in their mufflers, some manufacturers have been able to reduce sound insulation and muffler weight in their 1993 model automobiles while also eliminating back pressure from the exhaust system, thereby gaining performance while improving fuel efficiency.

Now let us imagine how the cloud of smoke and odor released by each exploding firecracker might expand outward in a relatively spherical fashion (once locally heated air has cooled by expansion and the resulting turbulence has subsided). But expansion by diffusion alone is a slow process, whether it is smoke spreading out in air or a sugar tablet dissolving in water. So you are unlikely to notice firecracker smoke and smell unless you are quite near. Of course, if moved along on the breeze but not excessively stirred, the smell could be detectable at a considerable distance downwind, merely because odor expands so slowly— skywriting smoke or jet contrails can similarly retain their shape for many minutes while gradually shifting across the sky as an integral part of that moving mass of air above you.

Sound and light always move outward from their source at top speed, leaving no trace once they have passed (other than any damage they may have caused). On the other hand, vaporized or dissolved molecules spread but slowly from any temporary center of concentration, whether it be firecracker explosion, dissolving sugar tablet or fresh dog turd. In air, any spherically expanding chemical front may first become apparent at the most dilute concentration that you can detect. If it is crossing your path, the chemical concentration then rises steadily until the center passes by or you pass it. Thus smell is an expanding-volume process (spreading from higher to lower concentration) rather than a passing-spherical-surface phenomenon. The volume of a sphere is proportional to its radius cubed (radius multiplied times radius squared) so even within an ideal spherical and uniformly smelly odor cloud, every time that cloud doubles in diameter, its concentration of smell molecules ought to diminish by eight times. In general, expanding smells are not distributed evenly in this fashion, so they fade out rather more rapidly as we move away from their apparent source (current center of concentration).

We Are All In The Soup Together

It seems that every living thing is submerged within innumerable, endlessly complex and interacting patterns of chemicals spreading through land, sea or air. Also that each organism is ceaselessly illuminated and vibrated by expanding electromagnetic and mechanical disturbances on every scale from all different directions and even ancient times. Yet, simply to survive, an organism must swiftly separate out the most timely and significant information from amongst all that other overwhelming and endlessly varied, often chaotic, input. Which brings us to your senses—those specialized receptors that allow most multicellular creatures to detect and sort out relevant information. As usual, rather than having been designed on a blank slate, each of your senses was built up over billions of years via simple modular recombinations and modifications of preexisting receptor molecules, cellular structures and processes. Throughout that arduous but relatively random design and engineering process, the grading system has remained strict but fair, even if a bit sloppy. "Pass" meant your ancestor may have come up with something reproductively advantageous. "Fail" suggested some potential ancestor was either unfortunate or displayed some critical deficiency under those particular circumstances. Even though a design might seem superior or its function outstanding, the fact remained—being eaten before achieving reproductive success meant automatic exclusion from the line of succession. Although this aimless self-scoring incrementally progressive design system might seem hopelessly inefficient, it had the great advantage of being able to sort through countless trillions of contestants at the same time. So with no guidelines or goals other than ongoing reproductive success, natural selection has managed to come up with amazingly effective special senses for the mantis shrimp, the duck-billed platypus, the hammer-head shark and other weird and wonderful life forms such as you.

For example, a small abyssal shrimp can allegedly detect 10 different colors. Somehow this skill allows it to graze more efficiently on the bountiful harvest of bacteria busily oxidizing highly reduced chemicals at those deep dark oceanic hot water vents. Perhaps some of those bacteria produce visible light, maybe in relation to water temperature. Possibly mantis shrimp vision (photosensitivity) overlaps the infrared (heat) so the water itself or nearby debris can be displayed as a rainbow of temperatures to keep alert shrimp from being boiled. Or maybe these shrimp can detect the faint glow emitted by newly arising vents as older ones move laterally and become inactive with spreading of the sea floor. Being poor swimmers, vent bacteria display extreme heat tolerance (thermophilia) rather than trying to detect the hottest water in order to avoid it. In contrast, the delicate electrical sensors on the leading edge of hammer-head shark eye wings or those bordering a duck-billed platypus' bill are able to detect tiny electrical depolarizations of large or small bottom-dwelling life forms as such organisms rest or innocently move about in the muck seeking food and romance. This provides shark and platypus with efficient access to otherwise under-utilized reducing-power resources that happen to be cowering beneath the muddy waters.

Larger animals are usually most interested in or threatened by creatures of comparable size that disturb their environment mechanically as well as electromagnetically and chemically. So grazers and browsers tend to have tall, steerable sound collectors at the very back of their skulls. These receivers are held high even during eating, well away from self-generated noises caused by moving about and chewing. Grazers and browsers usually have symmetrical, outward directed eyes and nostrils, as well as extended jaws to support their long grinding teeth. This allows the widest possible field of vision and smell while still permitting food inspection. On the other hand, pouncers generally direct both eyes forward since *binocular* vision permits more precise three-dimensional pouncing (and also makes it easier to spot a tiger in the tall grass). Both pouncers and grazers tend to have very sensitive detectors for the few molecules of plant or animal scent that may waft past on the breeze. Perhaps you view yourself as part-pouncer and part-grazer but your eyes focus together and you cannot even wiggle your ears. And despite having comparatively few odor receptors, you do seem to have one of the finest analysers on Earth for evaluating what you consciously or unconsciously may detect. So you watch, sniff and taste your captured reducing power before you swallow. And if unsuitable food evades detection by your primary mechanical, electromagnetic and chemical input sensors, that food can probably be forcibly rejected from one or both ends of your gastrointestinal tract.

How You Taste

Although your nose is able to recognize thousands of volatile molecules, the chemoreceptors located within crevices of your tongue deal primarily with the four critically important basic *flavors* of non-volatile solutes. *Sweet* usually implies easily utilized reducing power. Those tongue surface sensory cells with especially many glucose (and other sugar) receptors are concentrated near the tip of your tongue so you can readily select sugars for quick energy. Non-sugar low-calorie sweeteners may fool some of the people's taste buds some of the time but the per-person consumption of sugar and other high-calorie sweeteners has gone up since no-calorie sweeteners were introduced to the U.S. (so no-calorie sweeteners may simply reset your preference to a sweeter flavor). Even lead acetate ("sugar of lead") tastes sweet—the Romans valued that flavor in the slightly vinegared wine that they drew through lead pipes. By muddling Roman minds, both wine and lead poisoning presumably contributed to the fall of the Empire.

Salty also resides at the front of your tongue where it encourages appropriate replacement of extracellular NaCl lost in urine and sweat—both KCl and NaCl solutions stimulate your salt-sensitive receptors while Ca^{++} leaves a salt/bitter flavor. Although many cations have a salty taste, the salty flavor of Na^+ is accentuated when Cl^- reaches the fluids between taste cells—sodium chloride tastes saltier than sodium acetate, for example. Apparently your salt-flavor-detecting cells discriminate by the way different anions alter transmembrane and trans-epithelial voltages. *Sour*, on both sides of your tongue, helps to limit your intake of acid foods that might alter your body pH. And *bitter*, at the very back of your tongue, is most likely to meet well-chewed food (your last chance to throw it out). Bitter food we correctly tend to avoid, for toxic plant chemicals (alkaloids) tend to be very bitter.

Various combinations of sweet, salty, sour and bitter can help you to discriminate among important food inputs, although your perception of those flavors will vary according to your needs. Thus a huge piece of cake with plenty of frosting (lots of sugar and fat) may seem delightfully sweet and satisfying at first, yet later when you have had far more than enough, it might develop a rather sickening taste. And on a warm day following vigorous physical activities, added salt can improve the flavor of your salty snack foods. Your tolerance for sour fruit or juices is similarly dose and need related. And let us not forget *pain*, the fifth flavor. Here we refer not to a tooth-ache or biting the tongue but rather to chilies and other sorts of spices that stimulate pain fibers and often make your face perspire and nose run by autonomic reflex. Pain of the sort induced by spicy foods from India, Korea or Mexico is an acquired taste that in mild doses accentuates the gustatory process. However, even Mexican children must reach 3 to 4 years of age before they can appreciate this fifth flavor.

Surprisingly little is known about how you taste or smell. But in general, your sensory cells convert light, chemicals, sound and other mechanical signals into altered transmembrane potentials for your brain to interpret. Even exquisitely faint visual signals (1 photon), smells and tastes (a few molecules) and sounds (that displace one hair cell by less than 0.1 nanometer or billionths of a meter—which is close to thermal vibration levels) can be detected and then amplified for your evaluation and response. While salty, sour and sound are amplified directly by their effect on transmembrane cation channels, the detection of sweet and bitter, smell and light involves activation of GTP binding enzymes (G proteins—see Hormones).

How You Smell

Located well beyond reach of your exploring finger tip and also somewhat aside from the direct flow of air to your lungs, a dime-sized patch at the top of each nasal cavity exposes about three million *olfactory sensory neurons* under a thin sheet of specialized mucus. These olfactory nerve cells are continually replaced by new neurons developing from the *basal* cells that also provide new support cells. In the process of maturation, each olfactory neuron extends a dendrite to the psuedostratified epithelial surface and an axon to nerve cells in the *olfactory bulb* via perforations in the thin intervening bony *cribriform* plate—your olfactory nerves are quite short since those two symmetrical olfactory bulbs are just inside the skull. Although other sensory inputs are pre-processed at lower levels, the information received by your olfactory bulb goes directly to your cerebrum—so smells (but not sights or sounds) tend to activate relevant memory-nostalgia pattern-recognition circuits. And that recalled ambience can contribute to your safety or dining pleasure as there are so many different smells that any person, animal or object closely associated with a smell on one occasion is likely to be so again. Since breathing never stops, the direct cerebral olfactory connection (along with your hearing) also helps you to keep track of nearby matters at night, even when asleep. Given these advantages, the ancestral olfactory-cerebral connection persisted as your tremendously expanded and modified cerebrum evolved out of the cluster of nerve cells initially dedicated to analyzing odors.

It seems that some scents can stimulate your memory or behavior directly without entering your awareness. We know that female moth scent can drive a male moth mad with passion, sight unseen and a mile downwind. Similarly, the perfume industry thrives by misleading non-allergic humans through unrelated nostalgia circuits when a good bath might be more appropriate. And human females seem to interact by means of some *pheromone* (a so-far undetected scent with hormonal effects) that causes the menstrual cycles of women who room together to coincide—a form of peer pressure presumably based upon some scent emitted by the dominant female. In earlier times, such menstrual synchronization may have enhanced reproductive success by encouraging simultaneous births when seasonal variations in resources (or predators and parasites) necessitated group travel—which could affect newborn survival.

A male pheromone is emitted by the truffle, a fungus that grows underground on oak tree roots, most notably in France. Presumably that pheromone explains why truffles have long been valued for their flavor and possible *aphrodisiac* effects (honoring Aphrodite, the Greek Goddess of Love). The complex flavor of a truffle reportedly mimics the scent of a "freshly screwed-in bed" (another acquired taste, perhaps). That same (truffle) pheromone is produced by the testes and secreted in the sweat of adult human males—another source is boar pigs, which explains why female pigs are such devoted truffle snufflers. Interestingly, while judging a long series of photographs, both male and female humans rated a particular female as more attractive if those viewers were simultaneously exposed to (under the influence of) that pheromone. This is a good example of how natural selection encourages useful signal-and-receptor molecule combinations, for regardless of whether the signal is directly released by male sweat glands or only develops when armpit-odor-causing bacteria break up less volatile sweat gland secretions, those female humans or sows attracted to such scents gain reproductive advantage.

The truffle could conceivably derive reproductive advantage by being a turn-on for humans and pigs but it is likely that truffles utilize those pheromone and receptor modules for some more mundane truffly purposes. The current great demand and high price paid for truffles might therefore be viewed as a hazardous, indeed possibly lethal (for the truffle) side-effect. However, if the French and others can learn to farm them successfully, truffles may yet gain fitness through that pheromonally induced process of selection. Even without perfumes containing pheromones (should they be illegal?), we respond to the odors of others from the time of our birth. Indeed, body odor is an important part of the bonding process (and occasionally of unbonding as well)—and new parents might encourage early bonding with their infant by avoiding perfumes and scented toiletries.

A thousand different genes have already been identified for *olfactory-receptor-proteins* (many of which include carbohydrate) and hundreds of different lipid species are found within olfactory nerve cell membranes as well. Apparently the mammalian olfactory system can identify over ten thousand different odor molecules in air. Perhaps you could also detect dissolved substances with your nose full of water, but that experiment would be unpleasant. Therefore you rely upon taste to analyse the few crucial non-volatile substances regularly encountered in solution.

What You Smell

With the actual mechanism of smell as yet unclear, we can simply view each additional sort of odor receptor as another in a long series of piano keys upon which an almost endless variety of chords or single notes can be played with varying intensities. Naturally some of these effects are far more pleasant than others. So horsemen tend to be tolerant of and even nostalgic about the scent of horse manure, and cow manure symbolizes prosperity in certain cultures, but few humans seek out the smell of human shit. Yet I have several times encountered individuals who were truly convinced that their own shit did not stink, as well as others who presumed that every shitty odor represented the airborne escape of tiny shit particles from their turd-of-origin—with those particles then cruising freely on the breeze until firmly lodged in some innocent smellee's nose (or even on their toothbrush!). These are understandable interpretations of ordinary experience, even if somewhat incorrect.

Those taking undue pride in their own unscented stools might be shocked to learn that odors often become undetectable by exceeding the maximum signal strength of relevant olfactory receptors. For while the initial impression made by an intense chemical signal tends to be unpleasant, nasal and other chemical sensors thereafter report only changes (just increases, actually) in concentration rather than absolute values. So those who swiftly desensitize relevant nasal receptors in their own bathroom or outhouse are then no longer troubled. And wherever they may wander soon thereafter, they will initially be moving from greater to lesser concentrations of stink. However, there is no sensory signal for "things are getting less bad" other than the absence of a sensory signal. Even if leaving and soon returning to a smelly, unvented bathroom, the source individual may remain at least partially desensitized by that recent severe exposure. Thus he or she may judge self-generated shit odors with unwarranted favor, especially in comparison to those that are equally noxious but other-generated—for when the latter already seem troublesome at a distance, they can only become progressively more foul as you approach.

On a winter visit to a small Alaskan community, I once asked where the creaky old sewage truck disposed of its "honey-bucket" contents after these were picked up in house-to-house collections. A bystander pointed out that formal disposal ought to be unnecessary, given the ongoing significant leak from the truck's tank. Even though I tried unsuccessfully not to breathe thereafter, I detected no odor out there in the cold. And those delicate souls who complained about the dusty roads of this town in summer also detected none. This despite the fact that each passing vehicle on a dry summer day would stir up an impressive new cloud of that shitty dust. So when oxidised *dry* shit does get up your nose, it may cause no detectable odor. This is not surprising, for each exposure to oxygen allows bacteria to recover more of the food value remaining in stool (see Digestion). On the other hand, archeologists have collected and analysed 800 year old air-dried feces from old Chinese fortresses in the Gobi desert. That shit had lasted many centuries in mint (fine) condition because bacteria also require moisture for their metabolism. It is reported that by heating such antique stools and blowing the fumes up a dedicated and highly trained volunteer investigator's nose, these scientists could determine important details of that long-dead soldier's diet. Information of that sort has helped to clarify historical trade routes, since the source region of caraway seeds or other partially digested foodstuffs thus identified could often be determined.

Furthermore, chemists who study and replicate odors have found that the various chords played upon your nasal sensor piano keys usually represent just a few *volatile* (easily vaporized) molecules, so shit smells merely represent a shitty ditty rather than a nose full of shit. Even rocks have an odor when struck sharply together, for that strike momentarily creates a great deal of heat, causing some rocky constituents to vaporize so they can activate a "rock" tune in your nose. Similarly, a cinnamon roll emits specific molecules into the air that play "roll" messages. Interestingly, not every nose gets the same shitty rock and roll message, for some individuals lack certain receptors and all individuals differ in their sensitivities and abilities to discriminate between odors. Such variations of sensitivities between individuals may, in part, account for different "tastes" (smells actually) in coffee and wildflowers as well as barnyards. And what one observer could only identify as "either lilac or clover" might be classified more accurately by another with additional receptors tuned to less dominant volatiles. Often those added notes provide the overtones that can enrich and complete an odor. Skunk cabbages purposely heat their florets to increase the number of scent molecules being vaporized (see Metabolism). Newborn animals commonly remain odor-free for some hours until freshly arrived colonies of skin bacteria can begin to break down surface organic materials into smaller, more volatile molecules.

It seems that noses have their greatest survival value when deciphering very faint chemical messages. At such times, the odorant molecule is quickly detected and rapidly deactivated, so its effect can be both prompt and brief. That is another good reason for locating your nose as far away from your *anus* (asshole) as possible. However, regardless of how sensitive your nose may be or how many more working receptors you may have than others, from the time you first encounter an undisturbed expanding sphere of smell, you will usually be moving against a rising concentration gradient. And if you move through very slowly you may not notice much (for no gradient = no message), while going through more rapidly can greatly increase the gradient and thereby send a stronger message. So walking past an apple orchard at blossom time is pleasant, biking through is really nice and motorcycling past (with no windshield or helmet visor to disturb that humid evening or morning air before it enters your nose) can be memorable.

On the other hand, auto exhaust is both unpleasant and dangerous to smell. Yet if you mix auto exhaust with lots of other chemicals including lovely flower scents, you may smell very little at all, even when the visibility becomes markedly reduced by the dense smog of odorant molecules. Similarly, if you stimulate most or all nasal receptors the way those obnoxious and possibly dangerous bathroom deodorizers do, your brain perceives only noise to ignore rather than information. Indeed, very strong smells can temporarily or permanently disable your nasal receptors, thereby rendering you *anosmic.* Which is how a criminal lawyer (aren't they all?) discredited an apparently reliable witness who had recognized a particular incriminating odor at the scene of a crime. Asking him to first confirm his nasal expertise for the jury by identifying a few common odors, the lawyer gave the witness gasoline as his initial test odor, thus rendering him temporarily anosmic. At the same time, even heavy smokers on the jury were able to recognize the other odor in question.

You can lose your sense of smell for three or four days after heavy exposure to wood smoke around an open campfire. For as you have noticed, if you come close enough to be warmed, the smoke tends to be drawn toward you even if you are directly upwind, since your wind shadow reduces air pressure between you and the fire. Exposure to strongly reactive odorous chemicals such as formaldehyde (formalin) and certain other chemical fumes or smokes can cause anosmia that may persist for a couple of years. Although they often seem indifferent to that loss, many older chemists and pathologists remain chronically anosmic from repeated occupational exposures to strong-smelling volatile chemicals. So while the regular replacement of old olfactory nerves ordinarily keeps your nose up to snuff, a severe chemical exposure, infection, tumor, skull fracture involving the cribriform area, or other known or unknown agent can render an individual chronically or even permanently anosmic. And anosmia may be distressing, for it means losing the nostalgic smells and flavors that so enrich an ordinary existence. Among disadvantages of the anosmic state are the inability to detect hazardous gas leaks, pesticides, smoke and rotten food, as well as the loss of pleasurable overtones when eating or while smelling the flowers. After all, the enjoyment of fine food and wine is mostly in the volatiles. That is why the temperature at which food is served makes such a difference, and why *oenophiles* (lovers of the fermented grape) are so fussy about using properly shaped wine glasses and cooling their wine just so. On the other hand, an anosmic individual does not have to put up with annoyances such as a freshly tarred roof or a poorly ventilated bathroom, or the many city smells that make country living seem so sweet.

We Seek Pleasure

On this approach to your special senses, we have not yet explored beauty and pleasure, nor ugly and annoying. But it is those commonplace interpretations of your ongoing sensory input that eventually guide your behavior. Thus an enjoyable taste, smell, sight, sound or touch whets your appetite for more of the same. And it should no longer surprise you that those sensory inputs most likely to give pleasure also tend to engage you in reproductively advantageous behavior. Such behavior may include anything from better nutrition to improved socialization with the group to actual sexual and reproductive activities. This generally favorable relationship between beauty, pleasure, survival and reproduction only breaks down under conditions that your ancestors were unlikely to encounter—such as the purified neuroactive drugs that bypass sensory input to alter brain function directly, or the abnormal technologically created situations involving sensory deprivation or sensory overload (a rock concert, perhaps). While such chemicals or conditions may give pleasure, they need not improve survival or bring consistent reproductive advantage. Certainly you tend to avoid the ugly, scary and annoying, such as food that tastes bad, or excrement, decaying flesh and volcanic gases. You probably find the sight of open wounds and bleeding very distasteful, especially when the injured is nearby or happens to be a relative or friend or even yourself. And loud noises over which you have no control can be extremely irritating, as can excessively hot, cold or continuously wet skin.

Yet no matter how much you might wish to shut out that sort of obnoxious sensory input, you cannot. It is not safe to be out of touch, even when the input becomes strong enough to damage your special senses. Ordinarily the wide gap between barely detectable and potentially damaging inputs allows you sufficient time to flee. Interestingly, the perceived intensity of a stimulus does not relate in linear fashion to the actual strength of the signal. Rather, it tends to vary as some power function of the actual stimulus intensity. So you judge the loudness of a sound to increase more slowly than its actual signal intensity while the apparent strength of an electrical shock rises very rapidly (with the square or even cube of actual intensity). But it is also important that many benign sensory inputs provide pleasure. For pleasure makes it more likely that you will stay tuned to the information that surrounds you, which encourages acquisition of additional useful information with less effort. Clearly if you used those receptors only at times of obvious opportunity, trouble or danger, you might suffer adverse consequences due to missed signals or inevitable delays while tuning in.

Tastes Vary

There are innumerable examples of beauty being in the eye, ear, nose or mouth of the beholder. People vary, tastes vary and the impressions created centrally by sensory inputs also are subject to ongoing readjustment based upon need and past experiences. Perhaps much of this complexity and variability brings little survival advantage under ordinary circumstances but at least it keeps most of us from being irresistibly attracted to the same special individual for a mate. In that way too, variability tends to maintain itself and increase, even when it brings little advantage other than perhaps raising the percentage of available resources that any population is willing to utilize. Of course, without variability there would be no choice, no survival of the fittest, no evolution and no you.

Some Inputs Are More Important Than Others

You see in red, green and blue. Superman's X-ray vision allegedly gave him great advantage. Military infrared goggles encourage all-night fighting. Small bats echolocate during flight for guidance and the capture of insects. Large bats and almost all birds rely mostly on vision rather than ultrasound to direct their flight and help procure food. Nocturnal birds have a better sense of smell. Some flowers put out lots of scent. Other flowers and even lizards may be bright in ultraviolet but not visible light or vice versa. Why so much variety? Why don't creatures and plants just do whatever is best and compete on that basis alone?

Once again, it is not that simple. Certain inputs or outputs may only bring benefit under some circumstances. Echolocation is energetically expensive except during flight—when breathing already is forcibly coordinated with wing strokes. Furthermore, any terrestial creature that emitted strong echolocation pulses would generate lots of confusing echoes from nearby while also revealing its own location to others. Although they cannot echolocate, birds and large bats do have good visual systems and effective central processing. By giving up some visual capacity, those birds and large bats could devote more central processing to echolocation. But having only intermediate levels of both skills would be reproductively disadvantageous. Clearly it is risky to switch away from something that works, especially with bird brains so strictly constrained in size by the rate of gas diffusion through their egg-shell pores (larger pores would allow intolerable water losses—yet another natural balancing act). And if certain flowers find seasonal success via pollinators they attract and those same pollinators find these flowers the most rewarding, both may then evolve together into a codependent unit that others with different skills, tastes and characteristics are unable to join. Many sorts of wild berries are poisonous to you but one assumes that some creature out there gains reproductive advantage by spreading their seeds as well.

Your own color vision is not just a matter of esthetics either, for the survival of your ancestors was enhanced by having photoreceptors sensitive to green—the color most reflected by chlorophyll and therefore the most universally present visible light frequency on the heavily inhabited lands of our plant-powered planet. So all rod cells and about one-third of the cone cells in your eyes detect green better than any other light frequency. A separate set of cone receptors especially sensitive to red makes sense as well, for red happens to be the color best absorbed by chlorophyll (as well as by Earth's seas—those seas are blue because unlike chlorophyll they absorb blue light very poorly). And just as anything reflecting more red than chlorophyll will stand out in marked contrast, having the best reception of your blue-sensitive cones located just to the far side of green from red further reduces the opportunity for your predators and predatees to avoid detection among all the green grass and leaves. In other words, your special photosensitivities optimize contrast within the predominant *visible light* portion of the electromagnetic spectrum that best penetrates Earth's atmosphere from outer space. A purely green background is rarely encountered by larger-than-leaf-sized creatures, so larger animals generally seek camouflage in colors other than green. As for infrared vision, that might not have helped your earlier ancestors underwater where all surfaces are uniformly cool, nor those who later emerged on land in the tropics where everything is hot (except at night when your ancestors mostly slept or huddled in fear rather than roaming). Because ultraviolet light is quite well blocked by the ozone layer, any cone cells in your eye that detected ultraviolet light would not see nearly as well as those sensitive to visible light frequencies—presumably those flowers that are brightest at ultraviolet frequencies have evolved in co-dependent fashion with particularly ultraviolet-sensitive insects.

All electromagnetic energy travels at the usual speed of light. So the more frequent the waves, the shorter the wave length of that higher frequency radiation and the more energetic each photon. Indeed the energy carried by gamma ray and X-ray photons can alter atoms and molecules sufficiently to cause biological harm. Fortunately these high-energy photons are usually blocked by air molecules while still passing through our thick atmosphere from outer space. Thus higher frequency electromagnetic radiation had little relevance for the vision of your ancestors. Similarly, microwaves (radar) and radio/TV frequencies were uncommon on Earth until recently. But even if radio waves had been encountered regularly from the time of life's beginning, they still would not have been useful frequencies for animals to detect. After all, when a single radio wave may be as long as a football field, you can hardly expect it to reflect or even be distorted by some object as small as a lion or rhinoceros. So if you could see at radio frequencies, you might get the big picture but the small stuff (other animals, rocks, trees) would be invisible and therefore more likely to get you. Furthermore, while visible light is conveniently subcellular in wave length, you might find it impractical to run about with antennae or satellite dishes sprouting from your forehead in the fashion popularized by (extinct) Irish Elk.

For similar life-and-death reasons, humans are most sensitive to air-conducted sound vibrations in the 1000–4000 *hertz* (1–4 KHz or 1–4 thousand *cycles/second*) range, although young adults can detect sounds at 15–24 K Hz (and loud inputs at higher frequencies up to at least 90 K Hz are even detected by some deaf adults via bone conduction—perhaps through pick-up at the saccule or another inner ear structure). Sound of any frequency ordinarily travels at about 350 meters/second (1150 feet/second) in air. Thus your best heard wave lengths are about $\frac{1}{10}$ to $\frac{1}{3}$ of a meter long (3–12 inches), a range that includes the distance between your ears. That wave-length-to-head-size relationship holds for all mammals. So elephants, rhinos and whales converse at inaudibly low (to you) frequencies while mice chat about you and also listen at far higher frequencies than you can detect. Of course, owls and cats had better hear up to 50 KHz in order to accept a mousy invitation to lunch. Furthermore, barn owl ears are asymmetrically mounted (left higher and pointing down, right lower than eye level and pointing up) in order to improve vertical correction for sound intensity when flying in total darkness. Although mice and elephants hear best at wave lengths comparable to their skull width, they also squeak or roar (and some whales even sing) at frequencies that you can hear.

A breaking twig or light footstep puts out a wide range of sound frequencies. Thus the usual problem is not which frequency to monitor because it happens to carry critical information for you. Rather it is which wave length provides the most information for your particular head size (or the most easily detected pressure changes in your swim bladder, if you happen to be a fish bearing one of those multi-purpose air-filled sacs). Except for the very energetic and focused echolocation sounds emitted (at up to 120 K Hertz) by bats, the higher frequencies generally emitted by smaller creatures transmit poorly through air. But small animals generally pose less danger to larger creatures than to others in their own size range. So larger animals need not monitor these higher frequencies for defensive purposes. Nonetheless, being able to hear over a wide range of frequencies is clearly advantageous. The concentrated beam of sound and far-higher-than-usual frequency of bat or dolphin sonar is particularly useful when targeting lunch since it produces an easily reflected wave length well below target size. But wave lengths comparable to your head width are usually most helpful for locating the sound source in three dimensions, even though you and/or the source may be in motion or out of sight of each other. For your head partially blocks incoming sound, especially at head-size or smaller wave lengths. The resulting sound shadow can provide information about where the sound source is located, especially if the sound has not been reflected sufficiently to cause misleading echoes. However, your ears must first funnel (gather and concentrate) those passing waves of air compression and rarefaction. Then your sound-detecting cells can report centrally on which ear first perceived the sound wave as well as how that incoming wave was distorted by your immobile funny-shaped (non-symmetrical funnel) ears.

Even when a sound reaches both ears at once, you can usually tell by its directional distortion whether the sound source is in front, above or behind you. And while the sound persists, you can confirm its origin with a slight turn or tilt of your head to reduce sound pressure on one ear and further delay that ear's reception. When you change sound intensities and arrival times at your two ears in this fashion, you also detectably alter sound-wave *phase* differences between the two sides (sound waves alternate maxima and minima just like water waves, except that sound waves cycle between higher and lower air pressures). Admittedly all of this seems even more technical, complicated and amazing than your first bicycle ride—but your hearing skills actually developed in the same way that you learned to ride a bike—by simple trial and error. Fortunately for you, your long-suffering and naturally selected ancestors already had evolved ears and bicycles. Your simpler task was to figure out how to use them (without or despite any instructions). And before long those detectable differences in sound intensity and arrival times at your two ears were helping you to visually locate the probably huge source of those early cooing sounds (undoubtedly some doting relative).

Your internal comparison and evaluation of sounds received by each ear requires the simultaneous analysis of a great many different inputs. Somehow that allows you to detect just a few microseconds difference in sound arrival—far less time than even the fastest neuron takes to depolarize. Of course, you need not respond within microseconds, you merely have to differentiate between sound wave arrival times on that scale. This far simpler task could have been achieved in many ways. For example, a slight priority of identical inputs from one ear or the other might alter the resting membrane potential of certain centrally located neurons. Perhaps a right ear impulse of a certain frequency could cause one of those neurons to depolarize unless the left ear signal from a comparable receptor arrived first to cause hyperpolarization (or something like that). But as it turns out, your brain seems to contain *coincidence-detecting* neurons that respond specifically to matching phase inputs from both ears. Such a neuron would be farther from the ear that heard the noise first, as determined by the conduction speed of the axon carrying that report centrally. The advantage of such an array of neurons ready to report coincident impulse arrivals at their location is that each neuron can directly represent a position in space from which the sound seems to arise. *Visual disparity neurons* serve a similar function in animals with binocular vision, reporting on specific differences in viewing angle between the two eyes and again localizing the object's position relative to you.

In other words, it seems likely that specific nerve cells are chosen to respond by the length of delay between signals arriving at your right and left ears. The firing of a *characteristic delay cell* can then be most vigorous when the sound arises from a particular place relative to the head—creating an audio-spatial map to help the brain localize sounds. Similarly, different nerve cells respond to different visual angles, allowing binocular vision to provide a three dimensional map comparable to that based upon binaural hearing. Obviously, it is not necessary for this entire complex mechanism to have been created at once. It could far more easily have evolved through many intermediate individually advantageous steps to reach the present level of complexity and acuity about which we still have so much to learn.

It Has Always Been Harder To Listen Than To Talk

It required increasing intelligence for your early branch-swinging and later cave-dwelling ancestors to determine the significance of each others grunts and howls as these slowly became modulated by greater information content. Quite obviously, talking and listening must have evolved together. And surely that prolonged progression involved a great deal of body language with simultaneous visual signals supplementing and confirming verbal meanings. Indeed, some people still say as much with hand and face movements (even over the telephone) as with their voice—but that does not prove they are less evolved than the rest of us.

Your Hearing Machine

Your flexible but immobile external ear concentrates sound waves onto a delicate *tympanic membrane* that stretches across and seals the narrow inner apex of your ear canal. Being located around a bend, that *eardrum* is well out of range for your exploring fingertip. The outer eardrum and external ear canal are surfaced by stratified squamous epithelium—the canal also contains wax-producing glands that help your ear canal to shed water and perhaps immobilize or otherwise defend against inquisitive insects. Adult human males sprout varying numbers of hairs on their external ears (women cannot compete on that trait since the hairy ears gene is on the Y chromosome—see Reproduction). A tiny air-containing *middle ear* cavity is located on the medial (inner) side of each eardrum. Each cavity is lined by a single-cell layer of secretory epithelium and drains into the nasal part of your throat via an *eustachian tube*—those tubes allow you to equalize middle ear and atmospheric air pressures. Air travelers suffering from a cold or allergic sniffles can develop painful earache and some temporary loss of hearing during descent due to obstruction of one or both eustachian tubes. Often these symptoms clear as the ear "pops" (admits air—which moves the ear drum) when a later yawn successfully reopens that eustachian tube. Your middle ears readily release air with little pops as you walk, drive or fly to higher elevations (lower atmospheric pressures). The eustachian tube is short and straight during childhood, which allows throat infections easier access to the middle ear. Children often encounter new-to-them infectious organisms so they frequently have swollen *adenoids* (lymph glands) compressing the entry of their eustachian tubes into the throat.

Earaches associated with middle ear infections and blocked eustachian tubes usually signify pressure differences across the eardrum. Inflammatory secretions accumulating during bacterial infection may bulge the eardrum outward. On the other hand, a blocked eustachian tube cannot replace air being absorbed into your circulating blood—the resulting subatmospheric pressures draw thick fluid from middle-ear secretory surfaces while also retracting the eardrum medially. In the middle decades of this century, most children with recurrent earaches underwent routine tonsillectomy (tonsils are large lymphoid structures at the back of the oral cavity, one on each side) and adenoid*ectomy* ("-ectomy" means removal of)—but antibiotics are now the primary therapy for middle-ear infections. When persistent eustachian tube blockage causes hearing loss in childhood, a small hollow tube inserted through a thin part of the *tympanic membrane* (eardrum) can allow ongoing equalization of air pressures until inflammation subsides and the eustachian tube reopens. Small holes of this sort usually heal swiftly without residual damage after that flared-tip tube falls out or is removed. Having atmospheric pressure on both sides of your tightly stretched tympanic membrane permits incoming air compressions and rarefactions (sound waves) to freely flex that eardrum in and out. Every tiny displacement of the eardrum is amplified by a series of three little interconnected bony levers *(hammer, anvil* and *stirrup)*—the footplate of the stirrup vibrates the small membrane-covered *oval window* that separates your air-containing middle ear from your fluid-filled *inner ear.* Your eardrum has a surface area 25 times larger than that of the oval window so those three little levers crossing the middle ear vibrate the oval window about twenty-five times harder than the eardrum was moved by incoming sound. Two tiny skeletal muscles can somewhat stabilize your tiny middle-ear *ossicles* (bones) against anticipated strong noises but unexpectedly loud sounds exert their full impact on the delicate inner ear.

An eardrum is readily ruptured by rapid one-sided air pressure changes (by overly-swift descent when scuba diving or a flat-handed hit on the ear or strong pressure waves from a nearby explosion) or by the fluid pressure that slowly builds within an infected middle ear. Rupture of an eardrum by injury or infection reduces hearing about 25 decibels—a handicapping hearing loss since sound wave pressures must then be over an order of magnitude stronger for detection at the inner-ear windows. Stratified squamous ear-canal epithelium may grow through a widely perforated eardrum into the middle ear. This soon leads to an accumulation of cast-off skin cell debris (wet dandruff) within the middle ear cavity—left untreated, such chronically infected material may erode middle ear ossicles and adjacent mastoid bone air-cells or even extend onto nearby brain. Ordinarily, however, your air-containing middle ear nicely separates and insulates those delicate fluid-suspended inner ear contents from the outside world. And while the occasional middle-earache can be painful, it is far less troublesome than the deafness and dizziness that may follow apparently minor inner-ear injuries or infections.

Inner Ear

The inner ear serves all vertebrates as an organ of *balance* (the *vestibular apparatus*) and *hearing* (the *cochlea*). Both functions rely upon groups of specialized *hair cells*. Each hair cell is capped by a *hair bundle* (graduated series of cell-membrane-enclosed hair-like extensions) based upon an actin framework that is narrowest at its hair-cell-surface site of flexion. Movement of the hair bundle in specific directions causes the hair cell to change its transmembrane voltage. A positive displacement tugs on tip-link fibers that open cation channels and depolarization follows. Displacement in the opposite direction closes the 15% of channels that normally are open, leading to hyperpolarization of the hair cell. Stimuli at right angles to these directions have no electrical effect. Thus a hair cell acts as a *transducer*, converting tiny mechanical hair bundle movements into electrical signals for your central neurons to interpret, just as a phonograph converts needle motion into electrical signals destined for the amplifier and speakers. (In older spring-driven phonographs, the needle bumping along in a record's grooves vibrated a thin paper or tin diaphragm directly—the tinny or nasal quality of the recording was little harmed when a worn diaphragm was replaced by the thin top of a food can. Of course, frequent slow-downs meant one of the younger listeners had to hand-crank regularly.) One mammalian hair-cell may connect to over twenty nerve cells. Presumably that copious innervation takes full advantage of the rapid response rate of hair cells. As with transducers based upon crystalline piezoelectric materials, the inner ear may emit click sounds spontaneously or when electrically stimulated, although the mechanism of this reversed electrical-to-mechanical transduction is not yet clear.

Balance

Each of your three inner-ear semicircular canals lies at a right angle to the other two. Thus any acceleration of your head displaces fluid (by inertia) within at least one of these canals. Fluid emerging from any canal stimulates the group of hair cells near one end of that semicircular canal by displacing the flame-shaped *cupola* (gelatinous mass) in which those cell hairs are embedded. Once your own movements have been sensed in this fashion, copious neuronal connections to and within the brain stem swiftly redirect your eyes so that they remain centered upon the lady or the tiger. Since you must take your own movements into account during both square dancing and mortal combat, your voluntary motor nerve cell discharges are simultaneously reported back to the sensory areas of your brain—allowing you to anticipate the effects of your own muscle contractions so that you can integrate the next moves of your eyes, arms and legs.

Ice water entering the external ear moves semicircular canal fluid in similar fashion since cooled (more dense) inner ear fluid sinks and displaces warmer fluid upward (just as cold milk initially settles through your hot coffee). This *caloric* stimulation of your hair cells is interpreted as a violently spinning acceleration, so you become dizzy and unable to maintain balance or direction (while your eyes move back and forth rapidly—a movement known as *nystagmus*). If such a loss of orientation should occur to you while swimming with one ear dipped in very cold water (as when doing the sidestroke), simply dip the other ear until things stop spinning. Then keep your head out of that cold water until you are safely back in a boat or on shore (the caloric response may explain why so many individuals who fall into cold water then seem disoriented and swim away from the dock or in aimless circles until they drown).

In addition to the three bunches of hair cells associated with your three semicircular canals, you have two other clumps of inner ear hair cell sensors reporting from the *utricle* and the *saccule*. A gelatinous mass—with many small calcium carbonate crystals embedded on the side opposite to the hair insertions—lies horizontally in one of these chambers, vertically in the other. Those rocks in your head weigh down their watery gelatinous mass so that embedded hair cells can detect changes in position with respect to gravity. Similar *otoliths* in fish show microscopic daily-growth rings that can provide information about chemical exposures, temperatures and nutrients encountered during that day (thereby indicating migration routes while reporting on pollutants encountered en route).

Hearing

The sixth and by far your largest collection of hair cells is located inside your *cochlea* (a snail-shell-shaped structure found within each tiny inner ear). Your 15,000 *cochlear* hair cells sit upon a curving *basilar* membrane with their hairs embedded in a *tectorial* membrane. Each cochlear hair cell is tuned to a specific sound frequency by the length of its hairlike appendages. The basilar membrane becomes looser at its narrow upper end, thereby exposing auditory hair cells farther up the cochlea (away from the oval window) to progressively lower frequencies of sound (looser membranes and guitar strings vibrate more slowly). Your high-frequency hair cells near the oval window are first to die from the noisy input of rock concerts and jet engines. (Aboriginal populations not exposed to the noises of our civilized world often retain acute hearing into old age. And recent experimental evidence suggests that hair cell regeneration may someday become possible with appropriate local applications of cell growth factors.) A separate membrane-covered *round* window bulges toward the middle ear each time the stapes pushes the oval window toward the inner ear. That pressure relief allows easy passage of each sound (pressure) wave through your incompressible inner ear fluid from the oval window up one side of the spiral cochlea and down the other to the round window. Since both oval and round windows face back into the middle ear, a widely perforated eardrum should bring sound waves to both membranes simultaneously (although the stirrup would still affect oval window motion).

Cells Surrounded By High-Potassium Solutions Must Accumulate Sodium Ions

Your entire inner ear has a unique double-fluid-layer suspension that reduces extraneous bone-conducted vibrations (from chewing, head scratching and so on) while also damping flow noises from blood circulating nearby. Thus the delicate and intricate bony outline of the inner ear is separated from its identically shaped membranous contents by a thin surrounding layer of *perilymph*. Being ordinary extracellular or tissue fluid, perilymph is high in Na^+ and low in K^+. In turn, the inner-ear membranous structure is completely inflated by *endolymph*—a solution that resembles *intracellular* fluid by being high in K^+ and low in Na^+. Your hair cells live within and are nourished by that endolymph so they have had to reverse the intracellular/extracellular cation ratios maintained by other body cells. Since hair-cell cytoplasm is high in Na^+ and low in K^+, hair cell depolarization is associated with an inrush of K^+ toward that high-in-Na^+ intracellular fluid. Most likely it was not difficult to reverse the transmembranous ATP-driven Na^+/K^+ exchange pump under these unusual circumstances (in order to produce and maintain high intracellular Na^+ levels) since hair cells could not function without that reversal. So any construction defect that led to accidental pump reversal here would persist because it brought reproductive advantage while elsewhere such a defective and non-functional cell would vanish without a trace. Once again we see how certain spontaneous variations in modular constituents such as ion pumps and transmembrane channels may find beneficial applications under altered circumstances. And how cell membranes must separate fluids of different ion concentrations if those membranes are to have any useful purpose (see previous chapters).

CHAPTER 16

IN YOUR EYE

Vision Dominates Your Internal Display Of Reality;... Two Eyes Can See Faster Than One;... A Pinhole Camera Makes Sharp Images;... Pinhole Dynamics Depend Upon Slivers Of Bright Light;... Fingerhole Cameras Require Lenses To Bring Incoming Light To A Focus;... Refraction Of Incoming Light;... Light Can Be Focused By Refraction;... Focusing On Nearby Objects;... This Rainbow Is A Sign That A Flood Of Different Frequencies Has Been Refracted;... Darwin Was Puzzled;... Following The Light Path Through Your Eye;... Keeping An Eye Full;... Accommodation;... Your Retina;... Right Brain Sees To The Left And Left Brain Sees To The Right;... Your Retina Is Not Very Well Organized;... Making A Spectacle For Yourself;... Fish Eyes;... The Retina In More Detail.

Vision Dominates Your Internal Display Of Reality

People ordinarily believe what they see, even though seeing often leads to inaccurate or grossly incorrect internal displays of reality. And when all senses are involved, we usually trust our visual input far more than our senses of smell, taste or hearing. So if you eat something that resembles a rich dark red steak but find that it smells and tastes like fish, you will probably conclude that there was a storage or preservation problem or that you are eating seal meat, but clearly, despite the smell and flavor, it is meat and not fish. Similarly if it sounds like an elephant coming down the stairs but it looks like one of your favorite relatives, you will not be tempted to shoot. Normally, of course, your various senses tend to enrich and fill out your internal display of reality rather than disagree with one another.

Two Eyes Can See Faster Than One

Contributing to the speed and accuracy with which you evaluate a scene is the major overlap of your two visual fields. Early comparison of two separate eye inputs may allow those using both eyes to interpret a scene and develop an appropriate response more rapidly than when they have one eye covered. Furthermore, each eye performs a considerable amount of preliminary processing, filtering out or enhancing a great deal of the raw data arriving at 100 megabytes per second before the resulting more relevant input (with reduced noise and redundancy) is referred centrally. Within your brain, this information stream then diverges to allow separate analyses of positions, outlines, movements and colors. The outcome of that intermediate evaluation then creates the what and where of your next fleeting moment of visually defined internal reality. But before looking more deeply into your eyes, let us review how a pinhole camera works and how light can be focused. That may help us to see how the eye could have evolved.

A Pinhole Camera Makes Sharp Images

The pinhole camera is an internally blackened (to avoid reflections) light-tight box. A light-sensitive film is held against the back wall inside. There is no glass or plastic lens, just a covered needle-sized opening (measuring about 0.4mm in diameter) passing centrally through the camera's front surface. If that pinhole is momentarily uncovered and all goes well, both the rich man and his camel will pass easily through that eye of a needle to expose the film, leaving a clear upside-down left-right-reversed likeness of themselves. When a baseball player faces the camera with ball in right hand and glove on left, his image too will reach that film upside down. Fortunately, the ball and glove remain attached to their same (correct) hands—simply turn the exposed and developed film right side up and view it from the exposed side (but not through the back of a transparent negative film, for that would reverse left and right as in a mirror image). Apparently ball and glove have switched sides in the same fashion as head and feet. How is this possible?

Pinhole Dynamics Depend Upon Slivers Of Bright Light

In order to expose the film and thus be visible, every tiny part of the scene (which happens to contain no lights or mirrors) and ballplayer must be illuminated sufficiently to become an independent source of *reflected* light. One can therefore view light reaching the camera as tiny segments from sequentially expanding spherical surfaces of light being reflected by each point of the scene. Since the surface area of any sphere increases with its radius squared (here radius equals the distance between source object and camera) and the total output of light from an object is unrelated to the distance from which it is viewed, a camera twice as far away from an object will receive only one-fourth as much of its reflected light. And because a more distant object has a smaller percentage (or angular dominance) of the camera's field of view, that distant snow-covered mountain peak could appear smaller and less bright than the shiny new baseball nearby. As mentioned, the baseball in the player's right hand reflects light in all directions, even toward the cheapest seats in the bleachers. So almost all of the baseball-reflected light misses the camera—only the tiniest bit of it is properly aligned to pass through that pinhole. And with the player centered in the scene, that sliver of light from the player's right hand will pass through the pinhole heading toward the player's left. Similarly, a tiny bit of all light reflected from the left-hand glove passes through the pinhole toward the player's right. And those shiny white teeth (this player does not chew tobacco) send slivers of light angling downward into the camera, while his once-white socks send their light slivers through the pinhole toward the top of the film. If there is sufficient light and nothing moves, both the nearby scene and its distant background will remain in focus no matter how far back from the pinhole the film has been placed. But the farther back the film, the more those light slivers spread to form a *magnified* (larger) but less clear and intense image. Of course, total light entry depends only upon overall brightness of the scene and how long the pinhole remains uncovered, regardless of film position. The illumination can be increased by waiting until the sun comes up, or by turning on more lights, or by discharging well-timed flash bulbs somewhere behind the pinhole (in order to illuminate the near surface of objects close by) while the pinhole is uncovered. Flash bulbs are especially helpful when an image would otherwise be too faint or where the pinhole can only be uncovered briefly. For example, during a longer exposure, the moving ballplayer might travel across the scene, thereby blurring out his image.

Fingerhole Cameras Require Lenses To Bring Incoming Light To A Focus

It seems logical that "not enough light for a good picture" could also be overcome by enlarging the pinhole to something more generous—say a circular opening 10 millimeters in diameter. Indeed, when it really is important to capture some dimly lit scene, why not momentarily uncover the entire front of the film? After all, a much larger camera opening intercepts far more of each expanding sphere of light which should brighten the image in the same fashion as moving the pinhole camera much closer to the action. Unfortunately, a film exposed through a simple larger-than-pinhole opening turns out to be a total smear. That occurs because light from each spot in the scene—baseball, teeth, socks, grass—that passes through such an enlarged hole now overlaps across much of the film, thereby exposing the entire film more or less uniformly. So when developed, the film exposed through a

10mm fingerhole, or another film that was momentarily uncovered while facing the scene, will be relatively uniform in shade or color, revealing no objects let alone details. Even so, every part of that film did receive far more light than would have passed simultaneously through a similarly placed pinhole. And that increase in light represents a much larger (though still minute) fraction of all light reflected from every object in the scene.

It seems that fingerhole camera pictures might be useful if we could somehow sort out the additional light slivers entering that fingerhole and redirect them back to where each tiny sliver of advancing wave front would have hit the film in our pinhole camera. That apparently complex task can be simplified if the fingerhole is round, but before we consider how light is focused by a circular glass lens, we had better review the more general topic of how light can be bent and redirected.

Refraction Of Incoming Light

To simplify this exercise, let us assume that all objects to be imaged are located more than 20 feet from the camera opening. That means photons reflected from the target area will approach in essentially parallel fashion—or you could view those incoming spherical wave-fronts of light as almost flat surfaces more or less parallel to the camera front. But regardless of whether you prefer to view approaching photons as parallel rays or arrows or as a series of flat wave fronts, all light penetrates clear glass or water more slowly than it travels through air. So visible light that does not enter a glass or water surface perpendicularly will slow more on the side that first enters the denser medium. Thus light photons encountering a clear glass or water surface at other than a right angle are *refracted* (their direction of travel is altered—that is why a straight but partially submerged straw seems bent or displaced at the surface of your soda). You experience a comparable diversion when your right front car wheel wanders off the paved road. As that right side is slowed by snow, sand, grass or water, it will pull the car to the right (unless you can skillfully overcome this deviation). On the other hand, if your car were to strike a strong guard rail at a minimal (glancing) angle, it might simply bounce back onto the street. Some of the photons approaching a water or glass surface are similarly *reflected* (bounced away). A mirror is simply a super-strong guard rail that reflects all photons regardless of the angle at which they strike. And it is possible to curve a guard rail or mirror so that it will bounce all parallel, incoming, perfectly elastic cars or balls or photons back toward a single point where they can be collected. Even a small properly curved mirror can easily focus enough sunlight at one point to start a fire—another gentle reminder of solar-driven life's precarious balance between fire and ice.

Light Can Be Focused By Refraction

A transparent glass camera lens can also be curved in such a way that all light collected from a single object (e.g. the baseball) is redirected by its passage through the lens to produce a sharp image of that object on the film. Such a refractive gathering of parallel (or even diverging) light rays is readily accomplished by a doubly *convex* lens (one that bulges on both surfaces). Let us now sample a few light rays from the target scene directly ahead as they pass through that glass lens. In the simplest case, a light ray strikes the exact midpoint of the convex lens and proceeds perpendicularly through the glass. So with parallel incoming rays, the center point of the symmetrically curved front-of-lens surface becomes equivalent to the former pinhole, as light passing through it will not change direction. However, all incoming photons from the target that strike our convex lens surface off-center must enter that glass at an angle other than 90°, since the center is the high point of that lens. And since such off-center photons are initially slowed on their first-to-touch central side, all off-center photons are redirected slightly toward the lens center. Photons passing straight through centrally will not deviate as they escape from the symmetrically convex back surface of that lens into air. However, all light rays emerging at a non-perpendicular angle from the bulging back face of the glass camera lens will first enter air on their peripheral side. That extremely brief center-sided drag (longer glass path) again causes each off-center photon to deviate centrally as it speeds up on escaping from the glass lens. In this fashion, a lens with proper convex curvatures can concentrate all entering and essentially parallel rays of light from any part of the baseball onto the same spot of film that a mere sliver of that light would have hit after passing through an identically centered pinhole. But a lot more light has passed through this larger glass lens so the image is much brighter (providing better detail with a far shorter exposure time).

As with the pinhole, light passing through the lens from the right side of the scene exposes the left side of the film and the image will be upside down as well. Unlike light passing through a pinhole, however, the light gathered through a lens creates a clear image only at the one point where it comes into focus. Before or behind that focal point, the projected image of an object remains blurry. Furthermore, by collecting more light from a narrower field of view, a lens focusing on a distant object creates an image that seems much closer than it would appear within the wider-view pinhole camera at the same distance. Once again, even doubling the diameter of the tiny pinhole and inserting a tiny lens allows one to collect four times as much light during the same brief exposure—equivalent to locating the camera only half as far from the object. Note that we have not considered the apparent size of the image on the film, as we can easily expand or shrink that image by altering the lens focus while at the same time moving the film away or toward the lens. However, we obtain a sharp image only as long as the film remains precisely at the focal point of the camera lens. And if we wish to make an enlargement for better evaluation of detail or so the image (or some part of it) seems closer, the grain size of the film and lens quality as well as original illumination will determine how much magnification can be achieved without blurring the image. Additional complex computer enhancement methods can slowly deblur an image by increasing contrast while also filling in what ought to have been there. Your eye continuously performs that same trick on a real time basis. For example, whatever dominates the adjacent image at any moment is routinely expanded to fill in for information not received from the blind spot of each retina (see below).

Focusing On Nearby Objects

Light rays reaching your camera lens from nearby objects are still diverging noticeably (rather than being almost parallel). So in order to focus the light from nearby objects with the usual (distance) lens, the film must be moved back from the lens (to allow those diverging light rays a longer post-refraction path over which to come together). Of course, light rays from very nearby objects may continue to diverge even after refraction by the usual camera lens if that lens was not strong (curved) enough to bend them back into focus (the image of that nearby object then remains blurred at every film location). On the other hand, a blurry background results when the camera has successfully focused onto a nearby baseball, for if those diverging baseball rays are sufficiently bent to reach a focus at the film, all parallel light rays from the background must already have come together and then spread out again before striking the film.

This Rainbow Is A Sign That A Flood Of Different Frequencies Has Been Refracted

A prism separates white light into its constituent colors. Of course, there really is no such thing as a color. Colors just happen to be the graphic and effective way that you picture various wave lengths of visible light. So by stimulating various photoreceptors in accordance with their sensitivity to that specific wave length, any visible light frequency will be recognized and displayed internally as light of a particular color. Most of us would insist that colors really exist—or at least that they provide real information about a real object that (like any other attribute of such an object) might be subject to change. And most of us seem to experience colors in a relatively similar fashion. Or at least we have learned to label certain frequencies as red, green or blue and so forth whenever those particular frequencies show up on our internal reality display. But electromagnetic frequencies differ markedly in how readily they penetrate clear glass. Even within the narrow frequency range of *visible light* (which has a wavelength of 400 to 700 *nanometers* or billionths of a meter), the longer waves (carrying less energy and thus making fewer zigzags within a given distance) are refracted less than shorter light waves as they enter and leave clear glass. Therefore any spot of incoming light will spread into an orderly *spectrum* (consecutive display of all component frequencies) on passing through the non-parallel flat glass surfaces of a prism. Just as sunlight passing through certain size water droplets (on the side of you away from the sun) is naturally spread into a rainbow projected onto droplets behind them. In addition, a colored halo (almost rainbow) may surround the image of a bright point-sized light source such as a star after its light has passed through a single lens or inexpensive lens system. That *chromatic aberration* can be eliminated by passing the light through matched corrective lenses of proper curvature and composition.

Darwin Was Puzzled

Before we apply some of this optical information to the human eye, let us consider how a complex mechanism such as your eye could possibly have evolved. Darwin puzzled a great deal over that. He could not imagine how this delicately adjusted optical system with so many critically interacting components could possibly have come about gradually. For had the eye evolved over a long time as he suspected, it would have had to pass through an endless number of intermediate stages. However, each of those many stages could only persist and spread if its particular modification of the far more primitive eye was reproductively advantageous in comparison to the extant form. Yet it seems intuitively obvious that any partially evolved eye should bring little benefit since even minor modifications or injuries so easily disable our own fully developed eyes. But fortunately, as Dawkins has demonstrated, simply knowing that evolutionary progress is incremental and cumulative allows one to figure out how the fully developed eye must have come about.[1] Indeed, knowing from whence one came often identifies the path actually taken and shows why the path not taken was selected against. Furthermore, any likely (or even unavoidable) event can be made to seem most improbable if one focuses overly much on details. Thus one might forget that all roads formerly led to Rome if sufficiently distracted by the number, variety, source and style of cobblestones encountered along the way. Or one might recklessly conclude that an arrow

---

[1] Dawkins, Richard. 1986. *The Blind Watchmaker.* New York: Norton. 332 pp.

traveling at finite speed could never reach its target if advised that its flight path was divisible into an infinite series of points, each of which the arrow would have to pass in turn.

Still it is fair to say that only the most extraordinary series of unlikely coincidences could ever account for the fact that you, of all people, are sitting in just that chair, wearing those exact clothes and scratching that particular place while reading this fascinating paragraph at this very moment in the history of our Universe. But this too has come to pass—although mathematically speaking, it is so extremely improbable that it ought never to have occurred. Indeed, had anyone predicted all of these details some years ago, they might have been viewed as indulging in magic and perhaps burned at the stake (see Metabolism). On the other hand, even the most average and mathematically disinclined dolt of one thousand years ago would not have been astounded at the sight of you scratching, or by books and clothes and chairs. For clearly these were probable (rather than miraculous) developments over the following thousand years. Such a common sense approach to current events makes the pathway along which the eye must have developed seem almost inevitable. Indeed, the eye of an octopus appears to have progressed independently along a very similar route.

As a point of departure, remember that light sensitivity must be as ancient as life itself, having developed before the first photosynthetic bacterium absorbed its first photon nearly four billion years ago. Furthermore, any particularly light-sensitive surface spot might bring real advantage to bacteria, protozoa, algae or multicellular organisms seeking to remain near the water's surface where food was most plentiful. And if such a photosensitive spot happened to be located in a surface indentation, that dimple might sufficiently improve information about light direction to help "Dimples" become queen of all the beasts (because of her increased efficiency in moving toward the surface and avoiding predation). Of course, her advantage would only last until some more progressive multicellular descendent came up with an even deeper dimple.

Thus deeper-dimple-with-a-photosensitive-bottom probably ruled until almost-covered-deeper-dimple, having accidentally invented the pinhole camera concept, became first to clearly image the world. Her reaction was not recorded but we can hope she found it good. In any case, clear-covering-over-pinhole was perhaps the next big advance, a covered pinhole being less easily obstructed by debris. Sudden advantage then came to those with a larger aperture (opening) when an unusually concentrated protein solution enclosed within that primitive orb just happened to crystallize. The resulting clear glob of protein (consequent to accidental local overproduction of some easily crystallized enzyme) most likely took on a relatively spherical shape as it rolled about within the eye, bringing the world in and out of focus. Presumably a relatively spherical pre-lens with focus dependent upon position eventually gave rise to the modern spherical fish-eye lens with its attached fibers that move the lens back and forth to focus light at the photosensitive retina. There must have been an endless succession of other helpful mini-modifications before your own wonderful eye lens finally became available with its perfect shape and appropriate index of refraction. So it came to pass that those creatures with minimal chromatic aberration saw the light most clearly and ruled more or less wisely over the rest.

Following The Light Path Through Your Eye

Light *photons* must penetrate several different layers in order to activate *photoreceptors* in the *retina* at the back of your eye. The first and outermost layer is your clear *cornea*. This principal refractive element of the eye is surfaced by a layer of stratified squamous non-keratinized epithelium that is 4-6 cells thick. Between those *epithelial* cells and the single-cell-thick *endothelial* lining of the inner corneal surface lies a thin tough crystalline collagen-and-proteoglycan matrix that remains transparent only because of its very low water content. In contrast, tiny droplets of trapped water are primarily responsible for the light-reflecting quality of your *sclera* (the tough white cover that surrounds your eye except at the cornea). Colloidally trapped water similarly accounts for the lustre in your pearls, which can dry out and be ruined during prolonged storage in a low-humidity bank vault—another good reason to flaunt those pearls against your moist skin. Anyhow, that thin endothelial cell layer is responsible for keeping your corneas dehydrated—the active pumping of water from those collagen layers also requires the enzymatic breakdown of large extracellular water-loving proteoglycan molecules such as hyaluronic acid (so hyaluronidase is released into your corneas under the developmental influence of thyroxin—see also Collagen and Hormones).

Superficial corneal *abrasions* (scratches) are rapidly resurfaced (covered over) by inward migration of nearby epithelial cells. More persistent corneal damage can result from significant injury, infection or inadequate access to oxygen—conditions occasionally related to the use and abuse of contact lenses among other things. Damaged or deficient corneal endothelial cells sometimes allow water to infiltrate corneal collagen bundles and proteoglycans, thereby producing milky-looking corneal opacities. Any persistent loss of vision due to corneal scar or fluid infiltration, or by corneal weakening with loss of appropriate curvature (a condition known as *keratoconus*), may necessitate central corneal replacement with clear healthy live full-thickness corneal grafts from a recently dead individual. Despite the usual major immunological differences between host tissues and donor cells, over 90% of such corneal grafts survive because the cornea has no blood vessel connections and therefore no direct exposure to circulating host cells or antibodies. Indeed, cornea resembles cartilage in the way its gas exchange and nutrition depend upon diffusion through adjacent fluids. And without exposure to host cells that can recognize donor corneal cells as foreign, there is little likelihood of a host immune response (see Immunity). The occasional impending rejection of a corneal graft can usually be reversed with *topical* (locally applied) *glucocorticoids* (see Hormones).

After passing through the cornea, light enters the clear nourishing liquid of the anterior eye chamber. This *aqueous humor* (humor means "fluid" in Latin) is produced by a *choroid plexus* (secretory cells on capillary tufts in the *ciliary body*) located farther back within the eye. "Used" aqueous humor is primarily removed by absorption across a trabecular meshwork into a vein (*canal of Schlemm*) that circles the junction of cornea and sclera—as well as through exposed iris and other cellular surfaces and directly into various internally exposed venules (your aqueous humor circulation resembles your cerebrospinal fluid circulation because your eyes originated as outgrowths of your brain when you were just an ugly little embryo).

The central *pupil* (aperture) in the pigmented iris muscle narrows to about 2 mm in bright light. Thus bright light reduces any refractive errors consequent to incorrect off-center curvatures of either your cornea or your clear, doubly convex *lens*—that pupillary constriction in bright light improves visual acuity by taking you back toward the always-in-focus pinhole camera concept of your ancient ancestors. In addition, a narrowed pupil directs all light onto your closely packed color-sensitive *foveal* photoreceptors (discussed below). On the other hand, an adrenalin-producing fight-or-flight reaction will maximally dilate your pupils to about 8 mm diameter (allowing 16 times more light to enter regardless of illumination) since full peripheral vision in and out of the shade is then more important than ideal retinal illumination or perfect focus. Ordinarily, however, pupil size is adjusted by autonomic nerve reflex according to light intensity.

The lens of your eye is held in place just behind its pupil by fine *suspensory ligaments*. These ligaments attach to a circular *ciliary muscle* that rings the anterior inner surface of each eyeball, close by the ciliary body. The clear flexible well-ordered *crystallins* (crystallized proteins) that make up each lens are produced within the eye in large amounts—often by the combined efforts of repeatedly duplicated genes. Elsewhere in your body these same proteins appear only in appropriately tiny quantities as enzymes, heat shock proteins and so forth. That portion of your eye behind the lens is referred to as your posterior eye *cavity* while your *anterior* and *posterior* eye *chambers* are both located in front of the lens with the iris as their common border. Your lenses yellow and become increasingly opaque during a lifetime of use (in part due to the increased cross-linkage and disorder of lens proteins that have reacted with glucose in a *non-enzymatic glycation*). An early sign of lenticular opacities may be fine radial light rays that form a halo about a bright object or light at night (an effect often attributed to a religious experience or dirty windshield). Eventually a lens may become so opaque (it is then considered a *cataract*) that it must be removed surgically and replaced by a clear plastic *prosthetic* (artificial) lens of appropriate curvature to restore useful vision. Life style and occupation often determine how much lenticular opacity can be tolerated—a long-haul truck driver might opt for cataract extraction earlier than a retired person who mostly watches television and avoids driving at night.

Keeping An Eye Full

The jelly-like *vitreous* (or "glassy", as opposed to *aqueous* or "watery") *humor* fills the posterior eye cavity behind the lens. This relatively permanent posterior eye stuffing consists of collagen fibers stretched out within a clear gel. Along with your ongoing secretion of aqueous humor, that vitreous humor helps to maintain constant eye inflation (hence size, shape and focus) regardless of your position or state of hydration. Note that the back of the eye is gently curved rather than being flat like the back of your camera. As in a camera, the image at the point of focus is left/right reversed and inverted. Because you customarily view the world in this fashion, you compensate for it without noticing. One can even get used to seeing everything right side up (after some months of wearing inverting glasses), just as American pedestrians in London eventually learn to look right for traffic approaching in the curb lane—if they live long enough. Anyhow, your sclera and cornea form a tough fibrous envelope around your eye—which allows *glaucoma* (a condition generally associated with elevated aqueous humor pressures) to cause progressive loss of vision through destruction of delicate retinal cells (usually first detected as a loss of peripheral vision, or sometimes as trouble seeing at night—see below). Although there is a marked overlap between probably healthy and most-likely harmful eye fluid pressures, effective glaucoma therapy generally includes reduction of preexisting intraocular pressures.

Accommodation

Light passes through two lenses (your cornea and your lens) before reaching the retina. This allows some correction of chromatic aberration. En route that light traverses air-cornea, cornea-fluid, fluid-lens and lens-fluid interfaces, being refracted at each. Normally your eye adjusts its lens curvature as needed to bring light from nearby or far away objects into focus at your retinal photosensors. To *focus* clearly on *nearby* objects, you must *contract* the *circular* ciliary muscle which *loosens* suspensory ligaments, thereby allowing the lens to bulge. On the other hand, *relaxation* of the ciliary muscle that encircles your eye *flattens* the lens by increasing tension on those suspensory ligaments about its periphery. Such flattening brings the more parallel light rays from distant objects to a focus at the retina—but then nearby objects become blurry since their diverging light rays only reach a focus behind the retina (if at all).

Your Retina

The photoreceptor layer of each eye includes about 3 million cone cells and 100 million rod cells. When sunshine causes your pupils to constrict, the narrowed iris directs light centrally onto the densely packed color-sensitive *cones* of your *fovea*. In contrast, pupillary dilation in dim light brings the less closely packed but highly light-sensitive *rod* cells into play. Cones perform poorly in dim light so you lose color vision at night (except near colored or bright lights). Thus a faint star may only be visible if you look next to it rather than directly at it—since starlight too dim to excite foveal cones can still excite your rods. That is why persons who are losing their peripheral vision from glaucoma or other cause may have trouble seeing at night. You also have an off-center blind spot in each retina that corresponds to where optic nerve axons gather to exit the eye en route to your brain.

Right Brain Sees To The Left And Left Brain Sees To The Right

The two optic nerves meet just in front of your pituitary to exchange medially originating axons. Thereafter the right nerve only delivers information centrally from leftward looking photoreceptors of both eyes and vice versa. This useful arrangement allows continuous mapping of any object as it crosses your field of vision (e.g. from far left to far right). Otherwise, if each eye mapped only onto its same-side hemisphere, there would be computational discontinuity as an object appeared first on only one side and then later became visible to the other eye as well. Such a crossover of ganglion cell axons is especially important for the visual efficiency of predators with overlapping (binocular) fields of vision. Since binocular vision aids depth perception, it helped your primate ancestors grab tasty insects and swing off safely through the trees (see Growth and Development).

Your Retina Is Not Very Well Organized

Rather than being located up front where they could intercept all incoming photons, the photoreceptors of your beautiful eyes lie at the very back of your retinas beneath several layers of nerve cells, axons and blood vessels. In addition to reducing your visual sensitivity, this arrangement requires your rods and cones *not* to notice all of those overlying cables and pipes. Such not-noticing is achieved through the continuous minor oscillations of your eyeballs that allow each retina to discount any retina-stationary objects. Of course, those *saccades* (swift voluntary eye movements between selected targets in your visual fields) serve additional important functions—for like other sensory systems, your visual system is organized primarily to detect and report upon changing inputs. Thus a truly steady gaze at a boringly stable scene would soon cause the scene to fade. But to reduce computational complexity and confusion, each saccade is slightly preceded by an appropriate minor shift of the central image display that brings it into register with the anticipated change in peripheral input.

Nonetheless, a more efficient design would certainly have placed your light receptors at the front of the retina with the neural and circulatory support systems behind. That would also get rid of those blind spots and eliminate the tiny, faint, circular spots that you sometimes see pulsing across your visual field as you gaze at a bright but featureless sky. These spots represent red cells chugging through tiny retinal surface blood vessels that have been magnified by reflection from the convex inner (back) surface of your lens. But, of course, there was no design. Instead, your eye represents a currently competitive outcome selected from a long line of chance improvements that started with photosensitive spot in dimple and evolved through the something-like-a-pinhole-camera stage.

Most likely the posterior placement of those photosensitive cells reflects an important long-standing relationship with their underlying black (light-absorbing) pigment cells. Indeed, cells of this pigment layer appear to provide support services to their adjacent photoreceptor cells much as astrocytes do for neurons. So when additional layers of processing neurons upgraded the function of that primitive retina and the blood supply correspondingly increased within the enclosed orb, the photoreceptor layer gradually became buried under those other more-or-less transparent add-ons. Rather in the way old English homes display much of their plumbing amongst the ivy on their outside walls—functional and easier to get at when it freezes, they say, but certainly not ideal. The point here is that as eyes and older buildings evolve, every intermediate stage must have advantages over other available choices, including starting over or making more major renovations. So once again, evolution seems the better explanation for your eye—or for that inconvenient old English home you may just have inherited. Such less-than-ideal arrangements often remain quite functional as long as local competitors are subject to similar limitations. But if an invasion of American plumbers and plumbing appliances should ever gain a foothold in England, they will spread over that island kingdom just as Africanized honey bees have displaced domesticated bees throughout Latin America.

Making A Spectacle For Yourself

Being far-sighted can be advantageous for an Inuit (Eskimo) since their traditional prey is distant and wary. In contrast, being near-sighted would lead to failure as a seal hunter. Inuit children often seem less reckless in their early explorations and less verbal, which is appropriate to their traditionally dangerous surroundings and cramped housing. Indeed, these children are encouraged to watch from the sidelines until they can contribute to group activities that they already have observed and learned. While that approach makes them less likely to stand out in a typically competitive American classroom, one can imagine that any Inuit who talked too much risked getting heaved out of the old igloo during those long depressingly dark winter months (a major reproductive disadvantage). Such speculation aside, we know that the eyeball is relatively too short in any far-sighted individual. Diverging light rays from nearby sources will therefore reach focus well behind the retina, providing only a blurry view of those nearby objects. Far-sightedness (or near-sightedness) can be corrected by placing an artificial lens (spectacle) in front of the cornea or, under bright light conditions on snow and ice, by peering through two tiny, properly centered openings in otherwise opaque bone "glasses" (the pinhole camera concept again, this time utilized to avoid snow-blindness from excessive glare). Note that far-sighted and near-sighted describe distances at which seeing is clearly in focus rather than blurry—but these terms do not imply hyperacute vision at any distance since a normal eye will see as clearly at a distance as a far-sighted one. In fact, being far-sighted is disadvantageous, for a far-sighted person will have progressive difficulty with near vision as the maturing lens becomes less adjustable (flexible) with age—a very far-sighted older person may eventually be unable to focus clear images on the retina at all. Such individuals then require corrective lenses for all distances (bifocals or trifocals—but those bifocals you use for reading can cause you to fall down the stairs, since stairs are only seen clearly through the upper prescription). The corrective glass lenses for far-sighted (relatively-too-short eyeball) people are predominately convex in order to bring light to a focus within a shorter distance.

On the other hand, near-sighted individuals have relatively-too-long eyeballs. Although light from nearby objects is easily focused on their retina, light approaching in parallel rays from distant objects will come to a focus in front of the retina, causing far away objects to remain blurry. Older near-sighted persons without glasses are less visually disabled than older far-sighted persons—at least the near-sighted can still see nearby objects without corrective lenses. But even with relatively normal fluid pressures, the larger eyes of very near-sighted individuals are under increased tension (just as a water pitcher requires thicker walls than a water glass) which may raise their risk for developing glaucoma. Corrective lenses that allow a near-sighted individual to see clearly at a distance will be *concave* (thicker edges than middle). Such heavy-looking concave lenses cause parallel light rays from distant objects to diverge slightly (as if from a nearby object). Those artificially divergent rays are then more readily focused on the too-far-away retina. Of course, younger individuals can be quite near-sighted or far-sighted and still have a good range of focus. That ability to accommodate explains why the *ophthalmologist* (eye doctor with an M.D.) must strive to prescribe corrective lenses that return such a young individual to a condition comparable to the mid-normal visual range. And if that opthalmologist first paralyses the iris and ciliary muscles with belladona-like eye drops (their parasympathetic blocking action dilates the pupil), she should obtain a better view of the retina as well. In contrast, if you really strain your ciliary muscle for the best possible view of the eye chart while your refractive powers are being tested, then the glasses that are prescribed could represent a correction suitable for your eyes when really straining rather than for your eyes with ciliary muscle more or less relaxed. Such glasses could then become a possible cause for headaches and other symptoms.

An *optometrist* (a non-MD) is trained to refract and prescribe glasses but not necessarily to diagnose or medicate eye conditions or even dilate the pupil for a more satisfactory eye examination. *Opticians* prepare and fit glasses—they do not refract, diagnose or medicate ("optic" refers to seeing in Greek). In any case, once an appropriate corrective lens brings a young person into the normal visual range, secondary (bifocal) correction should be unnecessary. Interestingly, some animal studies suggest that both near-sightedness and far-sightedness may be modified by the way those eyes are used during childhood and adolescence. That again shows how your ancestors found reproductive advantage in adapting to the real world rather than through applying strict standards for the world to meet.

Astigmatism is the term used when some parts of the visual field are in proper focus but others are not. One test for astigmatism involves looking (with one eye covered) at the center of a circular pattern that resembles a bicycle wheel on side view. If some spokes of that bicycle wheel appear to be faint or missing (even though present) while the rest remain clearly in focus, one can diagnose astigmatism—meaning more or less radial segment (or other off-center) imperfections of the lens or, more usually, corneal curvature. In other words, if you have astigmatism of either eye (which is quite common), that cornea probably has one or more slightly high, low or otherwise misshapen spots or sections.

Fish Eyes

You become terribly far-sighted and everything appears blurry when your face is underwater because your submerged corneas no longer bend light the way they did in air. So you must wear an air-containing helmet, face mask or goggles in order to retain useful vision while diving. With such equipment, your hands, feet and any nearby fish will appear larger than they would on land, having been magnified by the additional refraction at the water-to-glass-to-air interface (incoming light rays approaching your face mask at an angle are bent centrally as they enter air, making teeth and tail seem farther apart than normal). Some diving birds achieve clear underwater vision without goggles by markedly enhancing their lenticular and corneal curvatures during each submergence. Fish that remain underwater have flattened corneas and a highly curved (spherical) lens. One surface-feeding fish that needs to see both above and below the surface has an hourglass-shaped pupil with a bulging backward-tilted lens that therefore presents its flattened aspect to the upper (air) view, while waterborne light entering the lower portion of the hourglass is bent more sharply by passing through the more spherically curved portion of that same lens.

The Retina In More Detail

The photoreceptor function of both rods and cones depends upon a molecule of *retinol* (a derivative of Vitamin A) changing its shape when struck by a single photon. Vitamin A is important for your skin development and metabolism and also closely related to retinoin (a tissue organizer chemical)—another case of multiple applications for slightly modified molecules. Rods and cones are most sensitive to photons of specific frequencies. Those frequencies are determined by the exact amino acid sequence of certain retinol-associated photopigment molecules known as *opsins*. The genes for these photopigment opsins are located on your X chromosome, which is why males (who inherit only one X chromosome) tend to have more color vision deficiencies than females (so color-blindness is a *sex-linked* trait—see Reproduction). Many vertebrates and invertebrates orient retinol's crucial double bond within thin organized layers of *rhodopsin* molecules (and rotate that common orientation between photoreceptor cells). This allows the birds and the bees to navigate according to the *polarization* of daylight when the sun is not visible. Presumably it also helps fishing birds to ignore reflected glare from the water's surface as they size up their next victim. Similarly sensitive fish are able to detect polarized reflections from other silvery well-camouflaged fish surfaces. Apparently there was no reproductive advantage for your ancestors in retaining this skill, so you must purchase Polaroid glasses to spot fish in those pools. Incidentally, the rhodopsin molecule also occurs in bacterial and other cell membranes—single-cell paramecia (protozoa) and algae seek or avoid light in response to light-induced changes in the electrical polarization of their ciliary or cell surface membranes. So rhodopsin is yet another competitively successful molecule that has found widespread utility among many life forms. The purple bacterium living so colorfully in the San Francisco Bay salt works captures solar energy by expelling a proton for each photon of light absorbed. Their purple membrane *bacteriorhodopsin* can be deposited in dry heat-stable layers. Its ability to change color repeatedly in response to light may find use in optical information processing. That light-driven proton pump could also be used to produce chemical or electrical energy directly from sunlight.

As mentioned above, your eye includes three sorts of cone photoreceptor cells—some see best at red, others at green, and still others at blue light frequencies. All rods see best at green. Through its effect on a molecule of retinol, a photon of properly colored light brings about closure of leaky photosensor cell Na^+ channels. The result is hyperpolarization of the rod or cone cell (as Na^+ continues to be pumped out). In this fashion, incoming light produces variations of electrical potential across the photoreceptor surface (retina) and those electrical variations correspond to both the intensity and color of light from the scene being viewed. Interestingly, this means that photoreceptor cell messages are not passed onward as action potentials. Among advantages of reporting each photon by hyperpolarization rather than through ordinary depolarization is the way stimulus intensity shows immediately in higher transmembrane voltages rather than being reflected in the rate of cell discharge—the latter would force you to wait until enough photoreceptor spikes had accumulated to establish rate and pattern. And speed is often essential, since an airborne fly can make a visually directed mid-course correction in 30 milliseconds. That means you cannot wait for a whole series of nerve cell discharges before deciding where to swat (you will probably miss anyhow). Furthermore, the *horizontal* cells that report on the *average charge* of a local group of photoreceptors provide a background against which sudden new photoreceptor information becomes dramatically evident to the *bipolar* cells that in turn keep *ganglion* cells informed (such a system avoids sending low-value "more of the same" messages centrally). Thus it is helpful for horizontal cells (which also communicate with adjacent horizontal cells via gap junctions) to be relatively slow in their cumulative reports of average photoreceptor charge. For that allows the bipolar cell (which only detects and responds to charge differences between photoreceptors and horizontal cells that it picks up at special *triad junctions* where all three meet) to more easily track moving objects and also detect sudden movements despite changes in background illumination that may vary by many orders of magnitude (e.g. as you leave bright sunlight to enter your dark and presumably unoccupied cave).

Overall then, the hyperpolarization of a rod or cone struck by light of appropriate frequency is reported centrally to the thalamus by *ganglion cells* after some preliminary processing. That preliminary processing is designed to extract important information and still avoid overloading your central processing capacity with continuously modified individual reports from the million ganglion cells of each eye. One useful built-in mechanism that helps to reduce information overload and the need for excessive computation is the *scale invariance* of your eye. That allows more and more of the image of any rapidly approaching object to be projected laterally where rods are less closely spaced—which maintains the internal display of that approaching object at the same level of detail except for important new information received by the closely packed cone cells of the small central fovea. The greater visual acuity of the fovea thus allows one to concentrate upon and get progressively more detailed information about fascinating items such as teeth and claws. Messages from rods are sent via party lines that allow any of a group of rods to speak through their shared bipolar and ganglion cells—which also makes things seen out of the corner of your eye more difficult to evaluate. On the other hand, each cone usually has its own dedicated *bipolar cell* reporting directly to a ganglion cell. Since *horizontal* cells help to modulate reports at the output end of rods and cones, they also enhance contrasts between light and dark as well as shapes, color and movement. And the *amacrine* cells (which occupy a more forward position) also help to readjust your interpretations of color as illumination varies—so that you still see blue in dim as well as bright light. But only ganglion cell axons pass out of the eyeball through the optic disc (your blind spot) in order to distribute their partially preprocessed messages to different parts of the thalamus for separate handling of information about color, shape and movement. In addition, the hypothalamus is kept informed of day/night and seasonal light changes so that it can adjust your various internal rhythms in an appropriate fashion (see Hormones).

The six muscles externally attached to each eyeball coordinate their gaze at the target with some *redundancy* (spare capacity). Thus the four up, down and sideways muscles are supplemented by two more (up-and-out and down-and-out) that improve control in the more important lateral directions (the other eye sees mostly nose anyhow). Eye muscles are reproductively advantageous because they allow quick changes of eye direction without moving the head. That allows you to remain completely still while keeping an eye out for predators, prey and reproductive opportunities. In contrast, an owl must fly in the dark using subtle visual and auditory cues to capture its wary victims. By eliminating eye movement, the owl keeps its visual and auditory displays of the target location in permanent register, while compensating for the resulting fixed stare with 14 hypermobile cervical vertebrae that allow 270° of neck rotation in either direction (for a total of one-and-a-half full turns).

The thin conjunctival layer covering your exposed anterior scleral and inside-of-eyelid surfaces is kept moist by fluids from the tear gland located laterally and deep within each upper eyelid. Your upper eyelids are also stiffened internally by tarsal plates. The eyelash hairs that sprout from your upper and lower eyelids are associated with sebaceous glands that grease your eyelid edges to keep tear water from flowing over the cornea except when you blink. An ordinary rate of blinking automatically moistens and refreshes your outer corneal cells and wipes corneal surfaces. A higher rate of blinking suggests an affectation, a tic or corneal irritation. Any minor excess of tears will drain via ducts at the inner corner of the eye into the nose. More than that will wet your face, which can be a reproductively advantageous way of demonstrating your strong and sincere emotions in a non-threatening fashion, or of washing away irritating substances.

CHAPTER 17

YOUR ENDOCRINE SYSTEM SENDS SIGNALS

All Molecules Carry Information;... Life Cannot Avoid Signaling;... A Signal Carries Information;... Signals Tend To Fade;... Circumstances Can Alter Any Message;... One Small Step At A Time;... A Signal Molecule Speaks To A Receptor Molecule;... By Its Response, The Receptor Amplifies The Signal;... Hormones Are Blood-borne Signal Molecules With Specific Targets;... Responses Must Be Timely And Appropriate;... Your Death-deferring Design Is Exceedingly Complex;... Suicide Is Usually Counterproductive;... Your Less Urgent Hormonal Signals Are Fat-Soluble And Act Via Intracellular Receptors;... More Urgent Signals Are Water-Soluble And Act At The Cell Surface;... Hormones Are Ordinary Signal Molecules Sent Out By Your Endocrine Tissues Via Body Fluids;... Cooperate Or Die;... Thalamus To Hypothalamus To Pituitary;... Your Thyroid Regulates Your Metabolic Rate;... Your Parathyroids Stimulate Your Osteoclasts;... Hormonal Control Of Blood Glucose;... Adrenal Steroids Can Make You Strong And Sexy, Or Cause You To Retain Salt Or Fall Apart;... Intermittent Release Improves Regulation.

All Molecules Carry Information

Information is carried on, in and about everything, everywhere. It is displayed through order, disorder, continuity, change, construction, destruction, concentrations, concentration gradients, energy and forces, input and output, presence and even absence. Furthermore, all matter and energy can carry information about origins, orientations and processes. Dead or alive, you do not escape from, nor can you stop producing, information. And even when you cannot understand it, you ignore information at your own peril. *Life cannot avoid signaling* its presence because every living thing continuously and unavoidably alters the energy flows and chemistry of its environment—to others those signals may mean food or competition or danger or an opportunity for reproductive success.

A Signal Carries Information

Noise is the background from which you must separate the signal. There are many ways to improve the signal-to-noise ratio. For example, a bright star may become visible in broad daylight if viewed from the dark bottom of a deep well, provided the sun is not also directly overhead. Of course, one person's signal may be another person's noise. Thus a passing airplane can drown out the most interesting conversation. Or a totally boring conversation may prevent a mechanic from detecting informative engine sounds. Furthermore, *any* particular *signal may have* a great *many different* interpretations, meanings and *effects*. Thus bats and owls retire at dawn's early light while robins consider first light their cue to start singing. And a police or fire siren may bring good news, bad news, both or neither. A shrill siren could also be annoying, or inaudible because your radio is on too loud or you are far away. The particular message that a siren carries for you might be "stop", "run", "call your lawyer", "hide" or "shovel snow off the nearest fire hydrant". Or that siren might cause you and others to turn on many more sirens, which would then *amplify* the initial warning of the volunteer fire brigade.

Signals Tend To Fade

The metabolic waste of any living thing contains stored reducing power and other useful materials but eventually these products become too dilute for detection or profit. Similarly, the information content of a signal tends to degrade as it melds into background noise—then the detection of its message becomes increasingly uneconomical and generally of diminished relevance as well. At one time, so many discarded brown beer bottles littered Alaskan roadsides that a legislator nominated them for Official State Flower. Some persons view cast-off empty beverage containers as pollution and waste. Others see them as a business opportunity. But most tend to lose interest when not enough bottles remain to cause ongoing offense or support profitable recycling. It becomes a matter of more important priorities and diminishing returns. On the other hand, scientists sometimes find it professionally and scientifically rewarding to identify a signal that is even weaker than the random background noise. Very distant stars that are less bright than the darkest night-time sky can still be identified and characterized by their consistent signal and spectrum. All it takes is an appropriate investment of scientific skills, telescope time and ultrasensitive detectors to extract and report upon (amplify) such otherwise hopelessly dilute messages.

Circumstances Can Alter Any Message

Our beautifully intricate information-laden languages and emotion-inducing songs represent endlessly elaborated primal screams, grunts and thumps—many songs continue to rely heavily upon those basic inputs. But complex languages and songs could only develop, advance and persist because they brought reproductive advantage to their users at every step. The progressive utilization of sound waves to carry ever-increasing loads of information depended upon simultaneous advances in sound reception and comprehension, as well as locally under-utilized frequency bands where significant sound interference (loud noise) was infrequent. The advantage always went to those who could send, receive and utilize larger quantities of more relevant information but there never was a conscious sit-down preplanned expansion of vocabulary and syntax. Rather language came about through the inevitable association of sounds with activities, events and objects. For example, under different circumstances "YeeHah!" might come to signify stumbling into a hornet's nest or perhaps copulatory success or "Look out for that tiger behind you!" Even today, the information content of any speech is only partially in the word sequence. It is also carried separately by intonation and inflection and by when and why who says what to whom—which may be even less important than what was not said and why. Or who reports on what has been said, and how they repeat, amplify or rephrase it. Also important to any message is how much of it was received and how much comprehended, versus how much was missed or which parts misunderstood. In addition, any response elicited and what that response might be depends upon still other factors such as whether the actual audience was the intended audience, the current condition and circumstances of that audience and so forth.

One Small Step At A Time

No biological process is completely reliable, so life forms and their component molecules tend to vary as widely as the competition will allow. Far greater variations occur, of course, but by the above definition, more extreme variations fail to reproduce since they are no longer competitive. As with our ever-evolving languages, those ever-evolving molecules occasionally and progressively develop significant modifications or elaborations that enhance their ability to transfer information.

A Signal Molecule Speaks To A Receptor Molecule

Even minor alterations in signal and receptor molecules may greatly expand the number and variety of possible interactions, depending upon which version of the signal is sent out and which receptor molecules then respond most vigorously in what fashion. As competing receptors become more efficient and complex, they necessarily alter the degree and even type of response that a signal may elicit. Not surprisingly, these sorts of evolving molecular processes have produced a few widespread (because effective) *superfamilies* of important signal and receptor molecules. Endless minor modifications within those families give rise to progressive variations in signal-receptor interactions that can transfer increasing quantities of ever-more-specific information. Some of that information improves detection of predators or prey. Other signal-receptor advances might upgrade chemical communications between and within various cooperating life forms.

By Its Response, The Receptor Amplifies The Signal

No matter how well signal and receptor may fit together, a signal cannot bring about a response unless it sufficiently alters the configuration of its receptor molecule to initiate some new chemical activity. That new chemical activity must then directly, or indirectly via intermediate molecular messengers, bring about *amplification* of the information delivered by the faint chemical signal in order to elicit an appropriate cellular response. At the start, of course, signals and receptors had no predetermined meaning or significance. Inevitably, however, there were ongoing interactions, some sort of molecular cause and effect. Then, during each molecular round, those responding less appropriately were trimmed from life's evolutionary tree. Eventually life's increasingly complex organizations came to rely upon highly specific and sensitive signal-receptor activations, although the particular significance and function of those specifically activated receptor molecules changed and expanded repeatedly during that ongoing evolutionary process.

Hormones Are Blood-borne Signal Molecules With Specific Targets

A hormonal signal may alter one or more sorts of receptor molecules. Each receptor affected then initiates a multi-step cascade of the molecular interactions requested. This cascade amplifies the faint original signal, thereby (directly or indirectly) bringing about an appropriate response to that specific hormone-borne information—subject to further modification by other inputs and control. Of course, even within your own body, what might be an appropriate response by a liver cell could prove fatal for a cell in your heart or brain—some signals are bound to demand urgent action from certain cells or life forms while remaining totally irrelevant to others. And different cells picking up an identical signal on an identical receptor molecule may still react in entirely different ways—each according to its own kind and condition. In other words, information that stirs your gall bladder need not exercise your gonads. So a great many urgent molecular signals simply pass by or are wasted upon cells not bearing appropriate receptors. And the fact that these specialized cells have flourished without those particular receptors makes it likely that they can safely continue to ignore this particular input.

Responses Must Be Timely And Appropriate

All creatures seek to dine without becoming dinner. Intra-species rivalries and the conflicting needs of predator and prey force continued improvements in matters such as speed, degree, duration and appropriateness of response. Some responses must be swift, sure and brief. Other readjustments better serve if gradual in onset and prolonged in effect. Where speed is of the essence, the entire pathway from receptor through amplification to response must be ready and waiting, chemically primed and energized to go. Yet such extensive preparation would be wasteful when a sustained adjustment or other gradual effect need not begin at once. Then it becomes practical to wait until a specific signal has placed its order with the appropriate receptor, since that allows the cell to avoid tying up costly resources in excess inventory. Furthermore, the timing of your response can be yet another critical variable in determining its effectiveness—responding too slowly to a predator is clearly disadvantageous but responding too quickly under less urgent circumstances can easily initiate serious oscillations or imbalance of output. And obviously there must be safeguards that prevent any response from inadvertently being triggered or inappropriately amplified.

As signal molecule concentrations rise, more general ancestral relationships often allow cross-reactions with less exactly matched receptors of the same superfamily. The products of such secondary interactions may then increase, decrease or otherwise modify the original signal-receptor effect. And that is very important, for in this difficult and dangerous world, "more of the same" may be a reproductively disadvantageous response to a stronger signal since the signal may be getting stronger because previous responses were ineffective or counterproductive. Under abnormal circumstances, the signal could also be getting stronger because there is no effective connection or feedback—as when we speak more and more loudly to foreigners who don't understand English. In view of the relatively endless possibilities for fine tuning signals and receptors, it is not surprising that your hormones and other molecular messengers generally adjust the "how much" and "how fast" of your biological rhythms rather than being primary determinants of whether a particular process will take place. Even those eventually irreversible growth and developmental changes brought about by your sex or thyroid hormones can be viewed as degree and duration responses to multiple hormonal interactions rather than as unique processes turned on or off by unique signals.

Your Death-Deferring Design Is Exceedingly Complex

So how in Heaven's Name did all of these *iterative* (repeated reapplications of output to input) and chaotic interactions and feedbacks ever become organized into an effective whole? How could something so very much more complex than any possible description develop so successfully without a careful and complete design? Surely the spark of life is far too precious, too basic and too well-integrated to have come about without a plan, by mere chance alone? Well, for the sake of this brief discussion, let us agree that overwhelming reproductive potential is inherent in all living things. Furthermore, we can reasonably accept the likelihood of randomly beneficial variations arising occasionally and even being transferred about within that vast but closely integrated system we know as Life. Let us also consider it evident that such variations can be recombined in uniquely effective fashions by sexual reproduction. And that some amongst those endlessly varied individuals will reproduce successfully while others will not—so that each new generation represents only the progeny of successful reproducers. Still, there *must* be more to life than that!

Actually, there is more to life's design than mere reproductive pressure or those infrequent random improvements and redistributions of molecular modules. Of course, without overwhelming reproductive urges and capacities there would be no subsequent generations and thus no life. And without variation there would only be stagnation and again no life. But a third essential contributor to evolutionary progress has always been destruction, not creation. For death wields the big eraser across life's blackboard, methodically making room for new calculations and formulas, always clearing away obstructions that might prevent the latest seemingly miraculous improvement in design from flourishing. Indeed, death never tires of wiping out life's errors, misjudgments and inadequacies. Death then cleans the erasers, recycles the chalk and conscientiously arranges an immediate play-off among all winning calculations at every step. Only death can usher ever-better solutions forward for their hard-earned moments in the winner's box. Because it simply erases faulty or outdated work without malice, death is an objective criticism that we take to heart. Death selects and later retires each player in life's great symphony. In the interim, death makes sure that they stay in tune. The absence of life is nothingness. Death is far more. Life requires and depends upon death for its sustenance as well as its design. We are entirely indebted to death for our existence and ongoing survival, for all that we are or ever can be. However, if death is so wonderful, why don't we just go out and kill ourselves in order to upgrade the world more quickly? Fortunately or unfortunately, it is not that easy to upgrade life, for it is life's competition that is so basic, essential and unavoidable.

Suicide Is Usually Counterproductive

Most likely we could agree on certain persons—presently alive or historical—whose early suicide would have been an unmitigated blessing. Regardless of that possibility, human evolution has now veered sharply away from the brief and brutish survival-of-the-fittest pathway that brought your ancestors such pain, grief and tears. In comparison, your personal stroll through the Valley of Death has been altered beyond ancestral recognition by the exponential advances of modern science with its hard working non-human machines and their outside-of-life information processing capabilities. It is those extraordinary and increasingly computer-based talents that have freed humanity from further dependence upon physical evolution to maintain reproductive success in the face of ever-changing reality. But your ongoing existence still depends entirely upon your urge to survive. Without that crucial motivation there could be little reproductive success. Our ancestors included only those most intent on survival despite overwhelming hardships and in the face of incredible odds. So although we recognize death as the inevitable end to all joy and suffering, we generally feel no desire to hasten our own demise. Indeed, it is evolutionarily and therefore emotionally correct behavior to defer death for as long as possible. That allows us to fight death with all of our might while still recognizing that if we ever really won that fight, the boring and risk-free life style then required would soon end the human race.

Your Less Urgent Hormone Signals Are Fat Soluble And Act Via Intracellular Receptors

Your growth, differentiation, metabolism, sexuality and responses to major stress are guided by fat-soluble signals acting upon receptors located deep within your cells. Those receptors differ markedly in their affinities and responses to various relevant signals that penetrate your phospholipid cell membranes—but the information represented by such intracellular signal-receptor complexes generally initiates gradual adjustments that take place over subsequent hours. Despite their many differences, intracellular hormone receptors all seem to belong to a single ancient superfamily (evidence for ancestral relationships between molecules includes common shape, conserved amino acid sequences at important sites, relatively unique conformations related to metal ions and so forth). Once formed, the signal-receptor complex of that family activates or turns down the transcription of perhaps 50–100 different intranuclear genes. The impact of the many new messenger RNA molecules thereby produced (or eliminated) is further amplified by their potential for translation into many identical proteins. Thus intracellular signal-receptor complexes bring about progressive adjustment of a cell's chemistry in obedience to incoming orders. And since you are composed of over 250 sorts of cells, a certain signal can require quite different genes to be transcribed in very different cells for entirely different purposes (although most genes are somehow blocked or otherwise misfiled so that they cannot be expressed in a cell where their product would be inappropriate). In any case, this intracellular receptor superfamily of yours stays tuned for specific *steroid* (modified cholesterol molecules that include your sex and adrenal cortical hormones as well as vitamins A and D) *announcements* and also *metabolic change orders* passed on *by thyroid hormone*.

More Urgent Signals Are Water-Soluble And Act At The Cell Surface

Unlike the rather slow adjustments in protein production that result when fat-soluble hormones activate intracellular receptors, the alterations caused by water-soluble signal molecules at the cell surface tend to be swift and brief. Although there are many different sorts of signals and each speaks to (alters the shape of) its own specific cell surface receptors, the resulting signal/receptor combinations generally act through relatively standard intermediate molecules to bring about the expected wide variety of effects. Thus the eucaryotic cells of slime molds, rats, lawyers and politicians rely heavily upon *G Proteins* (guanine nucleotide-binding proteins) to pass cell-surface signal/receptor instructions onward into the cell interior. These G proteins are attached to the inner aspect of the cell surface membrane where they can readily be activated by adjacent cell surface receptors. Activated G proteins rapidly turn on adjacent *effector enzymes* that then create many copies of some specific *second messenger* signal that diffuses through the cytoplasm to enhance or suppress the activity of still other enzymes. Since every enzyme brought into play (or turned off) could easily produce a great many product molecules, the result is a rapid and appropriately directed intracellular amplification of the original extracellular signal.

G proteins are especially handy at forwarding cell-surface signals because each consists of three modules and the three sorts of modules have enough variants to allow over a thousand specific interactions by uniquely constituted G proteins.

Furthermore, being a *GTP*'ase, each G protein ordinarily works for only a few seconds before shutting itself down to await the next signal/receptor interaction. This ability to serve as a timer (as well as switch and preamplifier) comes about because an activated G protein alpha module quickly releases the *GDP* it was carrying, which allows the more abundant GTP to enter. In turn, GTP encourages the alpha module to separate from its beta and gamma G protein modules and sidle along the inner membrane surface to activate a nearby effector molecule. But the alpha unit soon breaks its GTP down to GDP and the resulting change of shape causes it to leave the activated effector and return to beta-gamma—which may not have been idle during this period since it has its own range of effectors to activate or co-activate along with alpha. This clever method of limiting activity to the time required for self-hydrolysis of GTP is utilized by many other intracellular proteins—some members of that *GTP'ase superfamily* are involved in ribosomal protein synthesis, others regulate the rate of cell division.

Adenylyl Cyclase is a common membrane-bound effector molecule that produces *cyclic AMP* when activated by G proteins. Cyclic AMP is not an ordinary intracellular metabolite—being neither frequently encountered nor readily consumed makes it a great signal molecule for achieving intracytoplasmic activation of *protein kinases* (enzymes that add phosphate groups to certain other enzymes in order to turn those enzymes on or off). Other useful second messenger molecules include *cyclic GMP* and *inositol triphosphate*. Thus your cells utilize appropriate combinations of G protein modules to convert specific extracellular signals into appropriate cascades of intracellular enzyme reactions. The G proteins in your retinal rod cells amplify the impact of a photon on the rhodopsin molecule by turning on *phosphodiesterase* (the effector enzyme in your retinal rod cells) which converts cyclic GMP to GMP so the Na^+ channels being held open by cyclic GMP will close and effectively hyperpolarize your rod cell. In similar fashion, an epinephrine molecule interacts with a cell surface receptor to alter an adjacent G protein that in turn stimulates adenylyl cyclase to convert ADP to cyclic AMP which eventually allows the breakdown of glycogen and causes *phosphorylase* to release glucose (minus its phosphate) into the blood stream. G proteins, cyclic AMP and protein kinases are also involved in the production and release of steroid hormones from your adrenals and gonads. And *phospholipase C* cuts a plasma membrane phospholipid into two second messengers, one of which then releases intracellular Ca^{++} as a signal to open or close various ion channels.

Note that the term "second messenger" only makes sense if the first messenger is defined as the signal molecule that combines with a cell surface receptor. In actuality, a truly original signal is about as rare as a truly original sin, since any signal can also be viewed as the response to prior signals in life's ongoing process. Of course, at each step of any multistep cascade, molecular support groups that favor passing a signal along must contend with adverse information that would tend to suppress or reverse that signal's effect. And enzymes always lurk about ready to destroy the various enzymes or second messengers involved in such *regulatory* and therefore closely regulated reactions (see Metabolism). Amazingly enough, the lovable result of all of this incredibly intricate often-chaotic yet beautifully integrated chemical activity is the real you.

Hormones Are Ordinary Signal Molecules Sent Out By Your Endocrine Tissues via Body Fluids

The functional status of every cell in your body is ordinarily determined by its water, solutes and structural components. Different cells are affected in varying degree by all manner of alterations in those chemistries. For example, the lactic acid released into your circulation by hard-working muscle cells informs your liver cells of a job they must do. Similarly, CO_2 put out by those same muscle cells advises local blood vessels to dilate while also encouraging your brain stem neurons to alter the depth and duration of each breath. Even though numerous lactic acid and CO_2 molecules carry such specific messages throughout your circulation and even though those messages require distant cells to alter their activities in some important and predictable fashion, lactic acid and CO_2 are not classified as hormones. Presumably this is because lactic acid and CO_2 are common products of ordinary metabolism rather than designated chemical messengers not otherwise involved in normal cell processes.

Your elongated nerve cell axons transfer information much faster than your blood flow ever could deliver chemical messages. But even nerve cells must convert their electrical signals into urgent chemical messages and then wait for those molecules to diffuse across synapses to inform other cells. Yet the fifty-odd sorts of chemical messages known to be released by axons are considered neurotransmitters rather than hormones because they exert only brief and localized effects on one or just a few cells. On the other hand, blood-borne hypothalamic messages have acquired hormone status even though they only affect a few anterior pituitary cells. And the larger quantities of ordinary neurotransmitter released directly into your body fluids and circulation from the adrenal medulla (epinephrine and norepinephrine) or bowel wall (cholecystokinin) are viewed as hormones because of their more distant or widespread action. Both retinoic acid (derived from Vitamin A) and the activated form of Vitamin D are designated hormones since they only are released into your body fluids after chemical modification within your cells in order to induce specific responses at distant sites.

A large variety of other molecules distribute important information between your different cells. Some already have been granted the title and many others will eventually be recognized and added to the list of 50 or so messenger molecules currently classified as hormones. Some of the better known chemical messengers released into your blood by well-established *endocrine* (hormone producing) *tissues* such as your *hypothalamus, pituitary, thyroid, parathyroids, pancreas* and *adrenals* are reviewed in this chapter. Significant hormonal interactions involving your heart, kidney, GI tract and pancreas, lymphatic and thymic tissue, bone, bone marrow and blood, gonads and reproductive systems are considered under those headings as well. In addition to their hormone products, several of the above-mentioned organs or tissues also release *exocrine* secretions (onto body surfaces via ducts).

Cooperate Or Die

One of the major headaches of multicellularity is getting all cells and tissues to cooperate for the common good. As your central nervous system consciously and unconsciously sets its goals, the more urgent instructions go out electrically via motor nerves to your skeletal muscles. Housekeeping advice travels somewhat more slowly through sympathetic and parasympathetic nerves to regulate your smooth muscles, internal organs and sweat glands. But that still leaves a silent majority of your cells sitting out there without a synaptic clue as to what is currently expected of them. So most of your cells seek guidance from information-bearing solute molecules that diffuse out of your circulating blood. Some cells are specifically tuned to respond to certain chemical changes—a drop in the ionized calcium level of your extracellular fluids stimulates your parathyroid cells to release additional parathyroid hormone. However, those cells not individually *innervated* (taking their cues directly from a nerve cell synapse) as well as many that are, generally stay in touch by exposing a wide variety of specific receptors for relevant hormonal announcements. Those receptors affect the crucial chemical processes regulating cellular behavior—controlling such mundane matters as when to absorb excess glucose from the blood or how much to alter other important cell functions.

Thalamus To Hypothalamus To Pituitary

Your thalamus receives and analyses sensory input from all over your body. Your *hypo*thalamus (*just below* thalamus) then brilliantly coordinates your vegetative (unconscious or automatic) responses to that overwhelming sensory barrage. Not only does your hypothalamus control your sympathetic and parasympathetic outflows, it determines the hormonal outputs of various important endocrine glands through close regulation of its directly subjacent *pituitary* or *master gland*—and while your hypothalamus produces certain hormones, it also remains subject to feedback control by many others. Your entire pituitary gland is exposed to hypothalamic suggestions but only the *posterior pituitary* gets *direct* neuronal commandments from above—indeed, particular hypothalamic neurons even provide *oxytocin* and *antidiuretic hormone* to posterior pituitary storage axons for release whenever those same hypothalamic neurons are depolarized (see Kidney and Reproduction). In contrast, your hypothalamic neurons signal *indirectly* to your underlying *anterior pituitary* cells by dumping hypothalamic hormones into local capillaries that briefly enter small venules before branching out again as the capillaries that nourish and inform those pituitary cells. Such a *venous portal system* (capillary beds usually receive blood from arteries but here we have two capillary networks in sequence with the second set supplied by venous drainage from the first) allows the faint hormonal signals from specific hypothalamic neurons to stimulate or suppress specific anterior pituitary cells before those few hormone molecules are diluted to insignificance within the general blood circulation.

The separate and complex regulatory arrangements for hypothalamic supervision of your posterior and anterior pituitary reflect the separate origins of the embryonal *posterior* pituitary (from your hypothalamus) and the embryonal *anterior* pituitary (as an outpouching of your pharynx)—they also demonstrate how evolution regularly realizes very serviceable mechanisms through repeated modifications (see Eye). However, needless complexity inevitably raises cost and reduces reliability. In this case, your double-capillary arrangement slightly increases the risk of pituitary *infarction* (pituitary cell death due to inadequate circulation) under very stressful conditions involving maximal pituitary stimulation and very low blood pressure (for example, consequent to severe maternal blood loss during childbirth). After all, given two directly sequential capillary beds without an intervening heart pump, the second bed clearly must provide lower blood pressures. But the fact remains that this Rube Goldberg (unnecessarily complex) arrangement almost always works, which is the hallmark of tortuous designs that have evolved under severe competitive pressures.

The anterior pituitary releases more than half a dozen different hormones into the general circulation. One boosts thyroid function (*thyrotropin*, also known as *TSH* or *thyroid stimulating hormone*), others accelerate growth and maturation of bones and muscles (*growth hormone*), affect female breasts (*prolactin*), stimulate reproductive organs and gonads in both sexes (*luteinizing hormone* or *LH*, and *follicle stimulating hormone* or *FSH*), regulate adrenal glucocorticoid steroids (*adreno-corticotropic hormone* or *ACTH*—which also encourages aldosterone release) and affect melanocyte function in the skin (*melanocyte stimulating hormone* or MSH, as well as beta-lipotropic hormone and ACTH). MSH and ACTH have binding sites throughout the brain—both can affect learning, memory and attention. MSH also readjusts the temperature controls in your brain. In response to stress, ACTH and beta-endorphin (an opium-like peptide pain-reliever) increase your metabolism and provide pain relief.

Incidentally, the *pineal* gland (a small and physiologically obscure endocrine gland located above the third ventricle in the roof of the midbrain) puts out *melatonin* (which has nothing to do with melanocyte stimulating hormone). Higher levels of melatonin in the blood are thought to suppress puberty. So rising blood levels of LH coincide with declining melatonin concentrations as humans enter adolescence. And throughout your life, blood melatonin levels will continue to rise and fall with darkness and light—melatonin plays a role in your *circa*dian (*about* a day) rhythm and has been used to treat jet lag (see below). Your lizard relatives expose a simple light-sensitive *parietal* or third eye on the top of their heads that seems to help maintain their circadian rhythm. Despite being located deep within your skull, your own pineal cells remain *photosensitive* (respond chemically to light). By reducing oxidative damage to DNA and cell proteins, melatonin may have played an important role in helping life survive Earth's higher atmospheric O_2 levels during the last two billion years, since melatonin is the best scavenger for hydroxyl free-radicals ($^{\cdot}OH$)—the major cause of intracellular oxidative damage.

Anterior pituitary hormones are subject to various positive and negative feed-backs from the hypothalamus, which in turn must evaluate and respond to many inputs. For example, stress, ACTH, MSH and repeated stimulation of the female nipple during nursing all elevate prolactin levels—clearly prolactin production is inducible by more than one pathway. And anterior pituitary hormones may be back-regulated in a closed loop fashion by blood levels of the thyroid hormone or other product that they call for or stimulate—numerous cross-reactions include inhibition of prolactin production by a peptide split off from *hypothalamin gona-dotrophin releasing hormone* when that *GNRH* is activated to stimulate LH and FSH production (and also alter your mating behavior)—and a single *proopiomelanocortin* molecule (expressed mostly in the pituitary) gives rise to a whole family of shorter peptides including ACTH, MSH, beta-lipotropic factor and beta-endorphin. Furthermore, LH, FSH and TSH are all glycoprotein hormones closely related to *chorionic gonadotropin* (secreted by the placenta). Each of these hormone molecules is made up of two chains with identical alpha and different beta sub-units—representing yet another functional series of modular modifications. So while we may view any hormone as a single-purpose chemical signal, hormonal functions often overlap and are not clear-cut. For example, growth hormone may act directly on some cells and indirectly on others via liver-and-other-tissue-produced peptides known as *insulin-like growth factors* or *somatomedins* (see insulin below). ACTH is produced independently in inflamed tissues where it appears to have direct local effects on the inflammatory process. ACTH and MSH both modulate your immune system responses. Some of these hormones also act as neuromodulators or neurotransmitters. Obviously your blood-borne signaling arrangements are far from simple but a simpler system would not be nearly as versatile. So take the following simplistic discussion of your various endocrine glands with a grain of iodised salt—unless you grew up on a low-iodide diet.

Your Thyroid Regulates Your Metabolic Rate

Your thyroid gland will not noticeably bulge the front of your neck under normal circumstances. However, it might be barely apparent in a slender woman with a long neck, just below the midline bulge of her *thyroid cartilage* (Eve's apple?). The two lobes of your thyroid are connected by a thin band of thyroid tissue that crosses in front of the *trachea* (windpipe). Your thyroid avidly collects *iodide* (I^-) which it combines with modified *tyrosine* (an amino acid) to create *thyroxine*. Your thyroxine is then stored in a great many microscopic-sized *colloid follicles* throughout the thyroid gland until needed (follicles that enlarge enough to be felt are considered *thyroid cysts*). Thyroxine is released by your thyroid gland in response to thyrotropin signals from your pituitary. The consequent increase in blood levels of thyroxine and its more active tri-iodothyronine derivative suppress further thyrotropin release (a negative feedback). Continuously high blood thyroxine levels cause a hyperactive or *hyperthyroid* metabolic state that affects all of your cells. One large Western U.S. slaughterhouse recently caused a multi-State epidemic of temporary hyperthyroidism by inadvertently including cow thyroid glands in hamburger-designated bovine neck muscle trimmings. Presumably all of those temporarily hyperthyroid hamburger connoisseurs were not only warm with rapid pulse, fine muscle tremor and speeded-up metabolism, but also irritable and losing weight. Being hungrier than usual, they may have returned more frequently to consume additional thyroid-containing burgers (a positive feedback). Of course, such an excess of thyroid hormone in the blood from external sources is associated with a marked reduction of circulating thyrotropin levels (due to suppression of pituitary TSH secretion by that extrinsic thyroid).

Noticeable *goiter* (thyroid enlargement) can have many causes including a diet that is very low in iodide. Goiter used to be common at inland locations where little seafood was consumed—in Switzerland, for example. Since developed nations now routinely enrich table salt (NaCl) with a little iodide, most people not residing in third world nations obtain plenty of iodide in their diets. However, persons coming from very low iodide areas may develop thyroid disorders—including the production of *auto*antibodies (against *self*) and hyperthyroidism—if their chronically overworked thyroid is suddenly presented with an excess of iodide. By definition, when it becomes independently hyperactive from any cause, a thyroid gland no longer depends upon stimulation by pituitary thyrotropin. A persistently overactive thyroid gland can often be suppressed functionally with radioactive I⁻ or other medication—and a very enlarged but not overactive thyroid can often be reduced in size with oral thyroxin supplements. Either sort of thyroid condition can also be treated surgically. Permanent thyroxin supplementation will be required if the entire thyroid gland is removed (as for a thyroid cancer) or chronically underproductive. With or without treatment, hyperthyroid individuals occasionally develop *exopthalmos* (protuberant eyes) due to an autoantibody stimulated growth of fibrofatty tissues behind the eyes.

Thyroid replacement therapy is an especially urgent matter for the *under*productive thyroid gland of a *hypo*thyroid newborn (also known as a *cretin*) who may already have inadequate body growth and poor development of the nervous system as well as swollen tongue and *myxedema* (puffy skin). Not surprisingly, there is a great loss of human potential and productivity in third world nations where the average inadequate diet does not include sufficient iodide. Inexpensive iodide supplements have measurably improved health, educability and economic performance, as well as reducing newborn mortality rates and the incidence of congenital deafness. Adults who develop hypothyroidism—perhaps long after partial surgical removal or radioactive iodide treatment, or following autoimmune thyroid*itis* (inflammation)—tend to become mentally and physically slow, have an abnormally low body temperature and gain weight—if the cause is thyroid gland failure, they may have elevated blood thyrotropin levels as well. Of course, one can be fat and slow without being hypothyroid.

Total thyroid*ectomy* (*removal*) also terminates thyroid cell production of *calcitonin*, a hormone that lowers blood calcium levels by reducing osteoclast activity and stimulating osteoblasts while also encouraging excretion of Ca^{++} and Na^+ in the urine. That loss of calcitonin production generally does not cause noticeable problems. Indeed, a great many total surgical thyroidectomies were performed long before calcitonin was even recognized, yet high blood calcium levels rarely if ever resulted, presumably because parathyroid hormone production responds so directly to blood Ca^{++} levels. But it is also true that lack of calcitonin can diminish symptoms of postoperative parathyroid deficiency. And thyroid surgery does endanger those four very-small-and-hard-to-recognize parathyroid glands since your parathyroids usually nestle snugly against or even within the back side of your thyroid gland where they are well protected from all but surgical attack. Thyroid surgery also endangers the *recurrent laryngeal branch* of the vagus nerve. If damaged on one side, the patient will be hoarse—injury to both recurrent laryngeal nerves results in paralysis of both vocal folds, airway obstruction and loss of voice.

Your Parathyroids Stimulate Your Osteoclasts

All cells require appropriate blood Ca^{++} levels for normal function but only osteoclasts (which increase bone resorption and thereby raise blood calcium levels) and osteoblasts need pay attention to your little parathyroids—in contrast, your far larger thyroid gland sets the metabolic rate for every cell in your body. *Vitamin D* raises blood calcium levels by increasing bowel and kidney Ca^{++} absorption as well as by stimulating osteoblastic and osteoclastic activity in the skeleton (thereby also lowering parathyroid hormone levels). Osteoblasts are able to draw upon bone-matrix-fluid Ca^{++} and transfer this to your extracellular fluids as needed. Vitamin D affects most of your cells in a variety of ways and interacts with many other hormones. Its active form, *calcitriol* (1, 25 Dihydroxyvitamin D_3) is only produced as needed in response to reduced blood Ca^{++} levels. Calcitriol administration will probably have an important role in the prevention and treatment of postmenopausal osteoporosis (many persons with reduced bone mass or osteoporosis may have abnormal Vitamin D receptors). Ethanol suppresses parathyroid secretion and increases urine losses of Ca^{++} and Mg^{++}, thereby temporarily reducing serum Ca^{++} levels.

Hormonal Control Of Blood Glucose

Insulin is one of a family of peptide growth factors known as *insulin-like growth factors* or *IGF's* (and related to relaxin—an ovarian hormone). IGF's are produced in many tissues. Because they often mediate the action of growth hormone, IGF's are also known as *somatomedins*. Not surprisingly, there is some binding of insulin to IGF receptors and vice versa. Insulin is enzymatically produced by chopping the longer IGF-like *proinsulin* peptide into two shorter peptides that remain linked together by two disulphide bonds—it is then secreted at greater than basal levels in response to a meal, especially as blood glucose levels rise. One important insulin effect is to stop liver cells from releasing glucose when there already is sufficient glucose in the blood—insulin also promotes glucose uptake by liver, muscle and fat cells. Therefore, glucose given intravenously along with insulin can rapidly reduce dangerously high blood K^+ levels since blood-borne glucose moving into muscle and liver cells drags along K^+ (see also Kidney and Heart).

Insulin helps to keep your blood glucose levels in the normal range whether you are eating, not eating or very hungry. This is especially important for nerve cells which ordinarily require an uninterrupted supply of glucose. Of course, your insulin secretion is closely coordinated with changes in the release of many other hormones as well as the function of your autonomic nervous system. Among its other effects, insulin stimulates differentiation of (makes more) fat cells, encourages kidney recapture of Na^+ and promotes the ovarian production of androgens (male hormones). All mammalian cells bear insulin receptors on their surface and respond to insulin instructions in one fashion or another. Females with very high insulin levels due to *insulin-resistance* may develop polycystic ovaries, stop menstruating and become more *hirsute* (hairy). But polycystic ovaries and also menstrual irregularities have multiple causes, and many women with *normal* hormonal function (which includes production of both male and female sex hormones) often grow more hair on their faces, arms and chest than they might wish—especially women of southern European ancestry (moderate female hirsutism was presumably associated with reproductive advantage under Mediterranean conditions).

Insulin is produced in the *Islets of Langerhans* (named for the German medical student who in 1869 first noticed these tiny clumps of endocrine cells among the exocrine secretory cells of the pancreas). Those endocrine islets contain four different types of cells and receive both sympathetic and parasympathetic innervation. The circulation of each islet consists of a mass of capillaries resembling a glomerular capillary tuft (see Kidney) with the incoming arteriole first breaking into capillaries among the central *beta* (insulin-producing) cells in order to provide those critically important cells with an accurate sample of current blood glucose levels. Islet capillaries then drain outward past *alpha* (*glucagon* secreting) cells and *delta* (*somatostatin* secreting) cells as well as islet F cells (pancreatic polypeptide producers—function uncertain) before entering venules that lead to the portal vein.

Your secretion of glucagon (which raises blood sugar) diminishes as insulin concentrations rise in those pancreatic capillaries unless your blood sugar drops consequent to insulin release, as it might do with a mostly protein meal (another form of glucagon is produced by cells of your intestine wall). Somatostatin suppresses both alpha and beta cells as well as the release of growth hormone, TSH and most of your gastrointestinal peptide hormones—but being last in line for capillary blood flow within each islet means your delta cells must signal to those adjacent cells by sending that somatostatin signal through the general circulation (sort of like sending an air mail letter to the house next door). Liver cells utilize about half of the insulin that arrives through the portal vein from the pancreas—half of the rest lasts only 5 or 6 minutes in the general circulation. Insulin reaches muscle and fat cells mostly by being taken up actively into endothelial cells and then exported to tissue fluids—a process known as *transcytosis*. There are at least five related varieties of *glucose transporters* that can be brought to the cell surface in response to insulin signals. These transporters then move glucose into the cell along its concentration gradient. In addition, intestinal and kidney cells have *co-transporters* that actively import glucose against a concentration gradient by simultaneously allowing a Na^+ ion into that cell (another example of your Na^+ gradient at work).

Diabetes Mellitus means "to pass sweet urine" (a description based upon an outmoded method of urinalysis). Diabetes occurs when insulin production becomes insufficient, perhaps due to the destruction of beta cells by a person's own antibodies (such *autoantibodies* are a common cause of diabetes in childhood or adolescence). Those with early onset diabetes represent about 10% of all diabetics and their disease is generally referred to as *Type I* or *insulin-dependent diabetes mellitus* (*IDDM*). Diabetes mellitus is sometimes associated with normal or high insulin production (for example, when insulin receptors are genetically defective, blocked by autoantibodies or otherwise damaged). Diabetics tend to have impaired glucose transport into their beta cells and these beta cells are less responsive as well—even while still producing insulin in adequate quantities (beta cells also produce *islet amyloid polypeptide* which reduces the sensitivity of many cells to insulin). The liver cells of diabetics continue to release glucose into the blood when they ought to be taking up glucose in order to produce glycogen. A person with severe untreated diabetes mellitus will therefore have high blood sugar levels. The osmotic effect of excess sugar spilling into the urine then causes copious urine production with dehydration and salt losses. Uncontrolled diabetics also have altered protein, fat and carbohydrate metabolism (see Metabolism). If untreated, these disturbances and their associated *acidosis* (low blood pH) can lead to diabetic coma and death.

Unlike thyroxin (a modified amino acid that is active if taken by mouth), insulin is an easily *digested* polypeptide (quickly hydrolysed into individual amino acids when eaten) so it must be injected in order to have any effect. Frequent insulin injections can regulate abnormally high blood sugar levels but many quite-well-regulated diabetics still develop problems with vision, sensory nerves, kidney function and arterial circulation in their later years. Perhaps the extra insulin that must be injected to mimic the full effect of naturally produced insulin from the pancreas (which initially passes through the portal vein to the liver) undesirably stimulates IGF growth receptors in systemic arterial wall cells (see Circulation). Or possibly those late complications reflect the adverse chemical effect of high glucose concentrations, including unwanted covalent bonds between glucose and various body proteins. Such *glycated* proteins then develop further cross-links within or between proteins. The resulting dysfunctional, misshapen and clumped proteins (*advanced glycosylation end products*) become difficult for the body to break down and eliminate—their accumulation in the tissues of diabetics and other elderly individuals without diabetes could contribute to some of the degenerative conditions that afflict diabetics and the elderly (see Eye).

Other hormones promote the release of insulin. These include gastrointestinal inhibitory peptide, glucagon-like peptide, gastrin, secretin, cholecystokinin, vaso-active intestinal peptide, gastrin-releasing peptide and enteroglucagon. Many additional hormones, including growth hormone, glucocorticoids and *epinephrine* (like thyroid hormone, a derivative of the amino acid tyrosine) will raise blood glucose—it is the epinephrine secreted in response to hypoglycemia that makes you so sweaty and tremulous. Thyroid hormone increases your metabolic rate and thus glucose consumption as well. Obesity tends to worsen *adult-onset diabetes* (also known as *Type II diabetes mellitus, non-insulin dependent diabetes mellitus* or *NIDDM*) which affects about 5% of Americans over 40 years of age. The severity of adult-onset diabetes can often be reduced by exercise which beneficially affects the frequently associated obesity as well as any *hypertension* (elevated arterial blood pressure) and *hyperlipidemia* (increased blood fats and cholesterol). Suffice it to say that the cellular interactions by which you extract and utilize your stolen reducing power are very complex and not yet completely understood. However, the effective regulation of blood sugar must have been critically important to the survival of your ancestors through good times and bad, and therefore a big factor in their reproductive success. Thus it seems that a great many similar molecules acting through similar receptors at a great many different sites to coordinate a great many different effects can regulate your blood glucose far more effectively than any single on-off signal.

Adrenal Steroids Can Make You Strong And Sexy, Or Cause You To Retain Salt Or Fall Apart

As previously mentioned, the adrenal *medulla* originates from and remains an extension of the sympathetic nervous system. However, the adrenal *cortex* is derived from entirely different embryonic tissues so adrenal cortex and adrenal medulla operate quite independently of each other, even though cortex surrounds medulla. That close relationship simply reflects the fact that embryonal sympathetic-ganglion cells exposed to high glucocorticoid levels became adrenal medulla cells while otherwise identical cells matured into sympathetic nerve cells under the influence of fibroblast and nerve growth factors (or died of apoptosis if not needed).

The adrenal cortex contains three concentric layers of cells, each producing typical *steroid* (modified cholesterol) molecules with overlapping but generally quite different functions. Thus the outermost layer of the adrenal cortex produces mostly *aldosterone*, a *mineralocorticoid* that helps you to reclaim sodium from your urine and eliminate K^+ (see Kidney). The middle layer of the cortex produces *glucocorticoids* (mostly *hydrocortisone* and *cortisone*) which raise blood glucose levels and affect your immune system. Glucocorticoids also reduce inflammation and help you to gear up for intermittent stress. Abnormally high blood levels of glucocorticoids can lead to osteoporosis, poor wound healing, increased susceptibility to infection, untoward mental changes and high blood pressure. Abnormally low blood glucocorticoid levels must be treated with supplemental *hydrocortisone* or a comparable synthetic steroid to prevent depression of circulation and death after even minor stress.

As textbooks usually explain it, the hypothalamic secretion of corticotropin-releasing hormone induces the pituitary to release ACTH which causes the adrenal cortex to put out glucocorticoid in order to deal with the *stress* (be it depression, anorexia nervosa, alcoholism, strenuous exertions, hemorrhage or whatever). But the actual version is far more complex, with corticotropin-releasing hormone also produced and acting elsewhere (via the cerebrospinal fluid, for example), subject to many other influences, such as epinephrine, posterior pituitary hormones and vasoactive intestinal peptide. And ACTH may be released by posterior pituitary hormones directly, or by *interleukins* (see Immunity), or epidermal or liver *cell growth factors*. Furthermore, there are both slow and rapid paths for partially inhibiting ACTH release as well as corticotropin-releasing hormone production.

Clearly stress causes complex body-wide adjustments that affect everything from eating to reproduction (by cessation of ovulation, for example). So while you may remember the simpler explanation in which ACTH stimulates glucocorticoid output by the adrenal cortex, also keep in mind that, as usual, there is a lot more to it than that. Furthermore, although steroids ordinarily act over hours after binding to intracellular steroid receptors, it seems that gonadal and adrenal steroid hormones (as well as those costly and illegal anabolic steroids that sometimes cause steroid rages) can also affect your thoughts and actions as soon as they reach certain nerve cell membranes (where apparent steroid receptors have been discovered). To confuse matters even more, some intracellular steroid-family receptors may be principally activated by unrelated molecules such as dopamine. And whether you are female or male, the innermost cell layer of each adrenal cortex produces *both* female and male sex hormones (as do your gonads—see Reproduction).

Intermittent Release Improves Regulation

Most hormones are released from endocrine tissues to body fluids in intermittent bursts or pulses rather than continuously at a variable rate. Pulsatile hormone release provides temporarily higher concentrations at a lower rate of production, which is properly frugal and therefore reproductively advantageous. Furthermore, both the frequency and total amount of intermittently released hormone can easily be controlled by feedback from previous releases, and there is less need for receptors to detect differences in the quantity of signal molecules being released in each pulse. Not surprisingly, the production and availability of receptor molecules often varies in a way that allows an appropriate response by every relevant cell—after all, a non-responsive cell might as well not be there. Thus regularly administered drugs may lose effectiveness if they encourage a reduction in relevant receptor molecules. Intermittent activation is a common and prudent way to regulate the output of hearts, home-heating furnaces and water-well pumps. Such machines often deliver to some sort of reservoir that stabilizes output within an appropriate range for average conditions. Furthermore, if intermittent hormone pulses can do the job, that surely suggests you need not struggle to maintain steady blood levels of thyrotopin, ACTH, testosterone or whatever.

Indeed, your blood hormone levels vary widely during the day in an economical and predictable fashion most suited to your average daily needs—as part of your circadian rhythm. Gradual changes in that daily rhythm play an important part in the seasonal cycles of algae, flatworms, insects and vertebrates such as yourself. The melatonin levels in all of these life forms seem responsive to the relative length of daylight and dark, with vertebrate pineal glands and circulating blood both having their highest melatonin levels at night. So without apparent effort, you secrete hormones in an appropriate fashion for one who sleeps at night and is active by day, just as you regularly undergo seasonal variations (perhaps spring fever will eventually be recognized as a legitimate reason for time off). And even though you can always produce additional hormones as you gear up for unusual demands, it is far less stressful to cooperate with your internal clock. That is why a brief nap after lunch may boost your productivity more than a pot of coffee, while costing you less in wear and tear.

Further Reading:

Gilman, Alfred Goodman et al. 1990. *The Pharmacological Basis of Therapeutics.* 8th ed. Elmsford, N.Y.: Pergamon Press. 1811 pp. (useful for this and other chapters)

Linder, Maurine E. and Gilman, Alfred G. July 1992. "G Proteins". *Scientific American.* pp 56–65.

YOUR BLOOD

Life's Waters

The construction, utilization and biological breakdown of life's organic molecules all depend upon their solubility, ionization and reactivity in liquid water—so living things that become seriously dehydrated must either suspend their metabolic operations or die. Since all life is solar-powered, non-photosynthesizers generally survive by stealing organic molecules from plants or each other. The solar energy recovered when that *food* (stored reducing power) is oxidised can be used for construction and maintenance or to increase energy reserves. But fresh food is not always readily available and food storage poses problems since externally stored food soon spoils, while eating in excess of current requirements results in growing fat deposits that reduce mobility and increase risk. Therefore you must rely upon the living or recently dead tissues of plants or other animals to keep you running after more of the same.

Food spoilage shows that bacteria, fungi and other life forms grabbed theirs before you were ready to eat yours. To store food successfully you must somehow protect it from bacteria, flies, mice and the occasional elephant. As a start, you often try to inhibit or kill all incoming or resident microorganisms except those whose metabolic wastes can improve flavor or bring other benefit—as in bread and cheese. Naturally dried seeds (corn, wheat, beans, etc.) are quite resistant to micro-organisms. Direct dehydration by exposure to relatively dry air or even a partial vacuum (after slicing to expose raw surfaces) is often used to preserve fruit, vegetables, muscles and other foodstuffs. Surface applications of dry salt or sugar can rapidly remove water from sliced foods by osmosis. Boiling and then ongoing storage in animal fat (or vegetable oil) murders microbes while simultaneously dehydrating (and adding calories to) your food. Cooking in large amounts of sugar similarly *preserves* some foods by boosting total food *solutes* (ions and molecules dissolved in solution) to levels encountered in dehydrated foods.

Acidification suppresses most bacteria so you can sometimes preserve food for yourself while also enhancing flavor by sharing it with the microorganisms that produce yoghurt or sauerkraut. Higher concentrations of *ethyl alcohol* (a yeast metabolic waste that represents opportunity for other life forms) also suppress bacterial growth so there is no need to refrigerate your whiskey. During the slow freezing of animal or plant tissue, ice crystal growth removes water from increasingly concentrated adjacent fluids—similarly multiyear polar sea ice becomes desalinated as each freeze-thaw cycle forms new salt-free ice crystals while the less easily frozen *brine* (concentrated salt solution) drains away. But no matter how it is achieved, marked dehydration and *hypertonicity* (high solute concentrations) deprive micro-organisms within that food of access to usable water—forcing them to suspend metabolic activities or die—even you cannot benefit from the reducing power stored within your own dried or salted foods or preserves without supplementary water.

You Really Are Quite Wet

Most humans barely float so their overall density approaches that of water (one kilogram per liter). Fat cells contain little water so fat folk float freely. The very fat may carry half their body weight as water while truly skinny people are 75% water. So perhaps *two thirds of your total body weight is water* that you must haul about to keep your cells alive and working. And *two thirds of that* total body water *is intracellular* in location—the remaining one third fills your *extracellular space*. (A biolinguistic reminder: *extra-* means outside, *inter-* means between, *intra-* means inside). We already know that intracellular water contains very little Na^+Cl^- but quite a lot of K^+ (along with many other protoplasmic ingredients). On the other hand, extracellular water is almost 0.9% Na$^+$Cl$^-$ and includes little K$^+$. That *salty extracellular fluid* provides your cells with a metabolically comforting reminder of the ancient seas in which unicellular life first developed. Perhaps surprisingly, Earth's oceans have remained just about 3% salt by weight for billions of years despite new salt always arriving by river and ongoing evaporative water losses. Thus there seems to be a rough balance between salt being added and the superficial and deep processes leading to salt *burial* (in thick deposits wherever landlocked seas evaporate completely, or as ocean floor Na$^+$ and K$^+$ silicates are carried deep into Earth's mantle by subducting oceanic plates). The rhythm of Earth's major evaporative episodes seems to reflect the gathering and fragmentation of continents that occurs every 500 million years or so. Despite the fact that similar quantities of Na$^+$ and K$^+$ are held within Earth's rocky crust, the oceans contain thirty times more of the smaller Na$^+$ ion (which dissolves more readily because it fits so well between water molecules). In addition to being more tightly bound within Earth's minerals, K$^+$ is preferentially extracted from puddles, streams, lakes, rivers and seas by algae and other plants busily producing new high-in-K$^+$ cells.

An average 70 kilogram (150 lb) adult American male contains over 45 liters (more than 47 quarts) of water. Of those 45 liters of water, perhaps fifteen are extracellular in location. Interestingly, most other animals carry similar ratios of intracellular to extracellular water, regardless of whether they travel by land, sea or air. Quite clearly, if 15 liters of extracellular water can maintain the average male in a competitive state, any additional extracellular fluid would be unnecessarily burdensome while any less might represent *dehydration*. Although active multicellular ocean-dwelling animals generally haul along comparable quantities of extracellular salt water—and many sea creatures must even excrete salt in order to avoid dehydration—the more sedentary underwater sorts may get by quite nicely with cumbersome flow-through designs like that of a sponge.

Blood Is Your Medium For Exchange

An average male's many dozen *trillion* (thousand billion or million million or 10^{12}) cells have about the same total surface area as half a dozen football fields—his 15 liters of extracellular fluid spread evenly over all of these cells would form a fluid layer only one or two *microns* (millionths of a meter) thick. But your extracellular fluid is not that evenly distributed since your brain cells are tightly packed with little intervening fluid while your blood, cartilage and other connective tissues ordinarily include more extracellular than intracellular water. Evidently your cells can prosper with just a thin rim of fluid over some portion of their surface as long as the water, oxygen, nutrients, ions, current molecular messages and metabolic wastes of that extracellular fluid are continuously renewed or exchanged against supplies from the outside world. In comparison to that total cell surface area, your outer body surface is tiny—only about two square meters (20 square feet). Because your skin has to be quite impervious to water, gases, wastes and nutrients, your very competitive water-conserving design depends upon several extraordinarily large water-saturated internal surfaces for the exchange of gases and other solutes. Not only do those huge inner surfaces place major constraints upon the size, location and function of your lungs, kidneys, gastrointestinal tract and liver, but it is also necessary for these crucial surface-expanded organs to then interact efficiently with all of their distant dependent cells—which explains your flowing *blood*, the fluid organ that passes within easy diffusion distance (20 microns or less) of every healthy body cell except in your cornea, bone, cartilage, inner-ear and testicles.

Thus your cells all depend upon and interact via thin surface films of extracellular fluid that are continuously renewed by the blood being pumped endlessly through your *endothelial-cell-lined arteries, capillaries* and *veins.* Although many smaller and simpler creatures get along splendidly with far less, only the superb efficiency of your wonderfully integrated circulatory infrastructure could ever support your enormous cell population and the extraordinary specialization of your tissues. Your blood is the final common carrier that transports great numbers and varieties of cells, cell fragments and solutes in endless circuits between each of your tissues and your lungs while providing reliable pickup and delivery at all sites along the way. But unlike the securely positioned cells of your more solid organs, individual blood cells must travel alone and drift freely in order to squeeze several times per minute through your narrow capillaries. For it is across your thin capillary walls that all exchanges between blood and tissue fluids eventually take place. Since the transport function of blood cells requires them to keep moving, blood is the only organ that has no use for collagen. Therefore, any exposure of blood to collagen or some other *foreign* (non-endothelial cell) surface becomes a useful signal that blood has escaped from its proper endothelium-lined channels and is of no further use. Indeed, most cells outside of the circulatory system display *tissue factor* upon their surfaces—and the local generation of *blood clot* upon direct contact of blood with collagen, tissue factor and other foreign surfaces then seals off the leak so the rest of your blood can continue to circulate.

Your Just-in-Time Inventory System

To remain competitive, you avoid investing productive assets in excess inventory. While it certainly pays reproductive dividends to maintain a moderate energy surplus in the pleasing form of lightweight fat, too much fat soon becomes a burden. And your supportive bones ordinarily provide adequate supplies of calcium and phosphate at little extra cost. On the other hand, water is generally available and you are surrounded by atmospheric oxygen as well. So it makes sense to accept delivery of those crucial bulk supplies only as needed. Which is just another way of saying that you must live where there are adequate supplies of water and oxygen. Furthermore, the more active you become, the more water and oxygen you will require. And to maintain a stable level of oxygen and CO_2 in your blood, both your total blood flow and *respiration* (breathing rate and depth) must increase in accordance with your physical activity. Perhaps surprisingly, that breathing requirement would not be reduced even if you carried additional internal oxygen reserves, for any marked buildup of CO_2 soon stops your metabolism anyhow. Indeed, your arterial blood CO_2 levels ordinarily receive far more regulatory attention than your generally adequate arterial blood O_2 levels (see Respiration). But just as a crash-proof airplane could never get off the ground, your optimized design reflects a huge number of ancestral compromises between safety and practicality. So it generally takes your entire blood volume (about five liters of fluid blood) circulating continuously within your cardiovascular system to maintain your other extracellular fluids in a state that can support vigorous activity. And when that blood flow finally stops, the specialized cellular functions that depend upon timely pick-up-and-delivery of gases and nutrients also grind to a halt.

Nature selectively amplifies life's best responses to past challenges, but evolution cannot anticipate radical environmental changes yet to take place. Your ancestors were all short-timers. They could not afford metabolic investments with little likelihood of early payoff so such investments remain a minor part of your inherited survival strategy. However, a minor strategy can always become dominant if changing conditions favor its selection. For example, those individuals who readily accumulate a great deal of body fat are more likely to survive repeated feast-famine cycles. Famine has played such a persistent role in human history that most humans quickly gain weight when overfed. That is why inefficient metabolizers (individuals who can eat as much as they wish without gaining weight) are so unusual. Being highly efficient and no longer subject to such cycles has caused the Pima Indians of the dry American Southwest to suffer a high incidence of obesity and related disorders (e.g. diabetes and hypertension). Many insects, some rodents and bears *hibernate* as another way of extending seasonally limited reducing-power resources, for only during hibernation can they safely reduce their basal metabolic rates to any significant degree. Interestingly, certain hibernating rodents must occasionally rewarm during the winter at a significant cost in stored reducing power, merely to sleep and dream a bit (which they cannot do at subnormal temperatures)—another demonstration of the importance of sleep, and of life's delicate balance of urgent priorities.

Moderation Pays

Too much, not enough. Too soon, too late. Too big, too small. Too complicated, too simple. Too independent, not productive. Your entire beautifully integrated body is a very effective response to an enormous number of relentless pressures encountered by your ancestors. As a result, every one of your organs, and every tissue within each organ, and every cell within each tissue, and every organelle within each cell has been endlessly optimized—and all remain subject to ongoing interactive controls. When viewed from your twig on the evolutionary tree, it may all seem impossibly difficult to understand, but given the options available to First Life, your current complexity was inevitable. For example, always responding in the same fashion to a certain chemical or other stimulus could easily be a less successful ploy than sometimes not responding, or sometimes responding in a different fashion. Those who "do it the same way every time" may be skilled at what they do but they are unlikely to become leaders in innovation. So when conditions are changing and only the most effective responder gets to reproduce, life inevitably becomes more complex. In turn, life's increasing complexities unsettle the environment for others, so all living things eventually face increasingly difficult times—a minority meets the newest challenge while the majority disappears. As individuals evolve, their group behaviors and structures also must change or risk becoming increasingly irrelevant and burdensome.

It was the terrifying and bloody process of natural selection that eventually led to fine upstanding folk such as you, as well as every other currently competitive and thus equally efficient living thing. And all of you made it over the dead bodies of that forever silent majority of the less efficient, less adaptable or simply less fortunate who unwillingly gave up their tediously gathered reducing power so that the more competitive might endure. Yet in every age, an exemplary few, by their way of life and manner of death, somehow evaded routine recycling in order to send forward tantalizing fossil declarations of a time when they were among the most successful and efficient life forms under the warm noonday sun. Evolution is a process of trial and error that inadvertently adapts life to current conditions, so it is meaningless to claim that humans are more highly evolved than algae or worms or dinosaurs—each seems to have done very well in its own time and fashion. But if one could somehow reintroduce a dozen pairs of healthy, reproductively mature Tyrannosaurs to Brooklyn or Beirut or Bosnia on some lovely spring morning, it is highly unlikely that they would last even until nightfall. A dozen well-matched pairs of lovable adult humans might find themselves equally unsuited to the ambience of some dinosaur town. At times, being highly evolved may simply reflect an individual's ability to adapt, endure and avoid attracting undue attention from current holders of power or even their well-adapted parasites.

Uncontrolled population booms enhance social instabilities, exhaust environmental resources, disturb ecosystems and encourage migrations, thereby increasing opportunities for old and new predators and parasites and preparing the way for population crashes. Even within stable populations there is great variability. As individual variations merge into maladjustments, we refer to them as *diseases*. A diseased person may simply have insufficient control over some ordinarily minor aspect of growth, development or function. Although such disturbances often endanger survival and reproduction, normally detrimental diseases can become key to evolutionary advancement under markedly altered circumstances (see Sickle Cell Disease below). Some theorize that we differ so markedly from chimpanzees despite sharing almost identical DNA sequences because some subtle mechanism has disturbed and slowed human growth and development—and in many ways, you rather do resemble a cute baby chimp. Perhaps that prolonged infantilization has provided more time for development of your human brain and information processing capabilities in this ever more challenging world.

Here and in other chapters we consider a few of the complex controls that regulate your blood organ. Yet your other tissues and organs labor under similarly strict stimuli and restrictions. In general, increasing demands encourage growth while overgrowth triggers apoptotosis (programmed cell death). Although individual humans may vary greatly, their tissues and organs tend to assume individually appropriate sizes and shapes. So the fisherman who regularly pulls nets by hand eventually develops huge strong fingers of quite ordinary length. And the extraordinary ability of a liver to regenerate within weeks (another sign of its adaptability) is only rarely called upon following subtotal surgical removal. Even that impressive pregnant uterus soon shrinks to its usual small size after delivery. Similarly, your blood volume returns to normal within hours after you donate blood (see Fluid and Electrolytes). The accessibility, uniformity and fluid nature of blood make it the easiest organ to sample and study. And blood provides yet another demonstration of how closely controlled moderation within the current range of competitive possibilities tends to be the most successful reproductive strategy.

Your individual organization is a dictatorship that allows a cell to survive only while that cell has value for the group. Indeed, most or all of your cells carry instructions that cause them to die unless needed. That is exactly what happened to cells in the tail formerly decorating your embryonic backside, or the skin that once connected your embryo's webbed fingers and toes. Furthermore, the same strict rules apply to the white blood cells sent out as suicide troops, the red cells that die if not released into your circulation by a hormonal request for more red cells, the early nerve cells that die after providing a living pathway for other neurons to follow and any other misconnected nerve cells or nerve cell branches that receive signals unrelated to their designated function. Such a prevalence of apoptosis again shows how it can be cheaper to operate according to general rules that include the production and recycling of spares than to start and stop assembly lines in response to changing needs—furthermore, a needs-based response inevitably involves delay and only prompt responders get to reproduce. Interestingly, there is considerable anecdotal evidence that individual humans are more likely to die if not needed/ wanted/nurtured/involved; even negative attention (or a pet) can be healthier than no attention at all (see Stress, Immunity). But clearly there are important differences between the operation of an individual who has been organized in strict accordance with inherited ancestral instructions and the organization of fair and effective human societies by mutually beneficial rules only now coming into view.

Blood Plasma Is Extracellular Fluid— Red Cells Contain Intracellular Fluid

All organs contain both extracellular and intracellular fluid. About three liters of *extracellular water* circulate within your *intravascular compartment* (here defined as the blood channels of an average American adult male). The remaining twelve of your 15 liters of extracellular fluid are located outside of your heart and blood vessels in the *extravascular compartment* of various tissues and organs. Other than at the blood-brain barrier, there is relatively free trans-capillary exchange of water, gases and smaller-than-protein solutes between blood plasma and tissue fluids (between your *extracellular intravascular* and your *extracellular extravascular* fluids). Indeed, that is the way those five liters of blood continuously refresh the extracellular fluids of your various tissues. It would take relatively huge quantities of salt solution (as blood substitute) to distribute oxygen, nutrients and messages while also eliminating the CO_2 and metabolic wastes produced by your other 65 living liters. For among other factors, efficient pickup and delivery requires proper packaging to minimize handling and many essential molecules hardly dissolve at all in salty water. Consequently that much larger volume of circulating salt solution would need to flow at far higher speeds while moving lots of poorly soluble molecules about in particulate form—leaving most cells hard pressed to select essential supplies from amongst the solutes and small particulates whizzing past. With such a circulation, multicellular life would have had minimal mobility especially on land. In contrast, if you examine the blood flow within a typical capillary under the microscope, you will see a few red blood cells barely squeezing through the slender capillary channel between flattened endothelial cells. Especially considering the magnification, those red cells are not moving at all quickly. Indeed, that capillary blood flow often stops for a while as certain capillary cells constrict individual channels. Apparently it takes very small quantities of blood to serve all tissue cells dependent upon any particular capillary. In turn, that leisurely capillary flow surely improves the pickup and delivery of wastes and supplies.

Very Slowly, Blood Became Thicker Than Water

The complex and efficient blood organ that now keeps you alive evolved over hundreds of millions of years through an endless series of miniscule improvements upon the simpler salty solutions that once sustained your early multicellular ancestors. Every advance in the transport capacity of their blood brought your ancestors closer to a durable leak-resistant low-volume *cardiovascular system* (heart and blood vessels) that could function efficiently at low flow rates and low blood pressures—until eventually those incremental circulatory improvements made possible such amazing designs as the giraffe, the snake and you.

Capillaries Help Blood To Serve

An almost continuous sheath of flattened endothelial cells encloses your blood as it moves endlessly around your circulatory system. These endothelial cells vary considerably in form, function and fit, as appropriate to the tissue being served. Thus the endothelial cells of many tissues actively trap, ingest and transfer blood-borne proteins and small particles. Although the loose fit of most endothelial cells permits easy movement of fluids and white blood cells into surrounding tissues, some kidney endothelial cells also display large pores that expose their basement membrane filters—and liver cells are in direct contact with blood-borne nutrients as well as the blood cells passing through those partially endothelialized liver sinuses. In contrast, the tight fit of brain capillary endothelial cells enhances your blood-brain barrier. Endothelial cells also secrete lipid substances (prostoglandins) that increase or decrease local channel size and alter the likelihood of nearby clot formation. And they produce *tissue plasminogen activator* (TPA), a molecule that helps to dissolve adjacent blood clot.

The pathway of any one capillary may seem random but capillaries somehow end up passing within easy diffusion distance of almost all tissue cells, for tissue growth and metabolic factors tend to regulate capillary formation and thus distribution. Of course, every new capillary expands the interconnected capillary network through which your blood courses from tiny incoming arterioles toward tiny collecting venules. Furthermore, single-capillary blood flows can change rapidly in response to local tissue pH and CO_2 levels, waste and nutrient concentrations, hormonal and sympathetic nervous signals, blood volume, blood pressure, physical activity, body temperature and so on. Higher blood flow rates tend to deform endothelial cells or at least subject them to shear stresses that can affect branching angles and vessel growth—persistently elevated flows may injure endothelial cells.

Under steady flow conditions, the same volume of blood passes through every section of your closed circulation. Therefore blood flow slows where the total cross-sectional area of all parallel channels is greatest—just as a river slows where it widens markedly or breaks into numerous parallel meanders in the delta. That inverse relationship between flow rate and cross-sectional area explains how blood can reach a tissue quite rapidly through its narrow incoming artery, yet pass slowly through the relatively huge number of capillaries connecting artery to vein within that tissue. The slightly negative electrical charge of many endothelial-cell surface molecules helps to reduce adherence and drag as negatively charged blood proteins flow past. Blood viscosity rises when red cells represent over 55% of the total blood volume or with certain blood protein abnormalities. But increasing the flow rate of blood generally reduces blood viscosity by moving red cells and proteins away from channel margins and into more parallel orientations—such non-Newtonian behavior occurs because blood is not an ideal homogeneous fluid.

Within most tissues, cells of identical specialty tend to be uniform in size and function. The standard size of endothelial cells helps to maintain capillary diameters at 6-7 microns. With the diameter of your red blood cells also about seven microns, those red cells barely can squeeze through any capillary. Yet even while you rest, an average red cell must pass through at least two capillaries per minute, so there is a lot of direct red-cell-to-capillary-endothelial-cell contact. That continuous contact is probably of importance since capillaries could easily be somewhat larger. But with any slight narrowing of the capillary orifice able to block incoming red cells, local conditions can control local blood flows. Furthermore, while endothelial cells and red cells both have relatively huge surfaces, these cells still are small enough for diffusion to bring about rapid equilibration of trans-capillary gas pressures and solute concentrations. Indeed, molecular diffusion is so swift at these scales that even the inevitable momentary contact between a red cell and each capillary endothelial cell allows significant direct gas exchange between them (in addition to the indirect transfer of gases through blood extracellular fluids). Having red cells brush past regularly should also clear endothelial cell surfaces of any minor projections and loose debris, thereby helping to maintain capillary patency and the negative endothelial cell surface charge. And each gentle compression by a passing red cell expedites the transfer of nutrients and wastes across the endothelial cell.

Your Red Blood Cells May Be Red But They Are Not Cells

Properly speaking, your 20 trillion red cells should be considered formed elements of blood rather than cells, for a red cell ordinarily discards its nucleus as it enters the circulation from your bone marrow at the normal replacement rate of over 150 billion new red cells per day. Mature red cells are homogeneous biconcave discs carrying about ⅓ of their weight as *hemoglobin*. Hemo*glob*in is a roughly spherical (thus *glob*ular) oxygen-carrying protein molecule that consists of four interlocked polypeptide chains. Individually these chains closely resemble *myo*globin (myo- means muscle) in shape and function, although they bear quite different amino acid sequences. For with twenty varieties of amino acid available for protein construction, different amino acids can sometimes replace one another without affecting the final protein structure, provided the side chains of both amino acids are similar in size, shape and polarity. Thus the great variety of amino acid sequences found in hemoglobins from different species of vertebrates (and any differences between hemoglobins and myoglobins within the same species) simply document insignificant copying errors made during DNA replication over the ages. Since such innocuous substitutions tend to occur more or less regularly by chance, any inter-species differences in the amino acid sequences of such molecules can help to determine ancestral relationships between species as well as between particular molecules. On this basis, it seems that hemoglobin and myoglobin modules first entered divergent evolutionary pathways about 700 million years ago, while the two principal myoglobin-like (alpha and beta) subunits of hemoglobin only separated about 500 million years ago.

Every myoglobin molecule includes a single "hindered *heme*" group bearing a solitary iron atom. Although even an isolated heme group in the Fe^{++} (ferrous iron) form attaches willingly to oxygen, the Fe^{++} of a free heme is rapidly oxidised to Fe^{+++}

(ferric iron) which does not bind electron-hungry O_2. So isolated heme units have little value for oxygen transport or storage until *hindered* by a surrounding globin molecule which only allows oxygen to reach the Fe^{++} through a cramped *globin cleft*. This cleft also reduces the otherwise great affinity of heme ferrous iron for the tiny quantities of *carbon monoxide* (CO) that cells ordinarily produce, encouraging the eventual displacement of such CO as well. Your myoglobin binds and stores molecular O_2 for release to your muscle fibers as tissue oxygen levels fall. However, when four of those myoglobin-like units (and it has to be two alpha and two beta units) are appropriately associated as a complete hemoglobin molecule, they become an *allosteric* protein—a smart molecular machine capable of altering its own shape and function (including oxygen and water affinities) in response to various environmental conditions. Thus the increased CO_2 levels and declining pH commonly encountered in exercising muscle rapidly alter hemoglobin's configuration and adherence to O_2. This allows hemoglobin to make a fast dump of O_2 right where it will do the most good. At the same time, the ability of hemoglobin to bind both H^+ and CO_2 (hemoglobin's *buffer* effect) minimizes local pH changes consequent to lactic acid and carbonic acid (H_2CO_3) accumulation. This allosteric process reverses within the lungs, thereby expediting oxygen pickup as well.

Furthermore, the four oxygen molecules transported by each hemoglobin molecule load and unload *cooperatively*. This means every oxygen added makes it easier for subsequent oxygens to come on board, while each oxygen released reduces the affinity of hemoglobin for those that remain. Of course, hemoglobin always binds oxygen less avidly than myoglobin does for otherwise oxygen could not transfer from blood to tissue stores. Similarly, adult hemoglobin holds oxygen less tightly than fetal hemoglobin does, which encourages swift oxygen transfer as maternal and fetal circulations pass close by each other in the placenta. Hemoglobin clearly dominates red blood cell function so one might expect to find little more except water and ordinary intracellular solutes within those tiny biconcave hemoglobin packages. However, a small, highly anionic molecule known as 2,3 bisphosphoglycerate (BPG) also resides within red cells in equimolar concentrations to hemoglobin. BPG plays a crucial role as hemoglobin's assistant by preferentially attaching to less oxygenated adult (but not fetal) hemoglobin, thereby encouraging that hemoglobin molecule to release more O_2 into tissues as appropriate. BPG is readily displaced again at the higher oxygen levels prevalent in lung capillaries.

Free Hemoglobin?

A biconcave disc is what you get if you squeeze a spherical basketball from both sides while letting much of the air escape. That shape provides the red cell with an unusually large and pliable cell surface membrane around its relatively small content. Such a large surface area not only enhances gas exchange, it also permits red cell flexion and distortion during each passage through a capillary. In addition, having spare surface membrane allows a red cell to imbibe extra fluid when blood solute levels fall without rapidly reaching capillary obstructive size or being so liable to burst. Several different proteins including actin and spectrin contribute to the fibrous submembranous skeletal structure that keeps your red cell membranes in proper shape. However, if a red cell does rupture for any reason, it releases free hemoglobin into the blood stream. When out of its red cell, the hemoglobin molecule tends to split in half, which increases its toxicity to the kidney. In addition, free hemoglobin becomes even more adept at picking up oxygen but, lacking BPG (free BPG is easily lost across capillary walls into tissues and urine), 26 times less willing to release that oxygen at the tissue capillaries. Were it not for that risk to kidneys and the important BPG effect, one might reasonably ask, "Why bother with red cells at all? Why not just circulate free hemoglobin molecules in the bloodstream?" Indeed, many experiments suggest that despite its tendency to split in two and higher oxygen affinity (both of which can be modified chemically or by genetic manipulations), free hemoglobin can temporarily replace all red cells in some animals. But there are good reasons for retaining your hemoglobin within red blood cells.

For one thing, were your hemoglobin free, it would become your dominant circulatory solute, exerting significant osmotic and chemical effects (unless you could cross-link a great many individual hemoglobin molecules into large polyhemoglobin units still able to capture and release O_2 efficiently). Your usual fifteen grams of hemoglobin per 100 ml. of *whole blood* far exceeds normal blood *albumin* concentrations since that next-most-common blood protein is only present at 3 to 4 grams per hundred cc of *plasma* or *serum* (see below). And we know that even small alterations in your serum albumin level can cause major osmotic shifts of fluid across your capillary walls. Furthermore, the hydrophobic cavities in albumin molecules transport many of your less-soluble nutrients (fatty acids, some amino acids, divalent cations such as Ca^{++}, Cu^{++}, Zn^{++}) and molecular messages (such as steroids) in an easily released fashion that high plasma hemoglobin levels might well disturb. In addition, all of those newly exposed molecules of free hemoglobin would increase blood viscosity and alter blood flow characteristics (like herding along a great many swimmers rather than moving one boatload of passengers). And while free hemoglobin could easily pass through much narrower capillaries, the resulting faster capillary blood flow might interfere with gas and nutrient extraction. The blood chemical and pH controls that co-evolved with red cells might also be subject to great disruption if red cells were eliminated and hemoglobin set free. For example, it could be difficult to maintain *carbonic anhydrase* at levels adequate to dissolve CO_2 into blood plasma. And any benefit derived from repetitive contact between red cells and capillary endothelium would be lost.

Free hemoglobin would probably be less convenient for liver and spleen macrophages to grab and recycle than entire red cells are. The tendency of free hemoglobin to filter into the urine might lead to major losses of free hemoglobin while also endangering kidney function. Of course, a larger polyhemoglobin molecule would be far less likely to enter the urine or escape into tissues between loosely connected endothelial cells. And that is important, for free (or broken-down) hemoglobin enhances iron levels in blood plasma and tissue fluids. Ordinarily your *haptoglobins* and *transferrins* (specialized proteins that bind free hemoglobin or free iron tightly until delivered to liver cells where that iron is then stored on *ferritin* molecules) deprive potentially invasive bacteria of the iron they must have for growth and reproduction. But were significant quantities of free hemoglobin to overwhelm those circulating iron-scavenger proteins, you would become far more susceptible to blood-borne infections—yet the principal call for a hemoglobin-based blood substitute might come during major surgery or with severe injuries, where the risk of bacterial infection already is high. So rather typically, your dependence upon red blood cells instead of free hemoglobin has no single or simple explanation. Perhaps these difficulties can be overcome in time, for a useful blood substitute would have great value. But almost surely it won't be easy, given the close integration (codependence) of red cells with capillaries that has come about as a result of endless interactions and compromises between a great many factors over a very long time—some of which might have greater impact than any problems mentioned above. Once again we see how the reproductive success of the few survivors in every large and complex generation (natural selection through survival of the fittest) has been an exceedingly effective process for choosing between competing designs and optimizing the functional result. For regardless of how clever or durable any winning or losing design may otherwise be, some eat while many are eaten. Performance rules.

Your Inherited Blood Lines

Blood has long been recognized as a vital substance. "Blood lines" provide the legal basis of inheritance and relationship. And in tough times, families and tribes all too routinely revert to an evolutionarily correct, "us against them" mind-set that is likely to bring reproductive advantage for those "of the same blood" as they cooperate to take the lives and hard-earned property of others that they can overwhelm. When you must choose sides in a fight, blood relationships are often more important than right or wrong, or who is a better person or which side seems to be winning for "blood is thicker than water". That inherent tendency of relatives to help each other even when they dislike one another intensely simply increases the likelihood of passing along jointly shared genes—less reproductively advantageous behaviors might be far more sensible, fair and kind but they just die out. Yet only in recent decades have blood tests become available that could demonstrate the degree of relatedness among humans, or even reliably differentiate a sample of human blood from the blood of other mammals (except by some aspect of its cellular appearance under the microscope). Nevertheless your iron-based hemoglobin-containing red mammalian blood obviously differs from the copper-based hemocyanin molecules found in blue-blooded lobsters. That (and their inability to scream in air) may be why otherwise kind-hearted folk who no longer would consider burning a religious dissenter or even roasting a live red-blooded rabbit have no qualms at all about heating the old pot to a boil for inducing fatal thermal injury to an apparently unoffending live lobster. But different as they are, the desperate struggles of live lobsters to escape that hot pot certainly suggests that they find it unpleasant. Of course, few of us can imagine walking a mile in a lobster's claws, or indeed relating to them (uncooked) in any other fashion.

Your Blood Is Made Specifically By And For You

A brain in a pickle jar cannot convey the essence of its prior function or previous owner. Nor is a pint of blood in the blood bank any longer an integral part of a dynamic living organization. And when a relatively fresh pint of whole blood is transfused into a new owner, that blood and owner may take a while to get used to each other, even if no noxious DNA or RNA virus or other infective information or *allergen* (molecules to which the recipient is allergic) have been included. But transfused blood clearly is a lifesaver when you need it and even minor "transfusion reactions" are uncommon, usually brief and perhaps caused by the few donor white cells received (since these are not evaluated for compatibility when whole blood is cross-matched). Inevitably, however, some minor accommodation must occur between newly introduced blood components and their recipient. On rare occasions, donor white blood cells may even launch an immune counter-attack against their new host. So by definition the *homologous* pint (of another's blood) put in is less well integrated functionally within its new host than that *autologous* pint was whence it came. The volume of blood delivered during a single transfusion includes the preservative-*anticoagulant* (anti-clotting) solution with which blood was mixed while being drawn at the blood bank, so the donor may have contributed less than a *pint* (about 500cc).

No Needles! (The "N" Word)

We will consider the physiological aspects of blood loss later (Cardiovascular System, Kidney, Fluid and Electrolytes). For now, let us simply perform a small series of *completely painless* thought experiments. Our scientific studies will take place in some cozy, fun place like a blood bank or attractively decorated hospital laboratory where blood donations or samples can be drawn painlessly from your arm vein while you gaze out of the window and contemplate the many miracles brought about by natural selection, including sparrows and smog and gridlock. Just remember, "This won't hurt!" may be one of the three middle-sized lies.

Painless Experiment #1

We draw a simple glass tube full of your blood and watch it. Once in a while we tilt the tube. After a few minutes we note that the dark red liquid blood has become dark red jellied blood—this blood specimen has now *coagulated* or *clotted*. We continue to watch it. As pleasant hours pass, the clot shrinks away from the sides of the glass tube. The shrunken darker-red jelly that remains is now suspended within a clear white-to-yellowish fluid. Apparently the clot has *retracted* and thereby squeezed out that yellowish blood *serum*. As we continue to watch, the sun goes down. This is almost as much fun as listening to Iowa corn grow on a hot summer's night. When the morning sun next strikes our window, we discover that the dark shrunken jellied clot disappeared sometime during the night (clot *lysis* has occurred). Now we see a clear upper serum layer overlying a dark red layer of liquefied clot. If we invert that tube a couple of times, the red blood below mixes evenly with the yellowish serum above. Once again the blood sample has the uniformly red appearance of fresh liquid blood (or dilute ketchup). But this blood will no longer clot. Obviously (a) the red part of a dissolved blood clot includes red cells and is heavier than the clear serum; and (b) the regular clotting process traps all cells and cell fragments, for the serum is clear—which it would not be if it still contained intact cells and cell fragments.

Painless Thought Experiment #2

We draw fresh blood into a small glass beaker and stir it slowly with a thin glass rod for several minutes. The blood remains liquid and red as a cloudy pink-white layer builds up on the submerged section of the glass rod. If we now stop stirring, remove the rod and let the blood sit, we find that this blood sample doesn't clot either. As delightful minutes turn into hours, the red cells settle to the bottom half of the beaker, leaving clear yellowish serum above. Apparently (a) clotting need not involve red cells and (b) clotting cannot occur once that cloudy white layer (*fibrin*) has been removed by the glass rod and (c) red cells are always heavier than serum, regardless of previous clotting.

Simple Painless Thought Experiment #3—Variations On "Blood Thinners"
(Anticlotting Agents Or Anticoagulants) Played In Three Tubes

(A) We draw fresh blood into a glass tube that contains a little bit of clear *heparin* anticoagulant solution, then tilt to mix blood and heparin.

(B) We draw fresh blood into a second glass tube that contains a bit of another anticoagulant, *warfarin*, then tilt to mix blood and warfarin, and

(C) We draw fresh blood into a third clean glass tube containing a small quantity of any chemical that binds tightly to Ca^{++}, perhaps *citrate* or *EDTA*. Again we tilt to distribute the dissolved chemical through the blood sample.

As we observe these tubes, tilting occasionally, Tube B clots within a few minutes while A and C remain liquid. The clot in Tube B eventually retracts—later there is clot lysis. If left undisturbed, unclotted tubes A and C also change their appearance gradually, with red cells settling to the lower half while clear yellowish fluid remains above. Apparently we have successfully prevented clot formation in Tubes A and C. The clear yellowish *supernatant* (liquid above the settling red blood cells in those tubes) must therefore contain unused clotting factors. In fact, that never-yet-clotted overlying yellowish blood *plasma* will clot under appropriate circumstances. And once blood plasma has clotted (note that red cells need not be present), the clear fluid squeezed out during clot retraction or present after clot lysis will again be depleted of clotting factors—such a depleted fluid is referred to as blood *serum*.

Conclusion

Apparently heparin and citrate are the right stuff as far as anticoagulation is concerned but warfarin is a fake. Or possibly we were cheated by the supplier. Right? Wrong.

Warfarin was discovered when an alert dairy farmer noticed his cows toppling over dead after he fed them spoiled sweet clover. Autopsies revealed internal blood loss which eventually was attributed to the presence of warfarin in the spoiled clover. Warfarin rapidly became a popular and highly profitable rat poison—but when a reluctant military recruit ate considerable amounts of it without suffering more than a temporary bleeding tendency, this rat poison was investigated for use on patients. It turns out that warfarin only works indirectly to prevent clotting by blocking the activation of numerous *Vitamin-K-dependent proteins* in liver cells. Those proteins include the *prothrombin* from which *thrombin* is formed. And thrombin is the important *protease* (protein-digesting enzyme) in your blood that converts *fibrinogen* to fibrin. Fibrin was what collected on your glass stirring rod in experiment #2 above. More to the point, fibrin forms the tangled web of protein fibers that jellies your fresh blood, much as the gelatin from boiled horse scraps solidifies your lovely fruit salad (see Collagen). There is even more to this tale. Normal rats will bleed to death if you can prevent them from eating their own (or their neighbors) shit. For rats are *coprophagic* (eternally doomed to eat feces) because the essential fat-soluble *Vitamin K* produced by their resident colonic bacteria is not absorbed from the rat lower intestine. Fortunately for your more fastidious life style, the Vitamin K produced by bacteria within your own colon is readily absorbed during its first transit (see Gastrointestinal Tract).

Summary: Liver cells require *Vitamin K* (rat shit vitamin) in order to produce prothrombin and other *coagulant* proteins (and also *anti*coagulant molecules such as protein C). *Warfarin* (spoiled sweet clover anticoagulant) interferes with that function of Vitamin K. And that explains how warfarin became a clinically useful anticoagulant even though it acts indirectly upon prothrombin production (through its effects on your liver cells) and therefore cannot stop a tube of your own fresh blood from clotting. In other words, by competing with the *hepatic* (liver) regeneration of active Vitamin K, warfarin slowly and indirectly blocks formation of prothrombin and other clotting and anticoagulant factors. So to reverse that warfarin effect, inject lots of Vitamin K and wait a day or three (or if a rat, eat lots of shit and pray you didn't get the very-long-acting warfarin derivative only recommended for rats).

Commercial heparin is derived from the many heparin-granule-filled *mast cells* residing within beef lungs and hog intestines. Heparin stops blood from clotting by activating *antithrombin III* which then ties up and inactivates thrombin and other coagulation enzymes. Single-dose heparin anticoagulation of a patient ordinarily lasts just a few hours, due to normal heparin breakdown processes. Heparin's effect is rapidly *reversed* (within minutes) by the intravenous administration of *protamine*, a strongly basic molecule that quickly binds to heparin. Protamine is extracted from salmon sperm (see Reproduction).

Citrate or *EDTA* bind calcium tightly. Blood cannot clot without free calcium, for Ca^{++} ions have essential roles at several stages of the coagulation cascade. Although both heparin and warfarin can counter clotting within a live person, any attempt to anticoagulate an individual with citrate or EDTA would cause fatal *hypocalcemia* (low blood Ca^{++}) long before circulating Ca^{++} levels fell sufficiently to stop blood from clotting. Indeed, individuals receiving several blood transfusions within a few hours often benefit from extra Ca^{++} given intravenously (but *not* through the same tubing as citrated blood is being transfused for that would immediately clot any blood in the tubing) to help maintain appropriate blood Ca^{++} levels for nerve and muscle function. Even were blood not such an essential source of Ca^{++} for so many critical cell functions, there still would be no way to eliminate Ca^{++} from the circulating blood since the skeleton is such an inexhaustible source of replacement Ca^{++} (see Hormones, Bones).

EDTA binds strongly to some toxic heavy metals so it is administered to counter lead, mercury and other heavy metal poisoning. Intravenous EDTA administration, also known as *chelation* therapy, has unfortunately become a popular (unscientific) remedy for "hardening of the arteries"—this despite the fact that arterial wall calcium plaques lie deeply buried within your aging vessel walls (and are thus inaccessible to injected EDTA—although bone calcium is easily brought into your circulation). Surely one ought to avoid depleting the bone calcium of older possibly osteoporotic persons in this fashion. But on the other hand, an occasional individual with unrecognized heavy metal poisoning has undoubtedly shown marked symptomatic improvement following chelation therapy. For example, a menopausal woman developing osteoporosis can rapidly release some of the lead present in her bones. That lead probably accumulated over a lifetime—while drinking wine from lead-glazed pottery or ingesting lead paints or food contaminated by lead-containing solders (formerly used in food cans), or by drinking water delivered through older *plumb*ing or breathing leaded gasoline fumes and so on. The rapid menopausal release of bone lead (Pb) can undoubtedly produce many of the vague but unpleasant symptoms that come with lead poisoning. But relatively inexpensive blood tests will reveal which symptomatic person might benefit from chelation therapy for lead or other heavy metal poisoning. So it seems inappropriate to administer this costly and potentially harmful treatment to many unaffected old folks (who could easily achieve the same undesirable reduction in their skeletal calcium by eating spinach and rhubarb regularly, since both plants contain sufficient oxalate to reduce intestinal calcium absorption).

The clotting of your blood is a very complex process with dozens of sequential steps involving many different clotting factors. And as you would expect, that *clotting cascade* can be inhibited or amplified at each stage. Furthermore, no individual *activated clotting factor* (most of which are specific proteolytic enzymes resembling trypsin—see Digestion) lasts very long in the normal circulation—some are quite unstable, others are immediately attacked by enzymes lurking about or rapidly diluted to ineffective levels by the passing blood. That means clot usually forms only when and where it is needed unless there is some abnormality causing a *hyper-* (too much) or *hypo*coagulable (not coagulable enough) condition.

The heparin (a proteoglycan—see Connective Tissue and Cartilage) ordinarily present in your body may help to prevent excessive blood coagulation. A slightly different polymer, heparan sulphate, is found on the surface of endothelial cells and most other eucaryotic cells as well as in the extracellular matrix. On endothelial cells it too interacts with antithrombin III to provide a naturally clot-resistant surface. And various clotting factors can have quite diverse biological effects—for example, thrombin serves as a signal to white blood cells and endothelial cells while heparin affects several points on the clotting cascade as well as increasing capillary permeability, inhibiting platelet function, inhibiting proliferation of smooth muscle cells in blood vessel walls, inhibiting delayed hypersensitivity reactions (see Immunity) and regulating *angiogenesis* (new vessel formation). Presumably heparin plays so many entirely different roles because of the various ways its molecular shape and strongly ionized state may interact with other molecules. The great number of steps that precede blood clotting, with inhibitors and amplifiers at each step, provide ample opportunity for outside interference. For example, leeches produce *hirudin* which binds directly to thrombin and thus prevents fibrin formation while also keeping thrombin from initiating platelet clumping. Under some circumstances, leech squeezings can compete in cost and effectiveness with the heparin extracted from hog or cow viscera.

Normal Blood Clotting Requires Platelets

When *platelets* are few, blood coagulation is delayed and defective. Normally each cubic millimeter (mm^3) of your blood contains several hundred thousand of these approximately one micron diameter membrane-enclosed cytoplasmic fragments that have been released from the large *megakaryocytes* located adjacent to bone marrow blood sinusoids. There is increased risk of bleeding when that thrombocyte (platelet) count drops much below about 100,000 per mm^3—below 40,000 to 50,000 there will be a measurable defect in the ability of your blood to clot. A person with thrombocyto*penia* (*too few* platelets) tends to bruise easily and will continue to bleed from even minor cuts. In the total absence of platelets, such bleeding does not stop. And if flimsy clot eventually forms where blood has pooled upon the floor, that messy clot on the floor will not retract even if kept moist, for clot retraction requires platelets. Platelets respond rapidly to contact with foreign surfaces (extravascular cell surfaces, glass rods, collagen, etc.) by swelling, clumping together and releasing factors that activate the blood clotting cascade. *Serotonin* is one important factor released by activated platelets. Depending upon its local concentration, serotonin may directly constrict damaged blood vessels while dilating others nearby. Damaged vascular endothelium itself puts out increased amounts of *endothelin* (a vessel constrictor) and reduced quantities of *endothelial-derived relaxing factor* (nitric oxide or NO). *Macrophages* ("large cells that engulf") in your spleen and (sometimes) liver remove old platelets after they have circulated for four to ten days. Surgical removal of the spleen occasionally raises blood platelet counts to nearly a million platelets per mm^3—such high platelet levels can lead to a hypercoagulable state.

By inhibiting *prostaglandin synthase*, *aspirin* prevents platelets from sticking together or producing (blood vessel constricting) *thromboxane*. So if too many platelets or excessive platelet stickiness are a problem, aspirin can serve as an effective partial anticoagulant. Aspirin is taken daily by many older individuals in whom the likelihood of blood vessel *occlusion* (*blockage* that could cause heart attack or stroke) is considered a greater risk than the possibility of enhanced bleeding or other aspirin side effects such as allergy or gastric irritation. Even small doses of aspirin (less than one tablet per day) will permanently inactivate the prostaglandin synthase enzyme in all currently circulating platelets. Such an aspirin effect wears off in about four days, once most of the affected platelets have been replaced. Larger doses of aspirin may deliver *less* circulatory benefit since they also temporarily depress the production of endothelial *prostacyclin* (a local blood vessel dilator) in systemic vessels beyond the liver—in contrast, smaller doses only acetylate platelets passing through the small-intestine capillaries and portal vein en route to the liver. A great many other substances (foods, medicines, vitamins, spices) alter the activity of platelets. Even vigorous exercise by normally sedentary people can increase the stickiness of their platelets—an effect not encountered in the physically active. And the complex chemical processes involved in platelet activation may be initiated by diverse circulating substances such as epinephrine, ADP (adenosine diphosphate), thrombin and so on. *Southeastern pigmy rattlesnake venom* is a specific inhibitor of platelet aggregation—a *deerfly salivary protein* that competes for the fibrinogen receptor on the platelet surface may turn out to be an even more effective platelet inhibitor.

Cleaning Up Clots

Clot lysis is a routine part of cleanup and repair after your damaged blood vessels have been sealed by clot. Intravascular clot lysis is most active near endothelial surfaces since endothelial cells produce *tissue plasminogen activator* which converts the *plasminogen* present in blood clot to *plasmin*, an enzyme that chops long insoluble fibrin strands into soluble *split-fibrin products*. As noted in our painless experiments above, sufficient plasmin is produced in freshly drawn and clotted blood to bring about eventual clot lysis. Vampire bats secrete a very effective *plasminogen activator* in their saliva to keep their sleeping victims bleeding freely while they lap it up (that bat plasminogen activator may have commercial potential too). A large clot may persist for a long time under certain circumstances—especially within the *pleural space* after complete lung removal or within an *aneurysm* (significantly dilated segment) of the heart or large blood vessel. Such persistent clot may undergo gradual mechanical changes, sometimes in association with adjacent *fibrosis* (ongoing local scar tissue formation—see also subdural clot on brain). Nonetheless, an ordinary bruise goes through such a predictable series of color changes that one can confidently estimate how long it has been since injury by whether a "black eye" is dark purple, lighter in color, greenish, yellowish or faded. That reliable progression of color changes stops at death—which terminates removal of iron pigment by macrophages. Thus the interval from bruising to death also can be determined. Incidentally, you can get a black eye from a bump on top of your head. This occurs because blood escaping from damaged vessels beneath the tight fibrous layers of your scalp tends to flow toward locations where it is less constrained, such as the looser tissues about your eye. Similar minor bleeding into an inguinal hernia repair can cause the neighboring scrotum to turn a worrisome deep purple—this appearance may require strong reassurance that the testicles are not falling off. Such *hematomas* (blood collections or bruises) also clear up within a few days, although very large bruises disappear more slowly. A blood clot within a blood vessel is generally referred to as a *thrombus*. If that thrombus (or any other foreign object) breaks loose and moves with the blood flow to another site within the cardiovascular system, it has become an *embolus* (see Circulation). The process of clotting in vivo is *thrombosis*. Clot on an open surface tends to dry and retract into protective scab rather than undergoing lysis (lysis only proceeds in a wet environment).

Red Cells And Reticulocytes

Every cubic millimeter (mm^3)—not to be confused with a milliliter (ml) or cubic centimeter (cc) which are 1000 times greater—of your whole blood contains plasma and also formed blood elements. In addition to platelets, those formed elements ordinarily include 5000 to 10,000 white blood cells (see Immunity) and about 5 million red blood cells. These cells and platelets originate in your red bone marrow where progenitor or *stem cells* have established separate lineages for several different sorts of white blood cells, as well as for *erythrocytes* (red cells) and mega-karyocytes (your source of platelets). Normally a red blood cell circulates for about 120 days (during which time it travels about 300 miles) following release from your bone marrow. Then it is junked and recycled through engulfment by spleen or liver macrophages. A very young erythrocyte that has just entered the circulation may still show a partially visible internal framework on routine blood stains, in which case it is referred to as a *reticulocyte*. The percentage of red cells that are reticulocytes reveals how actively new red cells are being produced. When the *reticulocyte count* rises much above 1%, it hints at recent blood loss or abnormally rapid red cell destruction from any cause.

Painless And Neat Thought Experiment #4

Draw a sturdy glass tube full of blood. Mix with a trace of heparin to prevent clotting. Place in a centrifuge and spin rapidly for several minutes. That strong centrifugal force quickly packs red cells into the bottom of the tube, thereby saving you from waiting long hours for those cells to settle under gravitational pull alone. Now we see that your packed red cells occupy 45% of the entire blood sample volume. The thin whitish film above those packed red cells consists of white cells and platelets (lighter because they lack iron-containing hemoglobin). And the 55% of your blood sample located above this faint whitish film is liquid blood plasma. Normal plasma is clear and colorless to yellowish. If very yellow it suggests liver problems (*jaundice*). If very pink or red it indicates red cell rupture before, during or after that blood was neatly and painlessly withdrawn from your arm vein and centrifuged. If those red cells ruptured before that blood was removed from you (thus in vivo), you would expect to see reddish or black discoloration of urine by filtered free hemoglobin. More likely, however, some red cells ruptured in our syringe or tube (thus in vitro), not because they were abnormally fragile (as they might be in a person who inherited abnormal genes that produce spherical red cells rather than biconcave cells, for example) but rather due to the artificial surfaces those cells contacted or else our rough handling (too much syringe suction, perhaps) during their painless extraction from your arm vein.

For a direct determination of your blood hemoglobin content, we can easily rupture all red cells in a one ml sample of your freshly drawn blood by adding that whole blood to some standard amount of ordinary water in another tube. The water quickly causes every cell to swell and burst, leaving us with a clear red hemoglobin solution. The concentration of hemoglobin in this solution is then determined colorimetrically by comparing its shade of red to that of equally diluted standard hemoglobin solutions. Of course, since a red cell is ⅓ hemoglobin and your red blood cells occupied 45% of that spun centrifuge tube (hence a *hematocrit* of 45), we would expect to find about 15 grams of hemoglobin per 100 ml blood (15 grams percent). Hematocrit and hemoglobin often remain somewhat lower in adult human females who are still menstruating—there are disputed claims that such a slight chronic *anemia* (reduced hematocrit and hemoglobin) may lessen the risk of coronary artery disease and heart attack. Causes of significant anemia include blood loss, genetic abnormalities of red blood cells, infections (e.g. malaria, hookworm), nutritional deficiencies (insufficient iron, Vitamin B_{12}, folic acid, etc.), autoimmune disease, bone marrow problems and so on.

Although iron is plentiful upon Earth's surface, iron deficiency is the most widespread nutritional deficiency in the world. You absorb only a small fraction of the iron in your diet because you need very little iron and an excess is toxic (see below). Consequently, most third world populations do not get enough iron in their diet, especially when their diet is primarily vegetarian. About ¾ of your total body iron is present in your hemoglobin (at one gram of iron per four pints of blood), fifteen percent in myoglobin, 0.2% is bound to *transferrin* (which carries iron from point of release to site of need) and the rest is stored in *ferritin* and *hemosiderin* molecules (one ferritin molecule can bind 4500 iron atoms until needed). A final essential trace of iron serves as the electron donor and acceptor in many of your most important metabolic enzymes. The *lactoferrin* present on your mucosal surfaces and in your white blood cells and breast milk binds tightly to iron and thereby deprives (reduces the virulence of) bacteria and parasites that try to attack you. Humans also need iron to grow and thrive, and many poor people that previously were considered lazy and dumb have responded remarkably when their low iron levels were gradually built up (even before their anemia was corrected).

However, it can be fatal for a young child to swallow a few of mother's candy-coated iron pills. And adults with a chronically excessive iron intake (e.g. through drinking homemade beer stored in steel drums in Africa) or *hereditary hemochromatosis* (caused by excessive absorption of iron from a normal diet) commonly suffer from severe liver disease. Even the rapid correction of low body iron stores by iron injections has risk since it may encourage overwhelming infection in malnourished individuals as bacteria and other parasites suddenly flourish before the body can rebuild its immune system—iron deficient white blood cells often cannot destroy the bacteria that they engulf (see Immunity). The normal growth and development of your nervous system also depended upon adequate body iron reserves. Although those who are iron depleted generally secrete excessive amounts of epinephrine and thyroxin, they still cannot maintain ordinary body temperatures upon exposure to mild cold that others could easily tolerate. In addition to regular menstrual blood losses and the significant bleeding usually associated with delivery, women also contribute iron to the fetus during pregnancy (⅓ of a gram). So women are at greater risk for depletion of their body iron stores regardless of whether they live in a modern society or the third world. Since milk provides little iron, growing children require iron supplementation from other dietary sources.

When a person's hematocrit drops below 33 or so at sea level, the kidneys tend to boost normal serum *erythropoietin* levels (see Kidney). Erythropoietin must attach to a juvenile red cell before that red cell can mature—otherwise that cell soon disintegrates for lack of permission to survive (apoptosis again). Among immature red cell forms still in the marrow, the more mature require less erthropoietin to stay alive. Thus apoptosis selectively helps you to maintain a population of almost mature red cells in readiness for urgent release. Although mild chronic anemia is often well tolerated, a hematocrit of 20 or less reduces normal blood viscosity and increases the rate of blood flow—most people do not feel at all well when their red cells are reduced to that level. And death comes to individuals with healthy hearts and blood vessels (and adequate blood volume) when the hemoglobin falls to about 3 to 3.5 grams per hundred cc (equivalent to a hematocrit of 10–12). Of course, those in whom tissue oxygen delivery already is impaired by *arteriosclerosis* (narrowed arteries) or *heart disease* (defective pump) will die of heart attacks, strokes and so on, long before such low Hgb/Hct levels are reached (see Circulation, Fluid and Electrolytes).

Abnormally elevated body aluminum stores can cause anemia and bone disorders as well as fatal heart rhythm disturbances and brain dysfunctions. Apparently aluminum is toxic because the body treats it like iron but cannot wrap it in ferritin to reduce toxicity, even though transferrin carries aluminum as easily as iron. Aluminum interferes with the intracellular actions of calcium as well. Fortunately, the silicon in your diet can tie up and thus detoxify aluminum. Unfortunately, "soft" or acid water supplies carry lots of aluminum but not enough silicon.

Sickle Cells

Barring the occasional insignificant amino acid substitution, healthy human hemoglobin molecules are usually all alike. Of course, embryonic, fetal and adult hemoglobins all bear slightly different amino acid sequences in order to display the particular oxygen affinities that befit their station in life. But some blacks of West African ancestry carry a single nucleotide substitution in their adult hemoglobin gene that positions a hydrophobic amino acid at the molecular surface where it can stick adjacent deoxygenated hemoglobin molecules together. The long chains of deoxygenated hemoglobin that often result can distort red cells into elongated sickle shapes unable to pass through capillaries. Individuals carrying just a little of this abnormal *sickle-cell hemoglobin* (who have sickle cell *trait*) are generally *asymptomatic* (have no symptoms) because they also have a gene for normal hemoglobin. However, individuals bearing only sickle cell hemoglobin genes will suffer repeated life-threatening *sickle cell crises* whenever their blood oxygen levels fall sufficiently to provoke sickling.

Obviously that sickle cell *disease* is a bad thing to inherit, so the sickle hemoglobin gene ought never to have become common, let alone be present in up to 40% of certain populations in Tropical Africa. It turns out that the slightly fragile red cells of sickle cell trait are advantageous since they tend to rupture when invaded by malaria parasites, which therefore never gain a real foothold. So if two black parents, each with sickle cell trait, have four children, on average one of the four will die young of sickle cell disease while the two siblings with sickle cell trait will be protected from severe malaria. Of course, sickle cell trait is only advantageous in areas with a high malaria mortality rate. And even those with the trait may have a slightly increased susceptibility to sudden death when maximally deoxygenated (e.g. during competitive athletic events or high altitude mountain climbing). Under such strenuous circumstances, any minor sickling in small blood vessels feeding heart muscle might make these individuals more susceptible to fatal heart rhythm disturbances.

The severe anemia (due to early destruction of abnormal red cells) of sickle cell disease leads to narrowing of the blood vessels by scar tissue (in response to endothelial damage by high flow rates, abnormal red cells adhering to vessel walls and sludging of blood within the vessels). So younger children with sickle cell disease commonly develop strokes from blockage of brain arteries while strokes in older children are often due to rupture of thin-walled *collateral* arteries that develop to bypass such blockages. A fifth of such patients also develop kidney failure, presumably due to arterial disease of small vessels in the glomerulus (see Kidney). One interesting treatment for sickle cell disease (currently under study but not yet well tolerated or completely understood) attempts to dilute and thus separate adult sickle cell hemoglobin molecules by stimulating production of normal *fetal* hemoglobin (reactivation of fetal hemoglobin genes has been achieved with a number of chemicals including hydroxyurea and various butyrate derivatives as well as nucleoside analogues). Certainly your embryo produced *embryonic* hemoglobin without apparent difficulty in the *yolk sac* (see Growth and Development). And fetal hemoglobin was effortlessly produced during intra-uterine life. So why not just switch back to fetal hemoglobin if the adult hemoglobin molecule is defective (due to sickle cell or other abnormal hemoglobin), at least until more direct genetic repairs are possible?

Blood Types And Transfusions

Ordinary hemoglobin molecules may not vary significantly but human red cell surfaces do display hundreds of different molecules in an individually characteristic fashion (even molecules of bacterial origin may be absorbed onto red cell surfaces temporarily during the course of a severe infection). Since certain red cell surface molecules are both common and likely to be attacked by antibodies of the blood recipient (see Immunity), they become important factors to evaluate prior to blood transfusion. For when your blood enters the veins of another individual who does not share your particular red cell surface markers, blood-borne antibodies in the recipient may recognize the transfused cells as foreign (of course, the red blood cells from another animal species would be a far worse match). Such specific antibodies can cause the donated cells to stick together while other body defenses punch those foreign red cells full of holes, causing them to *hemolyse* (leak their contents). The resulting *hemolytic transfusion reaction* can be serious or even fatal for the transfusion recipient. So when you donate blood or receive a blood transfusion from another human, his/her red blood cells and yours must first be *typed* to detect any *A, B* or *Rh* molecules on the red cell surface. Small test samples of presumably compatible donor cells and recipient blood are then *cross-matched* (mixed together on a microscope slide) to rule out any uncommon antibodies in recipient blood that might attach to other surface molecules and clump those donor red cells—the few donated white cells may still be hit by such antibody attacks but that is generally of little consequence.

On rare occasions, transfused white cells can counterattack in a *graft-versus-host* reaction, especially with an immunocompromised host or one so closely related (or similar by chance) to the donor that donor cells have no detectable cell surface antigens not also present in the host (allowing donor white cells to avoid recognition as foreign by the host's immune system) although host cells still express antigens not present in the donor. That would leave host cells susceptible to attack by donor white cells but not the reverse. Graft-versus-host reactions can become a significant problem following bone-marrow transfusions which must introduce stem cells able to repopulate the damaged bone marrow and persist indefinitely. It is difficult to find a donor who will not elicit antibodies or attack the host under these circumstances, so recipients of bone marrow cell transfusions usually require ongoing *immunosuppression* to prevent attacks and counterattacks by host or donor cells (unless host and donor are identical twins and therefore match perfectly). Although the few mature white blood cells included in any blood transfusion soon die and cause little reaction, even peripheral blood may contain the occasional stem cell that could proliferate indefinitely after being transfused into an immunosuppressed or tissue-compatible individual—and following successful liver or other organ transplant, donor white cells disperse widely within the now *chimeric* (bearing cells of more than one individual) recipient.

All donor and recipient red blood cells can be sorted into a few major blood types. These types are defined by the presence or absence of common red-cell-surface *antigens* (sites liable to be attacked by a particular antibody). So if you have "Type A" blood, that simply means your red cells carry Type *A* surface molecules but not Type *B* surface molecules. Similarly a "Type B" person carries B but not A, while "Type O" carries neither A nor B. On the other hand, "Type AB" bears an A allele on one chromosome and a B allele on the homologous (matching) chromosome. This suggests that Type A can either be pure Type A (A and A) or A and O, just as Type B could be B and B, or B and O. However, you only receive one of each parent's two alleles via their *haploid* (carrying one-half of the usual chromosome count) sperm and egg, so an AO parent married to a BO parent could legitimately produce a child with AO, BO, AB or pure O blood type. The ABO system is important because those molecules are widely distributed in human secretions as well as among many other life forms including viruses and bacteria (those useful modules again). Indeed, your blood type may affect your ability to mount an effective antibody defense against an attacker with similar-to-your-blood-type cell-surface molecules—Type A individuals respond less effectively to smallpox infection, for example. Thus individual humans usually carry antibodies against any major red-blood-cell-type molecules not expressed on their own cells, and regionally common blood types tend to reflect the prevalence of particular serious infections rather than being evidence of recent shared ancestry.

In any case, you ought not transfuse B blood into an A or O recipient, for both A and O individuals would likely carry sufficient anti-B antibody to rapidly clump and destroy those transfused cells. On the other hand, AB individuals cannot carry either anti-A or anti-B antibodies. Therefore, at least in theory, an AB person could safely receive blood of any major type—so an AB patient is often referred to as a "universal recipient". In contrast, the small quantities of circulating anti-A and anti-B antibodies included in any single Type O blood transfusion will usually be diluted to insignificant levels in the recipient. So Type O blood (bearing a bit of anti-A and anti-B) can *usually* be given to any other human—which makes an "O" person a "universal donor". But here again, cross-matching remains very important for the detection of less common antibodies circulating within a potential recipient that might attack other surface molecules on red cells from that specific donor.

The Rh factor (first detected in your Rhesus monkey relatives) is the other common antigen that can cause transfusion incompatibility so Rh is routinely tested for during the typing of blood. If your red cells happen to display this monkey factor you are Rh positive (along with 85% of your fellow human primates). Thus an individual of AB type might be identified as either AB+ or AB- (depending upon his or her Rh status). Rh *negative* persons do not carry anti-Rh+ antibodies unless they have previously been exposed to Rh factor by transfusion or pregnancy. So an Rh negative person without clumping on cross-match can safely receive Rh+ blood (thereby presumably receiving an Rh positive blood transfusion for the first time). However, that Rh- person may subsequently produce enough anti-Rh+ antibody to affect a later crossmatch. Except in life-or-death emergencies, it is not advisable to give Rh+ cells to an Rh- female who might later become pregnant. For any Rh+ fetus of an Rh sensitized Rh- mother is likely to be harmed by maternal anti-Rh+ antibodies that cross the placenta and attack fetal Rh+ red cells. The resulting hemolytic disease of the fetus and newborn can lead to enlarged liver and spleen and diminished liver function or even cause liver failure as well as *bilirubin* damage to the brain of the newborn (see Digestion).

Even without a previous Rh+ transfusion, an Rh- woman may become sensitized by the few Rh+ fetal cells that enter uterine veins from torn placental vessels during forceful uterine contractions at the time of birth. Thus an Rh negative mother delivering an Rh positive child in a modern hospital routinely receives (intravenously) an antibody against the Rh+ antigen. That anti-Rh+ antibody rapidly eliminates any Rh+ fetal cells still within the mother's circulation after birth so she has little *post-partum* (following childbirth) opportunity to create anti-Rh+ antibodies of her own that might then harm an Rh+ fetus during a subsequent pregnancy. Apparently fetal blood cells can also be detected in maternal blood samples *during* pregnancy (which might some day allow maternal blood samples to provide sufficient material for genetic analysis of the fetus). The routine suppression of maternal antibody responses to fetal antigens during pregnancy seems to protect the few fetal cells that do get into mother's blood stream prior to childbirth (and also alleviate rheumatoid arthritis during pregnancy—especially when fetus and mother have dissimilar HLA antigens—see Immunity as well as Growth and Development). Of course, an Rh- mother who is ABO incompatible with the fetus is unlikely to become Rh+ sensitized since her already present anti-A or anti-B antibodies will rapidly eliminate any fetal cells entering her circulation.

Red blood cells tend to become BPG-depleted in storage—thus they may hold their oxygen more tightly than usual for a day or so following transfusion (until BPG levels again build up). Furthermore, such "tired" red cells may survive only a few more days or weeks within the recipient. But provided they did not deliver some viral or other unwanted DNA information, once those cells disappear from your circulation, your own personalized cells (say AB-) are the only sort remaining, no matter whether you received A-, B+ or O- cells. Except for a little wear and tear and perhaps some new antibodies, you have not changed a bit (see Immunity).

CHAPTER 19

YOUR CARDIOVASCULAR SYSTEM

Your Blood—Move It Or Lose It;... A Single Main Pulmonary Artery—A Single Aorta;... Your Beating Heart Drives The Circulation That Powers Your Beating Heart;... All About Bubbles;... It Can Make Your Blood Boil;... Collateral Blood Flow;... Arteriosclerosis, Atherosclerosis, Aneurysm;... An Embolus Can Only Go So Far;... All Muscles Pump Iron To Help Sustain Themselves;... Encourage One-way Blood Flow To Improve Circulatory Efficiency;... Murder?... Your Dynamic Veins;... Heart Muscle;... Heart Muscle Cells Squeeze Incompressible Blood—Heart Valves Maintain One-way Flow;... Things Really Were Different Before You Were Born;... Your Coming Out Led To Big Changes;... Your Inevitable Heart;... Capillaries And Lymphatics;... Heat, Rest, Elevation, Antibiotics;... Sometimes You Cannot Get There From Here;... X-ray Shadows;... Measuring Blood Pressure;... Pressures And Flows;... Your Parts Tend To Fit.

Your Blood—Move It Or Lose It

Blood vessels and neurons are easily lengthened so your size, shape and structure display the endless ancestral trade-offs made with respect to other limitations. For example, longer limbs increase running speed and reduce transportation costs while simultaneously slowing reflexes and enhancing body heat losses. But regardless of other design considerations, reproductive advantage requires that every structural change place vital organs, vessels and nerves where they are maximally protected. Naturally your central nervous system occupies the safest, least disturbed site in your body. From there it sends out the more or less rational requests that drive your musculoskeletal system. Of course, those neurons and muscle fibers can only do their job as long as the rest of your organs provide appropriate backup—so your lungs exchange gases, your gut absorbs nutrients, your kidneys stabilize the fluid and solutes of your circulation, your spleen retires worn-out formed elements from your blood and your liver maintains many of the important organic constituents of your blood. And every one of your tissues, organs and systems must support, cooperate with and even defer to your reproductive organs when these get aroused, or if there is even a chance that they might be called upon.

The continued survival of your many dozen trillion cells is entirely dependent upon your heart maintaining a satisfactory forward flow of blood through the enormous bed of capillaries that sustains each of your tissues. En route your blood unloads all sorts of supplies and picks up many wastes—yet only your blood gases are sufficiently disturbed by these exchanges to require reconstitution following every circuit through your tissues. Since your just-in-time gas exchange system obliges you to send all blood returning from tissue capillaries back through lung capillaries, the total blood flow through your tissues must equal the total blood flow through your lungs. Therefore the closed and circular *circulation* of larger land-dwelling creatures utilizes one muscle-powered pump to drive blood returning from lungs out through tissue capillaries and a second muscle-powered pump to push blood returning from tissue capillaries back through lung capillaries. Placing a muscle-surrounded blood-collecting chamber on the input side of each pump expedites refilling after each pump stroke and completes the simple logical efficient circulatory system based upon four contractile chambers that has so easily outperformed all competing models. For optimal protection and improved coordination of flows, those four heart chambers are gathered into a single central structure sheltered by your bony rib cage. There, unrestricted by your billowy lungs, your heart lies securely enclosed within its sturdy slippery mildly restraining fibrous *pericardial* sheath.

Your arms, legs and head are more or less circumferentially distributed around your body because that layout happens to be reproductively advantageous. As a result, your *left ventricle* (the pumping chamber that supplies freshly oxygenated blood to all tissues except lung) must move blood just back from your lung capillaries, lung veins and *left atrium* (the collecting chamber before your left or *systemic* ventricle) out in all directions at rather high pressures. Indeed, left ventricular pressures in an adult can intermittently raise a column of blood more than 150 cm above the heart, which exceeds the distance from your heart to your toes. On the other hand, little pressure is needed to push blood through those fluffy lungs on both sides of your heart. Therefore your *pulmonary* (lung) blood pump—which includes your *right atrium* (the muscle-enclosed chamber that collects your peripheral venous return) and *right ventricle* (the heart chamber pumping blood to lungs)—moves blood through your enormous pulmonary capillary bed at modest pressures that can lift your pulmonary blood up to 30 centimeters—about the height of your chest cavity.

It is possible to have adequate gas exchange without continuous blood flow through every capillary in your body. But capillaries in tissues other than lung do require ongoing access to blood, at least most of the time. Although the capillaries of your buttocks do not need continuous *perfusion* (blood flow) as you sit there fascinated by this book, soon enough you will become restless and shift about in order to bring new blood to *ischemic* (circulation deprived) portions of your "tired butt". Yet, you could continue reading for hours in a full bathtub without blocking those same capillaries, even though you would then become quite soggy. Having your brain suspended in cerebrospinal fluid similarly helps you to avoid circulation-blocking pressures at sites where your soft lower brain would press upon skull or tentorium cerebelli if not afloat (otherwise such pressures might oblige you to stand on your head every few moments).

Your lung capillaries become full only during the maximal blood flows associated with major exertion, so lung ordinarily offers little resistance to blood approaching through your pulmonary artery—only your dynamic regulation of lung-artery size (which controls resistance to blood flow and therefore the distribution of your right ventricle output) prevents that blood from simply soaking into your lung capillaries as if into a sponge. Close regulation of lung artery and vein capacities also eliminates excessive storage of blood within lung vessels while still allowing those wide fluctuations in flow between when you are at rest and exercising vigorously. Only the right ventricle (which sends blood to the lungs) and left atrium (which receives blood from the lungs) connect directly to your lungs. Having your left atrium located in the midline posteriorly while your right ventricle occupies the midline anteriorly presumably helps to balance blood flow between your two lungs. But such an arrangement necessarily displaces your right atrium toward the right heart border while your powerful left ventricle bulges out against your soft left lung. Two stubby (fat-finger size) *pulmonary veins* from each lung bring bright-red freshly oxygenated blood into your *left* atrium posteriorly. In contrast, the bluish deoxygenated venous blood returning from your upper and lower body capillaries enters the top and bottom of your *right* atrium via your two largest veins—known as *superior vena cava* and *inferior vena cava*.

A Single Main Pulmonary Artery—A Single Aorta

The task of your right ventricle is to intermittently squeeze deoxygenated blood into your *main pulmonary artery*, which then divides into left and right pulmonary arteries to supply both lungs. Your left ventricle similarly empties its freshly oxygenated blood into your single *aorta* for delivery to all systemic arteries. But while those two short low-pressure vena cavae reduce resistance to venous blood returning from your upper and lower body to your right atrium, your one elastic thick-walled high-pressure aorta fairly and economically serves your entire body. At its retrosternal (behind the sternum) origin from the top of your left ventricle, the aorta supplies fresh blood to your heart muscle via two pencil-sized *coronary* arteries. As the aorta then arches smoothly upward and backward over the root of your left lung, it gives off a large *innominate* (or *brachiocephalic*, meaning "arm and head") artery to the right side of your upper body. That innominate artery almost immediately divides into your *right subclavian and right common carotid arteries.* Next the aorta gives off the *left common carotid artery* to the left side of your head, then the *left subclavian artery* to your left arm. At that point, the *aortic arch* bends downward alongside of your vertebral column and becomes known as the *descending thoracic aorta.*

Multiple small pairs of *intercostal* (between rib) *arteries* that supply blood to both sides of your *thorax* (chest wall) arise from your descending thoracic aorta before that vessel passes through diaphragm muscle posteriorly near the midline (along your vertebral bodies and behind your guts) and becomes your *abdominal aorta.* That abdominal aorta then sends a branch laterally to each kidney (your two *renal* arteries) while a large, medium and small artery come off its front surface in sequence and several little branches emerge from its back surface. Those three anterior aortic branches (your *celiac, superior mesenteric* and *inferior mesenteric arteries*) distribute all of the blood supplied to your *intraperitoneal* abdominal *viscera*—in turn, the entire venous return from your stomach, small bowel, liver, spleen, pancreas and colon (large bowel) is directed through the *portal vein* to your liver for the initial processing of most nutrients. An additional supply of well-oxygenated arterial blood reaches your liver through *hepatic* (liver) branches of your celiac artery. Upon reaching your prominent lumbar vertebrae, your abdominal aorta divides into *right* and *left common iliac arteries*—these soon divide again to form your *internal* (to lower trunk) *iliac arteries* and *external* (to lower extremities) *iliac arteries.* Thus a rather long, gradually narrowing path is followed by those blood cells that pass through your ascending aorta and around the aortic arch into your descending thoracic aorta to reach the abdominal aorta. And that one large elastic aorta that so regularly changes its name but not its continuity or function, is solesource supplier of arterial blood to all *systemic* (non-lung) capillaries.

All blood vessels except your aorta and vena cavae have actively adjustable diameters. Systemic arteries, in particular, contain a great deal of smooth muscle in their walls. That smooth muscle is closely controlled by your sympathetic nervous system so that you can continue to *perfuse* (push blood cells through) the brain but not cheat the feet. The aortic wall itself has little need for smooth muscle, especially in its larger-diameter *intrathoracic* (inside of chest) portion, but it does have to be distensible and elastic so that it can easily accept the volumes of incompressible blood delivered during each heartbeat. Not surprisingly, your thick aortic wall includes a high percentage of elastin fibers as well as a lot of collagen.

Your Beating Heart Drives The Circulation That Powers Your Beating Heart

Your aorta passively accepts almost 100cc (about three ounces) of blood during each contraction of your left ventricle. The aorta's easy distensibility holds down the pressure at which blood can enter. And by deflating smoothly after each refill, that distended aorta maintains your systemic arterial blood pressure and thus the continuous forward flow of blood through all of your systemic arteries. However, elastin fibers undergo progressive covalent cross-linkage which (along with the buildup of calcium plaques and cholesterol deposits) gradually stiffens the average aorta in later life. Thus the aging left ventricle must work harder at higher pressures to force less blood through its increasingly rigid aorta.

If you could actually watch the normal flow of blood through some transparent aortic wall, you would probably be impressed by how slowly that blood moves ahead with each beat. A few ounces of blood surge in to stretch the aorta. Then the aorta gradually deflates as that blood streams in orderly fashion toward the most appropriate arterial exit. Another few ounces of blood are pushed in by the next heartbeat and so on. Certainly that blood is not just whizzing past. But to see blood really whiz, you might drop in on one of those realistic "blood and gore, stab 'em and shoot 'em" movies featured regularly at your friendly neighborhood family theatre. And do sit away from the front dozen rows where you seemingly risk being drenched or at least splattered by blood when the hero stabs the villain in the aorta (a lot more blood will whiz out of any hole in that stretched, pressurized, elastic aorta than you would anticipate from knowing that only three ounces of blood enter the aorta during every heart beat). Before it is over, and it will all be over very soon if one cannot stem the flow from that holey aorta, most of the villain's five liter blood volume will have gushed out in a rush. For as his aorta quickly undistends, blood also flows backward from all still-pressurized and reflexly contracting muscular arteries while his left ventricle faithfully forwards any blood it receives to aorta and thence out aorta's hole.

An equally large heroically placed hole in the villain's inferior vena cava (located behind the peritoneal cavity) releases a great deal of blood as well, since this hole allows blood from both iliac veins to escape forward without resistance while also draining blood back from the right atrium (and even superior cava). However, the inferior cava is thin-walled and not pressurized or elastically stretched. Therefore inferior vena caval blood loss does not squirt—it simply drains off in proportion to the size of the hole, as from a modest low-pressure faucet (indeed, if surrounding soft tissues and posterior peritoneum happen to block its low-pressure outflow, that hole may not bleed very much at all). But even a small vena caval opening can soon drain a villain's circulation unless he receives rapid blood replacement by transfusion. Of course, it does little good to pump blood into a villainous vein if that blood soon spills from some larger vein before even reaching the heart. Speaking of spilled blood or spilled paint, it is difficult to estimate how much blood or paint has been lost by viewing the area splattered.

All About Bubbles

The left and right sub*clavian* veins (located deep to the *clavicles*) and the left and right internal *jugular* veins (deep in the anterior neck) are large veins that return blood from your arms and brain. The subclavian and jugular veins of each side join to become your left and right *innominate* (or brachiocephalic) *veins*. Those two innominate veins at the base of your neck then merge into the well-protected superior vena cava within your chest. Now that you understand the anatomy, I will distract this new villain who is standing there quite relaxed and about to shoot our prostrate hero. Then you sneak over and plunge this short sharp open pipe into one of his internal jugular veins. Good shot! But there is hardly any blood loss at all! So that means we have failed! And after the villain calmly polishes off our hero with one careful shot, he will slowly turn to finish you off before disdainfully plucking that worthless weapon, your hollow pipe, out of his neck and tossing it aside. Right? Wrong!

The reason that your quick pipe in the jugular vein ploy did not release a lot of dark villainous blood (plenty of bright red blood would have spouted out had you missed the vein and drilled the adjacent carotid artery instead) is that blood within the jugular vein of an *upright relaxed* villain (or hero) falls freely toward the right atrium, delayed only by the wait for more venous drainage from the brain to replace it. But while the villain remains upright and relaxed, the continued inflow of air permitted by your pipe releases any blood still suspended in his jugular and innominate veins or superior vena cava. Of course, were he to tighten his abdominal and chest wall muscles and close off his airway as one does when lifting a heavy weight or taking a crap (a so-called "bear-down" or *Valsalva maneuver* often associated with a grunt due to compressed air escaping from the lungs), he would thereby squeeze his abdominal veins, inferior vena cava and right atrium. Indeed, a quickly performed villainous Valsalva maneuver would have forced blood back up the superior cava and jugular vein, pushing air and bloody froth out of your deftly placed pipe as well.

Anyhow, the evident lack of bleeding suggests that your pipe (accurately placed within the villain's largest neck vein) has permitted a lot of air to be siphoned into his villainous right heart. So now we have another worry. Does that air in the right heart simply invigorate him by pre-oxygenating his bluish venous blood before it can pass through those villainous black smoker's lungs? In that case the hero is done for and so are you. Right? Wrong. The hero may still get shot, of course, but that was in his job description. You should be O.K., however, because a) you cannot swim under a waterfall, b) you cannot swim when the surf is really high, and c) your car tends to vapor lock on a hot day if the gas line passes close by the exhaust system—as follows.

Briefly, you soap up and dive in to swim those few feet across to where that lovely large waterfall is pounding into the eight foot deep pool. When you have nearly reached "Mother Nature's personal shower" you suddenly sink from the sight of your admiring friends—soon they are also impressed by how long you can hold your breath. Unfortunately, that was not how you intended to impress them but never mind, since you are busily tumbling end-over-end in bubbly water. For a lot of air is driven into the eight foot deep pool by those falling sheets of water and that froth will not support your solid (water density) weight. It is not possible to walk out along the bottom either because you are continuously buffeted and turned over by the turbulence. Oh dear…

Or briefly, the surf is up, waaay up, and you lose your board—same problem of non-support by those foamy waters out where the big ones break. Oh well …

Or briefly, the gas line becomes hot enough to vaporize the gasoline within, which leaves your fuel pump pushing very compressible gaseous gasoline instead of incompressible liquid gasoline. Although, each surge by the fuel pump still increases gas fume pressures within the tubing, that push merely compresses gasoline vapor without moving gasoline along into the carburetor (or whatever those newfangled cars use instead). So the flow of fuel stops and your car stops. Just as our villain drops after sucking about 50 cc of air into his villainous pulmonary pumping chamber. For each heartbeat thereafter merely squeezes bloody froth while the forward flow of blood into those villainous lungs remains blocked by bubbles. His circulation stops—our minidrama ends. Now *relax!*

It Can Make Your Blood Boil

The amount of nitrogen, carbon dioxide and oxygen dissolved in your arterial blood plasma is determined by the individual pressures of those same gases in the air within your lungs (as the pressure of any gas rises, more will enter solution—see Respiration). But if surrounding air pressures suddenly drop, your blood could contain more gas than it can dissolve at these reduced pressures. So if you fall from a high-flying pressurized airplane or ascend too swiftly after helmet or scuba diving, your blood may fizz like a carbonated beverage. Furthermore, any bubbles that form on the systemic arterial side of your circulation (between lung and tissue capillaries) will tend to lodge in and block forward blood flow through your smaller arteries. This causes *ischemia* (inadequate tissue oxygen levels due to inadequate arterial perfusion) in tissues dependent upon those blocked arteries.

Of course, any CO_2 within such obstructive intra-arterial bubbles will tend to reenter nearby fluids and clear out quite rapidly, for red blood cells contain *carbonic anhydrase*, an enzyme that swiftly combines CO_2 with water to form very soluble carbonic acid as follows:

$$CO_2 \text{ plus } H_2O \underset{\text{anhydrase}}{\overset{\text{carbonic}}{\rightleftharpoons}} H_2CO_3$$

Oxygen, too, diffuses quite rapidly from small bubbles toward any nearby unsaturated hemoglobin or muscle myoglobin, but nitrogen gas (78% of Earth's atmosphere) enters and leaves your blood quite slowly. So even though N_2 ordinarily is present in far smaller quantities than O_2 or CO_2 (see Respiration), it is their nitrogen content that encourages air bubbles to persist long enough in arteries to cause a *stroke* (dead part of brain due to brain *anoxia*) or *infarct* (similar ischemic damage consequent to circulatory blockage in other tissues). The high nitrogen pressures encountered during deeper dives when using ordinary compressed air also can cause *nitrogen narcosis*, an intoxication that encourages unacceptable behavior such as light-heartedly offering your air hose to a passing fish.

Decompression disease with intra-arterial bubble formation can make its victims double over in pain. Known as *the bends*, that condition formerly afflicted many of those working under increased air pressure, such as the caisson workers building underwater bridge footings or excavating tunnels, or helmet divers who did not ascend (decompress) slowly enough. Fish-eating deep-diving dinosaurs also seem to have suffered from the bends (healed vertebral bone infarcts have been found in aquatic reptile fossils) but modern diving mammals have numerous physiological adaptations that help them avoid decompression problems (including rigid airways and collapsed lungs during the dive, a markedly slowed heart rate, circulatory diversions and additional oxygenated blood available for release from the spleen). Unfortunately, the bends can occur even when *helium* (less soluble, more swiftly exchanged, no nitrogen narcosis) is substituted for N_2 in the air supply that pressurizes your diving suit or scuba tank. The high-pitched quacking voice of a diver in a mostly helium atmosphere results from far lighter He atoms replacing the majority of N_2 and O_2 molecules. Of course, ordinary atmospheric air compressed to increasingly high undersea pressures soon causes *oxygen toxicity* (oxidative lung damage)—eventually it could even encourage spontaneous combustion. Similarly accelerated oxidation can result if you mistakenly grease the fittings on an oxygen tank with hydrocarbon molecules (ordinary grease or oil).

When intrapulmonary air pressures rise much above adjacent tissue fluid pressures, those over-inflated and unsupported lung air spaces may rip and release multiple air emboli into pulmonary capillaries and venules—and thence (via pulmonary veins and heart) to systemic arteries. Relatively compressed air enters the lungs when you breathe from an air bubble trapped within a sunken vehicle or from a scuba tank during a dive and then hold that entire breath while rushing to the surface in a panic (a very understandable reaction). But air *trapped* within your lungs must expand as surrounding water pressures fall, just as newly released air bubbles will double in size during a leisurely 10 meter (30 foot) rise to the surface. Under such stressful circumstances, disastrous *pulmonary air embolism* is easily avoided by purposely breathing out (exhaling) *continuously* during the ascent, rather than trying to hold that entire rapidly expanding breath until the surface has been reached.

Collateral Blood Flow

Certain of your critically important tissues are unusually protected from ischemic damage by having a more complete *collateral* (overlapping or alternate) arterial circulation. For example, your *Circle of Willis* provides sufficient collateral arterial inflow at the base of the brain so that you may get by without symptoms despite the gradual blockage of a carotid artery or slow arteriosclerotic closure of a *vertebral artery* (which normally delivers subclavian artery blood to your brain). Copious cross-connections similarly enhance the arterial circulation of your hands and feet. In addition, the persistent mild ischemia caused by progressive *arteriosclerotic* obstruction to blood flow encourages enlargement of any small arteries that happen to parallel (and therefore have the potential to bypass) the obstructed vessel segment. But regardless of the potential for such circulatory overlap, *sudden* blockage of a healthy artery by flow-directed lodging of an arterial *embolus* can cause ischemic damage to tissues currently dependent upon that artery.

Arteriosclerosis, Atherosclerosis, Aneurysm

Arterio*sclerosis* (*hardening* of the arteries) is a progressive life-long process based upon growing subendothelial deposits of *lipid* (fatty material), calcium salts and smooth muscle cells that gradually narrow the arterial *lumen* (channel). In certain susceptible individuals, comparable changes may sufficiently weaken arterial walls to allow *aneurysm* formation (marked arterial dilation—often in the face of quite ordinary arterial blood pressures). As they enlarge, such aneurysms become increasingly thin-walled and liable to rupture. But *atherosclerosis* (another medical term for crud-filled pipes) generally strangles blood flow through systemic arteries without causing aneurysmal expansions (some aneurysms may result from defective collagen or fibrillin fibers—see Collagen).

Apparently the adherence of blood platelets to abnormal (no longer endothelial-cell covered) vessel surfaces and the release of platelet-derived growth factors that stimulate smooth muscle cell growth, as well as the invasion of fat-laden macrophages into such areas during the effort to help clear up debris, are among important factors contributing to the atherosclerotic process. *Hypertension* (abnormally elevated systemic arterial blood pressures) and flow-related damage undoubtedly contribute to vessel wear and tear as well. Poorly supported endothelial surfaces (e.g. overlying the cheesy or even liquid arterial wall lipid intrusions known as *atheroma*) are also more easily torn. But at least the *fractal* distribution (giving a similarly random appearance at various magnifications) of your smaller blood vessel branches prevents your repetitive and forceful heart beats from setting up destructive resonances (damps any tendency to harmonic oscillations or chatter) in your arteries. Strokes and heart attacks often result from the sudden occlusion of a cerebral or coronary artery by clot formation on abnormal or damaged arterial *intima* (the endothelial surface) in the presence of arteriosclerosis. Even a temporary blockage of their circulation may damage or kill your continuously active brain and heart cells. Low-dose aspirin therapy (½ tablet or less per day) can reduce the incidence and progression of arterial wall disease by decreasing platelet stickiness and thereby limiting abnormal intra-arterial platelet deposits (see Blood).

An Embolus Can Only Go So Far

An embolic event occurs when intravascular gas, thick fluid or solid is passively borne through the circulatory system by the blood stream until finally it becomes lodged where larger vessels subdivide into numerous smaller arteries prior to the next capillary bed. Your *left*-sided (systemic arterial) flow of *oxygenated* blood begins in tiny *pulmonary* capillaries and ends in tiny systemic tissue capillaries, while the *right*-sided (systemic venous return to lungs) *deoxygenated* path returns blood from peripheral tissue capillaries to your pulmonary capillary bed. Although those huge numbers of tiny systemic or pulmonary capillaries certainly provide a far greater *total* cross-sectional area than any intervening portion of your circulation, only your larger vessels and heart chambers permit easy passage to an embolus moving with the flow of blood. And once such an embolus becomes firmly wedged into a systemic or pulmonary arterial branch (or branches), the tissues formerly provisioned by that newly blocked artery must either obtain adequate arterial inflow through some collateral pathway or do without (die).

So if a clot that has formed within a leg vein happens to break loose, that thrombus (now an embolus) will go with the flow through ever larger veins into the vena cava and thence through the heart before arresting as pulmonary artery branches become too small for that embolus to proceed. Whether or not that newly blocked pulmonary artery then causes local ischemia or symptoms depends upon the condition of patient and lungs, as well as exactly where and for how long that embolic clot then remains lodged (before clot retraction allows it to float ahead into a smaller vessel or clot lysis eliminates it entirely). An occasional embolus encourages further clotting locally where it becomes stuck, causing progressive difficulties. And if anticoagulation is ineffective or the blood remains hypercoagulable (sometimes a tendency toward hypercoagulable blood is inherited—which is hardly surprising, given the complexity of the clotting process), the initial pulmonary embolus may be followed by others—small or massive. Under such circumstances, one can markedly reduce the risk of subsequent clots passing through the inferior vena cava (the vast majority of pulmonary emboli originate in leg or pelvic veins) by insertion of some intracaval filter-type device.

When a person collapses and becomes pulseless (or develops very low systemic arterial blood pressure) consequent to incomplete or total embolic blockage of the main pulmonary artery, unusually forceful *CPR* (cardiopulmonary resuscitation) compressions on the lower anterior mid-chest can sometimes be lifesaving. For if an obstructing clot can just be driven into slightly more peripheral lung artery branches (where the *total* arterial cross-sectional area is greater) then a more adequate blood flow can resume through at least some *proximal* (closer to the heart) pulmonary artery branches and their lung capillaries. Of course, people collapse suddenly for many reasons other than pulmonary embolus, so one must use discretion as well as skill—especially as compressions that are too low could rupture the maximally distended liver (engorged by systemic venous return unable to reach the lung). It should not surprise you that systemic arterial blood oxygen levels may remain normal despite the very low arterial blood pressures that can accompany a massive pulmonary embolus. For even with minimal blood flows, any deoxygenated venous blood that does reach the systemic side will probably have passed through normal lung capillaries en route.

All Muscles Pump Iron To Help Sustain Themselves

Your heart muscle specializes in sending blood out to capillaries. Your voluntary skeletal muscles contribute to sending it back. And it all began when the very first group of muscle cells contracted and bulged out sideways in unison, thereby displacing nearby fluids. As time passed and muscles became increasingly organized, larger muscle groups often became enclosed by individual *fascial* (fibrous) envelopes. In addition to transferring locally generated forces, those partial or relatively complete collagenous sheaths kept muscles in place and reduced their impact upon passing nerves and vessels. But no matter how skeletal muscles might reorganize and specialize, every generalized contraction of a muscle inevitably squeezed all blood vessels within its compartment. That compression of their own blood supply (small blood vessels and capillaries) is one reason why voluntary skeletal muscles fatigue so readily during a prolonged or continuous forceful contraction— although a similarly persistent contraction would not tire slowly contracting smooth muscle or even require any effort at all from a dead person locked in rigor mortis (see Muscle).

Any work diverted to the compression of blood vessels necessarily reduces the external impact of a muscle contraction—an obvious reproductive disadvantage. Other than your heart, therefore, your muscles generally surround only their own personal blood vessels—the small nutrient arteries, capillaries and veins that service each muscle or muscle group within its own fascial envelope. And as you might expect, the effect of any major muscle contraction upon such enclosed vessels is to slow or stop arterial inflow, compress capillaries and speed venous outflow. Ordinarily, a muscle encounters little back-pressure when actively ejecting its own venous blood into larger veins outside of its compartment. With capillaries compressed and arterial inflow therefore minimal, the local effect of muscle contraction on circulation is necessarily one-way, moving blood out of local veins and away from that muscle.

Encourage One-way Blood Flow To Improve Circulatory Efficiency

Obviously, any thin wrinkles or infoldings of venous endothelium that interfered with easy venous blood outflow would be reproductively disadvantageous. However, there should be no adverse effect if such loose endothelial wrinkles moved aside readily to allow effortless egress of blood from all local muscle-enclosed veins during each muscle contraction. And if those non-obstructive endothelial wrinkles even partially prevented venous blood in larger proximal veins from refluxing back into just-emptied intramuscular veins during the subsequent muscle relaxation, then those flap wrinkles would thereby encourage forward blood flow through the muscle capillaries. Placing your immobile but very important brain up top similarly improves its venous drainage—only this time by gravity rather than through capillary and venous compression. Collagenous attachments (to surrounding bones and muscles) hold your dural sinuses and internal jugular veins widely open despite the strong suction effect caused by blood falling toward your heart within those veins. More generally, your venous drainage is enhanced by the intermittent respiration-and-exercise-related compression of all viscera within your chest and belly. Thus your voluntary muscles directly or indirectly expedite the return of venous blood from all tissues—even your skin, subcutaneous fat and bone benefit secondarily.

An artery brings blood to a muscle at relatively high pressures. By keeping arteries few and making their paths direct, your ancestors minimized their arterial blood volume, risk of arterial injury and resistance to arterial blood flow. But being few also obliged those arteries to remain quite adjustable in diameter. Not surprisingly, your arteries have rather thick muscular walls along with narrower channels than the more numerous thin-walled veins that drain blood away at comparatively low pressures. High arterial blood pressures and relative arterial incompressibility prevent the reversal of arterial blood flow during local muscle contractions. Therefore your systemic (and also low-pressure pulmonary) arteries lack internal valves. Your tiny thin-walled capillaries contain no valves either since simple compression easily boosts capillary blood into your low-pressure veins. Anyhow, capillaries are too small to risk hanging valve leaflets where red cells can barely pass.

With little blood momentum or capillary blood pressure remaining at the venous end of a capillary, it is important to encourage *venous return* (blood flow back to your heart). Hindrance, back-pressure, resistance to venous blood flow—

regardless of how you express it, there must always be spare volume capacity on the venous side. That implies many veins—indeed, a great many more outflowing veins than inflowing arteries. And each of those veins should potentially provide a larger channel than its corresponding artery. Furthermore, every vein should connect at many levels to many other veins so that returning venous blood can easily cross-over and follow the path of least resistance toward the heart—otherwise the inevitable compression of some veins as you stand, sit, cross your legs or lie down could obstruct blood flow through the particular capillary beds that they drained. So you have lots of veins and lots of collateral connections between those veins. And at every point where one vein enters another, delicate internal valves prevent the emerging venous blood from flowing back toward its most recently occupied capillary bed. In this way, each fragile one-way venous valve intermittently becomes a sturdy weight-lifter, supporting a column of blood that stretches up to the next valve.

Murder?

The neighbors heard screaming. Her apartment doors and windows were all locked when the police eventually broke in. The young woman lay dead in a pool of blood, still holding her bloody mop as a weapon. Police suspected foul work.

Autopsy revealed that her only injury was a torn *varicose* (markedly thinned and dilated) vein at the ankle. When her ankle struck the sharp edge of her kitchen counter and started bleeding copiously, she apparently screamed, panicked and ran for the mop to clean up the mess. Her damaged varicose vein continued to bleed severely as incompetent proximal venous valves allowed venous blood to drain all the way back from her pelvic veins and even right atrium. Of course, she could easily have stopped that frightening venous blood loss at any moment had she simply pressed a fingertip firmly upon the injured site or lain down with her leg above the horizontal.

Your Dynamic Veins

Varicose veins may result from chronically elevated venous blood pressures, as when a bulky pregnant uterus interferes with blood return from the legs until eventually those smooth-muscle-supported distal veins become temporarily or permanently overstretched—this anatomical situation is aggravated by the expanded blood volume of pregnancy. Sometimes a tendency to weak-walled veins is inherited. At other times, venous valves become stuck to the vein wall and rendered useless by an episode of *thrombophlebitis* (venous clotting and inflammation). But regardless of how or why proximal venous valves become non-functional, the net result is that distal veins bringing blood toward those incompetent venous valves are henceforth burdened at least intermittently by the great weight of a tall column of blood. This is especially true for veins of the lower leg where persistent venous back-pressure due to incompetence of more proximal venous valves can cause *post-phlebitic ulcers* if chronically high venous pressures prevent adequate blood flow through local skin capillaries.

The individual diameters of your many arteries and veins are ordinarily adjusted in accordance with orders sent out via tiny sympathetic nerve fibers distributed within the thin fibrous *adventitia* (outer envelope) of each vessel. The smooth muscle cells of your blood vessel *media* also respond to various circulating hormones that arrive through tiny adventitial blood vessels, and to local signals released by endothelial cells lining the vessel interior. As mentioned, all of your systemic arteries have a rather thick smooth-muscle layer except the proximal aorta which mostly lacks smooth muscle and utilizes a great many supportive elastin fibers instead. In contrast, your thin-walled veins are remarkably adjustable in diameter but they cannot continuously resist a lot of pressure. So you may faint if you stand very still for a long while, especially on a hot day when perspiration reduces your circulating fluid volume and sub*cutaneous* (beneath the *skin*) vessels dilate to radiate away excess heat. Fainting under such abnormal circumstances suggests that insufficient venous blood is being returned to your heart by peripheral muscle contractions. Consequently your heart pumps out less blood at lower pressures which especially affects higher parts such as your head.

A person will die of poor circulation to the brain (a stroke) if held upright in motionless fashion—one of several reasons that crucifixion gradually becomes fatal. However, an uncrucified person who faints will naturally fall to a more horizontal position that increases the blood pressure in his/her head arteries even before venous return improves. At that point, raising the feet and legs above heart level can further enhance venous return to the heart (see Fluid and Electrolytes) by mobilizing additional volumes of blood stored within leg veins. A reclining person can easily produce a similar surge in venous return by tightening leg muscles all at once. If you faint from an emotional shock (the term *shock* often refers to low blood pressure) your vessels probably dilated sufficiently to reduce your venous blood flow despite a normal blood volume. Area-wide venous compression improves venous blood return regardless of whether that compression is active or passive. But the nurse who conscientiously dorsiflexes each foot of a sleeping postoperative patient every five minutes also encourages more efficient emptying of all calf muscle veins and thereby reduces the risk of thrombus formation in any under-utilized lower leg veins where blood flow might stagnate. Even properly applied and carefully maintained elastic wraps or support hose are less effective than such passive muscle pumping in preventing venous stagnation under postoperative slow-blood-flow no-movement conditions. But external elastic support certainly can reduce fluid retention within limb veins and extracellular tissue spaces.

Heart Muscle

Unlike voluntary skeletal muscles, your heart muscle must contract rather slowly and regularly in order to squeeze incompressible blood out of those large blood channels (your heart chambers) through which blood passes to and from all of your tissues. So while your long slender multinucleated skeletal muscle fibers are efficiently shaped to exert their tension upon some distant bone or tendon, your short stubby heart muscle cells (most with just a single centrally located nucleus) are more suited for their combined concentric contractions by being securely interconnected at *intercalated discs* (along with the usual meshwork of collagen fibers to keep things in shape). Furthermore, those striated cardiac muscle cells need not wait for a nerve impulse to initiate each depolarization/contraction, for heart muscle cells have rather leaky Na$^+$ channels that eventually cause one or another of them to depolarize spontaneously (see below). Once initiated, the cell-membrane depolarization of each heart beat swiftly spreads between all atrial muscle cells through adjustable intercytoplasmic (thus direct electrical) connections known as *gap junctions*. This ensures that there will be no appreciable delay or discoordination of each cardiac contraction. So regardless of which atrial muscle cell may depolarize first, both atria contract as a unit. Similarly, both ventricles form a single electrically interconnected unit. Therefore as long as a regular heart rhythm ensures comparable filling (at rest), each cardiac contraction will tend to have much the same effect as any other, no matter where in the atrium that heartbeat may have originated. Of course, if the atrial beat is irregular (as in *atrial fibrillation*) or not properly timed with respect to ventricular contractions, the heart may fill less efficiently and therefore eject up to a third less blood per stroke. A great many heart rhythm disturbances can interfere with proper cardiac output but only *ventricular asystole* (total *absence* of systolic contractions) or *ventricular fibrillation* (chaotic ventricular muscle cell contractions that produce cardiac quivering but no coordinated heartbeat) will rapidly end all blood flow. Ventricular asystole can follow a heart attack (reflecting dead or injured heart muscle), or result from medication that suppresses heart muscle depolarization or prevents appropriate conduction of each heart beat from atrium to ventricle (representing an undesirable side-effect or reaction to that drug), or ventricular asystole can be due to some other abnormality such as markedly elevated blood K^+ levels.

Under some circumstances, a regular electrical impulse from an artificial *cardiac pacemaker* can stimulate regular cardiac contractions and thereby correct asystole. While ventricular fibrillation is a common cause of sudden death in the home or workplace, it often responds to temporary CPR followed by an electrical shock when a defibrillator becomes available. By causing all cardiac cells to contract (and thereafter relax) simultaneously, the direct-current electrical jolt from a defibrillator ends the chaotic feedback from still contracting ventricular muscle cells that can set off other cells as they once again become ready to contract. Fortunately, the long *absolute refractory period* of heart muscle cells (a time during which they will not respond to any stimulus) usually prevents fibrillation from getting started (see Muscle). In any case, the *sympathetic* (speed up!) and *parasympathetic* (slow down!) innervation of heart muscle normally regulates your heart *rate* as well as affecting cardiac contractility. And each heart beat ordinarily begins with *spontaneous* depolarization of a primary pacemaker cell in the *SA node* of your heart, for the cells of that *sinoatrial* node (located high in the right atrial wall adjacent to your superior cava and right auricle) usually depolarize most often and thereby dominate the heart rate—even at rest your SA node cells generally pace along at over 60 or 70 beats per minute. However, if the SA node does not depolarize appropriately, then atrial muscle, the *AV* (*atrioventricular*) node or even ventricular muscle will take over as your primary cardiac pacemaker. If SA node, atrial muscle and AV node all fail to discharge effectively, or if the AV node fails to conduct the atrial depolarization to ventricular muscle, the spontaneous ventricular rate usually takes over at a rate below 30 beats per minute, which is slow enough to cause fainting spells, poor cardiac output and even death.

In summary, each regular spontaneous depolarization of cells in your SA node initiates a wave of depolarization that passes rapidly over right and left atrial muscle cells. That electrical event (the *P wave* on your electrocardiogram or ECG) is soon followed by *atrial systole* (the coordinated contraction of these right and left atrial muscle cells). Atrial systolic contraction pushes much of the collected intra-atrial blood through your right and left *atrioventricular (AV) valves* (the *tricuspid* and *mitral valves* respectively) into the relaxed and therefore easily distended right and left ventricles. Although atrial depolarization precedes the onset of atrial systole, a slow wave of depolarization continues through your *AV node* delay circuit during the entire atrial systole. Those AV node cells conduct so slowly because they rely primarily upon Ca^{++} ion channels for their depolarization (rather than the usual rapidly opening Na^+ ion channels). But once that delayed electrical impulse emerges on the ventricular side of the AV node, your right and left ventricle muscle cells depolarize promptly (the *QRS complex* on your ECG—the subsequent *T wave* represents ventricular *re*polarization). The initiation of ventricular depolarization is expedited by *Purkinje* fibers containing specially conductive modified-muscle cells. *Ventricular systole* (muscle contraction) follows soon after the electrical depolarization.

Heart Muscle Cells Squeeze Incompressible Blood—Heart Valves Maintain One-Way Flow

As each new systolic contraction begins to squeeze just-collected intraventricular blood, ventricle chamber blood pressures soon exceed atrial chamber pressures and the flexible AV valves slam shut in response to the sudden reversal of blood flow—just as the first gust of wind slams a door. Ventricular cavity blood pressures continue to rise while *papillary muscles* (finger-like intra-cavitary extensions of ventricle muscle) pull ever more strongly on thin *chorda tendinae* (tendons) connected to the free edges of your AV valve leaflets, thereby preventing those AV valves from bulging back up into your atria (which would impede atrial filling as well as reduce the forward expulsion of blood by each ventricular contraction since atrial diastolic relaxation and filling must coincide with ventricular systole if the next atrial systole is to fill your ventricles during the subsequent ventricular *diastole*). Of course, your papillary muscles and chordae tendinae also keep those sturdy but delicate valve leaflets from being inverted into the relaxed atria. Your *pulmonic* and *aortic* valves swing wide open as intraventricular pressures exceed pulmonary artery and aortic pressures—thereafter pressures coincide within those great vessels and their respective ventricle chambers until ventricular diastole begins.

Your muscular cone-shaped left ventricle contracts with a noticeable twist that wrings out more of its enclosed blood while also expressing blood from the simultaneously contracting and thinner-walled right ventricle lying across its front surface. Nonetheless, your atria and ventricles never empty completely (variable but significant volumes of blood always remain within any heart chamber at the end of its systolic contraction). The onset of ventricular diastole (see octopus arm muscle) rapidly reduces ventricular chamber blood pressures below the blood pressures within your distended pulmonary artery and aorta, causing pulmonic and aortic valves to slam shut passively. Blood pressures within those *great vessels* (pulmonary artery and aorta) then decline progressively for the rest of ventricular diastole as their entrapped blood continues to flow out through all arterial branches. An interventricular *septum* (partition) of thick left ventricle muscle separates your left and right ventricle cavities (both atria share a much thinner septum). In addition to their papillary muscles, the *endocardial* (inside of heart) surfaces of both ventricular cavities are irregularly ridged by prominent muscle bundles. Presumably these *trabeculae carnae* ("beams of meat") buttress the inner ventricle wall against overstretching—perhaps they also initiate folding to bring heart wall muscle together in accordion fashion during each twisting ventricular systole. Moment-to-moment changes in *preload* (inflow) or *afterload* (output side) pressures often cause one heart chamber to pump a different amount of blood than another, but each chamber necessarily has the same output per minute as all others since any persistent imbalance of output would soon cause blood to accumulate in your tissues or your lungs.

Things Really Were Different Before You Were Born

Your superior vena caval blood flow is directed downward across the right atrium toward the tricuspid valve while blood moving up the inferior vena cava is aimed at the interatrial septum. Before your birth, that inferior vena caval flow was mostly well-oxygenated blood returning from your placenta via the umbilical vein. Upon striking the interatrial septum, that fetal inferior caval blood flow lifted a valve-like flap away from the left atrial side of your interatrial septum to open your *foramen ovale* (oval-shaped hole). This allowed most of the well-oxygenated fetal venous blood reaching your right atrium to pass directly across the interatrial septum into your left atrium. So your inferior vena caval blood flow mostly detoured around your right ventricle and lungs before you were born. With well-oxygenated inferior caval blood thus crossing over to your left atrium, left ventricle and ascending aorta, your "best" blood ended up supplying aortic arch branch arteries to your heart, brain and upper body. The rather deoxygenated venous blood returning from that upper body then passed through your superior vena cava, right atrium and tricuspid valve to reach your right ventricle, pulmonic valve and pulmonary artery. But even that reduced flow of deoxygenated right heart blood still exceeded the capacity of your not-yet-expanded-or-functional fetal lungs. Since your lungs offered such high resistance, much of your pulmonary artery blood flow ran off via the *patent* (open) fetal *ductus arteriosus* into your distal aortic arch just beyond the left subclavian artery. Therefore your fetal descending thoracic aorta bore mostly "used" superior vena caval blood that had arrived via the right atrium, right ventricle, pulmonary artery and ductus (bypassing your non-functional lungs).

Consequently the large volumes of refreshed blood from your placenta only mixed with venous return of your slowly growing lower body and diminutive lungs before being pumped out to upper body capillaries by your left ventricle, while your placenta and lower body capillaries received much of their blood supply from your right ventricle. Although your tiny non-aerated fetal lungs expanded at birth into light and compressible structures (just as a spoonful of sugar makes a lot of cotton candy), they needed very little blood for their routine growth, so completely separating your pulmonary and systemic circulations would have been disadvantageous prior to your birth. Furthermore, it made good sense to deliver well-oxygenated fresh-from-placenta fetal blood to your critically important, rapidly growing heart and brain rather than supplying all parts of your body (including those developing less urgently) with an equally deoxygenated mix of superior and inferior caval blood—given that your placenta already acted as a lung/digestive-system/kidney substitute and you had no need to move about right after birth anyhow. That left your most depleted blood flowing through descending aorta and internal iliac arteries to the placenta for gas exchange, nutrient uptake and waste removal. But the massive diversion of fetal blood into your placenta would have stressed a single systemic ventricle so both fetal ventricles were hooked up to the aorta in parallel.

Your Coming Out Led To Big Changes

When you took that first deep breath as a newborn and then hollered, you greatly expanded your lungs by inflating them with air. That sudden expansion of your newborn lungs reduced their resistance to blood flow (just as a compressed sponge becomes far more permeable when expanded). Your pulmonary artery and right-side-of-heart blood pressures gradually declined as your pulmonary arteries relaxed and dilated. The improved oxygenation of your systemic arterial blood (now based upon well-oxygenated pulmonary venous return) encouraged your still patent but now superfluous ductus arteriosus to shrivel and close down over subsequent days. Of course, the abrupt amputation of your placental circulation markedly restricted the outflow of blood from your aorta, thereby enhancing your systemic arterial resistance and boosting your aortic and left heart blood pressures. As rising left atrial pressures exceeded declining right atrial pressures, the left atrial flap came down upon and closed your inter-atrial foramen ovale (in over two-thirds of all humans that opening is thereafter completely and permanently sealed). Sudden cessation of their blood flow encouraged the umbilical extensions of your fetal internal iliac arteries and your entire *ductus venosus* (the venous bypass bearing placental blood from umbilicus to cava) to close. Note that *two* large umbilical arteries helped to maintain your placental blood flow and pressure while *one* large umbilical vein provided the least possible resistance to blood return (with no placental muscles to boost that blood back toward your heart). Before long your tied-off umbilical cord dried up and fell off, leaving your ordinary healed belly-button as the final memento of nine subaquatic intrauterine months. And another beautiful child was born who looked just like Dad and might possibly become President.

Your Inevitable Heart

No great imagination is required to see how your wonderful modern heart could easily have evolved through endless intermediate improvements from an ordinary muscle compressing its own blood vessels. Furthermore, your heart's slippery *pericardial* sheath is not so different from the fascial envelopes around many muscles. That your larger *coronary* (Greek for "wreath" or "crown") *arteries* usually lie within the *epicardial fat* also makes sense, for that protected yet outer-heart-surface location lets those arteries fill with oxygenated blood during ventricular systole. The smaller arteries and capillaries penetrating your heart muscle can then draw from already distended epicardial arteries as soon as diastole begins (when those smaller intramuscular branches no longer are compressed by contracting cardiac muscle). Even your *coronary sinus* (a large epicardial heart vein that empties heart muscle venous return into your low-pressure right atrium near the tricuspid valve) is where it ought to be. For were deoxygenated venous blood from heart muscle to drain into the *left* atrium (instead of circling around to the right atrium), that would annoyingly reduce the average oxygen content of left atrial blood just back from your lungs—such a reduction in circulatory efficiency would surely be a reproductive disadvantage. The low entry site of your coronary sinus into the right atrium is almost inevitable as well, since those alternating atrial and ventricular contractions are least disturbing to low-pressure structures that run in the fatty layer overlying the fibrous groove between your collecting and pumping chambers.

Capillaries And Lymphatics

The same arterial pressure that pushes blood through your capillaries also forces water out of the blood stream into surrounding tissues (except at the less permeable, highly controlled blood-brain barrier—see Nervous System). Many small solutes pass freely between (or through) capillary endothelial cells along with that water, leaving blood proteins behind. Serum *albumin*, a smaller and osmotically active (because more individual molecules per gram) blood protein molecule is present at over 3 grams per 100cc of your blood plasma—which allows albumin to draw almost as much tissue fluid back into your blood at the lower-pressure venous end of each capillary as leaks out into tissues at the higher-pressure arterial end. But tissue fluid accumulations increase with inflammation or low serum albumin levels, or during prolonged *dependency* (hanging downwards, especially below your heart and without much movement—hence high venous pressure) or consequent to high venous pressure from other cause. Tissues also accumulate more fluid as capillary walls become fragile with illness or advancing age, and blood proteins often escape from the circulation across abnormally leaky capillary walls—many other large molecules gain access to tissue fluids during the accelerated breakdown of local tissue and white blood cells that accompanies inflammation. But no matter how they reach your low-pressure extracellular tissue fluids, larger solute molecules cannot then cross the capillary wall into your circulation. So when your ancestors converted to a closed circulatory system, they avoided having chronically soggy tissues by also developing small many-valved endothelial-cell-lined *lymphatic channels* (that resemble and even respond to the same growth factors as blood capillaries). Your numerous repeatedly interconnected lymphatic vessels combine into ever-larger thin-walled conduits as the gentle flow of *lymph* drains excess fluids, solutes and particulate debris away from your tissues. That lymph undergoes repeated filtration through local and then regional *lymph nodes* (encapsulated structures that contain specialized collections of macrophages to clear particulate debris from the lymph as well as a great variety of immune cells waiting to detect and attack particular bacteria and other microinvaders—see Immunity). The lymph glands in your neck or armpit may become noticeably tender and swollen while dealing with an infection but ordinarily these smooth ovoid structures are most *palpable* (easiest to feel) in your groin.

Heat, Rest, Elevation, Antibiotics

Although the local production of lymph (excess tissue fluid) is enhanced by dependency, the flow of lymph occurs largely in response to your movements. Furthermore, mild heat improves capillary blood flow as long as there is no tissue ischemia. Therefore it is generally reasonable to treat an injured or infected extremity with "heat, rest and elevation" (except when cold reduces initial pain and swelling soon after an injury). But heat also boosts tissue metabolism so if the local arterial circulation already is *marginal* (barely adequate) due to injury, arteriosclerosis or other problem, increasing the circulatory requirement by heat without also improving that marginal circulation could cause *gangrene* (tissue death). Anyhow, within a day or so after a minor hand injury (even a pinprick) that introduces *pathogenic* (nasty) streptococcus bacteria, you might note a painful spot on the hand as well as a faint red streak leading from this site toward a swollen and tender lymph gland on the medial side of your elbow. Over the next day you might start to feel a bit ill or feverish as the red streak continued relentlessly toward your *axilla* (armpit), causing swollen tender lymph glands there as well. Soon thereafter you would likely develop a shaking chill and fever and feel very ill with the early stages of "blood poisoning" (bacterial invasion of your circulation) as tough streptococcal survivors of the hazardous voyage through your lymphatics gained access to your blood stream near the junction of your subclavian and jugular veins (where most of your lymphatic return finally pours into your venous blood).

Bacteremia (germs in the blood stream) was a common cause of death before antibiotics became available. It is again becoming popular as bacteria increasingly develop *antibiotic resistance*. Apparently transfers of plasmids and other genetic instructions occur frequently, even between unrelated bacterial species. And certain plasmids that happen to carry information on how to break down or otherwise cope with particular antibiotics can bring pathogenic bacteria increasing reproductive advantage as those antibiotics become an important environmental factor (for example, low-dose antibiotics have been widely incorporated in animal feed to enhance animal growth by conserving nutrients that otherwise might be lost through the bacterial production of methane). On the other hand, streptococci often blast their way through your initial defenses by utilizing enzymes such as *streptokinase* (which activates your plasmin to dissolve the protective fibrin strands that accumulate around an infection) and *streptodornase* (which dissolves the DNA tangles left as a final gift by your dying white blood cells). Were streptococci not such versatile chemists, those local tangles of fibrin and DNA would have helped you to wall off and thus localize their intrusion. We have seen that bacterial DNA information varies greatly and often includes DNA accidentally acquired from you or others. Indeed, inconceivably large numbers of microbes have been engaged in an unrelenting life-and-death struggle for almost four billion years—acquiring, recombining and exchanging endlessly varied RNA and DNA modules until a few lucky survivors once again uncover an improved solution for the latest life-threatening problem. Of course, that same basic competition also exists between all individual DNA modules in any individual bacterium or between combinations of modules that designate more effective plasminogen activators or improved antibiotic resistance— but it is early-loss-of-life versus survival-until-reproductive-success that drives evolution. In any case, such hard-earned microbial information now produces your bread, wine, cheese and antibiotics—and the genetic information of certain microbes has even been artificially altered to manufacture entirely new products. So while some *cardiologists* (medical heart specialists) still use streptococcal streptokinase to dissolve human coronary artery thrombi, others prefer to utilize the far more expensive and shorter-acting tissue plasminogen activator (a human-gene-transferred-to-bacterium product). Perhaps vampire bat plasminogen activator will eventually prove most useful of all—which would hardly be surprising since those vampires absolutely must keep the midnight snacks flowing or they don't live to reproduce.

Both your systemic venous blood return and forward flow of lymph are enhanced by one-way valves, muscle-generated tissue pressure changes and elevation above the heart (gravity again)—but normal tissue fluid pressures also contribute, as does the gentle massage given those tissues by huge numbers of simultaneously expanding capillaries during each heart beat. Because lymph ducts of all sizes interconnect copiously with one another as well as with small veins, lymph could reenter your circulation via a great many different routes. But your lymph primarily returns through right and (mostly) left neck veins where venous pressures are usually lowest (as blood hangs in arrested free-fall toward the heart—we have seen the same dynamics draw cerebrospinal fluid across your arachnoid villi into the sagittal sinus and produce venous air embolism).

Bacteria that gain access to your blood circulation are generally filtered out and destroyed at the next capillary bed. And as a bacterial infection advances relentlessly from injured tissues through your lymphatics into your venous circulation, the pulmonary capillary bed is next. Only rarely will a lung abscess or very severe lung infection release bacteria into the systemic circulation via the pulmonary venous drainage, for those mean-looking lung macrophages do not hang around inside of your post-capillary pulmonary venules as decoration. So when next you visit Drs. Paine, Slaughter and Coffin for some minor dental procedure, you can be relatively sure that any transient bacteremia consequent to jabbing at your gums and drilling about your bacteria-infested mouth will be cleaned up promptly in your lungs. Of course, some cardiac abnormality could still provide the occasional germ with a safe haven—perhaps adjacent to a damaged heart valve where turbulent blood flow has deposited a (here protective for bacteria) fibrin and platelet layer. Or the rare right-to-left shunt (see below) or even a persistently patent foramen ovale (when you are coughing or straining) might allow bacteria to bypass your lungs en route to your systemic arteries. Persons known to have heart valve problems or intracardiac shunts usually receive *prophylactic* (preventive) antibiotics before undergoing any dental work in order to minimize the chance that a transient procedure-related bacterial invasion could cause heart valve infection or left-sided (brain or other) tissue *abscesses* (toxic space-occupying collections of bacteria, dead white blood cells, tissue debris and so on).

Sometimes You Cannot Get There From Here

Cardiologists and *radiologists* (diagnosticians who specialize in using X-rays) often poke needles and tubes into various arteries and veins to obtain blood samples, administer inside-of-vessel treatments or inject radio-opaque dyes that can momentarily outline heart chambers, vessel interiors and blood flow patterns for recording on X-ray film. As with invading bacteria or embolic clots, the exploring *catheter* (hollow tube extension) of an invasive cardiologist or radiologist can only reach vessels and chambers that precede the next capillary bed. For example, a hollow needle stuck into your *femoral vein* (at the groin) can allow introduction of a blunt wire stylet for passage toward your right heart (if inadvertently threaded downward into your leg vein against the direction of blood flow, such a smooth and flexible wire would soon hang up in a venous valve leaflet). While moving proximally, that exploring wire from your femoral vein advances through your iliac vein and inferior vena cava (passing between your kidney veins and behind your liver) to enter the right atrium. From there it might continue straight up the superior vena cava (no venous valves there) or else turn to accompany your blood flow through the tricuspid valve into the right ventricle. Using X-rays, the cardiologist or radiologist can easily visualize and reposition that metal wire. Upon removing the original hollow needle, the flexible inside wire then provides internal guidance for insertion and placement of a slender and slippery hollow plastic tube (transvenous *cardiac catheter*). As it slides over and beyond the wire, that more limber catheter tends to go with the flow of your blood through tricuspid and pulmonic valves into your pulmonary artery and its branches.

On the other hand, a catheter similarly inserted and advanced through the femoral *artery* toward the heart will be moving against the flow of arterial blood. Here a guide wire is especially useful for stiffening the catheter as it is pushed up within the iliac artery and abdominal aorta, then up the descending thoracic aorta into the aortic arch. Once that catheter reaches the ascending aorta just above your aortic valve (hopefully without being delayed by repeated diversions into your head or arm arteries while traveling around your aortic arch), the guide wire is removed and radio-opaque dye injected through the catheter to visualize your aorta or coronary arteries. Or a floppy-tip catheter can be positioned against the aortic valve until it happens into the left ventricle during a subsequent systolic valve opening.

Here again, pressures can be measured, blood sampled and dye injected. It is generally impossible to advance a catheter farther upstream (across the mitral valve into the left atrium) from the left ventricle. The cardiologist may therefore choose to sample left atrial blood or measure representative left atrial pressures *indirectly* through *wedge pressure* measurements (by blocking a small peripheral pulmonary artery with a trans-venous balloon catheter that has an open tip beyond the small balloon). An alternative approach would be to advance a verrry long needle up the femoral and iliac veins and inferior cava, then across the right atrium to poke through the atrial septum (at the foramen ovale) into the left atrium. This *trans-septal* needle insertion carries obvious risks such as inadvertently poking a hole into the nearby high-pressure aortic root (relatively small volumes of blood escaping from an aortic needle-hole into the pericardium could easily compress the heart sufficiently to block venous return and thus stop the circulation—just as bird collectors quickly kill small birds by squeezing their heart between thumb and forefinger). Fortunately there are now several less invasive techniques (sonar, radioactive isotopes, X-rays, nuclear magnetic resonance and so on) that can provide useful images of heart chambers, heart valve function, blood flows and blood vessels.

X-Ray Shadows

All radio-opaque dyes contain heavy atomic elements whose numerous surrounding electrons deflect passing X-rays. When an X-ray beam is partially scattered by such materials before reaching the film, fluorescent screen or other detector, the remaining X-rays produce an outline of how that radio-opaque dye was distributed (just as light blocked by your hand does not strike the wall so your hand casts a shadow). However, an X-ray film taken of your body without such dyes will still display tissue shadows—for hardly any of the diagnostic X-rays passing through you are scattered by the lightweight *air* molecules you contain, a few more are deflected by low-atomic-density *fat* molecules, a moderate number will be deflected by closely packed *water* molecules (the major constituent of your muscles and other soft tissues) and many are scattered by heavier elements in the calcium phosphate crystals of your *bones*. So a list of your ingredients from least to most effective in blocking an incoming X-ray beam would be air, fat, water and mineral—not too different from the air, fire, water and earth that some ancient Greeks decided were Earth's basic elements. And the darkest (most exposed) portions of your routine chest X-ray film are where the X-ray beam passed primarily through air while the lightest areas show where the X-ray beam was almost entirely obstructed by bone minerals (or by all of the barium sulphate that you just swallowed or the iodide commonly incorporated within radio-opaque dyes for intravascular studies). Of course, an X-ray beam only casts useful shadows if organs or tissues of contrasting density are (or are not when they ought to be) silhouetted against each other. Often an outline is seen (a kidney, perhaps) only because a thin layer of fat comes between two layers of water density (separating kidney and back muscle, for example). The importance of contrasting densities can be demonstrated by directing an X-ray beam downwards through a *radiolucent* (transparent to X-rays) plastic cylinder or jar that contains a fatty layer above a water layer that in turn overlies a smooth layer of mineral (bone) density. Such a view provides no contrast and thus little information of value. On the other hand, an X-ray beam directed at that same jar from the side would show all 4 layers (air-fat-water-mineral) very clearly in silhouette. Incidentally, even the most powerful X-rays cannot penetrate Earth's entire atmosphere from outer space for the huge numbers of atmospheric air molecules encountered en route by those high-energy stellar photons provide as much shielding as a three-foot-thick wall of lead.

Measuring Blood Pressure

A needle and catheter measuring the blood pressures inside of your radial artery and aortic arch respectively will show quite similar numbers and wave forms since both tubes lie within the same pressurized systemic arterial compartment. However, a third catheter simultaneously measuring your left ventricular cavity pressures will differ significantly from the others during ventricular diastole—for as the surrounding ventricular muscle stops squeezing and ventricular cavity blood pressures rapidly drop toward zero, the blood just forced into your elastic aorta during ventricular systole is trapped there at near-systolic blood pressures by aortic valve closure. So the outflow of blood through your smooth-muscle-regulated arteries determines how much blood pressure remains within your arterial tree between heart beats (based upon how much blood still stretches your elastic aortic wall). That is why your aortic and radial artery blood pressures (in this particular example) only fall from 120mm Hg at the height of ventricular systole down to 80mm Hg by the end of ventricular diastole, even as your left ventricle cavity pressure falls nearly to zero.

Blood pressure is commonly expressed in millimeters of *mercury* (Hg)—the specified blood pressure will balance or support a column of mercury exactly that many millimeters tall. Mercury is about 13 times heavier than water so an equivalent column of water (or blood) would be 13 times taller (1.3 centimeters of water is required to balance every mm of mercury). A mercury-filled blood pressure *manometer* (a vertical fluid-filled glass tube often utilized for measuring pressures—or the *aneroid* type of air pressure gauge occasionally recalibrated against a mercury column) is generally more convenient for taking blood pressure measurements than a 13 times taller water column.

When you pump air into a blood pressure cuff wrapped snugly about your upper arm, you squeeze everything enclosed by that blood pressure cuff. If you pump the cuff up to 60mm Hg, that is much higher than your usual venous blood pressure but still less than the standard 120/80 blood pressure within your arteries that we will use for this example. So arterial blood continues to enter your arm even though it cannot then escape and return to your heart. This continued inflow soon distends your arm veins and makes it easier to remove venous blood samples through a hollow needle into a tube or syringe. Such an uncomfortable and artificially elevated venous pressure also increases venous blood loss if the hand or forearm has been wounded. This explains why an insufficiently tight *tourniquet* often *increases* blood loss from a *distal* injury (an injury farther out on an extremity than the tourniquet).

If you now pump the blood pressure cuff up to 150mm Hg, that obstructs all further arterial inflow to the arm. Any distal wound therefore stops bleeding (perhaps after draining much of the blood still stored in arm vessels). But if that tourniquet effect is maintained, the resulting painful total ischemia of the arm will lead to permanent damage within hours and then gangrene (death) of the entire arm beyond that cuff. So local bleeding is best controlled by local pressure—direct finger pressure or compression by a soiled cloth dressing in the dirty wound may be messy but it is far better than bleeding to death or tissue loss by gangrene.

As we gradually reduce the uncomfortably high 150mm cuff pressure on your (fortunately uninjured) extremity to reach 115mm Hg, we note the return of a weak *radial* (wrist artery) pulse. At 115mm Hg cuff pressure, an electronic monitor sensing blood pressure within a radial artery needle on the same arm would show a series of regular little bumps, with each bump reaching 5mm above zero as your systolic pressure reaches 120, then returning to zero as your arterial blood pressure drops below 115 (the current air pressure within the inflated blood pressure cuff). And if we open the radial artery needle to the atmosphere while maintaining that 115mm cuff pressure, we see intermittent small spurts of blood (you can cause a similar reduction in water pressure by standing upon or kinking your garden hose). When listening over the lower brachial artery with a *stethoscope* (this passive sound amplifier picks up sound vibrations via a comparatively huge funnel-shaped opening or even larger flat diaphragm and transmits that sound energy through hollow tubing into two rather narrow earpiece openings) at the 150mm cuff pressure, we heard nothing. At 115mm of mercury, however, we detect faint intermittent thumps (as blood starts to flow through the brachial artery when systolic pressures exceed 115 and then stops abruptly when diastolic pressures again drop below 115).

If we now reduce the cuff pressure to 90mm, the radial artery needle electronic monitor reports larger but still intermittent pressure bumps cycling from zero to 30 to zero, while the radial artery squirts blood somewhat farther when the needle is opened. And at 90mm cuff pressure the stethoscope still transmits those intermittent thumps (sound vibrations generated as arterial blood flow surges forward and then suddenly stops) while a finger feels the radial or *ulnar* (other wrist artery) pulse coming in a bit stronger. When we further reduce the cuff pressure to 60mm Hg, the digital pressure readout on the radial artery needle reports 60/20 even though the *actual* aortic blood pressure has remained 120/80 throughout this imaginary exercise. At a cuff pressure of 60mm, the radial pulse remains *palpable* (can be felt) and is even stronger, but the stethoscope no longer detects thumping pulse sounds because blood flow has become *continuous* (rather than intermittent) through those still-somewhat-compressed arteries under the blood pressure cuff. Incidentally, the best place to detect faint blood pressure thumps with your stethoscope is above the elbow on the medial side of the arm where the *brachial artery* runs in the groove between your biceps and triceps muscles.

Thus a simple stethoscope and slowly deflated blood pressure cuff on the upper arm allowed you to establish the highest cuff pressure at which intermittent blood flow could commence through the brachial artery. That intermittent heartbeat-like (water-hammer) thump sound first appeared as the cuff was slowly deflated past the peak-systolic level (highest arterial pressure) of 120mm Hg. Furthermore, the heartbeat sound then disappeared as the cuff pressure fell below the end-diastolic (lowest arterial pressure) level of 80 because below that cuff pressure the distal brachial artery blood flow no longer was interrupted (so no more turbulent start-and-stop flow thumps). On the other hand, you could feel a pulse (intermittent systolic filling followed by diastolic deflation due to radial artery blood run-off) as soon as the cuff pressure dropped enough to let some arterial blood through with each heartbeat. And that radial (or ulnar or brachial) artery pulse remained palpable regardless of whether arterial blood flow was continuous or intermittent. But since your ear could not detect sound waves with a frequency as low as your pulse, any heartbeat sounds heard while listening with a stethoscope over an artery when there was *no* blood pressure cuff inflation presumably represented higher frequency heart (valve closure) sounds transmitted along stiff artery walls or else turbulent blood flow within a nearby partially *occluded* (blocked) artery.

Electronic blood pressure sensors are expensive, strictly temporary and often unreliable monitoring devices. On the other hand, the live blood-pressure-sensitive cells of your aortic wall and in the upper reaches of each common carotid artery are extremely reliable and remain on duty full-time. Those carotid artery *bifurcations* (branching sites) also bear chemoreceptors that monitor arterial blood pH and CO_2 levels. Other relevant pressure and chemical monitoring activities occur in the brain stem, heart muscle and the inner lining cells of your heart (*endocardium*) and arteries (*endothelium*). Indeed, these continuously accurate lifelong readouts remain essential for the ongoing regulation of your blood vessel smooth muscle tone—otherwise you never could maintain appropriate regional blood flows and cardiac output with your constantly changing positions, activity levels and *blood volume* (consequent to beer parties, dehydration, diarrhea, tissue edema and so on—see Fluid and Electrolytes).

Pressures And Flows

If your aortic valve did not close properly, the resulting regurgitation of pressurized aortic blood into your relaxing left ventricle would reduce your systemic arterial pressures more rapidly than usual. Such an *insufficient* (leaky) aortic valve would therefore detract from the usual forward flow of aortic blood into downstream arteries and widen your *pulse pressure* (the difference between systolic maximum and diastolic minimum pressures measured in a systemic artery). A very wide pulse pressure produces a slapping pulse and generally implies a low diastolic blood pressure—which reduces diastolic blood flows through your coronary arteries into heart muscle. And that leaves your harder-working heart (pumping a lot more blood than usual) with subnormal blood flows to its own muscle—not a good situation. On the other hand, if your aortic valve were to become *stenotic* (narrowed by defect or disease), then your left ventricle would have to produce much higher intraventricular blood pressures to move enough incompressible blood through such a diminished valve opening—yet despite those elevated left ventricular systolic blood pressures, your peak arterial systolic blood pressures might remain normal or even low. Furthermore, the chronic effort of working against a constricted aortic valve would probably lead to *myocardial hypertrophy* (thickening of heart muscle)—again with a relative reduction in diastolic blood flow to the harder working and enlarged heart muscles.

Any of your four heart valves may become stenotic, insufficient or both. Mitral valve malfunction due to rheumatic fever, ruptured papillary muscle or torn chordae tendinae can cause lung *congestion* (too much blood in the lungs) while a *congenital* (before birth) narrowing of the pulmonic valve may interfere with proper pulmonary *perfusion* (not enough blood reaching lungs) and a markedly narrowed tricuspid valve can resist venous blood access to the right ventricle. As you might expect, moderately leaky heart valves are more easily tolerated on the lower-pressure right-heart side. We often speak of *right heart failure* when deficient heart muscle or defective heart valves prevent easy return of systemic venous blood to the lungs, or *left heart failure* when the valve or heart muscle problem interferes with blood movement between pulmonary veins and systemic arteries. Thus *peripheral edema* (fluid-filled legs) and liver swelling suggest right heart failure while fluid-congested lungs may represent left heart failure (or perhaps pneumonia or other lung inflammation).

Septal *defects* (openings between the two atria or between the two ventricles) or a persistently open ductus arteriosus are among the more common congenital cardiac abnormalities. Obviously a simple atrial septal defect or ventricular septal defect ought not cause any problems before birth when fetal heart chamber blood flows mix anyhow. But if such openings or a patent ductus arteriosus persist long after birth, they can cause noticeable signs of excessive (hence turbulent) flow such as *murmurs* (many heart murmurs are *innocent*, meaning they have no effect on health) or heart chamber enlargement as well as lung congestion. Left-heart-chamber blood pressures of children and adults usually remain higher than corresponding chamber pressures on the right side so any diversion of blood through an atrial septal defect, ventricular septal defect or patent ductus arteriosus should be from left to right (systemic to pulmonary) side. This means some (often much) of the newly oxygenated pulmonary venous blood will pass from left to right atrium through an atrial septal defect, or from left to right ventricle via a ventricular septal defect—the forward-directed systolic jet of left-ventricle blood spurting into the right ventricle often produces a localized *systolic* lift and *thrill* (vibration) on a child's soft overlying chest wall. On the other hand, the uninterrupted flow of blood via an open ductus arteriosus from aorta to pulmonary artery produces a continuous (but two-tone or machinery-like because systolic/diastolic) murmur.

In each of these cases, freshly oxygenated blood reappears in lung capillaries without ever having passed through systemic capillaries. So a major *left-to-right shunt* is an obvious waste of pumping effort that also tends to overwork the heart and overload the lungs with blood. Lung arteries may then respond by narrowing, which raises pulmonary artery and other right-sided pressures. Under some circumstances, those right-sided pressures eventually equal or even exceed comparable chamber or vessel blood pressures on the left side. Deoxygenated blood then spills into the systemic circuit across any septal defect or patent ductus. Major *right-to-left* shunts cause the peripheral blood to take on a *cyanotic* (deoxygenated) hue—leading to bluish discoloration of cheeks, lips and other unpigmented surfaces with a superficial circulation. Right-to-left shunts also allow clumps of bacteria and other particles arriving in the systemic venous blood to bypass lung capillaries en route to systemic arteries—so embolic strokes and abscesses of brain and other tissues are among the many serious problems associated with right-to-left shunts. A temporary right-to-left shunt may also develop through a still-patent foramen ovale if right-sided pressures are sufficiently elevated by a large pulmonary embolus or even by the many tiny emboli encountered in post-traumatic *fat embolism syndrome* (which might then allow later-arriving venous fat emboli to bypass the lungs and directly enter systemic arteries).

Sometimes the fetal aortic arch narrows markedly or becomes totally blocked near the opening of the left subclavian artery (perhaps this congenital condition is an abnormal expression of the obliterative process activated in the nearby ductus arteriosus soon after birth). Such a *coarctation of the aorta* leads to reduced arterial blood pressures and flows in lower body tissues so the kidneys constantly demand more blood pressure (see Kidney). An affected individual will have faint or absent pulses in the femoral arteries despite markedly elevated aortic arch pressures that can rise sufficiently to rupture a brain artery (cause a stroke) in childhood. The exact site of coarctation with respect to the left subclavian artery will determine whether left arm blood pressures are low, normal or high—right arm pressures remain high.

These elevated aortic arch blood pressures overwork the heart and can lead to early blood vessel aging (atherosclerosis) in the upper body. But even the severe proximal aortic hypertension resulting from coarctation barely maintains adequate lower-body arterial blood flows so this high blood pressure must be corrected surgically by removal or bypass of the obstructive aortic segment. Coarctation of the aorta usually creates an impressive collateral arterial circulation in the chest wall as enlarged and hypertensive subclavian artery branches feed blood through maximally distended intercostal arteries in order to fill the hypotensive descending thoracic aorta beyond the coarctation—this represents a reversal of the usual outward-bound intercostal artery flow from aorta into chest wall tissues. Since arteries grow in length as their diameters increase, the *expanded* and *tortuous* (elongated, therefore following a winding course) intercostals-as-aortic-collaterals can often be felt pulsating between the ribs (they may even wear away the bony under-surfaces of those ribs)—yet despite that major arterial enhancement, chest-wall venous blood flows remain normal.

A great many other anatomical and physiological abnormalities can alter blood flow and tissue oxygenation. The amazing thing is how often nothing goes wrong with such a complex, interactive system of pumps and pipes. But perhaps it is not so amazing when we remember that our ancestors could not have survived with substandard or non-competitive systems. Furthermore, those standards were continuously being raised by the competition. Modern medical and surgical advances have allowed many individuals to survive and reproduce despite serious congenital heart abnormalities. Luckily, many of those patients had *non-hereditary* congenital abnormalities (caused by viral infection, chemicals or other unknown influences) rather than misprinted genes that might lead to similar difficulties for their descendents.

Your Parts Tend To Fit

Your heart size and growth tend to be controlled by ongoing loads and demands. Any persistent stimulation of your sympathetic nervous system or continued chemical requests for "more blood pressure" from your kidneys will cause cardiac growth. Similarly, your skin, blood, lungs, gastrointestinal organs and other tissues generally remain appropriate for your size and the burdens that they must bear. Apparently there are no rigid preset sizes for the various parts of your body—had you been malnourished or chronically ill during childhood, that smaller you would still be properly proportioned. Thus the endless positive and negative feedbacks between your various tissues and organs seem to keep you just right with little conscious effort. But do watch those desserts.

CHAPTER 20

IMMUNITY

*Know Thine Enemy;... Be Strong;... To Thine Own Self Be True;...
Prevention, Localization And Delay;... The Battle Plan: Your Shock
Troops;... The Follow-up: Choosing The Perfect Weapon;... Command And
Control;... Preparing For The Next Attack;... Us Against Them;...
Monoclonal Tools.*

Know Thine Enemy

You are tall and strong and fast. You are good at detecting, analyzing and responding to information. So you generally can recognize a good lunch, evaluate a reproductive opportunity and stay out of serious trouble. However it takes a whole lot more than that just to get by. For in addition to the lions, leeches and litigators always lusting after every delectable asset of your eminently recyclable being, a host of other beastly challenges continuously arise to test your fight-or-flight, shoot 'em, flick 'em or buy 'em off capabilities. Furthermore, through no fault of your own, you have inherited the same ancient and implacable foes that eventually overwhelmed your ancestors. And it is those microbes and parasites that will continue to attack relentlessly in wave after wave, undaunted by their losses, until finally your defenses crumble and your repeatedly stolen reducing power is once again liberated. Yet their final big victory hardly differs from defeat as most of those invisible ones are consumed by other opportunists drawn to the feast. So life progresses, fueled by the fallen.

In this unending and unequal contest, the very large and complex never finally defeat the very small and numerous. True, the large win many more battles and outlast innumerable generations of the small. Huge specialized organizations such as the Soviet Union, General Motors, IBM, a blue whale or you, may even appear increasingly invincible as they accumulate power and resources. But being few also means that those who are so well adapted to the status quo necessarily lack variability. And as conditions continue to change and eventually change drastically, the same specialization of molecules, organelles, cells, tissues, organs and individuals that originally enhanced their acquisition of information and resources now makes them increasingly vulnerable to breakdown in some vital department. This sort of ecological fragility also allows our booming human populations to inadvertently deplete or extinguish other species of large plants and animals (by mistake and greed rather than intent). In contrast, their unlimited numbers, frequent inherited errors and regular viral, plasmid and other DNA and RNA attacks, prepare at least some microorganisms in every population for almost any eventuality—which also explains why despite moderately valiant efforts, humans have eradicated only one virus from the world.

Even that deletion of inanimate information was not based upon finding or fabricating a powerful cure. Rather, scientists and politicians were able to bring about a simultaneous world-wide human immunity to *smallpox* (an age-old often deadly human-dependent virus) by applying a closely related, naturally occurring, hardly ever fatal packet of *vaccinia* information to the broken skin of almost every susceptible human—thereby bringing about a protective counter-epidemic. Presumably the ancestral *cowpox/vaccinia* virus gave rise to smallpox as humans first came into regular close contact with infected ancestral cows ("vacca" means cow in Latin). In any case, being deadly and human dependent was the smallpox virus' fatal weakness. Even the more dangerous form of smallpox that killed 10%—30% of susceptible old-world humans was soon eliminated from the survivors by their immune responses, leaving no chronically infected individuals to distribute that highly infectious virus. But despite its high virulence when inhaled and the brief duration of illness before recovery or death, smallpox often persisted in the environment for 10 years or longer (and possibly far longer in permanently frozen northern soils or permafrost). Apparently some early European explorers found smallpox-contaminated blankets the perfect gift for highly susceptible Native Americans that they planned to loot, convert and destroy (smallpox eventually killed 90% of East Coast Native Americans). In contrast, cold viruses do not last long in the environment so their continued propagation depends upon a high attack rate with sufficiently limited *morbidity* (illness) so that current cold sufferers can remain up and about to sneeze upon others.

Vector-borne diseases have still other priorities since disease-spreading mosquitoes or other vectors feed upon sick and well alike. Here the virus, bacteria or other parasite (e.g. malaria) often finds reproductive advantage in keeping the infected vector relatively healthy and mobile while saving its highest replication rate and virulence for the human host—since converting as many human assets as possible into new parasites enhances the likelihood that the next blood-sucking vector will acquire some of those parasites for delivery to a new host during a subsequent blood meal. Then again, an infected and infectious louse need only stay in sufficiently good shape to stagger from its last lousy host (now deceased) onto a new one—there is no need for the parasite to keep that louse at its best. The so far intractable viral plague known as AIDS emphasizes how far we are from really understanding viral (or most other) infections, though it appears that even AIDS infections persist and spread according to similar evolutionary rules—where human behavior results in frequent changes of sexual partner or commonly shared needles among infected addicts, the virus enhances its spread through rapid replication (associated with increased virulence)—but when sexual partners only switch occasionally or needles are rarely shared, the advantage goes to a more chronic or indolent form of infection that can keep the patient up and about in the meanwhile (being terminally ill makes one an unlikely partner for sex or needle-sharing).[1]

Be Strong

As you sit there quite calmly reading these words, many trillions of microorganisms are vying desperately for any chance to live in, on or off you with no regard for how that might suit your purposes. Yet you remain blissfully unaware of the ceaseless highly lethal battles raging within and about you—unaware, that is, until things go wrong. And it does not take much. A sip of cool water from a beautiful mountain stream, a barefoot stroll on tropical sands, a delicious cream pie or lovely salad, a cough or sneeze in the subway, a passionate good night kiss or a stolen night of bliss and soon a whole new set of tiny invisible foes are fighting desperately to take your good health or even existence (after all, it is often yours versus theirs). So while individual microorganisms remain undetectable to your senses, before long the costs of that latest battle become all too evident. You just don't feel well, your glands are swollen, your stomach upset, red blood cell count low, white blood cell count high, diarrhea, swelling and redness, skin sores, a stiff neck, shaking chills, a raging fever, you cough and sneeze, you ache all over, your aorta or liver or heart or kidneys or salivary glands or testicles are under attack. You would rather stay in bed but reproduction is the last thing on your mind. And that probably doesn't matter because other humans are unlikely to view your obviously parasitized state with reproductive interest anyhow. You simply are not up to anything strenuous.

The arduous mating rituals of many complex multicellular life forms are disadvantageous to the weak, the ill and the poor. For an opportunity to reproduce, animals often must indulge in everything from back-biting, clawing and butting of

---

[1] Ewald, Paul W. April 1993. "The Evolution of Virulence". *Scientific American*. pp 86–93.

heads to disco dancing, nest building, the Easter Parade and a whole range of ridiculously gaudy or dangerously cumbersome forms of sexual ornamentation ranging from high heels and peacock feathers to the costly paraphernalia of the mighty hunter. These rituals and structures persist primarily because they identify and bring reproductive advantage to the most fit, healthy and/or wealthy. For it makes evolutionary sense that those who best resist bacterial and other parasitic infections, or who accumulate the most reducing power or wealth under current conditions, are likely to have healthier or at least richer offspring. Even your earliest bacterial or single-cell ancestors usually reproduced more successfully if not already under attack or parasitized by usually unhelpful viral or bacterial information.

The release of toxic molecules to which the source bacterium or cell is not susceptible can enhance its survival. For example, the penicillin-producing fungus is not itself endangered when a short segment of its penicillin molecule product permanently clogs the enzyme that strengthens bacterial outer walls by creating vital cross-links. Yet any bacterium thereby deprived of its sturdy cross-linked *peptidoglycan* overcoat is readily ruptured by its high internal osmotic pressure unless it resides within a hypertonic medium while the penicillin effect persists. So penicillin reduces microbial competition for available reducing power as well as bacterial predation on the penicillin-producing mold (which itself has no use for outer-cell-wall peptidoglycans). A huge assortment of other nasty products put out by bacteria, plants and your gardening friends can similarly deter or harm many nearby life forms, perhaps even dissolving those neighbors in preparation for lunch.

In general, the dangerous digestive chemicals routinely utilized within your own cells are produced in preliminary form and only activated after they have been isolated inside digestive *lysosomes* released from the secretory Golgi apparatus. Your cells can then treat any microorganisms that they may engulf within a bit of outer cell membrane (through *phagocytosis*) to a medley of their favorite lysosomal chemicals by fusing that bacteria-laden vacuole with a similarly enclosed lysosome. Many of your *small* or *large white blood cells* (*micro*phages or *macro*phages) are *phago*cytes (*eating* cells) that routinely recycle tiny captured organisms and debris by utilizing some of the 50 known lysosomal chemicals (which include several sorts of strong bleach) to advance their digestive process. Others of your cells dump lysosomal contents and various bioactive molecules externally onto larger prey or parasitic invaders (see below).

These never-ending small scale battles are no less real for being invisible. Indeed, with the reproduction rate of microorganisms so many orders of magnitude greater than yours, selection pressures at that "micro" level are correspondingly more fierce.

So it becomes important to defeat the more dangerous of those attacking microorganisms quickly since otherwise the more effectively you fight, the more fierce an enemy you could thereby create (because only the best-adapted-to-you of each attacking generation gets to reproduce). But the reverse also holds true, namely the better you get along with less dangerous assailants, the more likely they are to benefit by keeping you "on tap" for spare reducing power, perhaps without noticeable harmful effect. Of course, both of the above scenarios depend upon other factors too, such as the manner of spread and whether humans are the main species upon which this infection depends or only incidentally involved (as in tetanus). Eventually you and some of your microorganisms may even develop a *symbiotic* relationship and come to rely upon one another for certain services. The same spectrum of evolving relationships (predator, prey, parasite, symbiont) can be seen in most other interactions (business, politics, religion, friendships and the domestication of crops or animals).

Ultimately you are an exceedingly complex aggregate of specialized cells that represents a long-time and therefore tremendously evolved (greatly changed) cooperative of former bacteria. However, being a slow-growing rarely reproducing multicellular creature that cannot defend itself against such invisible invaders with spear, club and cannon, you are forced to get down and dirty, to meet or match the attacking micro-enemy's every move on their own terms. So like Commanders-in-Chief everywhere, you delegate the front-line dangerous stuff to your own microtroops—those white blood cells that endlessly patrol your perimeter, ready to repulse invaders that may breach your formidable surface defenses before those attackers can gain a foothold (so to speak). Of course, you must be ready to defend those troops and your civilian cells against toxic chemicals that such invaders may release into you. And it is critically important to continuously monitor and update intelligence reports about the identity of each invader, no matter how tricky that invader may be at evading your defenses or camouflaging itself as one of (or even inside some of) those many dozen trillion cells that are you. Not only must you keep up your guard, you also must minimize injuries to your civilian cells. For winning the next encounter is useless unless you survive as a healthy and functional entity. Furthermore, you must remain ever alert for any sign of weakness or corrupt tendencies among your countless civilian cells—almost inevitably, some of those cells will selfishly and foolishly place their own individual growth and replication before the welfare of the group. You even have to watch the behavior of your troops continuously and check their loyalties most closely. This is a dangerous world so you must run a tight ship. Thus you deliver overwhelming force and swift death to all invaders, death for disloyalty and cellular corruption, even death for suspicious behavior. Although the authorization to kill is delegated under standing orders, it too must be monitored and restricted if you are to survive and prosper.

To Thine Own Self Be True

Quite obviously your immune system represents an embattled dictatorship facing an impossible task. And sooner or later things do get out of hand. Perhaps certain of your cellular warriors are themselves secretly weakened, parasitized or cut down directly (as in AIDS). Sometimes the enemy simply overwhelms or outflanks your best surface defenses and localizing tactics, then rampages through your organization more swiftly than you are able to respond (the most virulent forms of hemolytic streptococcus-Group A). Other bacterial terrorists may offer little threat as invaders but are capable of releasing toxic chemicals or byproducts that could hurt or kill you (tetanus, diptheria, botulism). And on occasion, your cellular warriors and their high-tech specific attack molecules (*antibodies*) make targeting errors that misdirect some of their weapons against your own troops or ordinary civilian cells—the so-called "friendly fire" known as *autoimmunity*.

Not surprisingly, after a lifetime of unceasing warfare, the damage resulting from non-acute autoimmune diseases such as diabetes, rheumatoid arthritis, multiple sclerosis, glomerulonephritis and rheumatic heart disease tends to become increasingly apparent. As usual, the possibility for corruption among your civilian cells parallels their length of time in office (your advancing age), especially in tissues pre-authorized and already provisioned for ongoing growth and reproduction. So you are particularly threatened by insubordinate growths (tumors and cancers) of your continuously replicating reproductive system or your regularly replaced blood cells and the surface lining cells of your skin, urinary bladder, gastrointestinal system and respiratory tract. Since their ordinary function requires the continued production of cells, these tissues are especially susceptible to external disruptions by unwanted viral information or by mutagenic chemicals (e.g. the carcinogens in cigarette smoke) and energetic electromagnetic radiation at ultraviolet, X-ray or gamma ray frequencies.

No biological system can afford to be 100% reliable. Even coming close could be reproductively disadvantageous since outstanding reliability often implies excessive rigidity of purpose (over-specialization) as well as the inherent inefficiency that comes with many back-up systems. So eventually the controls of any immune system are evaded or disturbed—just another of those many "natural causes" for body failure associated with aging. But nothing is wasted, since all assets acquired in every lifetime are entirely recyclable and biodegradable. Although death generally has its greatest influence on the makeup of the next generation when it selects among individuals with future reproductive potential, the biased distribution of resources and information among ants, elephants and primates also influences who will survive to reproduce—thus those humans who are infertile by choice or happenstance may still alter the reproductive success of a great many others (including their own relatives) through warfare, persecution, slavery and the control of information and other resources including science and technology (e.g. by developing new weapons and defenses when those that were inherited falter in the face of the latest onslaught).

Prevention, Localization And Delay

The complex multilayered ever-adjusted blend of defenses that you have inherited usually protects you from unwanted viral information, significant bacterial attacks, troublesome single-celled protozoal infestations and small multicellular invaders. Indeed, only the tiniest fraction of those attempting to share your accumulated reducing power can ever survive upon you, let alone penetrate through your inhospitable borders. Of course, there will be days when you might prefer to seal your borders entirely in the manner of a bacterial spore but that would obviously be impractical for your moving, breathing, eating and otherwise functional life-style. So instead you must rely upon the fatty acids broken free from your greasy skin secretions by mostly friendly (saprophytic) bacteria to provide a mechanically and chemically inhospitable, acidic surface that repels many potential boarders. And the frantic competition for your scarce surface nutrients brings out nasty interspecies rivalries and defenses that further restrict your skin surface *flora* (population of microinhabitants). Even the ammonia released by urea-splitting bacteria feasting upon urine spilled into your vagina may help to suppress other potential invaders, as does the rinsing action of your perspiration and the continued export of your bacteria-laden respiratory tract and digestive mucus. In addition, it generally takes more than successful invasion of your most superficial cells to cause ongoing problems since those cells are continuously shed and replaced from below anyhow.

Along with their other antibacterial ingredients, your tears, nose drippings, perspiration and saliva all contain *lysozyme*, an enzyme that disrupts bacterial-wall polysaccharides—those unclothed bacteria are then susceptible to osmotic rupture. Lysozyme's DNA has gained wide distribution by being such a dandy antibacterial weapon—it is even found within chicken egg white. Furthermore, the *lactoferrin* in your milk and most other body secretions binds scarce iron and makes it un-available for bacterial growth while the *lactoperoxidase* of milk, saliva and tears also helps to suppress bacterial multiplication. Skin grease (sebum), perspiration and mucus not only kill and wash away—they also repel and reduce adherence of all sizes of potential predators—just as its especially thick and impenetrable slime coat-ing protects the slow moving jellyfish-eating ocean sunfish. Slimy surface layers are never sterile, of course. And the temporary truce between a fish and its own slime bacteria does not extend to wounds in your own skin that become contaminated by fish slime. Nor would those bacteria enjoying the sumptuous saprophytic life-style (while producing locker room odor) in the sweat of your own hairy armpit be equally benign elsewhere under other circumstances—*pneumocystis carinii* was not even known to be a common fungal inhabitant of human respiratory surfaces until it began causing fatal pneumonia in immunosupressed individuals, especially those suffering from AIDS.

As with other trespassers, your micro-predators often discover reproductive advantage in quiet coexistence rather than by bringing about overwhelming in-fection and early death. The same logic suggests that lynx would suffer great harm if they could catch and eat every last rabbit. Although your own food is generally cooked (or at least washed and no longer struggling), countless millions of living bacteria and other microorganisms are included in every yummy mouthful that you swallow (some of those bugs come with the yoghurt, sushi or salad, others from your own gums and teeth). But few are the micro-creatures that can pass through your seething cauldron of warm corrosive stomach acid unscathed—and those that successfully resist acid hydrolysis must then endure the gastrointestinal enzymes that so routinely cleave the carbohydrates, proteins, fats and nucleopro-teins of your unwilling victims.

Nonetheless there are many occasions when a surface injury or penetrating attack opens the way into your nutritious tissue fluids and cells, or even provides easy access to your blood and lymph capillaries. However, just as it takes more than one sperm to prepare an egg for fertilization, some minimum infecting dose or number of micro-invaders is generally required to bring off a successful infestation. For your white blood cells are always on patrol, ready to engulf and devour or at least localize any invading microorganisms. A more specific counterattack can then be prepared and launched once the enemy has been identified and its weaknesses become apparent. The delay involved in such careful identification prior to counterattack obviously carries some risk in the face of the more serious sorts of attacks that must eventually come your way. So would it not be better to counterattack at once, utilizing your entire arsenal—or better yet, to simply blow them away with that anti-microbial super-weapon undoubtedly developed by your long-suffering ancestors over long eons of painful trial-and-error? After all, aren't ancestors supposed to suffer, toil and improve themselves so their descendents can have it easier?

Unfortunately there is no such super-weapon. Nor is one possible in theory. For your own cells have far more in common with any single invader than the huge variety of potential invaders could ever have with one another. In other words, anything that could kill every potential invader would kill you as well (a problem that the military cannot solve either). So the broad-spectrum chemical herbicides and pesticides being used in such huge quantities are as dangerous to unintended victims such as yourself as they are to their proposed targets (your distant weed and varmint relations). In comparison, the more accurately targeted biological pesticides like penicillin are effective in tiny quantities and have far fewer side-effects for non-combatants. But such distantly related biological entities as streptococci, schistosomes and the AIDS virus still share various distinctive molecules with you that make them difficult to attack without also targeting yourself. Quite clearly, your body cannot afford to release poisons internally except very discreetly and carefully. So you target your enemies as precisely as possible in order to minimize collateral damage to your own civilian cells and structure.

The Battle Plan: Your Shock Troops

It takes time to develop, produce and deliver specifically appropriate responses to the enormous number and variety of possible attacks and attackers. So the first line of defense against bacteria (or other microparasites) that get into your flesh is a localized inflammatory response associated with increasing capillary permeability. Among blood proteins crossing those leaky capillary walls are specific *antibodies* that may recognize and fasten onto the new invader, thereby labeling it for destruction, as well as sufficient fibrin to tangle things up and help localize the invasion. This tissue inflammation is partially mediated by the first phagocytic white blood cells to arrive on the scene, usually within minutes. These cells are your *granulocytes*, 6000 to 7000 of which normally patrol within every cubic mm of your blood.

During bacterial invasions your blood granulocyte count usually rises rapidly as some of the vast ready reserves of granulocytes (microphages) stationed within your bone marrow are called up. But a granulocyte freshly released into your circulation only carries enough glucose (combat rations) for a day or so. There are good evolutionary reasons for not allowing granulocytes to compete with your other cells for the glucose within your circulation. Presumably these include the limited capacity of microphages for ingesting bacteria and their valuable role as first-attack suicide troops. In any case, those short-lived roving granulocytes respond with great sensitivity to local chemical signals from injured or inflamed tissues. Granulocytes also release chemical signals as they adhere to specific adhesion molecules displayed by disturbed endothelial cells. Then these mobile granulocytes push and squeeze their way between capillary endothelial cells into your inflamed tissues where they engulf and devour as many invaders as possible before dying. Their programmed cell deaths allow granulocytes to release additional digestive enzymes and toxic compounds as well as great tangles of DNA from their conspicuous multilobed nuclei (your personal version of entangling those attacking bacteria in red tape—another unexpected use for your DNA information).

Granulocytes are a little larger than red cells and come in three basic varieties. The common *polymorphonuclear neutrophils* make up 60%–70% of your normal *leuco*cyte (*white* blood cell) *count* in peripheral blood. Your *eosinophils* make up just 1%–2% unless a parasitic invasion or some *allergic* (*hypersensitivity*) process is underway, while *basophils* usually form only 0.5% of your total white blood cell population. These three varieties of microphages (their strange names derive from the sorts of stains that their granules attract) represent your cellular first line of defense. While their duties overlap to some degree, it is of and over the dead bodies of your neutrophils that you form much of the thick whitish content of a zit or pimple—which also includes blocked sebaceous gland secretions, entangled bacterial invaders (live, dying and dead) and a dense network of fibrin. From their subsurface storage sites in the respiratory, gastrointestinal and urinary tracts, your lurking eosinophils specialize in harming larger prey. Even parasites that are far too large for ingestion by your eosinophils still find the contents of those eosinophilic granules highly toxic—as do bacteria. Your circulating basophils and the functionally comparable (also bone-marrow derived) cells known as *mast cells* (which are distributed more like eosinophils) play important roles in tissue inflammation as well as your allergic responses (with the *IgE* that they pick up and carry on their surfaces—see below).

It is reproductively advantageous that localized infections cause pain (due to locally released inflammatory chemicals) since vigorous muscle activity or increased tissue pressures could easily push live bacteria into surrounding tissues—as you probably first discovered when the small painful pimple that you squeezed before the prom became a far nastier boil (abscess). Neutrophil phagocytosis is enhanced if bacteria to be ingested are first rendered sticky by specifically matched antibodies or circulating complement (see below). The inflammatory fibrin network also helps neutrophils to move about and pin down those invaders. But not all microorganisms that reach intracellular sites are doomed. Indeed, certain ones like malaria, tuberculosis and gonorrhea actually find red cell or phagocyte cytoplasm the perfect place to settle down and reproduce—for these parasites and bacteria have evolved ways to escape into the cytoplasm or otherwise prevent or inactivate chemical attacks (that might be brought to bear through fusion of the vacuole containing the ingested organism with a digestive lysosome).

As their description suggests, the remaining 20% to 25% of your circulating white blood cells generally do not have large light-microscope-visible granules (the granules of stained granulocytes are primarily lysosomes packed with nasty chemicals). These *agranular white cells* are your *lymphocytes* and the somewhat larger *monocytes* (the latter representing 3%–8% of your circulating leukocytes). While your neutrophils reach damaged or infected tissues within minutes, those monocytes arrive more slowly—but as they then advance out into those inflamed tissues, they gradually enlarge into the fearsome-looking *wandering macrophages* that play such an important part in signaling between cells, identification and cleanup of bacteria, and the elimination of accumulated cellular debris. Monocytes are close relatives of osteoclasts (see Bone) and also of the *dendritic* (many fine branches or arms) phagocytic cells located in your skin, liver and lymphatic tissues.

The Follow-Up: Choosing The Perfect Weapon

After your macrophages and dendritic phagocytes have methodically ingested and ripped apart their bacterial victims, they display a wide assortment of bacterial fragments on their outer cell surface. These fragments are firmly framed for presentation within special *major histocompatibility complex* (MHC, also known in humans as HLA or "human leucocyte associated") *antigen display molecules*, much as successful warriors used to hang scalps and other victim body parts from their belts. In addition to identifying all of those hard-working and victorious macrophages, such bacterial body part displays allow any of your hundreds of billions of tissue-based and circulating *lymphocytes* to drop by and try out these new *antigens* (short segments of molecules to which *antibodies* can be matched) for fit. Sort of the way aspiring princesses crowded around to try on Cinderella's little glass slipper—except that here, the more numerous the bacteria, the more bacterial parts are available for simultaneous display by a great many macrophages to a great many lymphocytes. It is very good news for any lymphocyte that truly fits such a bacterial part, for that lymphocyte is thereby stimulated to create a whole bunch of identical descendents (a *clone*). And soon those clones are hard at work, producing large quantities of the exact antibody that can best grab those annoying microbes by the particular anti-genic part being displayed—which usually allows you to hurt those specific microbes very badly and thus defeat their attack.

Your cells actually utilize two sorts of MHC complexes for their peptide dis-plays. MHC-1 presents short segments of ordinary cytoplasmic proteins that have been broken down by the usual proteosomes and other intracellular processes. MHC-2 displays somewhat longer peptide fragments derived through the lysosomal processing of bacteria and other ingested particles. While MHC complexes vary widely between individuals, they are identical on all cells within any one indi-vidual. And each of the dozen MHC components in any one person (six from each parent) is available elsewhere in dozens or hundreds of different forms. Some MHC complexes are better at fitting to and displaying certain peptides than others, so individuals vary in how effectively they can identify and respond to specific pep-tides from particular invaders. That is why an infection passed along through a blood relative is likely to be more troublesome than one caught on the subway. Since epidemics caused by serious and successful invaders often harm or kill (hence reduce the reproductive success of) those who are most susceptible, any prevalent infectious disease tends to enhance the percentage of individuals bearing MHC's that are more effective for displaying (and thus fighting) that same prevalent dis-ease (malaria in West Africa for example). This presented a particular problem for American native peoples whose closely related and therefore very similar MHC's made them uniformly susceptible to many "Old World" diseases that therefore could wipe out an entire community.

The large number of possible MHC's makes your personal major histocom-patibility complex display relatively unique and allows MHC recognition to serve as an important identity check on your own civilian cells. So while bearing the wrong MHC can get a cell killed in a hurry, you still avoid attacking your own cells as invaders whenever they display invader parts upon your personal MHC struc-ture. Although this helps you to deal with cellular invaders, it is a big problem for those receiving replacement organs or tissues from others. And the worse the MHC match, the greater the likelihood that a transplanted organ will be rejected by your roving antigen inspectors who "go on just doing their job" like well-trained bu-reaucrats even though their usual actions no longer make sense and, indeed, bring harm to the whole enterprise. So one must generally resort to poisoning or mis-leading those inspectors with *immunosuppressive* medicines or more specifically targeted molecules in order for foreign tissue grafts to survive.

Lymphocytes are slightly larger than red blood cells and easily halted within a capillary wherever necessary. You probably have 10 to 15 billion lymphocytes roaming through your blood stream at any one time, each desperately seeking the one perfect antigen that could grant it long life—for otherwise they must commit *apoptosis* (programmed cell death) within days. A great many billion more lymphocytes are positioned within your *spleen, thymus* and *lymph nodes.* (Your soft purple blood-filled and encapsulated *spleen* is located on the left upper side of your stomach while the soft yellowish and slightly nodular *thymus* gland with its rather diffuse margins and limited blood supply lies deep to your sternum on the front of your pericardium.) Smaller lymphocyte collections can be found within other tissues all over your body. The patches of lymphocytes within many of these tissues contain concentric cellular formations known as *germinal centers* where antibody-producing cells of appropriate lineage and specificity undergo final accelerated revision and retesting for improved antigen fit before entering production or undergoing apoptosis.

Command And Control

Almost all of your lymphocytes fall into one of two major groups, depending upon where they attended school. *T lymphocytes* are destined to become the officers of your immune system if they survive to maturity within your thymus. But first they must reliably recognize friend from foe. Any T's still in training that respond in hostile fashion when nudged by a friendly cell are allowed to kill themselves (apoptosis). Those T cells that recognize "self" MHC's (and hold their fire when only self peptide fragments are displayed therein) eventually graduate with honors and spend the rest of their careers politely checking MHC displays for the fragments of foreign molecules that inevitably betray invading life forms. Mature T cells are so good at this that they can specifically detect literally billions of slightly different foreign molecules.

Your lymphocytes vary in their ranks and duties. *Natural killer* cells (non-T, non-B, bone marrow origin) are authorized to punch all invaders full of holes, thereby killing them directly. Natural killer and your *cytotoxic (killer T* or *CD-8) lymphocytes* dispense that same harsh justice to any civilian cell that shows corrupt cancerous tendencies or harbors subversive infectious viral information—such traitorous or diseased cells are often betrayed by the abnormal or foreign peptides displayed on their MHC-1 complex. As mentioned, peptides for display are picked up on the basis of their fit in the MHC groove, but different versions of *transporter molecule* tend to deliver different antigen peptides to each cell's endoplasmic reticulum where its MHC-1 awaits. Even though self-incrimination carries a death sentence, your cells cannot avoid making such surface displays on their MHC structures. Not bringing freshly constructed MHC's to the cell surface is no option either since natural killer cells swiftly execute any cells not making enough MHC's— more often than not, such cells are cancer cells anyhow. Furthermore, cytotoxic T's appear able to induce apoptosis in their targeted cells—a useful skill since an appropriately directed cell can tear itself into many nourishing components within its own plasma membrane before being ingested whole by a macrophage or in parts by nearby cells—all without causing unnecessary disturbance or inflammation. Your usual (considerable) turnover of cells ensures that those many off-the-cuff death sentences delivered for presumed cause have little body-wide impact—except that your long-lived nerve cells must be protected from blood-borne white cells by the blood-brain barrier (neurons also make minimal MHC displays). Nerve cells can sometimes clear themselves of viral infestations when appropriate antibodies become available in the circulation (how they do that is unclear)—but intermittently activated viruses often persist within nerve cells (e.g. herpes cold sores and venereal sores and also the chicken pox/zoster viruses of the herpes group) for many years or life.

Killer and other T cells (and also glial cells in the nervous system) release a great variety of intercellular signals or *cytokines* including all sorts of *lymphokines*, *interferons* and *interleukins*. In addition, they send and receive a great many other chemical signals as they activate and instruct macrophages and also the *B lymphocyte* (antibody producing) troops on when, where and what to attack as well as when to stop fighting and mostly undergo apoptosis (note the lack of retirement benefits—these soldiers come cheap) after the latest battle has been won. The various grades of T cells include *inducer T's, suppressor T's* and *helper T's* (also known as *CD-4* cells) that fit the foreign antigen and then secrete *helper factors* which encourage (specifically-stimulated-by-perfect-fit) *B lymphocytes* to produce (antibody-secreting clones of granule-containing) *plasma cells*—all of these T's require the presence of macrophages to function. *HIV* (human immunodeficiency virus) attaches to CD-4 surface receptors in order to enter the cell so helper T cells become severely depleted during the course of AIDS—which opens such immunosuppressed individuals to all sorts of serious infections as well as malignant tumors. That gradual depletion of helper T cells apparently takes place mostly within regional lymph nodes prior to the development of any AIDS symptoms. The immune response is further weakened when AIDS patients also are infected by herpes viruses (which directly attack natural killer cells). Other T's include *delayed hypersensitivity T's* and *memory T's*. As you can see, it is all very complex and closely regulated.

Furthermore, this entire system must be partially but carefully deactivated during pregnancy to prevent immune rejection of the fetus (who clearly is other than mother), especially if some fetal cells enter the maternal circulation (see Blood). So the *placenta* (which is primarily based upon *paternal* instructions delivered by sperm—see Growth and Development) produces a protein that prevents proliferation of helper T cells while a factor secreted by the kidneys of pregnant females also blocks interleukin 1. Furthermore, the pregnant uterus contains a great many suppressor T cells. A better understanding of this natural version of an undisturbed nine-months-long foreign tissue transplant could eventually contribute to the prolonged survival of surgically transplanted tissues.

B lymphocytes graduate from B school. In birds their classroom is the *Bursa of Fabricus*—a lymph node located next to the *cloaca* (the combined rectum, urinary passage and egg duct). Having no cloacal gland within which to matriculate, human B-cells (that like all lymphocytes derive from bone marrow stem cells) must mature in other lymphatic tissues. Note how your lymphatic tissues occupy strategic sites for knowing thine enemy—in your mouth and throat (those tonsils and adenoids), in your small bowel wall and appendix where the highest intestinal bacterial counts occur, along mesenteric lymph drainage pathways from that bowel wall and along lymph drainage routes from your lung and all skin surfaces. Lymphocytes also pack heavily into your spleen and somewhat less so in liver (two spongy organs where macrophages filter out old blood cells, bacteria, cellular debris and so on).

Fully trained B cells can detect literally billions of different-shaped *foreign* molecules by fit while ignoring the equally huge variety of *self* molecules within or adjacent to your circulation (your corneal cells and sperm remain potentially effective antigens due to their lack of circulatory access). To interest B cells, soluble foreign molecules must usually have *molecular weights* (the sum of atomic weights of all atoms in the molecule) greater than about 10,000 *daltons* (about 10,000 hydrogen atoms—honoring John Dalton who developed atomic theory while trying to explain the partial pressures of gases). Perhaps that size limit keeps your immune cells from being distracted by fragments of food molecules that occasionally drift by—you would not want to attack your food with antibodies. Indeed, one way to encourage *immune tolerance* to specific proteins (e.g. myelin basic protein in multiple sclerosis or certain collagens in rheumatoid arthritis) may be to feed them to the patient. Hopefully that will not become the main method for organ transplant acceptance—somehow a bit of raw organ donor each morning for breakfast does not sound too appetizing. Smaller circulating molecules such as penicillin may occasionally activate your immune system, probably by adhering to some larger molecule that phagocytes can then grasp and break up for presentation to passing lymphocytes—but even here, oral penicillin appears less likely to induce an allergic response (in someone not previously known to be allergic) than the same antibiotic by injection—while placing penicillin ointment on the skin is quite likely to cause penicillin allergy.

In any case, if a certain B cell encounters a perfect fit for its one specific "foreigner detector" (surface antibody molecule) *and* is stimulated by a similarly boosted T cell, that B cell will give rise to a whole bunch of identical (thus a clone of) *plasma cells.* Those plasma cells then churn out great quantities of their particular antibody—the one that best fit the displayed foreign peptide. As you can see, B cells do their grunt work under the strict guidance and control of T cells. Note also that *both* must agree that there is an enemy worth fighting before antibody production can begin, just as the military divide essential firing codes between two or more carefully trained and selected individuals to prevent some unrecognized nutcase from starting a nuclear war. Many small businesses similarly improve their bottom line by sharing bookkeeping and cashier duties among two or more individuals to reduce both errors and theft.

Preparing For The Next Attack

In summary, B and T schools provide "discipline through apoptosis" to ensure that surviving student cells will not respond to any of your innumerable "self" molecular fragments. Yet those remaining billions of B and T lymphocyte surfaces still display over a billion different antibodies (foreign-molecule grabbers)—usually at a rate of one possible fit per cell. That means every B lymphocyte is always on the lookout for a single *epitope* (the one particular portion of an antigenic foreign molecule that its surface-borne antibody molecules will ever match). And when such a perfectly matched foreign molecular fragment is encountered, it can activate those specific surface detectors to grant that lymphocyte long life and many hardworking descendents. The usual epitope is a polypeptide 8 to 15 amino acids in length, and millions of such epitopes are likely to be available during any significant infection. So by supporting billions of different lymphocytes you virtually guarantee that any possible segment of foreign peptide will find a matching surface detector on at least one of your B cells.

When that special B cell finally is stimulated to produce antibody (after a frenetic last-minute re-selection contest among its almost identical offspring in the germinal centers of a lymph node during which the losers undergo apoptosis), each antibody molecule produced will replicate the cell surface detector that best matched this one epitope. So your antibody response must be delayed while one specific B cell multiplies into a large number of identical antibody-producing plasma cells because no other B cell detector has exactly the right configuration. Thus no other B cell can produce an antibody so well matched to that particular epitope of the foreign antigen. Of course, any antigen may be broken into several or even a great many different epitopes (separate or perhaps even overlapping antigenic portions), any or all of which could then stimulate antibody formation through the coordination of their matching B and T cells. As mentioned, most antigens are polypeptide portions of a protein, but even short carbohydrate-containing molecules can have great complexity (due to the many different ways that simple sugars can be hooked together with covalent bonds), which allows carbohydrates to be antigenic as well.

B and T cells all differ because there are hundreds of millions of different ways to construct polypeptides six to twenty-two amino acids in length out of the 20 available amino acids. So the variable, indeed slap-dash production of each cell's unique surface antibody virtually guarantees that every possible epitope will have at least one good B and T fit. As previously noted, many B and T epitope detectors fit "self" antigens or fail to recognize your personal MHC complex and therefore have to be destroyed or deactivated while still in B or T school, long before they reach the real world of antigen/antibody combat. Interestingly, one can sometimes add foreign molecules to the self category of newborn mice by direct intrathymic injection. But for the present, if you suffer from a disease like *hemophilia* in which blood clotting factor VIII can be absent, you may not have had an opportunity to eliminate B and T cells that fit factor VIII—so you may attack even that essential factor as foreign when it is repeatedly administered. And as you might expect, when your thymus (T cell schoolhouse) eventually *atrophies* (shrivels) during your golden years, you become increasingly susceptible to some sorts of infection and have a higher likelihood of autoimmune diseases and malignancies as well.

Autoimmune problems can arise at any age, of course, leading to antibody attacks on your own joints, heart, kidney, nervous tissue, endocrine tissues and so forth. Often such attacks follow some otherwise insignificant infection in which certain viral or bacterial surface epitopes happen to resemble some self epitope normally displayed by the cells of a certain tissue. For example, your immune response to a minor *strep throat* infection (pharyngitis due to Group A beta-hemolytic streptococci) could lead you to attack your own heart cells (*rheumatic fever*) or kidney cells (*glomerulonephritis*). Even AIDS may only become symptomatic when the immune system attacks a particular viral surface epitope that is identical to a helper T cell surface epitope. Such confusion is hardly surprising in this world of interchangeable DNA modules. Indeed, it is amazing that this sort of problem is not more common. Cell surface hormone receptors seem especially prone to autoimmune attack by *blocking* antibodies or *activating* antibodies (antibodies that so closely resemble the ordinary hormone signal that they serve in its stead but then do not quit acting when the hormone would). Even if they do not lead to endocrine cell or hormone target cell destruction, those receptor-antibody complexes may permanently deactivate or activate important cell receptors, leading to various endocrine cell dysfunctions such as diabetes mellitus and hyperthyroidism.

Like other proteins, antibodies carry a net electrical charge due to their ionization in water as well as their close association with various smaller cations and anions. Therefore antibodies and other serum proteins will migrate through an appropriate solution or gel when driven by an electrical field. The speeds at which such molecules travel is determined by their size, shape and electrical charge (the pH and other characteristics of the solution can alter protein ionization and configuration). Under appropriate circumstances, your blood serum proteins will separate electrically into four major fractions—*alpha, beta* and *gamma globulin,* and serum *albumin.* We have already considered albumin's contribution to your blood osmotic pressure and have seen that it carries all sorts of molecules through the blood stream in its hydrophobic clefts. Certain blood globulins serve similar and often highly specific roles in the transport of particular steroids (see Hormones). And antibodies are *glob*ular proteins that tend to travel together in the *gamma glob*ulin fraction—so those impressively large (several milliliter) gamma globulin shots deliver a wide assortment of currently relevant antibodies (extracted from banked blood) that provide immediate *passive protection*. The diptheria immune serum (antitoxin) rushed to Nome by dog sled in 1925 (and commemorated by the annual Iditarod sled dog race from Anchorage to Nome) represented a dramatic delivery of passive immunity to quell an epidemic. In former times, horses were actively immunized against tetanus, diptheria and other toxins. *Antitoxin* antibodies extracted from an appropriately immunized horse's blood serum could then provide specific passive immunity when there was insufficient time to develop a useful response within a person's own immune system. But after one or two horse serum injections, a person was likely to develop an antibody response against (become "allergic" to) subsequent *horse serum* injections. The resulting *serum sickness* was common and could be serious (especially in the days before cortisone became available to suppress the immune response and reduce inflammation—see Hormones).

Active immunization occurs when you are stimulated by independently attacking or medically administered antigen to produce your own antibodies via your own clones of plasma cells—you can actively produce specific antibodies in response to actual infections—or to the circulating by-product of some minimally invasive bacterium (e.g. the *tetanus toxin* produced by the tetanus bacillus)—or to a chemically altered (in order to make it non-toxic but still antigenic) version of such a toxin (e.g. tetanus *toxoid*)—or to some weakened whole version or antigenic fragment of an unwanted infection (measles, polio, whooping cough and so on). In all such situations, useful immunization depends upon at least one specific epitope of an antigenic molecule finding the perfect B and T match that will encourage an effective antibody-producing response by the resulting clone of plasma cells. More commonly, several different epitope portions of a particular antigen or group of antigens will cause several different clones of plasma cells to initiate antibody production. In general, antibodies are most helpful when they target surface molecules on the living invader. Sometimes it takes repeated exposures to build up *adequate resistance*. Thereafter a swift and successful *anamnestic* (secondary) antibody response should protect you from becoming ill again due to that particular cause. But while some exposures provide lifelong protection, others require frequent restimulation of the immune system to keep them in effect.

In addition, being successfully immunized by tetanus toxoid only protects you against tetanus toxin—the rusty barnyard nail that penetrates your foot could still introduce streptococci or other bacteria against which you have insufficient resistance (then you would probably require antibiotics to help you overcome the resulting infection). Similarly, flu shots or a recent attack of *influenza* (the old Italian term for an illness thought due to the *influence* of a particular malalignment of the planets) can protect you from another attack by that particular version of virus, but influenza still makes repeated sweeps through the same human population by being assembled in a variable fashion—often utilizing different parts acquired during cross-infections involving some of the dozen subtypes endemic to ducks, gulls and other water-birds as well as pigs and humans (so new flu strains often originate in association with Chinese rice paddies where there are lots of ducks, pigs and people in close proximity). That continuous process of trial and error eventually results in another flu epidemic when a form of flu virus exposing only new-to-you antigen finally arises. Thus flu counters your variable defenses by its own naturally selected variable surfaces.

Us Against Them

The live foreign lymphocytes included in a transfusion of fresh whole blood are generally attacked and destroyed before they can do harm. However, lymphocytes given to an immunocompromised host (e.g. included in the bone marrow infusions and blood transfusions received by a cancer patient after total body irradiation) may survive in sufficient numbers to attack host cells. Transfusion-associated *graft-versus-host* disease can follow intra-uterine blood transfusions to the not-yet-immunocompetent fetus or a blood donation from some close relative. In the latter case, donor lymphocytes exhibiting only shared MHC antigens will escape attack although they can still detect and respond to other MHC surface antigens on recipient cells. Fortunately for the effectiveness of such blood transfusions, lymphocytes are quite sensitive to X-rays and other high energy photons—so the incidence and severity of graft-versus-host disease can be reduced by irradiation of blood destined for immunodeficient hosts (as well as by avoiding non-essential transfusions between close relatives). Under some circumstances, preliminary blood transfusion from a potential donor can improve host acceptance of subsequently donated tissues. This effect may be based upon lymphocyte deactivation through overwhelming contact with an appropriate antigen in the absence of necessary associated signals. Eventually it will become possible to reliably induce such *anergic* (unresponsive) states to specific antigens (e.g. donor tissues) while the patient still retains immune competence against microorganisms in general (the strong immunosuppressive medications often needed to prevent transplant rejection also increase the risk of infection or malignant growths).

All antibody/antigen complexes tend to stimulate or expedite macrophage and microphage phagocytosis. In addition, antibody attachment to a bacterium or cell can activate the *complement cascade* of your blood. Like the *perforin* secreted by killer lymphocytes, this defense system punches holes in the antibody-labeled invader (or mismatched blood cell or "foreign" tissue transplant cell). The complement cascade is a sequentially activated series of nine steps that eventually produces a persistent hole in the antibody-tagged bacterial or cell membrane through the transmembranous insertion of multiple barrel-stave-like modules. Any cell receiving one or more holes of this sort will soon equilibrate with the surrounding fluids by swelling and rupturing or else leaking its guts out—not nice, but effective (your blood cells wisely display complement-inactivating surface molecules). A clone of plasma cells can produce up to five varieties of antibody to the same antigen, each serving a somewhat different purpose. Those five classes of antibody all travel (in an electrical field) among the gamma globulin proteins of your blood and are the specific *immunoglobulins* (Ig) *G, M, D, A* and *E*. All share the same basic Y-shaped antibody molecule. At their tips, both arms of any particular Y display identical *variable peptide* sequences (specific to the clone of plasma cells producing that particular antibody molecule and not shared with any other). Each variable (tip-of-Y) peptide sequence is secured to its standard peptide handle by disulphide bonds. The stem of the Y and those two standard arm peptides are known as the *constant* portion of each Y-shaped antibody molecule (being constant for that particular form of immunoglobulin).

Following your first exposure to an antigen, it takes about 4 days for your hard-working plasma cells to produce and release useful amounts of IgM antibody molecules. Those four days were once a time of getting sicker and sicker (from pneumococcal pneumonia, perhaps). Then came the *crisis* during which everyone tiptoed about so as not to further stress the critically ill patient (remember those "Hospital Zone, Quiet" signs?). If the crisis preceded effective IgM production, it usually resulted in death. Otherwise, with specific IgM antibodies building up rapidly, one could hope for a marked *lysis* (drop) of fever and gradual or even swift improvement. That critically important IgM is the largest antibody molecule you can produce. IgM has a cartwheel shape and each molecule bears five separate Y-shaped components that allow it to grab and clump several bacteria by their tiny epitopes for easier disposal. Although IgM is not produced in large quantities, it is found in blood and also cerebrospinal fluid. IgM exists in both free and membrane-bound versions, the latter molecule being anchored to the cell membrane by a hydrophobic (thus soluble within the phospholipid membrane) tail section. Both IgM and IgG also activate complement.

Your plasma cells switch to producing IgG when the IgM is almost gone (or to IgA or IgE, depending upon the stimulating antigen and how it was presented). IgG therefore shows up in useful amounts about 10 days after first exposure to an antigen. However, upon re-exposure to the same antigen, IgG is produced within 3–4 days by an *anamnestic* (accelerated) response of long-lived *memory T* and *memory B cells* that may have been hanging around for many years. Those memory cells tend to gather in the spleen, lymph nodes, bone marrow and wherever else the antigenic action might take place—for example, beneath your skin or gastrointestinal mucosa. No IgM is needed or produced following such reexposure since effective immunity to that particular invading antigen already exists. IgG and the cells that produce it remain in the blood circulation for about a month. IgG molecules are small enough to cross the placental barrier between maternal and fetal blood streams, so IgG provides the fetus passive protection as the mother develops active immunity. Ordinarily your total blood level of IgG runs about one gram percent or one gram per deciliter, meaning one gram of assorted IgG antibody molecules are distributed in every 100 cc of serum or cerebrospinal fluid (it takes just tiny samples of fluid to make such determinations).

IgA is present in blood, and in even higher concentrations in milk, spit and tears, as well as in gastrointestinal and urinary tract secretions. IgA lasts 5–6 days and inactivates bacteria, viruses and toxins. Human breast milk provides both IgA and *mucin* to passively protect the infant's lungs (by spill-over of milk) and GI tract from viral and bacterial attack while also reducing the likelihood of food allergies. IgD lasts about 3 days after being released from its normal cell surface location by cell death. Free IgD apparently plays no significant role in the immune process. IgE levels are especially elevated in people with allergies or worm infestations. Apparently the occasional allergy simply reflects an unfortunate side-effect of fighting parasites. Since the IgE system developed under conditions where parasites infected a large fraction of the human population, even a partially effective but immediate response to those parasites would bring significant reproductive advantage despite having such undesirable side-effects. Serum IgE levels only remain up for 3 days but when exposed to its appropriate antigen, that specific IgE causes your circulating basophils and those mast cells lurking at sub-epithelial sites (that have picked up IgE for their surface displays) to dump histamine, serotonin, bradykinin and other nasty molecules that can bring about smooth muscle spasms in host and (hopefully mostly in) worm. The host is supposed to survive any resulting asthma attack or even *anaphylaxis* (catastrophic allergic response occasioned by eating shrimp, or the occasional penicillin shot, bee sting or whatever else a person may be highly allergic to) while the parasite entering your tissues or worm in the GI tract gets a severe cramp, causing it to die before maturing or fall away from your gut wall to a deservedly miserable death in the bucket or toilet (does anyone *really* believe that God made all of those disgusting intestinal worms, each according to their kind, and saw that they were *good?* Compared to what?).

Thus you initially respond to microbial attack with granulocytes, killer lymphocytes, macrophages and circulating non-specific agents such as complement. Thereafter—if the problem persists (and depending upon cytokine influences)—your immune system has two possible approaches to viral, bacterial or cellular attacks. One directed by helper T lymphocytes is *cell-mediated* and involves direct cell-to-cell attack by natural killer and cytotoxic T cells responding to unusual or non-self peptide epitopes (displayed by infected cells on their MHC-1 complexes). Your other helper-T-based response is *humoral* and more widespread, and it relies upon raising specific antibodies against epitopes of foreign antigens that are properly displayed by macrophage MHC-2 complexes. Those antibodies then seek out such foreigners, be they foreign molecules, invading viruses, bacteria or cells. The resulting antibody-antigen complex identifies that antigen for destruction by phagocytosis or other attack (complement, perforin, etc.).

Apparently lymphocytes find metallic surfaces and long-chain polymers too boring and non-specific for any response. And you continuously form new lymphocytes while the older ones soon die by apoptosis unless their unique surface-bound "test antibody" happens to have the exact size and shape necessary for dealing with current invaders—in which case there is a delay as the best-fitting T lympocyte "clones up" and stimulates the best-fitting B lymphocyte to form plasma cell clones for producing lots of copies of that particular competitively chosen antibody. All of which suggests that one cannot be allergic to penicillin or anything else without some previous exposure, although one may not always be aware of such earlier exposure in the diet or by inhalation, skin exposure or even injection. Of course, most people never produce antibodies to penicillin at all, despite receiving repeated penicillin shots or pills. But once a person does become allergic, he/she generally remains so, for IgE laden basophils and mast cells are granted long life (there is no reproductive benefit in thus protecting you from future penicillin attacks so the frequent occurrence of penicillin allergy simply indicates that penicillin injections played no role in the survival of your ancestors).

Specific memory T cells tend to be attracted to and stored within the most appropriate tissues, especially when these are inflamed. Apparently the endothelial cells of blood vessels in your different tissues carry more or less unique address molecules on their surfaces to which such memory T's adhere as they scrape by. That sort of selective localization increases the likelihood that appropriate memory T's will be rejuvenated by repeated exposures to their one particular antigen. Of course, the customary two-pronged immune response to invaders (killer cell plus antibody) does not always work out exactly right. For example, undesirably long-lasting B cells may result from and then sustain ongoing viral infestations by Epstein-Barr virus, which can eventually lead to uncontrolled B cell growth—one such B cell cancer is known as *Burkitt's lymphoma*.

Monoclonal Tools

Artificially produced antibodies have become important ultra-specific tools in biology and chemistry, as well as for medical diagnosis and treatment. Thus an antibody to the active part of a hormone fits that hormone exactly so antibodies formed against that first antibody may have part of the original hormone shape and perhaps the same information-signaling effect as that hormone. Or a specific antibody can be used to eliminate a particular autoimmune antibody or to bind-up and thereby deactivate some harmful bacterial toxin. Exquisitely sensitive antibodies are used within airport luggage "sniffer" machines to detect specific drugs or chemicals, and fluorescent-dye-labeled antibodies readily locate tiny quantities of antigen deep within a living cell. Antibodies that target a certain type of cancer cell can be modified to deliver toxic molecules or radioactive atoms, thus serving as carefully addressed letter bombs. And antibodies can help to identify and separate out important stem cells needed for the regeneration of bone marrow. The antibody to an antibody might serve as an alternate antigen for immunizations when that seemed safer than using the original antigen (there might be resistance to being immunized with an "almost surely deactivated" AIDS virus, for example, even if such immunization had no chance of stirring up an autoimmune attack on helper T cells). And an antibody constructed to fit some critical intermediate (transitional) molecule can serve as an enzyme to encourage a desired chemical reaction, even in solvents other than water.

With no end to such clever applications, the large scale production of specific antibodies has been expedited by *monoclonal* techniques. Simply trap that one perfect mouse plasma cell by fishing for its surface antibody using the antigen of your choice. Then *fuse* (combine two into one) that plasma cell with some rapidly growing rodent *multiple myeloma* (B-cell clonal cancer) and soon you can grow a huge clone of such fused cells, each pumping out that one special monoclonal antibody. Almost as impressive as what you must do every single day within your own immune system, simply stay to alive.

Suggested Reading:

Special issue on The Immune System: September 1993. *Scientific American.* pp 53–144.

CHAPTER 21

RESPIRATION

Life Is One Big Gas Exchange;… Partial Pressure And All That;… CO₂
Tension And CO₂ Content May Differ Markedly;… Gills And Lungs:
Everted and Inverted Gas Exchangers;… Pumping Air;… Conserving
Water;… Conserving CO₂;… Lung Should Be Elastic.

Life Is One Big Gas Exchange

It may take half a century for the cool surface of an abandoned iron implement to rust and flake away (4 Fe plus 3 $O_2 \rightarrow$ 2 Fe_2O_3). Yet within seconds, the purposely incomplete combustion of smoldering tobacco leaves ($CH_2O + O_2 \rightarrow CO_2 + H_2O$ + heat) begins to release *addictive nicotine*-bearing smoke, *poisonous carbon monoxide* (CO) and *cancer-causing tars*—a toxic burden that will likely cause one-fifth of all premature deaths in the developed world. Similar complex organic molecules once coated Earth's primeval surface under a reducing atmosphere and may have played an essential role at life's beginning, but even a heavy smoker wouldn't have liked it back then. Since oxygen releases its covalently bound electrons very reluctantly, tobacco plants and photosynthetic bacteria utilize a lot of sunlight to pry hydrogen *atoms* out of water for construction of life's substance and power—that solar energy can be released explosively when H_2 and O_2 recombine to form H_2O. In contrast, an especially vigorous intermolecular collision is enough to break H_2O into H^+ and OH^-—and pouring a small amount of strong acid into a concentrated alkali solution (or vice versa) only produces enough heat energy (as H^+ and OH^- recombine) to boil and splatter such nasty-till-neutralized solutions all around. Evidently it takes a lot less energy to ionize (pluck a proton from) water than to disrupt it entirely (yank two electrons and two protons away from the O). In other words, H_2O is at a lower energy state than H^+ and OH^- which in turn represent less potential energy than H_2 and O_2. Therefore, life's ample energy comes from carefully returning H to O in many small intermediate steps that include the utilization of proton gradients (see Metabolism).

Until about two billion years ago, any waste O_2 discarded during *photosynthesis* ($H_2O + CO_2 + light \rightarrow CH_2O + O_2$) soon encountered other elements willing to donate electrons. But photosynthetic life prospered and Earth's surface minerals (silicon, aluminum, iron, magnesium, etc.) eventually became fully oxidised despite the ongoing exposure of reduced materials by continental erosion—only then could that production of O_2 and predominance of photosynthesis over *combustion* (reversed photosynthesis) contribute significant quantities of free O_2 gas to the atmosphere. While such free oxygen expedited the oxidation of exposed organic molecules to CO_2 and H_2O, a small percentage of the organic material being produced by Earth's *biosphere* (living surface) continued to accumulate beneath Earth's oxidised surface as peat and humus buried in swamps, in thick organic sea-bottom sediments and within the great underground *hydrocarbon reservoirs* (coal, oil, tar and natural gas) that represent life's ancestral remnants following anaerobic microbial digestion. Also contributing to the build-up and preservation of atmospheric O_2 was the early burial (often in association with microbial metabolism) of reduced iron and sulphur compounds following their release from sea-floor hot water vents.

Only after Earth's atmospheric O_2 exceeded 15% by volume were open fires of organic materials possible—and if atmospheric O_2 ever exceeded 25%, any lightning strike would have ignited catastrophic conflagrations (including forest fires after trees finally evolved over 300 million years ago). In any case and for whatever reasons, Earth's atmospheric oxygen content appears to have settled between those extremes, holding steady at about 21% over recent hundreds of millions of years. This confirms an ongoing equilibrium of sorts between O_2 production by photosynthesis and O_2 consumption by oxidation, and between the burial of newly formed organic material (and other reduced compounds) and the ongoing exposure of oxidizable minerals through surface weathering or atmospheric releases of reduced volcanic gases. Some *purple* and *green phototrophic bacteria* utilize light and reduced sulphur compounds such as H_2S rather than H_2O—these bacteria excrete granules of sulphur rather than O_2 gas ($2H_2S + CO_2 + light \rightarrow CH_2O + H_2O + 2S$). And other purple and green non-sulphur bacteria utilize organic acids and carbohydrates to reduce CO_2. But cyanobacteria (and their chloroplast descendents within Earth's plants) remove electrons from always available water and thereby release the O_2 that you require to reverse that energy storing process for the production, maintenance and operation of your own complex biomolecules. To ensure that your biological oxidations proceed slowly at low temperatures, you "bleed off" and divert that plant-procured energy in many small controlled underwater steps. We have noted that water has a high heat capacity and puts out fires, but the avoidance of spontaneous combustion is just one of many reasons for life's chemistry remaining water-based—water is also a great solvent for polar molecules—breaking water apart is a good way to store solar energy—water reacts in many anabolic and catabolic reactions—water is plentiful on Earth—and water is the only liquid substance naturally encountered at Earthly temperatures (except for mercury). While life undoubtedly began in water, the need to limit oxidation rates still restricts biological processes to utilizing dissolved O_2 molecules rather than gaseous O_2 directly—that dependence upon abundant sunshine, water, dissolved O_2 and minerals explains the remarkable productivity of wetlands and estuaries along the sea shore.

Partial Pressure And All That

Our present atmosphere is almost 21% O_2 by volume, over 78% nitrogen (N_2) and about 0.03% CO_2, with argon (a noble gas) making up the remaining 1%. The average sea-level air pressure (the sum of all atmospheric gas partial pressures) defines *one atmosphere*. One atmosphere just supports or balances a column of mercury 760mm (¾ of a meter) tall. That pressure, also referred to as *760 torr* (1mm Hg at 0°C = 1 torr), equals 14.7 pounds pressure per square inch of exposed surface. So at sea level the partial pressure of O_2 is about *155 torr*, that of N_2 is about *600 torr* and that of CO_2 about *0.23 torr*. The *barometer* (invented by Torricelli) demonstrates changes in atmospheric pressure through changes in the height of a mercury column that initially filled a sealed-tip glass tube. The open bottom end of that vertical glass tube remains submerged in a dish of mercury to prevent air entry into the tube. Were there no atmosphere above the mercury dish (hence a vacuum), then the vacuum produced by mercury falling away from the closed tip of that tube would equal the pressure created around the mercury dish by removal of all atmospheric gases. In that case, the mercury level in both tube and dish would be the same. Thus an atmospheric pressure of zero defines a complete vacuum.

Regardless of its solubility, the equilibrium concentration of a gas in water is proportional to the partial pressure of that gas in the adjacent atmosphere—so if you double its partial pressure you will double the amount of that gas in the solution. The water solubility of any gas also falls with rising temperature (as increasing molecular velocity encourages gas escape) or increased salt content (less unoccupied space between water molecules). Nitrogen's water solubility is half that of oxygen but the atmosphere contains four times more N_2 so water carries twice as much dissolved N_2 as O_2. Interestingly, N_2 is five times more soluble in fat than it is in water so it takes much less time for a new partial pressure of N_2 to equilibrate with your blood than with your fat via that blood, and obese scuba divers run an increased risk of the bends even if following recommended decompression schedules. Of course, your blood serum holds relatively little N_2 or O_2 and almost all of your oxygen is delivered by the specialized oxygen-transporting molecules within your red cells. Indeed, hemoglobin allows a liter of your blood to carry about two hundred cc of O_2 while a liter of water only absorbs 4–10cc dissolved O_2 from the atmosphere (depending upon temperature and salinity). Admittedly 4–10cc does not sound like much, but it is enough for a 200 gram fish (0.44 lbs) consuming only 20–80ml O_2 per/hour (that limited solubility of oxygen also means our relatively huge atmosphere is hardly affected by the amount of O_2 dissolved in Earth's waters).

It is generally more convenient to quantify *dissolved gases* by reference to their *partial pressures* than to their equivalent gas volume. Thus the oxygen *tension* of a solution refers to the specific partial pressure of atmospheric oxygen that would account for this amount of dissolved oxygen at equilibrium. Of course, daytime photosynthesis by aquatic plants and algae consumes CO_2 and releases O_2 while aquatic life consumes O_2 and releases CO_2 during the night. Therefore the unchanging partial pressure of oxygen within the atmosphere may bear little relation to the variable oxygen tensions measured locally within Earth's surface waters. Like O_2, CO_2 is utilized or released by living cells in dissolved form. Since the water solubility of CO_2 is 30 times greater than that of O_2, a considerable volume of CO_2 can be present within your carbonated beverage without greatly raising its partial pressure of CO_2. Indeed, if your freshly opened liter bottle of soda water was equilibrated and bottled under a pure CO_2 atmosphere at 760 torr (one atmosphere), it contains over a liter of dissolved CO_2—for even though gas molecules take up lots of room when whizzing and banging about in the atmosphere, they fit right in amongst those vastly more numerous liquid water molecules—that is why dissolved gases hardly alter the volume of a fluid. After all, a cubic meter of water contains one million cc of H_2O while a cubic meter of room air becomes fully saturated by a mere 30cc (one ounce) of water in vapor form. In turn, the water vapor in 100% humid air contributes about 5% of that atmosphere's total gas pressure.

Some semi-dormant volcanoes (especially those near subducting oceanic plates that carry sea-floor carbonates down into Earth's molten mantle) release compressed CO_2 into the bottom waters of their crater lakes. Since their lake-bottom CO_2 tensions can equal lake-bottom water pressures, these crater lakes may dissolve enormous quantities of CO_2—and such highly carbonated water is significantly heavier (more dense) than otherwise similar but uncarbonated water. Now we know that one atmosphere equals 760mm Hg. We also know that mercury is 13 times heavier than water. Thus one atmosphere equals 13 x 760 or 10,000 mm of water, which is about 10 meters or about 30 feet of water. So if some hypothetical 300 foot deep volcanic lake were to become CO_2 saturated from its bottom, it would have a bottom water pressure and potentially a bottom CO_2 gas partial pressure of about 10 atmospheres. That means every liter of that bottom water could hold at least 10 liters of dissolved CO_2.

Such great quantities of dissolved CO_2 can cause calamity if some disturbance—a landslide or earthquake perhaps, or even the seasonal turnover of slightly cooler and therefore more dense surface water sinking through warmer gas-laden waters below—stirs up that heavy bottom water and thereby gets the lake all fizzy. For CO_2-laden bottom waters that are disturbed and begin to rise must release gaseous CO_2 as surrounding water pressures decline. That positively fizzy feedback stirs things up even more. Before long a major portion of those CO_2-rich lake bottom waters are suddenly carried to the top by the huge CO_2 bubbles that they now contain. Such a massive overturn of water in a volcanic crater lake may be associated with enormous releases of cool (because just expanded) and therefore heavy CO_2. When that CO_2 temporarily blankets the surrounding area, it can suffocate all nearby (and downstream) life that is dependent upon ongoing access to atmospheric O_2. A 1986 overturn of this sort at Lake Nyos in Cameroon, West Africa, killed over 1700 people and many animals as CO_2-laden waters frothed more than 75 meters (almost 250 feet) above mean water levels.

Similar events are expected there in the near future, both because this 200 meter deep crater lake can become CO_2 saturated within 20 years of last degassing and because its unstable outlet dam (40 meters high and composed of volcanic ash) seems likely to wash away soon—in which case the large size of this lake would put many people in both Cameroon and Nigeria at risk of drowning at the same time as its rapidly dropping water levels encouraged another major release of CO_2. Among remedies being tested are the placement of multiple vertical pipes far down into the water. Once upward flow has been established in such a pipe, it is maintained by gas bubbles escaping from the rising solution—prewarming that cold water by discharge onto nearby land might further reduce the risk of cold (and therefore heavy) surface waters causing an overturn.

CO₂ Tension And CO₂ Content May Differ Markedly

CO_2 combines with water to form carbonic acid as follows:
$CO_2 + H_2O \rightleftarrows H_2CO_3 \rightleftarrows H^+ + HCO_3^-$. Reading that equation from left to right, one can see that adding more CO_2 to water leads to formation of more carbonic acid (H_2CO_3). Carbonic acid is a *weak acid* that only partially breaks down into a hydrogen ion (H^+) and bicarbonate ion (HCO_3^-) at normal body pH. Such an *equilibrium reaction* can also be read from right to left—so adding extra H^+ would produce a more acid solution containing less HCO_3^- and more H_2CO_3. And that increase in H_2CO_3 would create a higher CO_2 tension in that more acid solution. On the other hand, a more alkaline solution (achieved by reducing the H^+ concentration or adding OH^-) tends to move the reaction equilibrium to the right, with the solution holding more of its total CO_2 as HCO_3^- (which does not directly enhance the measured CO_2 tension until considerable acid is added during a laboratory determination of total CO_2 content). Evidently the total quantity of *dissolved carbonates* in a solution at equilibrium with a specific CO_2 tension will vary according to the pH of that solution. Despite the significant impact of pH on total CO_2 content, near-surface sea waters generally are sufficiently stirred by wind and wave to equilibrate with overlying atmospheric gases so oceanic CO_2 tensions usually remain close to 0.23 torr. Indeed, Earth's vast oceans are not noticeably affected by local volcanic contributions of CO_2, nor by carbonate released from dissolving *limestone* (calcium carbonate), nor by bacterial releases of CO_2 from rotting vegetation and so on.

Sea creatures less than about one mm in diameter can absorb enough oxygen through their wet surfaces to lead very active, competitive, reproductively successful lives. Since CO_2 is a larger molecule than O_2, it diffuses through water more slowly (at 86% the rate of an O_2 molecule). But CO_2 is 30 times more soluble in water and tissue fluids so at the same pressure gradient about 25 times more CO_2 molecules than O_2 molecules will diffuse through an equally thick layer of water per second. Therefore CO_2 *retention* (accumulation) is rarely a problem for underwater creatures while they still can obtain sufficient O_2—except in stagnant waters where CO_2 levels may rise markedly. On the other hand, simple diffusion soon becomes insufficient for oxygen delivery as creature diameter increases, particularly if the entire organism is live cellular material rather than 99% enclosed and inert jelly as in a jellyfish. So enlarging organisms must either reduce their metabolism or expand their *respiratory surfaces* (those body surfaces that have a role in gas exchange).

Gills And Lungs: Everted And Inverted Gas Exchangers

Larger underwater creatures generally exchange gases across delicate everted respiratory appendages. Unlike ordinary surface wrinkles or other skin-covered structures, these specialized *gills* are *perfused* (supplied) by *desaturated* (low-in-oxygen but carbon-dioxide-burdened) venous blood. As it flows under those thin gill surfaces, this blood exchanges gases with surrounding waters on behalf of all internally positioned body cells. By pumping its blood through those fine layers of gill structure in a direction opposite to the passing flow of water, an undersea creature can extract up to 60% of total dissolved O_2 (for that *countercurrent* pattern of flow puts blood just exiting the gills next to newly entering water from which it can still acquire a bit more O_2, while very desaturated incoming blood gains its initial O_2 from already depleted waters most of the way through the gills).

Their endless quest for reducing power normally exposes sea creatures to an adequate unidirectional flow of oxygen-containing water—that water flows across their gills more rapidly as they swim faster and therefore require more O_2. Resting fish move oxygen-bearing water across their gill surfaces by intermittently expanding the mouth cavity to bring in new water and then compressing that water while moving gill covers back and forth to force it out sideways. A swimming fish builds up higher than average water pressures in front of the mouth and lower than average water pressures laterally where exhaust water escapes from under its gill covers—that unavoidable pressure difference markedly enhances water movement across the gills. An occasional flow reversal (brought about by sudden enlargement of the oral cavity in the presence of closed lips) is the fishy equivalent of a cough to clear particulate matter from the fine gill structure—fish also expand the oral cavity suddenly as they suck in their next victim. Fish vision is unaffected by swimming speed since the transition from high water pressure in front to low water pressure on the sides occurs just where fish eyes are located (see Eye).

Certain fish prosper in stagnant low-oxygen waters by enclosing usable air in their *swim bladders* (which can also serve to detect sound waves and control buoyancy). Some swim bladders contain swallowed air, others obtain their gas by a countercurrent blood-to-blood air-exchange mechanism (see Kidney). Clams, mussels and similarly stationary sea creatures ordinarily pump far more water past their gills while seeking nutrients than would be required for adequate oxygenation. Organisms fixed to the sea floor, sea vegetation or other stable structures additionally benefit from wind, waves and tidal currents that prevent local waters from becoming too stale or oxygen-depleted. Yet while very effective under normal circumstances, the thin-walled blood-inflated gill structure does have major disadvantages, as anyone who has ever caught a fish can tell. For if its desperate ungainly flopping does not quickly bring that landed fish back to water, it soon suffocates, gasping miserably. Air-based fisherpersons are often puzzled by the inability of fish to utilize the good fresh air that surrounds them on shore. But as delicately expanded gill surfaces dry and shrivel, no surface water remains in which atmospheric O_2 can dissolve. And any gaseous O_2 penetrating those relatively impermeable air-dried gills simply causes further drying. Of course, a passing fish might find it equally strange that air breathers are so intolerant of submersion in oxygen-rich waters. For if its desperate ungainly flopping does not quickly bring that submerged air-breather back to the surface, it soon suffocates, gasping miserably. Observant crocodiles have learned to hold their struggling land-vertebrate victims underwater, knowing that they will soon relax and more readily release their repeatedly stolen reducing power.

As with gills, lungs are mere respiratory appendages through which deoxygenated blood circulates for renewal. However, lungs are *inverted* (internalized) for conservation of water vapor while gills are everted into oxygen-bearing waters. Your thin pulmonary gas exchange surfaces are easily damaged by exposure to fresh or salt water. Fresh water does more harm than salt because fresh water causes osmotic rupture of lung cells as well as red cells. But since a kilogram of easily inhaled air delivers 100,000 times more O_2 than a kilogram of viscous water, it would clearly be impossible to move useful quantities of low-oxygen water in and out through your tiny airways, no matter how well your lung surfaces might tolerate such submersion. Furthermore, a gill-like oxygen exchange surface that brought your warm blood into close contact with cold sea water could easily cause fatal hypothermia. Nonetheless, certain species of tuna do warm particularly important muscles (as well as their eyes and brains) while swimming in cold waters by circulating the relevant part of their blood to and from the gills via closely mingled vessels that encourage countercurrent heat exchange (thereby saving and redirecting some of the heat produced during vigorous muscle activity).

Dogs have survived being submerged in warm isotonic saline solution at eight atmospheres oxygen tension—which ought to provide about 200 ml. of oxygen per liter of solution (comparable to your usual arterial blood oxygen content)—but even here, more gas exchange may have occurred through skin and outer mucosal surfaces than via any oxygen-enriched water that reached respiratory surfaces within the lung. So gills extract oxygen far more efficiently from water than lungs ever could. Incidentally, the oceans equilibrate with atmospheric O_2 so effectively that partial pressures of O_2 often do not differ a great deal regardless of depth (those CO_2 partial pressures only rose with increasing water depth in the above-mentioned volcanic crater lake because CO_2 was constantly entering its bottom waters at high pressure).

You expose over 90 square meters (the floor space in a small house) of moist gas-exchanging surface to the fully humidified air within your lungs—were that huge respiratory surface suddenly everted (gill-like) and exposed to the atmosphere, your evaporative fluid losses would kill you in less time than the average beached fish takes to stop flopping about. Of course, that cold-blooded fish consumes far less oxygen (and therefore requires a much smaller respiratory surface area) than a more rapidly metabolizing warm-blooded mammal or bird of comparable weight. Gills ordinarily encounter oxygen in solution so there is no risk of drying those highly vascularized (rich in blood vessels, hence blood-red in live or recently deceased fish) exposed gill surfaces. Having gills dense-packed with capillaries in this fashion helps to minimize their wetted surface and reduce drag while also limiting solute exchanges with surrounding waters. Your own respiratory surfaces are not nearly as red since lungs must store air as well as exchange gas efficiently. A resting man consumes about 14 liters of dissolved O_2 per hour. A maximally working man requires up to 280 liters of O_2 per hour. If dependent upon some sort of artificial underwater gill device, such a hard-working man would have to totally deoxygenate more than 56,000 liters of normally oxygenated water per hour. Thus the far greater work capacity of warm-blooded air breathers explains why birds and mammals now rule the seas (though they remain in constant danger of drowning).

Pumping Air

Just as "Fish gotta swim" (or at least gulp) in order to keep fresh water flowing across those gills, all but the smallest and simplest land creatures gotta *ventilate* (move new air in and used air out of) their lungs. Some snails and scorpions still rely upon gas diffusion without ventilation and many insects depend entirely upon diffusion to maintain useful gas mixtures in the *tracheal air passages* that approach within easy diffusion distance of all their underlying cells. Ordinarily, the tracheal openings through a wax-coated water-repellent insect exoskeleton are too tiny to admit water droplets. But insects soon drown when sprayed with soapy (hence reduced-surface-tension) water that readily penetrates the smallest openings through those *chitinous* surfaces (chitin, a polysaccharide found in the cell walls of most fungi, is also the principal structural material of insect and crustacean exoskeletons).

Frogs compress upper airway and buccal cavity air in order to *push* it into their lungs. You expand your chest cavity in order to *pull* atmospheric air into your lungs. Land mammals and birds generally coordinate breathing with limb movements to enhance efficiency. A galloping horse inhales during extension of the body and exhales as muscles flexing the body help to squeeze air out—the ventilation of a trotting dog is considerably more complex and relies in good part upon gait-dependent pendulum-like visceral movements as well as intermittent fore-limb impacts on alternate sides of the chest—both of which reduce the requirement for frequent diaphragm muscle contractions and make more effective use of air still within lung airways by mixing it with alveolar air (see below). Under these circumstances, visceral deceleration and acceleration move more air than active diaphragm contraction since a single diaphragm contraction (more a repositioning, actually) only occurs once per several breaths. It seems that the liver is especially effective in stretching the elastic central-tendon of the diaphragm (to which it is held by sturdy folds of peritoneum) while diaphragm muscle is relaxed—all of which enables the trotting dog to coordinate breathing and rapid leg movements without driving its diaphragm muscle at inefficiently high speeds. Despite their ungainly legs that stick out sideways, your reptile relations run as efficiently as mammals. But reptiles gain length of stride by flexing side-to-side as they advance. Presumably such fish-like swimming movements mix fresh airway air with adjacent alveolar air but they also interfere with expansion or compression of the chest—so reptiles must regularly stop running in order to breathe. Few air-breathing vertebrates also exchange gases effectively while underwater. Among those performing both activities reasonably well are *amphibians* such as the frogs and salamanders whose larvae have gills while the adult forms exchange gases via lung and skin surfaces. More primitive respiratory structures allow certain fish to undertake short strolls between ponds or even climb trees (using fins as legs).

Creationists often try to discredit evolutionary theory by pointing out that so-called "missing links" between species are missing because God never made them. But no play on words can obscure the clear continuum and obvious overlap of forms, abilities and requirements between totally gill-dependent and totally lung-dependent creatures. It is especially meaningless to demand additional "missing" intermediate forms when so many closely related, intermediate sorts of creatures already are known (e.g. fish that swallow air to supplement oxygen uptake at their gills or other surfaces, or that absorb O_2 through specialized air sacs or via primitive supplementary lungs). Furthermore, each new intermediate species discovered, declared or defined automatically creates additional "missing links" on either side of it. And the most heated disputes about whether certain fish should be assigned to one species or genus or even allotted their own species often revolve about the acquisition and presumed evolutionary importance of certain fishy structures or characteristics—sometimes the current assignment simply reflects an apparent absence of sexual reproduction among otherwise closely related forms.

So it is vague as well as circular reasoning (hence tautological) to insist that each creature must be "of its own kind" because God made them so, since species classifications are based upon mere human definitions (and often disputes) to start with. In other words, it defies logic to use any vague restatement of a vague definition as confirmation for any part of that definition. Nor ought one use the breath-holding and underwater swimming skills of whales as proof that whales were created "as is, where is" in the sea. For clear fossil evidence and the remaining *vestigial* (no longer needed or functioning) pelvic bones of modern whales confirm that whales were land-based with well-developed lungs long before they resumed a fully aquatic existence. Presumably they initially went down to the sea in trips because air-breathing gave these efficient warm-blooded mammals such advantageous access to the bounteous stored reducing power of gill-dependent sea life—then before long the most highly successful well-insulated whales became too fat, streamlined and otherwise specialized to compete on land. Many birds, from puffins to pelicans to penguins have similarly come to dominate underwater life forms while modifying their own structure and life styles to suit.

Conserving Water

So how do you prevent destructive drying and also minimize water losses from the delicate, hugely expanded gas-exchange surfaces exposed within your lungs? Clearly those internal lung surfaces must be kept moist in order to dissolve gases, yet remain dry enough for easy air access to lung capillaries. Quite logically, that balance is best maintained if your moist internal lung surfaces only encounter air that has already been fully warmed (cold air holds less moisture) and humidified within your nasopharynx. The next time you sneak a peek way up someone's nose, note the three folds of mucosa-covered tissue bulging in from the lateral side. Those *turbinates* help to warm, stir and moisten incoming air—being cooled by evaporation even allows them to recover some moisture from the warm fully humidified outgoing air. And the snot that often occludes your nostrils simply represents mucus proteins and other solutes left behind when your copious nasal secretions have given up fluid to incoming drier air—which is why Alaskans must pick and blow their noses more frequently than Hawaiians do. Quite obviously, you could not tolerate a similar build-up of dried mucus in the airways of your lungs (for one thing, your fingers are too short and fat).

Your tear glands respond rapidly to a cold and windy day—and as you blink away the tears that prevent cold air from drying your corneas, your nose receives its first shot of fluid to help compensate for the low water content of that cold outdoor air. Many other nearby secretory surfaces contribute moisture as well—so while your *lacrimal* (tear) ducts emerge just beneath the inferior (lowest) turbinate, your anterior sinus secretions (frontal, maxillary and anterior ethmoidal) drain below the middle turbinate, and your posterior sinuses (sphenoidal and posterior ethmoidal) drain under the superior turbinate. Still it takes an extensive superficial (submucosal) network of blood vessels to supply sufficient nasal fluid for evaporation—which is why unearned bloody noses (not due to injury based upon insult) are so common—those copious submucosal vessels often bleed as a result of minor abrasions by intranasal fingers and handkerchiefs.

Your lungs must maintain a moist surface, yet clear all airway secretions in order to keep smaller air passages open. Loose collections of mucus and debris are often broadcast widely by cough or sneeze but your pulmonary *toilet* (cleansing) is mostly based upon the slow steady upward movement of a moist and sticky mucus sheet that also transports trapped dust and debris. The routine accumulation of that mucus during times of gentle breathing explains all of the noisy throat clearing during quiet movements at the symphony, as well as the initial gargling sounds that you emit (unless you first clear your voice box) when answering the telephone if you have not talked for a while. This mucus is delivered up by the synchronized rowing movements of a great many *cilia* (that are easily paralysed by cigarette smoke—see Reproduction) projecting out from your *psuedo*stratified *columnar respiratory epithelium* (this epithelium is composed of cells that vary in height and thereby *misleadingly* give a layered appearance).

When you breathe out (exhale or expire) into a cold winter's air, your humid breath becomes visible as a rapidly dissipating cloud. This happens because each quickly cooled exhalation momentarily includes more water molecules than it can retain as invisible water vapor—cooling warm humid air also causes it to deposit dew on cold grass or the outside of your iced drink. Your lungs conserve water by humidifying as little outside air as possible. Larger lungs would take more time and effort to fill and empty—smaller lungs would force you to breathe rapidly and inefficiently. Dogs evaporate a lot of water from their upper airways when they *pant* (make rapid shallow breaths) in order to cool their fur-insulated inner selves without bringing too much fresh air into the lungs.

The soft billowy portions of your lungs are composed of *alveoli*—those end-of-airway sacs are where all respiratory gas exchanges occur. Intra-alveolar air dissolves into alveolar wall fluid and maintains an O_2 diffusion gradient toward the oxygen-depleted blood within your lung capillaries while carbon dioxide diffuses from that blood toward the lower partial pressure of CO_2 present within your aveolar air. En route to your alveoli, air from your nose or mouth and throat must first pass through your *larynx* (voice box) between your two *vocal folds* into the *trachea* (windpipe). When those vocal folds are tensed and held almost together, the passing air flow causes them to vibrate rapidly between open and shut—which provides the audible sound waves (vibrations or buzzing) that you then tune and modulate with the laryngeal and pharyngeal (throat, mouth and nose) muscles involved in talking or singing. Air within your trachea soon reaches your *thorax* (chest) where the trachea divides into your left and right *main bronchi* (the primary airways of your left and right lungs). These main bronchi subdivide repeatedly into ever-smaller *lobar, segmental* and *sub-segmental* bronchi before finally branching into tiny terminal *bronchioles*—each serving several alveoli. *Used* air remains within your air passages at the end of every expiration—that *dead space air* reenters your alveoli at the beginning of the next inspiration (which helps you to conserve CO_2—see below). In contrast, the fresh air within your upper airways at the end of each inspiration generally gets little use (but see trotting dog earlier this chapter).

There can be little air flow within an alveolus during *respiration* (breathing) since the total alveolar surface area is so great (that 90 square meters again). For even your full *vital capacity* (biggest in-and-out breath) only moves about 5 liters of air toward and away from that huge surface. Nonetheless, your alveoli certainly enlarge visibly with maximal lung expansion (at the end of a "deep breath in"—defined as *full inspiration*) and diminish in size during maximal lung deflation (at the end of a "deep breath out"—thus *full expiration*). Even after that best expiratory effort, another liter of *residual* air remains within your airways and alveoli—some of that residual air can be squeezed out by a forceful compression of your upper abdomen (the *Heimlich maneuver*, best performed by another person from behind you when necessary to dislodge food blocking your larynx).

At rest, you take about a dozen 500 ml breaths per minute—almost one-third of that half-liter *tidal volume* is stale dead-space air containing more CO_2 and less O_2 than atmospheric air. Reduction of dead space air is just another of many reasons (weight and balance, relationship to heart, breathing dynamics, skeletal considerations, etc.) for locating your lungs as high in your body (as close to the air intake) as possible. You breathe more deeply during exercise so your dead-space air becomes a smaller fraction of each breath. When still out of breath from running, you might wish for wider airways to improve your air flow but that would increase your dead space and require deeper breathing at rest—much narrower airways would minimize your dead space air and reduce your tidal volume but so restrict your airflows that you no longer could outrun the competition. Thus it is advantageous to have airways of moderate diameter that can adjust somewhat to suit your varying needs—those active alterations of your airway diameters rely upon contraction or relaxation of smooth muscle cells in the walls of your smaller airways (your *bronchi* and *bronchioles*).

The "adrenalin rush" of a fight-or-flight response opens your airways widely— those airways then narrow as you breathe at rest. However, when your airways already are narrowed by allergy or other inflammation, their smooth muscles may contribute to the decidedly unhelpful *bronchospasm* associated with an *asthma* (wheezing) attack. Non-mechanical (chemical) ways to combat such airway inflammation and its associated asthmatic bronchospasm include the administration of antibiotics (for any associated infection) as well as anti-inflammatory adrenocortical steroids *locally* (by inhalation) or *systemically* (by tablets or injections that give a body-wide effect). One can also inhale or inject *adrenalin* or similar substances that mimic the airway impact of a sympathetic fight-or-flight reaction but such medication tends to lose its effectiveness with regular use. Or one can try to block any parasympathetic nerve impulses that might contract those smooth muscle cells. Parasympathetic blocking drugs include *atropine* which is part of the *belladonna* (pretty lady) mixture extracted from the *deadly nightshade* plant (apparently dilated pupils were considered "sexy" even if the pretty lady then had blurred vision and visual distress in bright light—but, of course, no problem with candle light). Atropine-like molecules are also used in eye drops by the ophthalmologist (eye doc) to dilate your pupils and paralyse your ciliary muscle during an eye examination (you will be bothered by bright light and unable to focus your eyes until that effect wears off).

An ordinary expiration leaves some intrapulmonary air that can still be expelled through a more vigorous expiratory effort. That last gasp *expiratory reserve* measures about 1200 ml. So together with your dead-space air and the persistent one liter *residual volume* of air that you cannot exhale (unless external abdominal compression forces your guts up against your diaphragm), you retain over two liters of stale air within your lungs and airways after each normal expiration. We have already seen that each 500 ml breath at rest adds only 350 ml of fresh air to the stale air already collected inside. Therefore you could easily improve your breathing efficiency, enhance your blood oxygen levels and eliminate most of that useless waste CO_2 by *hyperventilating* (breathing more deeply and frequently). Right? Not really.

Conserving CO_2

It turns out that your inverted lungs, as opposed to everted fish gills, are especially good at conserving both water and carbon dioxide. You might ask, "Why save a useless waste product like CO_2?" But please don't. For we already know that CO_2 has important effects on your body. Indeed, your ordinary respiratory efforts occur entirely in response to your systemic arterial blood CO_2 tensions rather than to your systemic arterial blood O_2 levels—unless that arterial oxygen tension falls below 50% of normal. This physiological arrangement reflects the ancestral discovery that breathing to regulate arterial blood CO_2 levels within narrow limits automatically provides plenty of oxygen—an obvious compliment to the efficient hemoglobin molecules and wonderful lungs that you have inherited.

Normally the fluid layers lining each alveolus are thin and the underlying alveolar cells themselves are thin (less than 0.2 microns) and the pulmonary capillary walls adjacent to each alveolus are thin—so the O_2 and CO_2 tensions of your systemic arterial blood stay nearly the same as the O_2 and CO_2 partial pressures in your alveolar air. Thus your arterial and alveolar pO_2's average about 100 torr at sea level while your pCO_2's run about 40 torr (compared to 155 torr pO_2 and 0.23 torr pCO_2 in the atmosphere). Furthermore, since the partial pressure of water vapor in your fully humidified lungs at body temperature (37°C or 98°F) accounts for 47 torr of the entire 760 torr in one atmosphere, all other gases combined must then add up to 713 torr (the remaining 94% of that wet alveolar air). The atmospheric pressure reaches 47 torr at an altitude of 20,000 meters (65,000 feet) above sea level so at that elevation your lungs would hold only water vapor—at any higher altitude (hence even lower air pressure) water would boil from your lifeless lungs (see below).

Both your tissue fluid pCO_2 and systemic venous blood pCO_2 run about 45 torr so there is little difference between the CO_2 tension of your tissue fluids and that of the air in your alveoli. However, carbon dioxide is so soluble and therefore diffuses so efficiently through body fluids that even the 5 mm Hg gradient between tissue pCO_2 and lung pCO_2 allows your blood to relieve those tissues of their excess carbon dioxide. Of course, the *carbonic anhydrase* present within all of your red blood cells speeds the rate at which CO_2 gas dissolves into and is released from your blood by more than a million times, as follows:

$$CO_2 \text{ plus } H_2O \xrightleftharpoons[\text{anhydrase}]{\text{carbonic}} H_2CO_3 \xrightleftharpoons{} H^+ \text{ plus } HCO_3^-$$

Once again, if we read that reaction from left to right, we see that any increase in your blood CO_2 tension causes your blood to become more acidic (by raising H_2CO_3 levels and hence the concentration of H^+). But if we make changes from right to left (note again that these equilibrium reactions can go either way), part of any acid we may add will combine with bicarbonate (HCO_3^-) to become H_2CO_3 (for carbonic acid is a weak acid only partially ionized at body pH).

Your respirations closely control your arterial blood pCO_2 in order to continuously adjust this important *buffer system* and thereby stabilize your blood pH. Therefore, if you should happen to develop diabetic *acidosis* or become unusually acidotic from drinking a cupful of vinegar (acetic acid), you will automatically breathe more deeply to reduce the excessive acidity of your blood by blowing off more blood CO_2 than usual. The above equation shows that eliminating more CO_2 diminishes your arterial blood concentration of weak acid (H_2CO_3) and raises your blood pH back toward normal. Conversely, if you should develop *alkalosis* (a low-acid or more-alkaline-than-usual condition) because persistent vomiting has caused you to lose a lot of stomach acid, your breathing rate and depth will automatically slow and decrease so that you retain more of that same CO_2 (weak acid) than usual and thereby rebuild the acidity (reduce the pH) of your blood toward normal.

Thus an abnormally *high* pH (alkalosis) plus abnormally *high* CO_2 in a radial artery blood sample shows that you are automatically retaining CO_2 by *under*breathing to compensate for your *metabolic alkalosis*. On the other hand, if we find an abnormally *low* pH (acidosis) together with a *high* CO_2, you ought to have compensated for that acidosis by breathing more quickly and deeply. The fact that you have not blown off that obvious excess of CO_2 suggests some respiratory dysfunction (severe emphysema perhaps, or some problem with your nervous system, or simply too much sedation)—therefore such a low pH condition caused by CO_2 retention is known as *respiratory acidosis*. But if your arterial blood pH and CO_2 are both below normal (*acidosis* with *low* CO_2), you are appropriately blowing off more CO_2 than usual to compensate for *metabolic acidosis* (an acidosis not due to CO_2 retention), while *high pH* with *low CO_2* suggests that you are simply *hyperventilating*—so just relax and breathe in and out of this paper bag to enlarge your dead-space and those symptoms of nervousness, cramping and other funny sensations will soon stop. Hyperventilation is a common response to anxiety or unexpected (or anticipated) pain.

Note that these examples of respiratory regulation never even mentioned oxygen. And that is reasonable, since low arterial O_2 levels only begin to drive respiration (your so-called *anoxic drive*) long after increasing arterial blood CO_2 levels should already have encouraged more adequate ventilation. As usual, however, it is not quite that simple—although a mildly elevated arterial pCO_2 is a strong stimulus to your respiration, any major buildup of CO_2 also has a sedative effect. So if your pCO_2 rises rapidly for any reason, you are likely to "pass out" from those high arterial blood CO_2 levels at about the same time as your anoxic drive kicks in (those who prepare for underwater swimming by hyperventilating in order to reduce the insistent breathing signals from their rising arterial CO_2 levels may therefore faint, inhale water and drown).

The low pO_2 experienced on tall mountains (at 5500 meters or 18,000 feet elevation the sea level atmospheric pressure is halved to about 380 torr—the top of Mt.Everest at 8848 meters has a one-third atmosphere ambient air pressure which is about all that well-acclimated human mountaineers can survive even briefly—with an ambient PO_2 of about 50 torr and alveolar PO_2's down around 30 torr, their anoxic drive keeps them hyperventilating to obtain barely enough oxygen (which sufficiently reduces arterial blood CO_2 to inactivate the usual CO_2-based drive to breathe—such low-oxygen levels also cause metabolic acidosis which additionally stimulates breathing). You can easily become anoxic or accumulate dangerously high CO_2 levels after carotid artery surgery has damaged the delicate O_2 and CO_2 sensors of your *carotid body*—especially when rebreathing within a small closed space (e.g. with your head covered by a sheet). Some infant deaths from SIDS (sudden infant death syndrome) have been attributed to rebreathing stale air through a soft pillow in the prone position. Infants vary widely in how effectively they can regulate their respiratory gases.

Interestingly, birds maintain a one-way flow of air from airway to posterior air sacs on inspiration, to lung from posterior sacs on expiration, to anterior sacs on next inspiration (which also delivers new air to posterior air sacs) and out of the anterior sacs on next epiration (which also pushes air into lungs from posterior sacs) and so on. Such a "four cycle" breathing system with its improved (because flow-through and somewhat countercurrent) oxygen extraction permits adequate air storage and conserves CO_2 and water vapor despite using smaller lungs. This arrangement also improves breathing efficiency at high altitudes, so wild geese fly easily over human climbers struggling up Mt. Everest with their oxygen tanks—but perhaps goose lungs retain more dust on account of those circuitous air flows. Anyhow, if it still seems unnatural to ignore O_2 levels and let systemic arterial blood CO_2 levels determine your rate of depth of breathing, remember that a pO_2 of just 40 Torr will saturate three-fourths of your arterial hemoglobin with oxygen. So to answer the question posed so many paragraphs ago, storage of stale air within your lungs actually is good for you—up to a point. For by breathing to regulate your arterial CO_2, you ordinarily stabilize your arterial pH in addition to oxygenating your red cells and conserving water (you might save a bit more water by breathing only as needed to oxygenate your hemoglobin, but you would then require other mechanisms for regulating your pH in the face of those variable CO_2 levels—nature wisely selected your ancestors from among those who used rather than fought their CO_2 gradients).

Gaseous O_2 and CO_2 dissolve readily in the thin film of extracellular fluid lining the inner surface of each alveolus. And that flimsy fluid film remains intact because only warm air with 100% humidity enters your lungs. But the water layer lining each tiny air-containing alveolus can also be viewed as a bubble perforated by a tiny end-airway bronchiole. So why don't those wet stretched-out alveolar surfaces collapse like an ordinary punctured water bubble in order to form a more stable water droplet? It turns out that your lungs continuously counteract the alveolar surface tension (which is based upon hydrogen-bonding between adjacent water molecules) with a multi-component *surfactant* (detergent) secreted by cells near each alveolar entry. Temporary overexpansion of an alveolus quickly dilutes its surfactant and increases surface tension, encouraging that overstretched alveolus to become smaller. On the other hand, a just-collapsed alveolus should have high surfactant concentrations within its fluid lining and thus reexpand more readily (one end of each detergent molecule dissolves easily in water while the other end is attracted to the lipid membrane of those thin alveolar lining cells). Slightly stretching those surfactant-secreting cells with an occasional *yawn* or *deep sigh* helps maintain your surfactant production at effective levels.

The lung tissue of prematurely born infants tends to be low in surfactant so some premature infants benefit by having surfactant (extracted from raw chopped animal lungs or else artificially concocted) sprayed into the airways of their poorly expanded lungs. When a premature birth is anticipated, a day or two of adrenal corticosteroid therapy for the mother prior to delivery may improve the newborn's lung function as well (see Growth and Development). Liquid perfluorochemicals or PFC's have a great capacity for carrying O_2 and lungs filled with PFC's are readily inflated by gases despite having inadequate surfactant levels (PFC's have been evaluated as possible blood substitutes). The severe pulmonary artery spasm often associated with respiratory failure of the newborn and with Adult Respiratory

Distress Syndrome appears to respond favorably to nitric oxide (NO) inhalation. This *vasodilator* (blood vessel relaxant) has the great advantage of acting locally (because it is so rapidly eliminated by hemoglobin—see Brain). That allows tiny doses of inhaled NO to affect only lung-vessel smooth muscles, unlike many other medicines that dilate blood vessels throughout the body and thereby cause unacceptably low systemic blood pressures.

Lung Should Be Elastic

Normally expanded lungs are continuously pulled toward a smaller size by stretched *peri*-alveolar (*around* the alveolus) elastic fibers and by the surface tension within each alveolus. But a great many phagocytes (bearing nasty digestive fluids that are also capable of destroying pulmonary elastic fibers) hang out inside of your tiny pulmonary venules in order to intercept all manner of undesirable debris and bacteria arriving through the systemic venous blood as healthy cells brush on past. When those white blood cells die, their escaping digestive fluids can damage the adjacent lung, especially in individuals with insufficient blood levels of *alpha-1-antitrypsin* (antitrypsin here really means *anti-elastase*)—a glycoprotein produced in the liver. Thus persons inheriting an alpha-1-antitrypsin deficiency suffer progressive breakdown of their lung elastic fibers and alveoli—a process that leads to pulmonary *emphysema* whether or not they smoke cigarettes but at an especially young age if they do smoke (in part because smoke-induced inflammation recruits additional white blood cells to the lungs). Fortunately, a blood test is now available to detect such susceptible individuals and alpha-1-antitrypsin expensively extracted from pooled human blood plasma (that has been heat-treated to inactivate viruses) is available for weekly or monthly intravenous infusions. And even as you read these lines, bacteria containing transferred human alpha-1-antitrypsin genes are slaving away to provide (a hopefully less costly) human alpha-1-antitrypsin.

Now let us watch that older cigarette smoker try to breathe (but please don't stare). See how hard she/he must gasp to draw in small amounts of air and how he/she does not appear able to breathe very far out at all. Note also how she/he enlists neck, shoulder, chest wall and abdominal muscles for each breath (ordinarily those so-called *accessory muscles of respiration* are activated only during maximal breathing efforts) and how stiffly the head and shoulders must be held to help those accessory muscles enhance chest wall expansion. Normal expiration is impossible in advanced emphysema (as well as during episodes of *air-trapping* associated with asthma) for extensive alveolar breakdown leaves every emphysematous lung riddled with inelastic small-to-large pockets of stale air (known as *cysts, blebs* or *bullae*) that resist compression. When so much of the chest cavity is occupied by trapped air, any attempt to force expiration simply squeezes the heart and distressingly interferes with venous blood return—therefore each new inspiration necessarily begins with the thorax already at near-maximal size (ribs held up, diaphragm almost flat).

How You Breathe

During quiet inspiration, your chest wall muscles lift ribs up and out while contraction of your upwardly domed diaphragm muscle forces your abdominal viscera downward. The resulting expansion of your chest cavity and lungs reduces intrapulmonary air pressures and draws atmospheric air through your widening airways to your enlarging alveoli. But soon the inspiratory neurons of your brain stem are inhibited while neurons activating muscles helpful to expiration come into play—your thin, elastic and relaxed diaphragm muscle is then bulged back up within your chest by the abdominal wall muscle pressure exerted upon your guts. At the same time, your chest wall muscles relax so your ribs can move down and in. The consequent slow compression of your pulmonary alveoli and intrapulmonary airways pushes some of their air content outward to the atmosphere.

Note that your intrapulmonary airways are pulled open as the chest cavity enlarges during inspiration—they then narrow again during expiration. *Wheezing* (musical sounds made by high speed air flow through the inflamed, congested or otherwise narrowed airways of asthma, lung infection, left heart failure and so on) therefore becomes most evident during expiration (entirely blocked airways make no sound, of course). As your need for air exchange increases, you recruit additional neck, shoulder, chest wall and abdominal muscles into each inspiration and expiration in order move more air in and out of your lungs more rapidly. The elasticity of normal lungs helps them to become smaller during every expiration— indeed, were they not so adherent to the chest wall, your normal lungs would quickly deflate to the smaller *fully relaxed* size determined by unopposed lung elasticity and alveolar surface tension. Surprisingly, such a relaxed lung (within a widely opened chest at surgery, for example) still retains a moderate volume of (airway and alveolar) air and is rather resistant to further compression. In marked contrast, when persons with severe emphysema finally suffocate miserably (those who profit from tobacco sales ought to help care for these unhappy souls), their overinflated lungs will balloon forth if the chest is opened at autopsy.

Any portion of lung exclusively dependent upon a blocked bronchus soon becomes airless as the ongoing flow of blood readily absorbs all alveolar gas (since O_2 absorption into nearby capillaries leaves a persistent pN_2 and pCO_2 gradient toward the blood). An *atelectatic* or *collapsed* (completely airless) portion of lung is reduced in size, darker in color and rubbery to the touch. In contrast, a *consolidated* lung (or portion of lung) may retain near-normal size but resemble liver in weight and consistency. Lung becomes consolidated and more or less airless when its alveoli fill with fluid. With left heart failure (pulmonary venous congestion) or pulmonary edema from other cause, that alveolar fluid resembles extracellular fluid. But intra-alveolar fluid accumulations associated with lung inflammation or a bacterial infection (as in pneumonia—"pneuma" derives from the Greek term for breath) are likely to include more cellular debris, blood proteins and possibly bacteria.

Your chest cavity never becomes as small as your relaxed-size lungs so alveolar surface tension and elastic lung fibers are always pulling your smooth and wet lung surface away from the smooth and wet inner-chest-wall surface. But lung cannot separate from the chest wall unless free air or fluid are available to fill the space thereby created—just as you cannot pick up an inverted wet plate from a smooth plastic countertop unless the suction caused by your lifting attempts is first broken (by prying with a knife blade to allow air entry or else slipping the plate to an edge of the table). Air may gain access to the potential *pleural space* around lung by leaking from damaged lung or through a chest wall perforation (or via another more unusual route). Such *free intrapleural air* allows the slippery *visceral pleura* (over the lung surface) to come away from the equally moist and slippery *parietal pleural* layer (that covers the inner aspects of your ribs and intercostal muscles as well as the upper diaphragm surface and your *mediastinal contents*—your heart, blood vessels, airways, esophagus and everything else located between your two entirely separate pleural cavities). Although damaged lung easily leaks air into the adjacent pleural space, free intrapleural air cannot reenter that lung—just as you can blow air from a balloon through a hole in its tip but not suck air back. For soft rubber balloons (or soft alveolar walls) simply collapse (fall together) when outside of balloon (or outside of lung) air pressures exceed the air pressure inside. In this way, a ruptured balloon or torn lung acts as a one-way valve, permitting air leakage during distension but allowing no air reentry when collapsed. While it continues, such a one-way flow of air contributes to a growing *pneumothorax* (accumulation of air outside of lung but within a pleural cavity).

A major lung-surface air leak, or any *positive-pressure resuscitation* (which involves forcing air into trachea and lungs at above atmospheric pressure) in the presence of even a minor lung-surface air leak, can rapidly lead to *tension pneumothorax* (a high-pressure space-occupying intrapleural air accumulation that sometimes builds until it compresses the heart and lungs sufficiently to stop their function) unless the involved pleural cavity is swiftly deflated by creation of a substantial *trans-thoracic* (through the chest wall) opening to the atmosphere (most commonly via a large hollow tube). Once an open chest tube is properly placed to persistently vent the chest cavity, the life-or-death emergency is over—for heart and lungs are then no longer *compressed*. Of course, a *simple* pneumothorax (some intrapleural air but not enough to compress underlying lung during inspiration) will persist as long as an open chest tube still allows easy atmospheric access to the pleural space. But at least that underlying more-or-less relaxed lung can again exchange some air and the heart is no longer pushed against the opposite lung—and with intrapleural air pressures once again below venous pressures, the heart can easily refill after each systolic contraction.

Full reexpansion of lung generally improves pulmonary function so a simple pneumothorax is often treated by attaching a non-obstructive (to outward air flow) one-way valve or even a high-capacity air-suction device to enhance outward air movement at the atmospheric end of the chest tube. But as long as a simple chest tube merely remains *open*, every expiration moves air *out* of both mouth and chest tube. Similarly, each inspiration causes atmospheric air to flow *into* all holes (such as your mouth or an open chest tube) that lead to the chest cavity. For *intrathoracic pressures always remain above atmospheric during* normal *expiration* (otherwise air would not leave your lungs)—just as intrathoracic pressures always remain *subatmospheric during normal* (although not during positive-pressure respirator-assisted) *inspiration*.

Despite the occasional pneumothorax, *hemothorax* (pleural space blood accumulation), *empyema* (pleural space infection) and other pleura-related problems, those slippery pleural layers were essential when they first allowed your newborn lungs to inflate/expand freely—even now they continue to enhance the uniformity of lung ventilation within each asymmetrically expanding and relaxing hemithorax (dominated as it is by diaphragm movement at the lung bases). Those better-ventilated lung bases continue to get more than their appropriate share of blood under no-gravity outer-space conditions—which might have been expected, since it remains reproductively advantageous to direct more blood through the better-ventilated portions of your lung.

Hopefully this chapter on breathing has made more sense to you than preceding ones on hormones or the immune system. For the simple physical constraints imposed by solubilities, diffusion, partial pressures and flows make it unlikely that incredibly efficient mutant individuals sporting some radically new respiratory system will suddenly overwhelm all who stubbornly stick to the "tried and true" methods worked out over the past 700 million years. On the other hand, past billions of years have not yet exhausted possibilities for new signal-receptor interactions—for better methods of attack—for improved techniques in self-defense—for interesting ways of getting even or just jamming up another's chemistry. In your responses to parasites, anything goes and the nastier the better, but the latest innovation only brings reproductive advantage until the enemy catches on and catches up. That is why you must memorize the latest discoveries in endocrinology or immunology while still expecting frequent surprises, yet you can probably figure out basic respiratory functions once the parameters are understood. These might seem contradictory examples of natural selection at work but gas exchange is easily optimized while almost any molecule can carry information and fighting efficiency depends entirely upon the enemy and the terrain.

Further Reading:

Krogh, August. 1941. *The Comparative Physiology of Respiratory Mechanisms.* N.Y.: University of Pennsylvania Press (originally). N.Y.: Dover Republication. 172 pp.

Schmidt-Nielsen, Knut. 1990. *Animal Physiology, Adaption and Environment.* 4th ed. N.Y.: Cambridge University Press. 602 pp.

von Hippel, Arndt. 1986. *A Manual of Thoracic Surgery.* 2d ed. Anchorage, Alaska: Stone Age Press. 502 pp.

Little, Colin, 1991. *The Terrestrial Invasion, an ecophysiological approach to the origins of land animals.* N.Y.: Cambridge University Press. 304 pp.

Bramble, Dennis M. and Jenkins, Farish A., Jr. 1993. "Mammalian Locomotor—Respiratory Integration". *Science* 262. pp 235–40.

CHAPTER 22

DIGESTION

Overview;... Laundering Your Stolen Assets;... How You Eat;... How You Swallow;... It Ain't Necessarily So;... Dissolving Your Prey;... Dissolving Your Bacterial Competitors;... Osmosis Again;... Hormones And Enzymes;... Exposing Your Villi;... Fighting Over Your Lunch;... Another Noble Plumber Who Risked Plumbism;... Too Good To Waste?... Exploring The Human Abdomen;... The H Word;... Ostomies;... Soil Is Alive—It May Be Healthy Or Sick.

Overview

Early life bequeathed its DNA upon all living things. But your particular modification of those essential instructions can only be passed along if you continuously refuel and refurbish your DNA delivery system with the standard parts unwillingly relinquished by previous players in this Tinkertoy world. Many of your personal polypeptides and proteins are enzymes that direct life's usual molecules to interact in ways that best maintain your naturally selected and especially delightful design. With no obvious limit on the variety of possible enzymes, it seems that an equally endless variety of different life forms might have been constructed from those same involuntarily shared parts. Yet life is subject to many constraints—chief among which is the fact that it began in and still depends upon water—so the shapes assumed by complex biomolecules and life's membranes tend to reflect their solubilities in water, and most of life's chemical reactions take place in water. Thus complex proteins, fats and polysaccharide molecules are all built up through endothermic *dehydration synthesis* (removal of water). And polypeptides, neutral fats and carbohydrate polymers are ordinarily broken apart by catabolic solar-heat-releasing *hydrolysis* (disruption by water).

A bacterium often lives and reproduces within its relatively enormous food supply while dissolving and absorbing nutrients from the tissues and wastes of its living or deceased benefactors. In contrast, an amoeba often engulfs its tiny bacterial prey before reducing it to life's usual constituents within an intracellular *digestive* lysosome. Every cell of a simple multicellular creature such as a sponge must similarly capture and dissolve its own prey. But you are a more advanced killing machine, so some of your cells send messages while others carry out the attack by running fast and hitting hard. The close coordination of your highly specialized neurons, muscle fibers and other support cells carries a cost, however, for those finely tuned and highly competitive cells are no longer able to rip off and absorb their fair share of each victim. Therefore you require yet another group of highly specialized cells to break down, sort out and make available in standard form the construction materials and captive reducing power currently stored within a live moose or a chocolate mousse. Of course, the standard organic fuels (reduced molecules) that result from such preliminary digestion must then be further cracked and refined through your cytoplasmic reactions and mitochondrial oxidative-phosphorylation centers before finally they can power your essential enzymatic processes—those endosymbiotic mitochondria are life's true mercenaries, working within every eucaryotic cell on all sides of every bloody transaction for an ongoing share of the spoils.

Laundering Your Stolen Assets

The tissues and organs of your powerful *digestive system* are many and diverse—apparently it is not easy to keep your own tissues intact while routinely disassembling the rather similar tissues of those you have overpowered. The old alchemists puzzled endlessly over questions such as "If you discover a universal solvent, what will you keep it in?" Your own digestive system takes a more pragmatic approach by breaking down the organic materials of others in a careful stepwise fashion to avoid endangering itself. Thus your digestion proceeds rather slowly in discrete steps as your food advances through appropriately resistant segments of the long hollow *gastrointestinal tract* (*GI tract* or *gut*) that extends from your throat to your *anus* (asshole). En route, your *digestive system* (which includes that gut and its affiliated structures) mechanically breaks down and chemically hydrolyses your stolen reducing power. So when your most recent prey has finally been disassembled into ordinary amino acids, glucose, fatty acids and nucleotides within your gastrointestinal *lumen* (channel), there may be little to show whether a particular molecule derived from a yak or a yam.

But the variety and quantity of building blocks released by your internal hydrolysis (the presence or absence of essential dietary amino acids, fatty acids, vitamins, minerals and so on) certainly depends upon what you just ate. Being at the top of the food chain usually allows you to digest any adjacent life form that suits your taste. With most living things dependent upon the same mechanisms and biological processes, the more critically important bionutrients tend to be widely available in *ordinary foods* (your usual plant and animal victims). And there is little reproductive advantage in producing what is easily stolen (or as lobbyists say, "Why run for office when politicians come so cheap?"). One might therefore argue that any of the so-called *essential* amino acids, fatty acids and vitamins actually have minimal relevance to your everyday survival, since it generally takes long months of deprivation under unusual circumstances before specific dietary deficiencies become evident—apparently they became "essential" only because your ancestors had no reason to manufacture such easily stolen, hence low-priority items.

Although one might reasonably expect a proper digestive system to break down and absorb a whole duck (including guts, feathers and feet) or an entire apple tree, in practice you prefer to eat roast duck skin, fat and muscle—or Mom's apple pie (baked, sweetened, cored apples within a fat-laden crust). Indeed, you would find raw duck feathers or apple tree branches quite indigestible, and too many apple seeds might endanger your health (due to their cyanide content). Yet some among the simplest of bacteria, fungi and worms are able and willing to digest all duck and tree parts—otherwise the landscape surely would be piled high with feathers, beaks, seeds and knot holes. But just as it does not pay a gold miner to extract that last ounce from his claim or the bird to pick the last delectable berry or juicy bug from a bush, there is no reproductive advantage in developing a more complex digestive system that could degrade anything you might even consider eating into useful components—especially when it is so easy to construct an anus. For life's contests are rarely won on the basis of who ate everything put on their plate or who produced the least waste. Rather it is a matter of life or death to ensure that you consistently derive an energy profit from whatever you do eat. Not surprisingly, therefore, the cells of your gastrointestinal tract give preferential handling to high energy foods of the sorts your ancestors were likely to acquire. And your gut cells have little interest in digesting cellulose or the low-value keratin proteins in fluffy hard-to-chew-and-swallow feathers and hair. Similarly, crude oil is rarely available and it contains many toxic or otherwise indigestible hydrocarbons for which your ancestors never acquired a taste or the necessary digestive enzymes.

Nonetheless, within a surprisingly wide range, you can break down most proteins, carbohydrates and fats that you are likely to buy, gather, catch or kill. So you are *omnivorous* like a bear or pig rather than *carni*vorous (meat eating) or *herbi*vorous (vegetarian). And your front teeth grab and *incise* while your canines tear off bite-size morsels for your molars to grind. Still, you prefer rich (highly reduced) foods—those with the greatest caloric output when oxidised, such as the honey regurgitated by busy bees and muscle fibers fried in fat (hamburgers, deep fried fish) and starchy foods cooked in fat (the plant starch of your french fries and donuts is more easily digested after cooking). The internal hydrolysis of those foods leaves you with a more persistently satisfied feeling than any meal of lettuce or celery, baked bread or rice. And that good feeling inside comes from not only satisfying your own cells but also through pleasing a far larger population of other intestinal inhabitants—it pays to keep your guests happy as well.

How You Eat

In order to benefit from the food you have caught and cooked (now dead, denatured and partially decomposed), you still must eat and dissolve it. So after you inspect, smell and taste it, you begin by chewing, tearing, crushing and grinding it into smaller fragments while adding *saliva* (a complex fluid that contains *amylase* to speed the breakup of polysaccharides, as well as a slippery mucus binder) and continuing to taste, feel, sort, stir and squeeze it with your tongue. Before long, that delightful morsel of recently deceased rabbit flesh or chunk of heat-injured potato is converted into a moist mushy well-lubricated *bolus* (lump) that hopefully will hang together for transfer by sliding (swallowing). The salivary glands that produce your thickest, most mucoid secretions have short ducts that empty under and near the sides of your tongue. Your larger *parotid* salivary glands (that along with testicles seem especially susceptible to inflammation by the mumps virus—be sure to get your mumps shot) are located near the angle of your jaw bone below each ear. From there they send volumes of more *serous* (watery) fluid through the long *parotid duct* that enters your mouth via the mid-cheek mucosal surface on each side. Since those salivary secretions also play an important role in suppressing bacterial damage to your teeth, any medications or other conditions that produce a persistently dry mouth also increase chances for tooth decay. Under such circumstances, it might help to sip (allegedly antibacterial green tea or other low-acid sugar-free) fluids or chew sugarless gum (but too much chewing could cause sore jaw joints) in order to simulate or stimulate the secretion of saliva.

It is necessary for you to create a consistently mushy, stuck-together mucoid food mass (that you never would even consider eating if it came that way) prior to moving it past the opening of your *larynx* (voicebox) into the upper end of your *esophagus* ("to carry eating"), hopefully without dropping any portion into your airway to start a coughing spell or possibly even choke you. As you have learned, that transfer is not always so easy, since your own larynx occupies a lower, more hazardous position than that structure had in your early primate ancestors who preferred to eat and run rather than sit around discussing the meaning of life when lions or leopard might return at any moment to the kill being scavenged. In any case, your intricately coordinated swallowing mechanism lifts the upper airway (trachea and larynx) against the *epiglottis* (a curved mucosa-covered cartilaginous cap that fits nicely onto the larynx) as your food slips by—its high attachment to the back of your tongue ordinarily holds the epiglottis well away from your open airway. The extremely complex and agile muscles of your tongue are anchored to surrounding structures such as the *hyoid* bone. Although big and bulky, your tongue is adept at sorting and pushing food about within your mouth for selective grinding and processing—it performs that task with consummate skill and daring while dodging in and out amongst those unshielded choppers and grinders (such a design would never be allowed under current OSHA regulations).

Your tongue has a very sensitive surface which tastes, feels pain and even can detect, manipulate and extract the single hair that came in your soup. Indeed, a significant segment of your sensory (cerebral) cortex is devoted to analyzing signals from that sensitive tongue surface, which is why your eyes water and you feel such uncommon concern whenever you do happen to bite your own tongue (an example of how items destined for digestion are subjected to far different treatment than your own similar and adjacent tissues). As your tongue muscles finally collect and squish each moistened slimy *bolus* (clump) of chopped, ground and mixed food posteriorly past the epiglottis into your throat or pharynx, your upper esophageal *sphincter* (circular muscle that controls an opening) automatically relaxes to gratefully accept that welcome reducing power. If you enjoy comparing tongues, you will have noted that your dog's tongue is not nearly as bulky, strong or useful for chewing and swallowing as your own. Indeed, your tongue more closely resembles those beef tongues so tastefully displayed amongst other chunks of animal muscle, fat, viscera, brains, faces and feet on the meat counter of your kindly neighborhood grocer. Canine (dog) teeth are pointed and arranged for grabbing and slicing rather than grinding, so a dog's long and slender tongue (that curls so usefully for drinking) has little to do with sorting and pushing food about. Indeed, a dog cannot simply tongue-transfer the bloody ripped-off flesh that it so enjoys to where those concentric waves of posterior throat and esophageal muscle contraction can take over. So dogs must wolf their food, tossing their meat backward while moving their head forward to engulf it. Such inertial feeding is rapid, competitive and reproductively advantageous, although perhaps over too soon.

How You Swallow

Those body surfaces subject to abrasion (your skin, mouth, throat, esophagus and vagina) are covered by stratified squamous epithelium in which the outermost *cornified* (dead and shingle-like) layer of cells being rubbed away is continuously replaced by fresh cells moving up from deeper layers. Thus your esophageal epithelium resists wear and tear, especially from any hard or sharp items on your menu (bones, nuts, eggshells, etc.). And numerous mucus glands supported within the *sub-epithelial connective tissue layer* further slime your firm longitudinally folded esophageal surface to speed the food upon its way. That continuous sub-epithelial layer—which provides loose support for blood vessels and nerves throughout your entire intrabdominal gut—simultaneously frees your single-cell-thick intestinal epithelium to fold and wrinkle independently of heavier outer muscle layers. The additional epithelial wrinkling brought about by delicate subepithelial smooth muscles further slows food passage so that your intestinal absorptive surfaces are utilized more effectively. The entire long tube formed by your hollow esophagus and intraabdominal bowel is supported by circularly oriented smooth muscle cells which in turn are enclosed within an outermost layer of longitudinally arranged smooth muscle cells. Coordinated contraction (*peristalsis*) of those two sturdy muscle coats milks your food forward through the various processing stations of your gastrointestinal tract—the circular fibers constrict to trap or compress the bowel contents and prevent backflow (just as you bring your thumb and forefinger together when hand-milking a cow) while a segment of the outer longitudinal muscle shortens (the way your palm and fingers progressively expel milk trapped within the teat). Of course, your gut only changes position slightly and cyclically as those passing waves of smooth muscle contraction move intestinal contents along.

A large mouthful of food expands your empty esophagus the way a snake dilates when swallowing another struggling victim. While a snake has room to expand in most directions, your own esophagus cannot enlarge posteriorly against your solid vertebral bodies. So when your flat, wide and empty esophagus becomes full and round, it necessarily presses forward against the back wall of your always *patent* (open) trachea—which (like your bronchi) is reinforced by individual cartilaginous rings (much as a flexible vacuum cleaner hose may contain a wire or plastic spiral to keep it from collapsing). And unlike a cold-blooded snake that can exchange gases across its skin during the twenty minutes a large victim being swallowed may block its (the snake's) airway, your breathing must continue throughout each meal. But to ease the passage of larger bites from your latest victim, your U-shaped tracheal rings only surround the front and sides of the tracheal *lumen*, so a large intraesophageal bolus of food simply indents the soft posterior *membranous* portion of the tracheal wall in passing, rather than bumping down along an entire series of posterior tracheal rings.

Any food that becomes lodged in the upper half of your esophagus can narrow your airway by pressing against the soft membranous posterior tracheal wall. Such a blocked esophagus is uncomfortable, frightening and associated with copious drooling as large volumes of saliva are stimulated to help wash the food down. And you will be sitting up and gagging, not lying flat on your back—for in the latter position you rely upon the esophagus to drain fluid away from the back of your throat (another good reason to locate the trachea in front of the esophagus). But you should still be able to talk or at least whisper and you won't choke to death. Furthermore, a vigorous *Heimlich Maneuver* (compression of the upper abdomen) is *not* likely to relieve any esophageal obstruction. Save that dramatic effort for a person with food blocking the larynx who cannot move any air—such an acutely endangered individual will certainly be *unable* to talk, cough or even whisper (see Respiration). Food that remains lodged in the esophagus may require careful esophagoscopic removal to minimize risk of esophageal injury—often followed by *endo*scopic (internal) inspection or x-ray contrast studies to rule out esophageal damage and determine what caused the food hang-up—too large a mouthful and muscle spasm versus esophageal narrowing by scar or tumor, for example.

Food usually slithers from your *cervical* (neck) esophagus through your *thoracic* (chest) esophagus and on past the diaphragmatic *hiatus* (esophageal pathway or *opening* through diaphragm muscle) far more rapidly than you can read this convoluted sentence. Esophageal lumen air pressures may be reduced somewhat by the elastic tendency of adjacent lungs to pull away from (and so expand) that thoracic esophagus, but both your thoracic and abdominal contents (including the short final segment of esophagus normally located within the abdominal cavity) are alternately expanded (or at least decompressed) and then squeezed by the chest wall, diaphragm and abdominal muscle contractions associated with respiration and exercise. Your *distal* (farthest from its source—your mouth) esophagus usually shows increased smooth muscle tone near your entirely intraabdominal *stomach* ("stoma" means opening). This *higher pressure zone* or *lower esphageal sphincter* relaxes during peristaltic food passage into the stomach but subsequently resumes its more contracted state.

It is a mere fraction of a meter from the back of your tongue to your anus as the crow flies—yet your entire gut between those two points is about seven meters (over 22 feet) in length. Therefore your GI tract (below the relatively straight esophagus) necessarily follows a highly variable and tortuous path within the thin slippery *parietal peritoneal lining* of your *abdominal cavity*. Note that your enlarged and muscular *stomach* is but a relatively short *proximal* (upper) portion of your continuous hollow GI tract (gut)—and that many other organs such as liver, spleen, pancreas, uterus and ovaries also are located within your muscle-enclosed *abdomen*. Correct usage therefore suggests that when someone slugs you in the stomach, you actually were struck in the mid or left upper abdomen over your proximal GI tract—although that information may not make you feel any better. And a kidney punch could cause distress on either posterior flank just below the ribs—near one of those two *retro*peritoneal (*behind* the peritoneal cavity) kidneys that have your vertebral bodies, abdominal aorta and inferior vena cava between them in the back part of that same muscle-enclosed abdominal space. Although often transparently thin, your posterior peritoneum cleanly separates the kidney circulation from your digestive blood supply.

The mucus-protected columnar secretory cell lining of your stomach is less tolerant of surface abrasion than the stratified squamous esophageal epithelium, but it is quite capable of resisting the usual overlying acid-digestive brew. However, as your stomach kneads and compresses your food, it may occasionally force some acidified content back up into the esophagus. Such *gastric reflux* is more likely if the *proximal* (esophageal-entry portion of the) *stomach* has been displaced into the lower-pressure thorax—such a *hiatus hernia* is quite common in the elderly. Frequent acid reflux tends to irritate the esophagus and cause pain or a burning sensation under the sternum or in the throat—a condition commonly referred to as *heartburn*. These symptoms may be aggravated (and nasty-tasting stomach content is more likely to burp up into your throat and mouth) if you lie down soon after a large meal, especially if that meal included caffeine, alcohol, chocolate or mints. Usually you can eat *antacids* (buffers of acid such as $CaCO_3$ or $NaHCO_3$—as in "Tums" or "soda bicarb") to obtain temporary symptomatic relief from simple heartburn. More serious reflux problems are treated with medicines that reduce stomach acid production, or else by surgery to resecure the herniated stomach within the abdominal cavity and restore a *properly acute gastroesophageal angle* (with the esophagus entering the side of the upper stomach at an angle rather than diving directly into the top of the stomach—such an acute entry encourages stomach muscle contractions to block reflux as long as the lower esophageal sphincter also is supported externally by intra-abdominal pressures).

The muscle fibers of your tongue, pharynx and upper esophagus are of the voluntary skeletal sort, while lower esophageal and more distal gastrointestinal muscles are smooth (involuntary) muscles. Thankfully, your *external anal sphincter* (outermost locally prominent circular muscle band surrounding the anus) is once again voluntary striated muscle. So you exert a certain amount of voluntary control at both ends of your GI tract but have little power to intervene in between. The term *voluntary muscle* implies that such muscle is under your direct control. However, just as a hammer tap on a knee tendon makes the voluntary muscle of your anterior thigh contract (in an automatic or reflex reaction to that sudden stretch), you actually have little further control over your coordinated swallowing reflexes once food reaches the posterior third of your tongue. Of course, you can still reject swallowed food if it turns out to be unsatisfactory despite all of your inspections, tasting and feeling by tongue. Defective food is sometimes vomited from the *upper GI tract* (stomach and upper small bowel) by abdominal muscle contractions coordinated with *reverse peristaltic* gastrointestinal movements—or if such food only proves detrimental upon arrival in your lower small bowel, you can often rinse it out quickly by developing *diarrhea*. Many of your intestinal reflexes are regulated by the parasympathetic nerve fibers of your two vagus (tenth cranial) nerves. In contrast, your sympathetic signals mostly serve to shut down your viscera when the intestinal blood supply must be diverted to your fight-or-flight muscles during an emergency. Quite a few hormones also play major roles in the coordination of your gastrointestinal processes (see below).

It Ain't Necessarily So

For immediate accurate retaliation, you must determine the exact spot upon your surface that is being bitten. But before the advent of surgeons and antibiotics there was little benefit in knowing whether a pain originated in your heart, lung, pericardium, esophagus or appendix. So pain reports from intrathoracic or intraabdominal viscera tend to be vague, although often insistent enough to encourage relief-seeking behavior such as resting or breathing gently or not eating—sometimes it takes expensive invasive tests to determine whether your chest pain comes from heart, pericardium, lung, esophagus, chest wall or abdomen, or indeed, if it has any *clinical significance* (relevance to your health) at all. For example, cardiac *ischemia* (inadequate coronary arterial blood supply endangering some of the heart muscle) could cause typical *angina* (crushing mid-chest or left-chest pain radiating into the left arm) in one person while another equally afflicted might only experience vague chest discomfort or perhaps an ear-, jaw- or arm-ache.

Dissolving Your Prey

Your stomach interior is compartmentalized by *gastric rugae* (large mucosal surface folds) that delay food passage. The dilated upper portion of your J-shaped stomach is known as the *fundus*. Food that arrives there may remain unchanged for some hours, especially if a lot of food already is undergoing maceration and acid hydrolysis in the lower *antral* portion of your stomach. As each mushy food packet leaves your fundus, it is treated to a corrosive hydrochloric acid bath by the single-cell-thick gastric-lining of your stomach which pumps out H^+ in exchange for K^+. This *mucosa* extends down into a great many deep irregular crypts and crevices—an arrangement that maximizes your stomach's secretory surface while minimizing the area of exposed stomach that must be protected from its own acid digestive secretions by a thick layer of alkaline mucus. Your stomach has an especially rich blood supply (as back-up for its extensive secretory and muscular activity) and is the only segment of your entire hollow GI tract enclosed within three (rather than two) major muscle layers. Its job description includes kneading and diluting every soft slimy incoming bolus of food (and any drink) into a fluid mush and eventually the dilute suspension of nutrients known as *chyme*. That chyme is intermittently released in small squirts to your *duodenum* (the first segment of your *small intestine*) whenever the constricting *pyloric sphincter* (stomach outlet muscle) relaxes. Unlike respiratory tract epithelium, your GI mucosa has no need for ciliated cells since intestinal contents are milked along by peristalsis.

If eagerly anticipating this meal, your initial salivation and gastric secretion of acid and pepsin may have preceded the first bite. That rather obvious *cephalogastric* or head-to-stomach *reflex* was formally demonstrated when Pavlov's dinner bell caused his dogs to salivate and their stomachs to secrete digestive fluids. And when food polypeptides contact your distal gastric mucosa, *gastrin* (stomach hormone) is released into your circulation to encourage production of additional gastric secretions. Almost all vertebrate stomachs (including yours) secrete the protein-chopping enzyme *pepsin* in an inactive *pepsinogen* form, to be activated by acid hydrolysis once that proenzyme is safely beyond the overlying mucus barrier. Your stomach also produces *intrinsic factor* which expedites intestinal absorption of *Vitamin B$_{12}$* (a cobalt-containing pigment that prevents *pernicious anemia* and is structurally related to magnesium-containing chlorophyll as well as heme).

Dissolving Your Bacterial Competitors

Regardless of how much you may brush, floss, pick, rinse or gargle, your mouth remains a veritable zoo where one might easily encounter over 300 different species of bacteria and many other tiny life forms, each hoping to raise countless generations of happy descendents in this wonderfully warm, moist and nutritive environment. Human bites often cause serious infections by introducing an anaerobic sample of that zoo into the wonderfully warm, moist and nutritive tissues of the fist or other body part. Particular populations of these bacteria have earned special recognition—*streptococcus mutans* is frequently associated with tooth decay and gum disease while flourishing populations of *helicobacter pylori* (also referred to as campylobacter pylori) may inhabit the gastric, pyloric and duodenal mucosa in association with visible *gastritis* (gastric inflammation) or *ulcer* (an open sore in stomach or duodenal mucosa that sometimes penetrates the underlying gut wall). You are generally stuck with your own personal bacterial *flora* since these tough little warriors compete quite desperately for their spot in the gum. So no matter how clean the restaurant, how hot the soup or how often you have washed your hands, every bite of food includes a major burden of bacteria (millions per ml. of your saliva) and other microorganisms. But few of those tiny passengers will survive the wild ride through your seething cauldron of stomach acid—indeed, persons most troubled by an excessively sour stomach may have an offsetting advantage when it comes to destroying the tuberculosis bacteria that could come in that tasty but unpasteurized milk. Those unusually durable helicobacter attach most readily to mucosal cells bearing Blood Group O-associated surface antigens. Helicobacter then neutralize stomach acidity locally by secreting *urease* to release ammonia from urea ($NH_3 + H^+ \rightarrow NH_4^+$). There is some evidence that submucosal connective tissue immune cells play a role in the ulcerogenic process and are also affected by the anti-ulcer drugs that reduce gastric acidity and secretions. In any case, several weeks of combined antibiotic and bismuth subsalicylate (Pepto-bismol) treatment can usually eliminate the helicobacteria that seem to cause these and many other chronic stomach ailments (up to and including cancer).

All newborn mammals (including human infants) ordinarily subsist on breast milk. Although their stomachs produce little hydrochloric acid (H^+Cl^-), these newborns do secrete *rennin*, an enzyme that coagulates milk and thereby slows its transit and improves its digestion. Curdled and denatured milk protein can then be mixed, mashed, broken down, dissolved and gradually hydrolysed into its easier-to-absorb constituents. Human infants also secrete *gastric lipase* to initiate the breakdown of milk fats into fatty acids. Rennin and lipase are not secreted by your adult stomach, but commercial rennin (extracted from the stomachs of slaughtered calves) produces those *curds* (potentially tasty cheese solids) from milk, leaving *whey* (the fluid residue). Newborns do not digest proteins very effectively so maternal antibodies present in human breast milk generally reach the small intestine intact. There those antibodies (especially IgA) adhere to specific cell surface receptors and are absorbed by phagocytosis—thus intact breast-milk-borne IgA is transferred into the infant's blood stream after protecting GI tract surfaces directly and lung airways indirectly (via *aspirated* milk—milk that "went down the wrong way"). Human breast milk also contains *mucin* as protection against infectious infantile diarrhea.

There are many other factors favoring human breast milk over cow's milk in the early months of life—infant allergies and even autoimmune problems such as childhood diabetes could possibly result from early exposure to bovine milk albumen which, some work suggests, shares an epitope with pancreatic beta-cell membranes in genetically susceptible individuals—other evidence denies this relationship and places the blame for juvenile diabetes on an immune response to *glutamate decarboxylase* (which converts glutamate to GABA within the central nervous system and also in the pancreas—and this enzyme also resembles one fragment of a virus that has been associated with the onset of juvenile diabetes). Adults denature almost all food proteins entering their stomachs so they cannot acquire any useful level of antibodies by eating them (although human mothers still pass along minute traces of cow antibodies in their own breast milk—cow milk antibodies have been blamed for infant colic). The close physical contact between mother and infant encourages her resident bacteria to colonize the infant's inner and outer surfaces—which made your mother's personal milk-antibody production particularly relevant to your early need for passive immune protection as your newborn immune system was becoming competent (prior to your birth, a functional fetal immune system capable of attacking mother would have been disadvantageous). Mother's intimate exposure to your stools, vomit, skin flora, sneezes and coughs also encouraged her to produce antibodies against infections troubling you.

Osmosis Again

Salivary amylase becomes inactive when acidified so any further slow breakdown of carbohydrates or fats in an adult stomach reflects unguided acid hydrolysis. The dilute chyme that squirts from your stomach through the pyloric sphincter still has the same low pH (1 to 2) that encouraged protein breakdown by gastric pepsin. Chyme generally contains long polypeptides for *pepsin* is a proteolytic enzyme that hydrolyses polypeptides at just a few specific sites. Since chyme is relatively *isotonic* (has nearly the same osmotic pressure as blood) by the time it leaves your stomach, water must equilibrate rather freely in either direction across your stomach wall—as does alcohol (ethanol). A fatty meal tends to hold alcohol and pass through your stomach more slowly than a high protein meal, therefore a fatty meal may delay the transfer of alcohol across your stomach lining cells. However, a complex carbohydrate (polysaccharide) meal moves through the stomach quite rapidly—and eating several candy bars when rather dehydrated would temporarily make you even more dehydrated as each delicious sugar molecule in that hypertonic stomach content attracted additional water from your circulating blood. You might even become dizzy until the sugar-laden fluid accumulating within your gut lumen finally returned to your circulation across the small bowel wall. Such circulatory disturbances can be blamed upon the fact that candy bars only evolved quite recently, for your stomach ordinarily avoids drawing in a lot of fluid by carefully *not* digesting your dietary protein, carbohydrate and fat all the way down to amino acids, sugars and fatty acids (as it could easily do, even though thereby raising the risk that the stomach might dissolve itself).

The glass of water you just drank is rapidly absorbed into your stomach and small bowel lining by the osmotic concentration differences between that water and your body fluids, and soon eliminated as urine unless you are dehydrated. In contrast, water taken with a meal merely reduces the fluid contribution that your stomach must make before that meal can proceed as chyme. Thus what you eat and drink may be absorbed across your stomach or small bowel wall more or less rapidly—depending upon the order, rate and amount that you eat and drink them. An appropriate mixture of food and fluids inside of your stomach should therefore allow a canteen-free hike without repeated stops to urinate.

Hormones And Enzymes

Intermittent releases of acid chyme onto your duodenal mucosa stimulate the secretion of hormones such as *gastric inhibitory peptide, cholecystokinin* (especially with fatty meals) and *secretin.* These hormones respectively reduce gastric activity, cause the discharge of *bile* from the gall bladder and stimulate the release of *alkaline pancreatic secretions* into the *descending* (second) *portion* of the duodenum—where your *bile duct* and *pancreatic duct* enter the duodenum together at a tiny sphincter-controlled opening known as the *Ampulla* of *Vater.* Depending upon where it lodged, a gallstone from the gall bladder might therefore block just the gall bladder duct, causing chole*cystitis* (gall *bladder inflammation*)—or it could obstruct the common bile duct, causing *jaundice* (a yellowing of the sclerae and other tissues and body fluids due to retention of *bilirubin* which is a product of hemoglobin breakdown)—or it might become impacted distal to where the bile and pancreatic ducts join, causing jaundice and perhaps also *pancreatitis* consequent to the reflux of bile into pancreatic ducts.

Pancreatic enzymes ordinarily become activated only within the duodenum, far from their pancreatic cell of origin. Very severe *auto*digestion (*self* digestion) and inflammation may follow the abnormally early (pre-release or intra-pancreatic) activation of pancreatic enzymes. That variety of pancreatitis can cause serious illness or even death, with survivors subject to many possible complications such as retroperitoneal abscess or cyst formation at the pancreas (located behind the stomach). Pancreatitis is occasionally associated with viral infection, alcohol abuse and other factors. Cigarette smokers have an increased incidence of pancreatic cancer—this malignancy is usually incurable (due to invasion of adjacent tissues) when symptoms first develop.

Your ongoing intestinal food stream usually maintains a near-neutral pH of 7 to 8 once alkaline pancreatic secretions have neutralized the acid chyme. Pepsin becomes inactive at that pH and other proteolytic enzymes take over. When cholecystokinin causes the gall bladder to contract and release bile, the chyme becomes more greenish yellow or brown in color. Bile contains various cholesterol and fatty salts whose *detergent action* (with the polar end of the molecule dissolving in water and the other end in fat) helps your intestinal fluids to dissolve or emulsify ingested fats. As mentioned, bile also includes those potentially toxic bilirubin pigments derived from the breakdown of heme (at one stage in their complex life-cycle, malaria organisms eat a lot of hemoglobin—and the quinine-type antimalarial medicines poison these parasites by interfering with their ability to safely store discarded heme). Pancreatic secretions include enzymes that hydrolyse fats (known as pancreatic *lipases*), proteins (*trypsin* and *chymotrypsin*) and polysaccharides (*amylase*). Individual cells of your small bowel lining secrete other enzymes locally to digest adjacent food disaccharides and polypeptides in preparation for immediate absorption of their sugars and amino acids.

Exposing Your Villi

In addition to the numerous irregular surface folds that slow food passage, the small bowel absorptive surface is enormously enhanced by a great many tiny *villi* (finger-like projections) which give your small bowel mucosa a velvety appearance to the naked eye. Each villus has a central arteriole passing close enough by returning venules for counter-current solute transfers to maintain high ion concentrations at the villus tip. The many columnar cells exposed upon each villus additionally expand their intestinal surfaces with a great many *microvilli* (tiny actin-supported finger-like cell membrane projections) that together form the so-called *brush border* where food absorption as well as much final digestion takes place. These microvilli lie under a thin *glycocalyx* (a protective meshwork of digestion-resistant glycoprotein and mucopolysaccharide filaments). Food normally takes several hours to pass through your small bowel. The churning and mixing of small bowel secretions with bowel contents helps your *duodenal* and *jejunal* surface cells to actively take up glucose, fatty acids, amino acids and as many other essential nutrients as possible before that food stream reaches the *ileum*, the third and final section of your small bowel (these small bowel segments look pretty much alike, actually). At any point along the way, the small bowel fluid content reflects an ongoing balance between intestinal Na^+ absorption and Cl^- excretion as well as osmotic fluid shifts consequent to solute removal by intestinal epithelial cells.

Fighting Over Your Lunch

A large glucose meal (say soda pop and candy) should deliver plenty of glucose to your gut lining cells by passive diffusion. So why waste energy actively absorbing glucose, amino acids and other valuable molecules when you could simply wait for them to diffuse out of your continuously stirred intestinal contents? Wouldn't it be more economical to at least delay active absorption until you see what nutrients still remain farther on down your intestinal tract? Well actually, the sooner you withdraw all useful molecules, the sooner your precious water will follow along osmotically. And the other reason to hurry is that your lower small bowel (distal jejunum and ileum) and colon are literally teeming with uncountable hordes of hungry bacteria ready to grab your lunch if you do not absorb it first. These bacteria are fierce descendents of tough survivors who made it through your acid stomach. They remain locked in a life-and-death struggle with each other as well as your intestinal lining cells to see who gets what. Of course, your mucosal cells have many advantages including an adequate blood-borne supply of O_2, along with immune cells and antibodies for back-up. But conditions really are difficult out there in the intestinal channel with little if any free oxygen and usually no free iron, plus a great many enzymes and various antibodies mixed in with bile and all of the other toxins put out by you or your chaotically flourishing inner population of bacteria, protozoa, fungi and even worms.

Whenever you change your diet, you greatly affect those uninvited internal residents. Initially you may note few signs of bacterial distress (some hunger pains perhaps?) as your new or inadequate diet leads to wholesale death and destruction among previously successful microorganisms. Indeed, you may only become aware of having caused that devastation as new bacterial varieties take over and cause various temporary symptoms such as excessive gas production, cramps, diarrhea and so on. So it is easy to see why human herbivores (vegetarians) and carnivores both tend to get "an upset stomach" when they annoy their current intestinal inhabitants by trying the other diet. As in any zoo or prison, when your desperate intestinal inmates are upset, they will at least roar to let you know. Similarly, every time someone takes a crap and then makes you a lovely salad with their unwashed shit-contaminated hands, you will most likely be subjected to yet another fierce takeover battle between resident bacteria and those newcomers. The latter need not even be dangerous, although often they are. They merely need to be tough and different enough to upset your bowels. Or if your food contains milk and you are one of many adults who cannot digest lactose, that lactose is welcomed and soon digested by your bowel inhabitants. But as lactose-loving bacteria then flourish you will again have gas, cramps and diarrhea due to their additional metabolism and the unabsorbed solutes within your intestinal lumen.

So just as it makes sense for you to eat only what you need, it also makes sense for your intestinal epithelial cells to quickly grab all useful or high-energy solutes (and the water they attract) on your behalf. Once actively absorbed, those sugars and amino acids enter villi capillaries for transport through your portal vein to the liver—hydrolysed *fats* are absorbed passively through cell surface phospholipid membranes and then transferred into villi *lymphatics* (also known as *lacteals* because of their visibly milky content at surgery soon after a fatty meal) which eventually empty into veins at the base of your neck (see Circulation). The fatty acids, phospholipids and cholesterol absorbed from your diet generally travel within your blood stream as *chylomicrons* (tiny protein-covered globules) or within the hydrophobic clefts of your serum albumin molecules. Presumably the immediate diversion of dietary fat toward your systemic venous blood improves your portal venous blood flow and simplifies the processing of other nutrients and any toxic materials by your liver—dietary fat is mostly destined for deposit within your fat cells so it has little relevance to your liver anyhow.

Low-density lipoproteins (LDL's) are the major cholesterol transporters in your blood. Each LDL particle includes about 2000 cholesterol molecules and 1000 phospholipid molecules wrapped in a detergent-like (partly polar, partly fat-soluble) *apolipoprotein* molecule—one subform of that apolipoprotein may play a role in Alzheimer's disease (a degenerative brain disorder). LDL particles deliver the cholesterol that is so essential for construction of all cell membranes (see Cell) and for production of those steroid hormones (slightly modified cholesterol molecules) that have such a profound effect upon your growth, behavior, salt content, ability to tolerate stress and much more. But as usual, when some is good, too much is bad—so *LDL-associated cholesterol* is often referred to as your "bad" cholesterol since *elevated* LDL cholesterol levels are associated with an increased likelihood of heart attack due to coronary artery atherosclerosis. On the other hand, your high density lipoproteins (HDL's) tend to return surplus cholesterol from the periphery to your liver for disposal in the bile. High levels of one variety of *HDL-associated cholesterol* seem to be associated with a reduced risk of heart attack so HDL-cholesterol is often considered your "good" cholesterol (HDL's may also provide protection against certain types of serious infections by binding to bacterial endotoxins). Elevated blood LDL levels can result from a genetic deficiency in LDL receptors or from a modern high-cholesterol diet that stimulates your cells to reduce their LDL receptors once their cholesterol needs have been met (your ancestors produced most of their own cholesterol molecules in their liver cells since the plants and insects upon which they subsisted had little cholesterol on offer).

While persistently high LDL levels in the blood seem to encourage atherosclerosis, oxidised LDL's and some varieties of apolipoprotein are even more atherogenic (see Metabolism and Circulation). Oxidised LDL's stimulate macrophage scavenger receptors, and cholesterol-loaded macrophages (foam cells) are prominent components of *fatty streaks* (early subintimal atherosclerotic deposits in arterial walls). Furthermore, oxidised LDL's are *cyto*toxic (injurious to *cells*) and they stimulate white cells to release cytokines and growth factors that induce proliferation of arterial wall smooth muscle cells while also encouraging an autoimmune response to their own presence. Interestingly, apolipoproteins are about 80% identical to *plasminogen* molecules (the precursor form of the *plasmin* that causes clot lysis). And plasmin, like about a dozen other blood proteases closely involved in blood clotting, is related to the trypsin molecule produced in your pancreas (but trypsin is a more generally digestive protease). Once again we see closely related molecules finding many different uses. And as usual, it is all far more complex than the above discussion would suggest—with more being learned every day.

The osmotic pressure of your ongoing intestinal content gradually declines as smaller organic solutes (the ripped-apart molecules of your most recently digested victims) are progressively absorbed across your duodenum and upper jejunum. This encourages a steady passive recovery of residual intraluminal water as long as the passing flow is not speeded by any dietary indiscretions or bacterial toxins. Thus the bulk of your water intake is entirely mixed with *stool* (shit) before you finally "drink" it into your intestinal cells. Yet it still makes sense to drink only microfiltered, boiled, chlorinated or otherwise "clean" (not stool-contaminated) water so you can avoid ingesting diarrhea-causing or other pathogenic life-forms that recently flourished in the intestines of others.

By the time your intestinal content reaches the *ileocecal valve* at the far end of your small bowel, this material is recognizably stool (with a porridge-like consistency). The ileocecal valve is a floppy, redundant area that encourages one-way flow of small bowel content into the *cecum* (first portion of your large bowel). That cecum is probably located in your right lower abdominal quadrant and has a single tube-like sealed-tip *appendix* arising from its bulging lower end. In *post-gastric* or *hind-gut fermenters* such as the horse or rabbit, the greatly enlarged cecum is the site of vegetable matter digestion by cellulase-secreting bacteria and protozoa. Rabbits eat their own soft fluffy cecal content directly as it is released at the anus in order to recapture food value generated by hind-gut cellulose fermentation—they put this material aside for further fermentation in the gastric fundus before final digestion and elimination as ordinary fecal pellets on the second pass. In comparison, even the most confirmed human vegetarians only achieve partial breakdown of dietary cellulose within their GI tract, presumably because conditions there do not favor the prolonged storage, fermentation and recycling through bacteria and protozoa upon which the significant enrichment of cellulose-dependent nutrition relies—consequently higher-in-fiber diets tend to pass through you more quickly in order to allow an increased intake of such low-nutrient vegetable matter.

Your *colon* (*ascending* colon, *transverse* colon, *descending* colon, *sigmoid* colon and *rectum*) continues the drying-out process, moving bowel content forward slowly with the assistance of three prominent longitudinal *taenia* (external bands of muscle) until *defecation* (shitting) takes place. Your *external* (voluntary muscle) *anal sphincter* is usually in control of when you will defecate. However, there is a limit to how long your external anal sphincter can withstand those forceful sigmoid and rectal peristaltic contractions, particularly as your traitorous internal anal sphincter automatically relaxes during each coordinated contraction of your rectal smooth muscle. Anal *incontinence* (leakage of gas and diarrhea or even formed stool) can occur under many conditions, although it is more common in older women and after difficult childbirth. This condition can often be stabilized or improved by medication, diet control and sphincter exercises—surgery sometimes enhances control as well.

Another Noble Plumber Who Risked Plumbism (Lead Poisoning)

When next you take a *crap*—thereby honoring the memory of Sir Thomas Crapper who was knighted by the Queen for helping to develop the flush toilet[1]

---

1 Reyburn, Wallace. 1969. *Flushed with Pride: The Story of Sir Thomas Crapper.* London: Macdonald and Co. 95 pp.

(although she never made him a member of her Privy Council), note that your more or less solid smelly brown feces do not rapidly swell up or fall apart upon entering water in the toilet bowel. This suggests that your feces do not include many small solute molecules. A very foamy smelly floating stool may contain more fat than usual (not enough pancreatic lipase, perhaps). A black, tarry, really stinky stool may include slightly digested blood from a bleeding gastric or duodenal ulcer. Stools soon lose their odor as their surfaces dry and oxidise (although anyone cleaning up after the neighbor's dog knows that newly fragmented dog poop rapidly releases additional smelly vapors). But in the presence of adequate oxygen, those partially broken-down smelly molecules with high vapor pressure soon undergo further oxidation as one or another sort of bacteria happily acquires their residual reducing power. *Aeration* plays such a major part in *sewage treatment* because extra oxygen enhances bacterial oxidations. The eventual output is CO_2 and clean water as well as a residue of less-easily oxidised substances commonly referred to as *sludge* (which may include toxic heavy metals).

Too Good To Waste?

As you can see, a lot of life remains within any stool, especially herbivore manure. Back on the farm, fresh horse manure, still fermenting vigorously (unlike cow manure in which many of the microbes are digested *after* they have made their contribution to the fermentation—being more efficient digesters, cows have an advantage when forage is scarce and of low quality—and many weed seeds that pass unscathed through horses are easily digested by cows), was often packed into a shallow trench to warm the small greenhouse-like enclosure known as a *cold frame* (where seeds could be planted before outdoor spring planting was possible). Dogs, pigs, ravens and rats often eat feces for their nutrient value. Indeed, farmers have successfully raised hogs beneath chicken houses with those hogs subsisting for six months solely on chicken shit and any spilled chicken feed. Although hogs can prosper in such low-budget accommodations, they eventually should be finished on a grain diet to eliminate any "off" flavor before the slaughter.

No vertebrate (and few invertebrates) can digest cellulose unassisted. So cows and other ruminant herbivores maintain a warm and moist fermentation vat enclosing rechewed cellulose fibers thoroughly mixed with symbiotic bacteria secreting cellulase and other enzymes that release short-chain organic molecules for absorption through the rumen wall. As mentioned, those flourishing hordes of bacteria are then consumed by other bacteria and protozoa and a large fraction of the protozoal population is in turn digested and absorbed each day by the cow's intestinal cells. In some dry areas where wood is scarce, herbivore manure has value as a building material and fuel. Even the casually discarded output of your own intestinal tract retains lots of good reducing power for other creatures (and half of that bulk may be living and dead bacteria of over 500 different species—at an average concentration of 100 billion to 1 trillion [10^{12}] bacteria per gram)—it also provides important nitrogenous compounds and minerals necessary for healthy plant roots and shoots.

While a *high fiber* (high cellulose and lower energy concentration) diet reduces your intestinal transit time and increases your fecal output, a *low residue diet* does the opposite—allowing you to extract more of those more concentrated nutrients before your resident bacteria can get at them. So on a low-residue diet your fecal output (bacteria and other dietary debris) and *flatus* (farts, which represent swallowed air and also the gaseous metabolic wastes of your intestinal microorganisms) are markedly reduced in volume. Although the methane and other gases produced by those intestinal inhabitants can certainly cause discomfort when withheld, such high-pressure intraluminal gases are then partially absorbed into the circulation and expelled along with other gaseous wastes (onion and garlic sulphides, the various aldehydes and other breakdown products of your expensive alcoholic beverages and so on) through your lungs, hopefully causing less offense thereby—unless you are one of those close face-to-face communicators.

When a hunter kills an herbivore that has a full stomach and intestine (the standard condition), its live gastrointestinal residents will continue to generate heat and gases, ferment and digest, transfer DNA and reproduce, fight and die—so dead gut wall soon disintegrates, allowing widespread bacterial contamination and overgrowth *unless* that herbivore is *gutted* (undergoes total intestinal tract excision) soon after death. Undamaged meat (voluntary skeletal muscle) remaining on the carcass can then cool and *age* properly (decay more gradually in a drier, more bacteria-resistant mode). Interestingly, bears and other carnivores often consume the abdominal contents of their kill before eating the rest. And the Inuit (Eskimo) have known for millennia that caribou stomach content makes a tasty and digestible salad quite unlike its flavorless lichen source.

Exploring The Human Abdomen

A long vertical midline abdominal surgical incision through your skin, fat and the thick collagenous fascia (that extends laterally over your two paramedian rectus muscles) exposes your semi-transparent parietal peritoneum for division. Such a generous opening allows easy access to your entire abdominal cavity unless there are *adhesions* (intraperitoneal bridges of collagenous scar tissue marking sites of previous inflammation or injury). A gloved hand exploring the undersurface of your thin diaphragm muscle can feel your heart beat and your lungs being inflated by the anesthesia respirator. Your *omentum* (a mobile sheet of fatty tissue which includes easily dilated blood vessels and many cells involved in immune responses) ordinarily hangs down from the *greater* (lower and outer) *curvature* of your J-shaped stomach in front of the small bowel. When that omentum is lifted upward, its attachment to the transverse colon also becomes evident. Although flimsy and transparent in a newborn, the omentum may become inches thick with fat in a very obese person.

The fragile omental surface adheres readily to inflamed tissues and helpfully walls off any infected area where normal intestinal inhabitants (or even vicious newcomers such as typhoid bacteria) may have broken out through your gut wall. For regrettably, despite all of the good food and expensive wine that you may have shared, your intestinal bacteria and other parasites or symbionts would just as soon digest you as your lunch (yet another example of how life can only be productive and competitive at the edge of chaos). And their big break could come if your appendix becomes blocked and inflamed with its circulation compromised—or that small *diverticulum* (colon wall outpouching full of stool) springs a leak—or your ulcer finally penetrates completely through the stomach or duodenal wall— or your small bowel becomes sufficiently twisted, entrapped, obstructed and inflated so that its blood circulation is threatened or blocked locally (as in a *strangulated hernia*). Under these and many other circumstances, some of the stool bacteria (or even worms) from your bowel lumen may pass rapidly through your weakened or perforated gut wall and spread into nearby nooks and crannies of your peritoneal cavity to flourish in these potential spaces at your expense (breaking down your tissues, releasing toxins and overstressing your immune system to the point of collapse or self-injury).

Bacterial invasions of the peritoneal cavity caused a great many human deaths before modern antibiotics became available following World War II. *Peritonitis* is still a very hazardous condition, although it often remains *localized* (walled-off and restricted to a small portion of the peritoneal cavity) when omentum and nearby loops of inflamed bowel wall stick promptly together. But peritonitis also can become widespread—at worst, those stool bacteria may contaminate the entire abdominal space between diaphragm and pelvic floor, from in front of the (retroperitoneal) kidneys and great vessels to the inner aspect of the anterior abdominal muscles. Furthermore, infected or otherwise damaged peritoneal surfaces often become heavily scarred (infiltrated with collagen) during the healing process—making subsequent surgical access difficult or almost impossible while simultaneously increasing the likelihood of later intestinal blockage that might require such surgical attack.

An exploring hand moved gently to your left upper quadrant along the front surface of the stomach eventually encounters the soft, delicate, easily ruptured (especially if enlarged by malaria or other red-cell-destructive disease) *spleen* between stomach, diaphragm and colon. The liver occupies your entire right upper abdominal cavity, extending broad peritoneal attachments onto the overlying diaphragm (which may have relevance to breathing—see Respiration). Your small and ordinarily thin-walled gall bladder lies just under the forward liver edge. If the omentum is temporarily displaced onto the anterior chest wall, your colon (large bowel) becomes readily visible as a frame around those loose loops of small bowel. A flat mobile curtain-like *mesentery* suspends your small bowel along one side of its entire length. The two slippery peritoneal surfaces of that mesentery arise from your posterior parietal peritoneum and enclose a fatty layer traversed by a great many small bowel lymphatics, nerves, arteries and veins. The oxygen-depleted nutrient-laden venous blood from your surface-expanded gut (and from all other intraperitoneal tissues except uterus and ovaries) gathers within your *portal vein* and is routed to the liver. There that gastrointestinal venous drainage slows markedly as it enters a huge number of tiny tortuous (hence surface-expanded) and incompletely endothelialized *hepatic* (liver) blood channels known as *sinusoids* where phagocytes lie in wait for bacteria, worn blood constituents and other circulating debris while *hepatocytes* (liver cells, their exposed surfaces expanded by microvilli) deal with just-absorbed nutrients and toxins—a separate inflow of well-oxygenated hepatic arterial blood supports all of that complex metabolic activity. Processed sinusoid blood proceeds via your hepatic veins into your inferior vena cava just behind the liver.

This *venous portal* or *double capillary bed* system (the first set of capillaries in bowel wall, the second set of sinusoids in the liver) gives your liver cells first option on all absorbed nutrients (except for those fats being diverted through intestinal lymphatics). As the primary destination of reclaimed animal and vegetable parts, your liver clearly is the designated major metabolic organ of your body. In addition to storing spare iron, Vitamin B_{12} and your fat-soluble vitamins (A, D, E, K), liver modifies Vitamin D prior to its final activation in your kidney and stores excess glucose (as *glycogen* or animal starch) for release into your circulation as necessary. Your liver also produces a great many other important body chemicals, utilizing raw materials that you eat and digest—it breaks down and partially recycles old hemoglobin, deactivates and eliminates a wide variety of those toxic molecules always included in your diet (another good reason for varying your diet in order to avoid overloading any particular system) and produces blood albumin, blood clotting factors and so on. Therefore, if the next biology examination asks where an important non-hormonal substance is produced or some nasty chemical is detoxified and you don't know, liver would certainly be a good guess.

All liver products and wastes are exported through your hepatic veins or dumped into the bile. Much of that bile is modified by intestinal cells and bacteria before being reabsorbed into your intestinal capillaries and lymphatics or else discarded with your fecal output. Many medications are modified or deactivated by your liver cells so some medicines are best given intravenously or delivered by absorption through the oral, rectal or vaginal lining in order to reach the systemic circulation without first passing through the liver. Although medicine absorbed from under your tongue bypasses the liver entirely, only part of the medication delivered by a rectal or vaginal *suppository* will bypass the portal venous drainage (those huge torpedo-shaped suppository tablets are specifically designed for drug absorption through the rectal or vaginal epithelium—despite repeated instruction, some patients still complain that they are hard to swallow).

Its impressive response to growth factors allows liver to regain full normal size within two or three weeks after a major portion has been removed. But liver is often *acutely* (suddenly) damaged and sometimes permanently destroyed by certain chemicals (*toxic hepatitis*). Liver can also suffer acute harm or chronic injury with scarring as a result of ongoing *viral hepatitis* as well as bacterial or parasitic infestations, alcoholism and other diseases. Progressive *cirrhosis* (chronic scarring with hardening) of the liver interferes with portal venous blood flow through liver sinusoids toward the hepatic veins and inferior vena cava. The resulting *portal hypertension* (abnormally high blood pressures within the portal vein) demonstrates the near-total absence of collateral pathways through which portal venous blood might bypass the liver.

That naturally selected lack of collateral pathways around the liver reveals the importance of directing all nutrient-laden portal venous blood through those relatively obstructive liver sinusoids while larger volumes of venous blood returning from other tissues and organs flow easily through separate low-resistance channels toward the lungs—portal venous blood also would bypass liver sinusoids via such lower-resistance pathways, were they available. It turns out that portal venous collateral circulation is barely possible but only at both ends of the GI tract—in and around the lower esophageal wall and around the anus. The *esophageal varices* (chronically overstretched "varicose" veins within the esophageal wall) that sometimes develop in response to portal hypertension may eventually rupture and bleed severely into the upper stomach or lower esophagus. Any *hemorrhoids* (varicose veins around the anus) associated with cirrhosis are similarly high-pressure and liable to bleed.

The "H" Word

Hemorrhoids are very common, of course, and only rarely associated with portal hypertension and cirrhosis of the liver. Along with lower extremity varicose veins, hemorrhoids often swell and become symptomatic as a woman's blood volume increases during pregnancy and her enlarging uterus impedes lower body venous return through the adjacent iliac veins. But a great many people who are not pregnant also develop bulging symptomatic veins around the anus. It is often possible to prevent *progression* (worsening) of such hemorrhoids by keeping the buttocks firmly together during normal sitting, while also avoiding long unproductive periods or excessive straining on the toilet seat as well as overly persistent toilet paper wiping. It is better to catch up on your reading elsewhere while you sit with buttocks held together to support those veins. And rather than pull on and irritate the anal epithelium by wiping endlessly with toilet paper, consider washing the area clean with mild soap and water in the shower. An increased intake of fluids and dietary fiber (and prune juice) can often remedy mild *constipation* (the infrequent elimination of excessively dry and bulky feces) and thus reduce straining at stool as well as hemorrhoid symptoms. Of course, increasing constipation or rectal bleeding may have some other more serious cause that requires medical attention. Surgery is often effective when hemorrhoids become very troublesome. The occasional hemorrhoid that clots (a thrombosed hemorrhoid) can be quite painful to sit upon for several days. Incidentally, the clots that develop within hemorrhoids or in superficial arm veins after delivery of intravenous fluids and medications, cannot escape to become hazardous pulmonary emboli. Ordinarily, therefore, a clotted hemorrhoid is simply a pain in the ass. A very large hemorrhoid might include rectal mucosa coming out through the anus (a minor *rectal prolapse*), which raises the risk of anal incontinence.

Ostomies

Sometimes it becomes necessary to surgically bypass (defunctionalize) the rectum or even entire colon. Any diversion or decompression of gastrointestinal content (usually onto the abdominal skin surface) through a tube or via a side-opening in a bowel loop or an end-opening of a bowel channel can be referred to as an *ostomy*.

Unfortunately, the many delicate blood vessels entering your outer small bowel surface are easily damaged by drying, thereby depriving the adjacent small bowel wall of its circulation. So especially before collateral circulation develops during healing, one must keep any exposed small bowel moist or else folded back upon itself in such a way that only the more dehydration-tolerant mucosa is exposed to air.

A *gastrostomy* usually involves placing a hollow rubber or plastic tube through the stomach wall into its lumen. Stomach and *duodenostomy* drainage tend to be very watery unless the person is eating. A *jejunostomy* will put out thicker material that still passes readily through a tube. *Ileostomy* output is quite thick and more or less continuous. Dietary experimentation can reduce gas production and stool volume from an ordinary ileostomy but with little possibility of reliable control, an ileostomy (collecting) bag becomes a necessary precaution and nuisance. (However, some surgical specialists have developed techniques to permit construction of an ileostomy that can collect stool until emptied electively on a regular schedule—thereby eliminating the external bag for some patients.) A *colostomy* will release increasingly solid material, depending upon the disease or condition being treated and which part of the colon has been entered or interrupted. If much of the proximal (right and transverse) colon remains on line and in good condition, it is often possible to train that colon to empty at convenient times—usually through stimulation of colonic contraction by inflating the colon with a liquid *enema* of some sort. A trained colostomy can be quite trouble-free and merely covered by a small dressing—some owners have even declared their colostomy a welcome relief from chronic constipation. Interestingly, the lining of any defunctionalized portion of colon tends to deteriorate in the absence of ongoing exposure to stool—another example of disuse atrophy and a demonstration of how your more specialized tissues and organs can only adjust up to a point.

Soil Is Alive—It May Be Healthy Or Sick

Ordinary *soil* is more than just a random mix of clay and silt, sand and gravel, for any fertile soil also contains huge numbers of living things and all sorts of cast-off organic materials. The fine particulate organic content of soil is repeatedly passed through and broken down within the intestines of worms (unlike yourself, each of that endless variety of worms is just as long as its intestinal tract)—and that organic matter is continuously being degraded by bacteria, fungi and protozoa—with all of these life forms feeding upon each other as well. Except for deep sea rift colonies primarily energized by bacterial oxidation of the hydrogen sulphide (H_2S) emerging from sea floor hot water vents, all life relies upon and also (through wastes, death and decay) encourages plant growth. In addition to being invaded by an enormous density of live plant roots and their fungal extensions, soil is constantly replenished by rotting vegetation and decomposing animal matter. Furthermore, much of a soil's organic content arrives as herbivore manure (being fewer in number and on a rather restricted low-fiber low-residue high-herbivore diet, carnivores generally produce far less stool). Every farmer knows that soil is readily improved by manure and *compost* (composting refers to the accelerated biological digestion of waste organic residues). Similarly, letting land lie *fallow* for a year allows unharvested organic plant material to decay and fertilize the following year's crops. On the other hand, a bucket of rich soil gradually loses its organic content (hence fertility as well as bulk) through bacterial oxidation when left idle in a warm greenhouse.

It is likely that solar-powered bioregenerative modules will soon become practical for home food production—although crops of the future may bear little resemblance to their wild or domesticated agricultural antecedents. For the easily optimized conditions within such bio-ag units should permit a far more intensive, selective and efficient production of nutritive plant reducing power than our Whole Earth biosystem, which forces plants to do many other things for reproductive advantage besides converting solar power into glucose. An optimist might view the soil of such a self-regenerating food production and waste recycling biosystem as a marvelous living entity from which other life draws sustenance. A pessimist might call it shit. Both would be correct.

CHAPTER 23

METABOLISM

Life's Shared Complexity;… Life is sweetness and light;… Oxidation And Reduction—Electron Grabber Seeks Electron Donor;… Your Sweet Power Requires Free Oxygen As An Electron Acceptor;… Your Slow Burn;… Joan Of Arc Preferred Men's Clothes;… Glycolysis;… Kreb's Cycle;… Oxidative Phosphorylation—The Payoff Comes In Transferring Electrons To O_2

Life's Shared Complexity

An average bacterium can be viewed as the coherent sum of possibly 1000 different chemical reactions. Many of those reactions share common molecular motifs (for example, dehydration synthesis or hydrolysis) so an ordinary bacterium actually contains far fewer *types* of chemical reactions—and perhaps only 100 different molecules play a central role in life.[1] While some believe that these common characteristics and related molecular structures reveal God's unitary design and handiwork, others view that same complexity and those close molecular relationships as evidence of life's shared ancestry. But neither creationist nor evolutionist has ever caught any form of life in violation of the First or Second Laws of Thermodynamics. Life neither creates nor destroys energy. Every living thing eventually runs down and wears out. Only bacteria appear potentially immortal, since one or the other daughter cell could continue to subdivide indefinitely. Yet even here, each daughter is just half the bacterium its parent was, so it would be more correct to say that some tiny fragment of a bacterium's DNA might be passed along to descendents almost indefinitely—which applies to your own DNA as well. And although bacterial DNA occasionally lasts for millennia in spore form, a part of yours could go even longer if you are fortunate enough to become attractively fossilized.

Man was either created in God's actual image or he was not. But if man truly is in His image then God Himself must be subject to those same energy laws and all of the ordinary rules of physics and chemistry. For God (and therefore man-in-

[1] Stryer, Lubert. 1988. *Biochemistry.* 3rd ed. New York: Freeman. 1089 pp.

His-image) would have a utterly fantastic and far different design if He had no need for water-based cellular metabolism, respiration, circulation, digestion, excretion and reproduction—moreover, God could never have tolerated the conditions on prebiotic Earth unless free of such constraints. Nor ought man-in-His-perfect-image be ornamented in any fashion—especially not by useless nipples. Such attractive decorations belong on woman where they also serve an important mammalian function. This overlap of male and female decor seems especially inexplicable in light of biblical claims that woman was created as an afterthought from a different design for a different purpose (see Genesis for two conflicting versions of that transaction). But at least we all can agree that man truly is King of the Beasts. Right? For clearly, none can withstand his organized might. Except, of course, those pesky ever-changing and potentially immortal bacteria that are both essential components of all "higher" life forms and the ultimate victors at every death. Of course, the proposition that God constructed man in His own image out of bacteria is the ultimate theological non-starter in the unending race for a profitable pulpit. So since religion has nothing to do with science anyhow, rather than examining fundamentalism, let us examine something that really is fundamental to life such as life's ongoing need for energy. Perhaps a review of your basic energy-handling processes can cast additional light on whether man's design is truly divine or if man and woman are mere modern manifestations of mindlessly modified nucleotide sequences passed passionately forward from ancient times.

Life is Sweetness And Light

A tree is very solid and heavy, being constructed of cellulose and filled with water. Cellulose is a water-insoluble form of polymerized glucose—thus a polysaccharide. Only plants can make cellulose. With few exceptions, only bacteria and fungi can break cellulose apart. This may be another example of how Nature tends to separate anabolic and catabolic pathways. Especially within a cell, it rarely pays to build something that simultaneously is being deconstructed close by. Sometimes the pathways for building and breakdown can be separated physically. In other cases, the reagents differ sufficiently to prevent confusion. Thus your cells utilize their hard-earned hydride ion (H^-) reducing power for *anabolic* purposes only when that hydride is delivered upon *NADPH*—while *NADH* is the dedicated hydride carrier for your *catabolic* production of ATP (see below). Your cells are continuously rebuilt—on a daily basis, you replace about 400 grams (a pound) of your own cellular protein. Of course, a mere fraction of that replaced protein is structural. More consists of the many short-lived molecules that help you adjust to life's signals and instabilities. But even when your cells seem to be working smoothly, it pays to recycle no-longer-needed proteins before they can become *sufficiently oxidised* (often by highly reactive *free radicals* bearing unpaired electrons) or *glycosylated* (the likelihood of protein amino groups becoming covalently bound to glucose rises along with blood glucose levels) to bring about the early onset of cellular old age—those durable non-functionable glycosylated proteins often undergo further cross-linking with other proteins (just another example of how life's metabolic processes are forced to function at the border between order and chaos in order to compete). In any case, you cannot afford to organize or maintain a large molecular inventory so you make do with disposables and vigorous recycling.

Glucose is the water-soluble *hexose* (six carbon sugar) molecule that provides much of life's energy and structure. All simple sugars have an *-ose* ending to their names and are carbohydrates. *Carbohydrate* suggests "hydrated carbon" since early chemical analyses revealed a general formula of CH_2O (which implied a water molecule covalently bonded to a carbon atom). It is true that the atomic ratios in sugar are $(CH_2O)_N$ with the N of a hexose being six, for a pentose five, and so on—but as it turns out, each H and OH are attached to their carbon atom by separate covalent bonds rather than every carbohydrate carbon somehow being bonded to an intact water molecule. And that more-or-less repetitive structure is what allows any carbon of one simple sugar to form a covalent bond with any carbon of another simple sugar (or with protein amino groups as mentioned above)—so spontaneous or enzymatically driven dehydration syntheses can create a great many different, and very complex carbohydrate-based structures.

In essence, the apparently miraculous but actually ordinary (although complex) electrochemical process known as photosynthesis creates a lovely tree by first letting sunlight split a water molecule, then conserving that solar energy investment by a prompt separation of charges across an intracellular membrane. The inventory required for manufacture of a mole (180 grams) of glucose molecules is well known—just 48 moles of red photons (about 2300 kilocalories of light) can convert six moles of dissolved CO_2 plus six moles of water into sufficient sugar ($C_6H_{12}O_6$) to sweeten a lot of apples or provide a very solid cellulose hammer handle, while releasing six moles of O_2. If we now reverse the process and burn such a wooden hammer handle or its equivalent one-third pound of dried apple sugar in an ordinary oxygen-containing atmosphere, that newly created mole of glucose will release 686 kilocalories of trapped solar energy as heat, plus six moles of CO_2 gas and six moles of water vapor. Thus chlorophyll-based photosynthesis manages to store about 30% of intercepted red light energy within covalent chemical bonds, an impressively good yield. Even more impressive is the estimate that every 15 days our Earth intercepts as much solar energy as is trapped within all known coal reserves (thus it takes only minutes to melt your favorite chocolates all over the dashboard of your closed automobile on a sunny summer's day). More remarkable yet is the fact that essentially all of our sun's energy output heads off into space in every other direction, missing this tiny distant planet #3 entirely. But most remarkable are the 10^{22} stars of our visible Universe (many producing much more power than our sun) that undoubtedly hold planets of far greater interest under their sway. Just so you keep that 10^{22} number in proper perspective, it amounts to over a trillion stars per live human. So for Heaven's sake, what is everyone fighting about?

Oxidation And Reduction—Electron Grabber Seeks Electron Donor

Plant chloroplasts invest solar energy in order to split water—recombining of hydrogen with oxygen returns that solar energy investment. In other words, water is an extremely stable hard-to-split molecule and a considerable energy reward is always posted for the return of wayward hydrogen atoms to their eager oxygen partner. The covalent interaction that forms water is especially powerful because hydrogen is one of the strongest *reducing agents* or *electron donors* in living systems while oxygen is one of the strongest *oxidizing agents* or *electron grabbers*. Normally the biological *oxidation* of a reduced atom or molecule (oxidation of your stored reducing power, for example) is an *exothermic* (solar-energy-releasing) process involving a *loss of electrons*, or the *loss of a hydrogen atom* or the *formation* of a covalent *bond with an oxygen atom* (or some other avid electron acceptor). On the other hand, *reduction* is an *endothermic* process (requiring an external input of solar energy) that increases the chemical potential energy or reducing power of an atom or molecule with respect to oxygen. Reduction involves the *acquisition of electrons*, or the *acquisition of a hydrogen atom* (along with its electron) or the *rupture* of a covalent *bond to oxygen*. Since hydrogen, oxygen and other atoms or their electrons are neither created nor destroyed during oxidation or reduction, every oxidation of one atom or molecule must be associated with the reduction of another. Similarly, while an oxidation-reduction reaction neither creates nor destroys energy, every *spontaneous* (hence energy-releasing) oxidation-reduction reaction results in a loss of useful stored (potential) energy. So living things convert part of the solar energy trapped within covalent bonds into mechanical work (or use it to drive some anabolic project) and the remainder inevitably escapes as heat—thus all things run down.

Intracellular oxidation-reduction transactions usually involve the sale or purchase of a proton along with two electrons as a hydride ion (H^-) or else the simultaneous transfer of two hydrogen atoms (each a proton bearing its own single electron). The hydride ion is a very strong reducing agent since H^- desperately wants to be rid of both electrons. *NAD$^+$* (nicotinamide adenine dinucleotide— derived from *niacin*, a B Vitamin) and *FAD* (flavin adenine dinucleotide—derived from *riboflavin*, another B Vitamin) serve as your intracellular electron carriers. NAD^+ delivers a single hydride ion plus a spare proton (as NADH + H^+) while each FAD transports a pair of hydrogen atoms (as *FADH$_2$*). In their reduced (high energy) state, both NADH plus H^+ and FADH$_2$ provide reducing power to drive enzyme-directed anabolic reactions. Therefore NAD^+ and FAD are considered *coenzymes* (essential non-protein cofactors). Obviously NADH plus H^+ carries greater reducing power than FADH$_2$ since a hydride ion is a stronger pusher of two electrons toward oxygen than two ordinary H atoms could ever be. Indeed, the electron transfer potential of NADH toward oxygen is 1.14 volts or 53 kilocalories per mole of NADH. Biological systems tend to transfer electrons and H's in pairs. This relates not only to the formula of H_2O but also to the destructive hyperactivity of any O_2 atom that acquires only a single extra electron and thereby becomes a free radical. Such a *superoxide anion* (O_2^-) does not enter normal O_2 pathways—instead it attacks and disrupts many different molecules in its crazed attempts to pass off that spare electron or grab another (so that one of the two covalently-bonded oxygen atoms can complete its outer electron shell). Superoxide anions are occasionally produced when Fe^{++} is oxidised to Fe^{+++} plus e^-. Therefore your tissues normally contain an enzyme *(superoxide dismutase)* that can convert superoxide into H_2O_2 (hydrogen peroxide) for disposal by enzymes such as *catalase*.

Note that spontaneous *oxidation* (rusting) of an iron atom brings about an *increase* in its *positive* charge, while the reduction of iron ore (or rust) to metallic iron (which requires an input of heat energy to smelt the iron and drive off its O_2 in a reducing atmosphere) reduces Fe^{+++} to Fe. The uncontrolled oxidation of a burning building may burn (rust) through the large exposed surfaces of a steel I-beam even more swiftly than it can weaken and destroy a solid wooden beam of equal strength. In general, an atom whose charge becomes more positive (or less negative) has been *oxidised* while one whose charge becomes less positive (or more negative) has been *reduced*. Of course, the simple *ionization* of preexisting cations and anions (as when Na^+Cl^- entering solution separates into Na^+ and Cl^-) is a far less energetic process than an oxidation/reduction reaction in which one atom rips electrons off another (as when water's oxygen acquires an electron from sodium metal: Na plus $H_2O \rightarrow Na^+$ plus OH^- plus ½H_2 plus a lot of energy).

Your Sweet Power Requires Free Oxygen As An Electron Acceptor

Over the past two billion years, continuing enormous releases of toxic photosynthetic waste gas (free O_2) have irretrievably polluted Earth's atmosphere—driving flourishing anaerobic life from Earth's surface and ending the loss of Earth's hydrogen into outer space (see Chemistry). But that oxygen also opened the way for aerobic organisms to recover more energy (and therefore perform more work and further heighten competitive pressures) by completing the oxidation of reduced organic molecules such as glucose back to CO_2 and H_2O. Although reducing conditions still prevail deep underground where ongoing bacterial metabolism eliminates free oxygen, a few aerobic life forms (including blind catfish) manage to flourish along certain oxygen-bearing underground water streams that may run deeper than 600 meters or 2000 feet. Nonetheless, Earth's buried (reduced) hydrocarbons generally remain well-isolated from the free O_2 upon which your high-powered existence now depends. So while a complete combustion of all buried hydrocarbons might deplete much of Earth's atmospheric O_2, this is not a practical concern in view of the biosphere's ongoing processes of photosynthesis and the active burial of organic materials by sedimentation and continental drift—most buried hydrocarbons are so diffusely and deeply distributed that their recovery for combustion would cost far more energy than might be gained anyhow. Furthermore, the costs of solar/hydrogen power are nearing competitive levels and even the past century's massive fossil-fuel-powered industrialization has only reduced Earth's atmospheric O_2 concentration by a tenth of one percent.

Photo*synthesis* can be viewed as a multistep photon-powered *anabolic* process involving electron transfers and a separation of charges across membranes. Many photosynthetic bacteria utilize the resulting *proton gradient* directly to produce ATP, while plants and the cyanobacterial relatives of plant chloroplasts use electrons ripped away from water to produce both ATP and NADPH for the reduction of CO_2 ($6CO_2$ plus $6H_2O$ plus some NADPH and ATP becomes $C_6H_{12}O_6$ plus $6O_2$) into glucose. Glucose then serves in a variety of extracellular and intracellular roles (for example, as cellulose for support, or sap sugars to feed roots, shoots and fruits). Similarly the *oxidative phosphorylation* (glucose-for-ATP exchange) that takes place within each of your cells converts the *electron transfer potential* of glucose (via another proton gradient) into the more easily utilized *phosphate transfer potential* of high energy phosphate compounds like ATP (and also GTP, UTP, CTP and TTP). Note that adenine is one of the four nitrogen-containing bases that make up RNA. As previously discussed, it seems likely that early RNA-based life gave rise to modern DNA-and-protein-based life. Certainly the dominating presence of adenine at many critical stages of life's fundamental energy transfer processes—in NAD^+, FAD, ATP and coenzyme A—and also the primary role of RNA splicing enzymes, messenger RNA, transfer RNA and ribosomal RNA in the production of protein—all argue strongly for RNA involvement from the start. The particular task of your adenine-based *coenzyme A* is to shuttle two-carbon *acetate* groups about while grasping those acetates via a *high-energy sulphur bond*. This arrangement provides coenzyme A with a high *acetyl-group transfer potential*. And that is yet another important intracellular skill since those two-carbon acetate chunks have exactly the right size and shape for efficient "burning" to CO_2 in the *Krebs cycle* (Krebs is the chemist who deciphered this cycle) that powers each cell. The high-energy sulphur bond between coenzyme A and acetate takes on much of the energy released during each disruption of a carbon-to-carbon covalent bond (as longer carbon chains are chopped into two-carbon acetate units)—that salvaged energy is then transferred along with the acetate.

Your Slow Burn

Although life can neither create nor destroy energy, it certainly can conserve or waste it. Thus it is likely that you will become increasingly burdened by bounteous stores of body fat if you eat too much, regardless of whether your diet is primarily fatty, protein or carbohydrate. For those principal food groups (your basic building blocks) and nucleic acids all release energy when oxidised. The likelihood of becoming obese also rises in those who carry more fat cells, eat more fat in their diet, enjoy a higher caloric intake and have a sweet tooth (see Digestion). If you actually burn a gram of fat in an oxygen-containing atmosphere, you will release about 9 *kilo*calories (nine *thousand* calories) of trapped solar energy as heat. A gram of protein or carbohydrate only releases about 4 kilocalories when it goes up in flames (one kilocalorie will warm a liter of water by 1°C. Each of the *calories* that you count in your food is actually a kilocalorie with the kilo part deleted for convenience, and to reduce guilt?). Despite its lower-than-fat energy content, glucose is the most readily available source of intracellular energy since fats are stored as hydrophobic neutral fat while amino acids must first be *deaminated* (have their amino ends cut off) in the liver so those toxic *ammonia* (NH_3) molecules can be repackaged in pairs for elimination by the kidneys as soluble non-toxic *urea*

$$H_2N - \overset{\overset{\textstyle O}{\|}}{C} - NH_2$$

—only then can the non-amino remainder of each amino acid be chopped into two-carbon chunks for extraction of its stored chemical bond energy via the Krebs cycle.

Some years ago, the Inuit mayor of Barrow, an Arctic Sea village in northern-most Alaska, was consulted by U.S. researchers in Antarctica on how best to stay warm in very cold weather. In addition to fur clothing, his main recommendation was dietary as he pointed out that researchers would feel a lot warmer if they ate more meat and fewer vegetables. Consumption of animal muscle and fat certainly can warm a person—in contrast, subarctic human vegetarians tend to be uncommon and uncomfortable for unlike horses and hares, human vegetarians are not much warmed by their brief and inefficient internal fermentation of ingested plant matter. A purely vegetarian diet may also reduce testosterone production in males—possibly that finding is related to the fact that many plants (including wheat) produce enough estrogens to suppress reproduction in any herbivore that relies overly much upon them—just another example of how a varied and moderate diet can be healthier than a lot of any one thing (a varied diet also limits your intake of particular plant-produced or artificially disseminated pesticides, toxic heavy metals or *estrogen-like pollutants* such as *PCB*'s, *DDT* and some *petroleum by-products*). In any case, if you devour a nice thick juicy steak for supper this evening rather than picking at your customary veggies and pasta, you probably will sleep warmer tonight and so require fewer blankets (or open your bedroom windows more widely). Of course, confirmed vegetarians will also upset their intestinal microorganisms. Apparently the digestion and metabolism of proteins releases about 30% more heat than the breakdown and utilization of dietary carbohydrates. Presumably some of that excess heat production occurs in the gut as peptide bonds are hydrolysed and some in the liver where deamination takes place. But regardless of site or mechanism, your roundabout access to protein-stored reducing power is less efficient (every metabolic step loses some energy as heat) than the more direct conversion of intracellular carbohydrate to *ATP*.

Although some amino acids bear non-polar (hydrophobic) side chains, proteins and carbohydrates are polar molecules that dissolve readily in water. In contrast, fatty acids are primarily non-polar hydrophobic hydrocarbons except for their polar carboxyl tips. So while each hydrogen-bearing carbon of your fat is maximally reduced (packed as full of solar energy as possible), fat is less easily dissolved and brought into Kreb's cycle than carbohydrate or protein. Thus neutral fat must first be brought out of storage and broken into fatty acids before each linear hydrocarbon molecule can be chopped into soluble two-carbon units (a process known as *beta oxidation*) for transfer by coenzyme A. By the time your fatty acids are ready for insertion into Kreb's cycle, some of their caloric content has been lost. Furthermore, you metabolize fat at a leisurely pace, perhaps in part because the overall rate of fat digestion is limited by its relative water insolubility. Most of your fatty acids bear an even number of carbon atoms. In theory, the progressive beta oxidation of a fatty acid with an odd number of carbons should leave a single-carbon residue that might create all sorts of chemical havoc—as highly reactive formaldehyde, for example. So the beta oxidation of any fatty acid with an odd number of carbons stops at the three carbon stage. The three-carbon *propionyl* coenzyme A that results can be utilized elsewhere without risk. Many multistep intracellular biochemical reactions produce similarly useful molecules at various stages that can be diverted for other purposes. For example, one intermediate three-carbon molecule (not the usual 1, 3, bisphosphoglycerate of cytoplasmic glycolysis but 2,3, BPG) that accumulates in your red cells diminishes the oxygen-grabbing qualities of partially deoxygenated hemoglobin, thereby enhancing O_2 release into the tissues (see Blood).

Incidentally, only the small glycerol end-piece of a neutral fat molecule can be converted into glucose by animals. On the other hand, amino acids are regularly broken down and rebuilt into glucose through an intracellular process known as gluco*neo*genesis (*new* glucose creation). Gluconeogenesis is stimulated by growth hormone as well as epinephrine, thyroxine, glucagon and cortisol. Hungry humans and ravenous ravens can therefore produce essential intracellular ATP from their own stored proteins, carbohydrates or fats. But since glucose is the most readily utilized source of cell energy, bad men and loving parents (those who are not dentists) will generally treat little children to candy rewards rather than pork rinds. That preference for glucose could relate to the outcome of a winter helicopter crash in Alaska several years ago. The pilot sustained back injuries and was unable to move while the passenger walked about briskly, flailing his arms to keep warm. By so doing, the passenger may have used up his glycogen stores more rapidly. Although both men were inadequately clothed, only the passenger was dead of hypothermia when rescuers arrived the following morning. At some point during that frigid night, the passenger's obligatory switch to "burning" his fat reserves for continued shivering must have been associated with a fatal slowing of his overall metabolic rate.

For similar reasons, obesity is difficult to reverse. Indeed, your adult body weight generally remains relatively stable despite wide variations in food intake. Thus overeating stimulates heat loss (hence inefficient utilization of that excess reducing power) through increased norepinephrine secretion (which also encourages heat release from any brown fat stores—see below) while any attempt to lose weight through dieting is soon viewed with alarm by your regulatory systems, which respond as if to the onset of starvation regardless of how obese you may currently be. Therefore your initial efforts to lose weight actually reduce the mobilization of your fat stores (by beta oxidation into two carbon acetates for your Kreb's cycle) and decrease your production of the soluble four-carbon acetate precursor molecules such as acetoacetate (the energy sources preferred by tissues such as your heart and kidney). This decline in fat breakdown represents a direct response to reduced thyroid secretion as well as the resulting high serum levels of acetoacetate for your body counters your attempts to diet by slowing its metabolic rate. Furthermore, the well-insulated very obese overheat easily with exercise and require more time for rewarming after even mild hypothermia. They also have altered insulin and adrenal cortical hormone responses. These sorts of mechanical difficulties and physiological responses make it tough for fat folk to lose weight despite heroic limitation of their caloric intake. In any case, repeated cycles of fasting followed by overeating (especially of fatty foods) probably cause more health problems than remaining pleasingly plump. So while there are a great many more interactions involved in dieting than we have yet discussed or even understand, it clearly is not as simple as those fine folk who so readily deliver the free advice, "I don't mean to be critical but… if you would only eat less, you could easily lose weight". For in dieting you must evade or override important ancestral adjustments that once enhanced survival through seasons of plenty and long periods of starvation. The occurrence of diabetes in 5% of the U.S. population may in part reflect metabolic adjustments that once improved survival during recurrent famines. As the Pima Indians have found, maximal metabolic efficiency can be dangerous on a modern affluent diet. Interestingly, mice lacking in brown fat also have increased metabolic efficiency and soon become obese—and once obese they also become insulin-resistant and then *hyperphagic* (they eat too much). Apparently brown fat plays an important part in helping rodents (and perhaps humans) maintain a stable weight despite a variable caloric intake.

Joan of Arc Preferred Men's Clothes

An average 70 kilogram male who is ⅔ water will carry total body energy stores approaching 100,000 kilocalories in fat (11 kilograms), 25,000 kilocalories in protein (mostly in muscle), 600 kilocalories in gl*ycogen* (also known as *animal starch* and found mostly in liver and muscle tissue) and 40 kilocalories as glucose. Even though that represents an impressive concentration of reducing power, there have only been rare and unsubstantiated reports of humans undergoing spontaneous combustion (see brown fat below). Indeed, as in the well known case of Joan of Arc who was burned at the stake, it generally takes quite a bit of wood or other combustible reducing power to bring a person up to ignition temperature. The burning of wet wood poses a comparable problem, since the high-thermal-capacity water must first be driven off before the activation temperature necessary for ongoing *fuel combustion* (ongoing oxidation of vaporized fuel gases) can be reached. Thus a 70 kg. man would require an input of about 3000 kilocalories to warm his 45kg. of water to boiling from 37°C, and perhaps another 27,000 kilocalories would be necessary to vaporize that water—which still leaves a net gain of almost 100,000 kilocalories upon total combustion of that average man.

These inexact calculations may not fit Joan of Arc, of course, as her particular dimensions remain unknown. But she probably was a fairly husky farm girl for she swung a notably effective sword. Since her executioner alleged that her heart simply did not burn and was found intact among her ashes, we must subtract the reducing power represented by that noble heart from our estimate of the net heat release consequent to Joan's combustion. But it all turned out well in the end. Even though the Church had Joan of Arc burned alive in the 15th century, they compensated for that error by declaring her a Saint in the twentieth century—presumably because of all she suffered (at their hands). In some ways, Joan of Arc simply was ahead of her time since the last straw leading to involuntary combustion of France's greatest heroine was her reckless refusal to wear women's clothes—she insisted that men's clothes were far more comfortable (which may still be true of French fashions). However, that is religious history. Back now to how you extract energy in useful form from your stolen reducing power. Hint: Only enzymes can bring about useful reaction rates at more comfortable body temperatures. Indeed, without facilitation by those enzymes, the activation energy for your oxidative processes would lead to combustion as in the case of Saint Joan above.

Glycolysis

As blood glucose levels rise after a meal they cause insulin to be released from your pancreas which in turn encourages glucose uptake and anabolic activities by your cells (e.g. protein synthesis and fat storage). Upon entry into the cell, each glucose is promptly enveloped by a *hexokinase* molecule to expedite phosphorylation by ATP. In its ordinary unfolded (glucose-free) condition, hexokinase does not act as an ATP'ase (it is unable to hydrolyse and thus waste ATP). Indeed, hexokinase must change its configuration by slamming shut upon each glucose before it can attach an ATP to a non-polar corner of the glucose molecule. Once phosphorylated, that glucose-6-phosphate can no longer leave the cell. Only liver cells, which store much of your total body glycogen, contain a *phosphatase* to relieve glucose of its phosphate burden so the newly naked glucose can slip back out into the passing bloodstream. Thus you rely upon liver cells to raise your blood sugar by exporting glucose whenever pancreatic glucagon signals the onset of low blood sugar levels. Glucagon also stimulates *glycogenolysis* (glycogen breakdown), inhibits *glycogenesis* (glycogen production), increases *gluconeogenesis* (glucose production from amino acids) and blocks unnecessary *glycolysis* (glucose breakdown) while simultaneously inhibiting fatty acid synthesis and encouraging the mobilization of fat. The epinephrine and norepinephrine also secreted in response to significant *hypoglycemia* (low blood sugar) cause muscle glycogenolysis, inhibit muscle uptake of glucose and leave you shaky (as after a fight-or-flight reaction). In addition, they enhance glucagon secretion and mobilize fatty acids from fat while inhibiting insulin secretion. Consequently your blood sugar rises as glucose is released from liver cells while liver and muscles switch to utilizing *ketone bodies* like acetoacetate for fuel (in order to use less glucose).

The ten sequential enzymatically directed steps of *glycolysis* (glucose breakdown) occur in the cytoplasm of every cell, with each step controlled by a specific enzyme. Glycolysis begins with one six-carbon glucose molecule plus two ATPs and ends with a couple of three-carbon pyruvic acid molecules—each bearing two high-energy phosphate groups able to regenerate ATP. Thus glycolysis provides a net gain of two ATPs per glucose, plus two hydride ions that convert two NAD^+ to two NADH (with two H^+ ions tagging along for later use). But when excessive muscle exertion results in a temporary local shortage of *oxygen* (your final electron-and-hydrogen-ion grabber), the only way glycolysis can continue—in order to maintain cellular metabolism at *viable* (life-sustaining) levels—is by off-loading the newly formed hydrides-plus-hydrogen-ions onto those newly formed three-carbon *pyruvic acid* molecules to create *lactic acid* molecules.

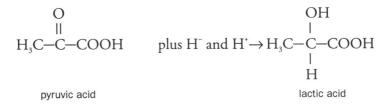

pyruvic acid   lactic acid

Lactic acid therefore accumulates in your muscles and adjacent blood capillaries whenever strenuous exertions momentarily deplete local oxygen supplies. And until that oxygen deficit is repaid, you will notice muscle fatigue and air hunger (panting). By reducing blood pH locally, lactic *acid* enhances oxygen release from passing red cells—your liver cells eventually utilize that lactic acid for energy or regenerate it into glucose. While this sequence of events may seem a bit roundabout and complex, it does maintain a flow of energy at comfortable body temperatures.

The release of three-carbon acids seems to have been how the energy-from-glucose game was played by your bacterial forebears (whose descendents—hence your distant cousins—include the bacteria that acidify your yoghurt) prior to the build-up of atmospheric oxygen. In the case of yeast (a fungus), the primary waste molecule that accumulates under no-free-oxygen conditions happens to be a two-carbon alcohol (H_3C—CH_2OH). It is hardly surprising that those who enjoy drinking yeast metabolic wastes (ethanol provides over 5% of the calories in the average American's diet, some alcoholics derive over half their caloric intake from ethanol oxidation) are rendered a bit dysfunctional. After all, unlike the comparable situation of rats eating feces (see Blood), alcohol provides no essential nutrients. Although ethanol intake enhances your energy expenditure—in part due to heat losses associated with *peripheral vasodilation* (flushing), it also reduces your oxidation of fats—so when alcohol ingestion raises your caloric intake above your energy needs, you noticeably accumulate fat (that beer belly).

Glycolysis provides only a modest increase in ATP energy stores but that sort of low return on the initial glucose investment kept your early ancestors in the competition until aerobic conditions developed. At that time, over two billion years ago, bacteria were the dominant life form, although eucaryotes (cells with nuclei) were developing from cooperative combinations of bacteria. Not surprisingly, some amongst that vast anaerobic bacterial horde accidentally uncovered the far greater reducing power available to those capable of an aerobic existence. The first aerobic (pre-mitochondrial) bacteria then prospered on the pyruvic and lactic acid wastes being released by early eucaryote hunter-gatherers—as did those early eucaryotes that engulfed (but could not successfully devour) some tougher-than-usual ancestral mitochondrion. Ever since then, mitochondria have had a monopoly on the complete internal combustion (to CO_2 and H_2O) of three-carbon waste products from glycolysis—releasing some high-energy phosphates in exchange. There are practical limits to all partnerships, of course, and extra mitochondria are routinely digested and recycled through nearby lysosomes to this day. So credit your early eucaryote ancestors for making that huge transition from a livelihood based upon bacteria-baiting to running the world's first organized farming enterprise that milked the most highly productive mitochondria of ATP while consuming those not needed. Successful eucaryotic farmers gained further advantage as rising atmospheric O_2 levels and enhanced mitochondrial aerobic skills delivered a more consistent ATP profit.

Dozens to hundreds of mitochondria now slave within each of your cells, converting products of cytoplasmic glycolysis such as pyruvate and NADH into energy-rich ATP molecules. Those mitochondria also play a part in the production of other important molecules (amino acids, haem, some nucleic acids and so on) needed by the cell. In turn, mitochondria import many of their essential peptides from the cell cytoplasm. And while fatty acid oxidation occurs within the mitochondrial *matrix* (inner compartment), fatty acid synthesis takes place in the cytosol of the cell—yet another effective separation of anabolic and catabolic pathways. Because your cells and mitochondria have so irretrievably intermingled their assets, both must suffer the consequences if either is afflicted by harmful genetic information. Since mitochondria are passed out to daughter cells on an "as is, where is" basis, any daughter cells that happen to receive mostly defective mitochondria will then replicate and pass along that disadvantageous burden—and the earlier such a problem arises during embryonic development, the more severe the eventual consequences.

Kreb's Cycle

Oxidative *decarboxylation* describes the process of shortening a three-carbon pyruvate to a two-carbon acetate while generating one energy-rich NADH and releasing one CO_2 molecule. The acetate is then delivered by coenzyme A onto a four-carbon oxaloacetate to create *citrate* (a six-carbon acid displaying three carboxyl groups). Thus the high-energy sulphur bond holding acetyl to coenzyme A transfers energy from one C-C bond (broken as pyruvate released CO_2) to another on the citrate molecule. Both carbons of each newly added acetate are oxidised sequentially to CO_2 in a series of enzyme-directed steps that regenerate oxaloacetate, leaving it ready to relieve yet another acetyl coenzyme A of its acetate burden. The nine steps from citrate to oxaloacetate thus complete Kreb's cycle, converting one acetate into two CO_2 molecules plus three NADH plus one *GTP* (equivalent in energy to an ATP) and one $FADH_2$. This carefully controlled multi-step combustion has been restricted to the mitochondrial matrix ever since it first occurred within some simple mitochondrial ancestor over two billion years ago.

Although molecular O_2 is not a direct participant in *Kreb's* (also known as the *citric acid* or *tricarboxylic acid*) *cycle*, this cycle cannot continue under anaerobic conditions since energetic electrons acquired by NADH and $FADH_2$ can only be off-loaded safely through a sequence of electron transfer molecules known as the *electron transport chain* or *cytochrome system* that in turn utilizes oxygen as the final electron and hydrogen ion receptor (thereby producing H_2O). By delivering its hydride ion to the electron transport chain, each *NADH* eventually gives rise to *3 ATP's*. Every *$FADH_2$* carries enough energy to produce *2 ATP's*. So this mostly mitochondria-based process of *oxidative phosphorylation* eventually produces 36-38 moles of ATP (equal to about 260 kilocalories of energy stored as high-energy phosphate bonds) from the 686 kilocalories of solar energy stored within each mole of glucose. Thus the conversion of each glucose molecule's reducing power (through glycolysis, oxidative decarboxylation, Kreb's cycle and the electron transfer system) into the phosphate transfer potential of three dozen ATP's is almost 40% energy efficient (of course, we started with nine times that much energy in the form of red photons, and there will be further unavoidable losses as this ATP is utilized—but there are plenty of red light photons out there so don't worry about how many you use, just worry about how you use them).

The *energy charge* of any cell is readily determined by its current ratio of ATP to *AMP* (the latter has no phosphate transfer potential). You utilize and regenerate about 40 kilograms of ATP per day at rest, or up to ½ kilogram of ATP per minute during maximal exercise. The average ATP molecule lasts only about a minute, for many enzymes are ATP'ases, capable of hydrolysing ATP to perform mechanical work or to drive *coupled* (otherwise energetically unfavorable) anabolic reactions. By maintaining low ATP levels, a cell reduces the risk of runaway oxidation and overheating—spare high-energy phosphates stored on phosphocreatine allow your muscles to recharge a few ATP molecules while Kreb's cycle and the electron transfer chain try to catch up with metabolic demands resulting from vigorous muscle activity. Reversible reactions such as enzyme phosphorylations (by specific *kinases*) and dephosphorylations (by specific *phosphatases*) are among the many ways that enzyme activities are regulated within the cell. Other controls include the rate at which enzymes are synthesized and degraded, inhibition of enzymes by the accumulating products of their reactions, and the limited availability of essential cofactors such as NaD^+, FAD, Fe^{++}, Zn^{++}, Mg^{++} and Ca^{++}. Like enzymes, none of these cofactors are used up in the reactions that they help to catalyse. Pb^{++} acts a lot like Ca^{++} but does not let go as well, so lead poisons calcium-activated cell processes.

Oxidative Phosphorylation—The Payoff
Comes In Transferring Electrons To O₂

Both NADH and FADH₂ bear a pair of electrons with a high electron transfer potential (a lot of reducing power). Unlike the well-organized Kreb's cycle which occurs within the mitochondrial matrix, the final sequential steps of oxidative phosphorylation are carried out in respiratory assemblies fastened to the deeply-indented inner mitochondrial membrane. The electrons flowing from higher-electron-potential to lower-electron-potential carriers within these assemblies push their proton companions out through the inner mitochondrial membrane, creating an outside-to-in H⁺ gradient across the inner membrane. Since the cations creating this *electrochemical* gradient across the inner mitochondrial membrane are protons, the outer surface of that membrane also has a lower pH than the mitochondrial matrix. That *proton-motive* force then generates ATP as protons force their way back in through *ATP synthase* molecules (which provide a proton pathway across the inner mitochondrial membrane), thereby releasing newly formed ATP molecules. Spare mitochondrial ATP is exported to the cell cytoplasm. The electron flow stops whenever all available ADP has been converted to ATP. Each ATP bears four negative charges (4-) and is exported in exchange for ADP (3-) in an electrically driven process that utilizes *ADP-ATP translocase*, so every ATP molecule exported reduces the inside-negative electrical gradient (voltage) across the inner mitochondrial membrane as well as the overall efficiency of ATP production.

Arsenic can substitute for phosphorus on ATP but high-energy arsenic is unstable and breaks down rapidly. Thus arsenic is a poison that uncouples oxidation from phosphorylation. Brown fat, which is plentiful in hibernating animals, human infants (especially between the scapulae) and cold-adapted mammals, is brownish in color because of its high mitochondrial content. Brown fat cells are capable of discharging their proton gradients directly or even hydrolyzing their ATP in order to produce heat. Brown fat mitochondria uncouple oxidation from phosphorylation by short-circuiting mitochondrial protons through a *thermogenin* molecule into the mitochondrial matrix (rather than forcing those protons to work their way back through an ATP synthase molecule). Such a short-circuit produces heat in the same way as mixing acid with alkali (don't try it, for it can splatter or explode) or accidentally short-circuiting your car battery with a metal wrench (this rapidly releases enough heat to melt steel, as in arc welding).

It is not too difficult to dream up some metabolic error that might allow overcharged brown fat cells (not normally present in significant amounts in the human adult) to ignite spontaneously, or to imagine how arsenic poisoning could have a similarly disastrous effect. An occasional unconfirmed report of a spontaneously combusting human has referred to heavy ethanol consumption. And a great many house fires that lead to combustion of an inebriated individual have been blamed upon untended cigarettes. But the negligible water content of fat cells might allow an unusually obese, arsenic-poisoned and/or alcohol-laden individual little opportunity for keeping the rest of his/her (perhaps abnormally brown) fat out of that fire, once it ignited from whatever cause. Skunk cabbages similarly short-circuit their proton gradients to heat their florets, vaporize more scent and thereby attract insects for pollination. One wonders if an excessively amorous skunk cabbage has ever overheated and started a forest fire? While that clearly could cause reproductive failure of the skunk cabbage in question, perhaps its nearby and slightly cooler siblings would gain offsetting reproductive advantage from their own sublethal hot scent? Just one more reason for skunk cabbages to prefer wet surroundings (and for us to stop speculating).

Proton gradients also generate ATP in bacteria and chloroplasts. Furthermore, these gradients power the active transport of amino acids and sugars into bacteria and run the motors that rotate some bacterial flagellae (at an estimated cost of two hundred protons per rotation). Proton gradients also help to move electrons from NADH to NADPH for anabolic purposes. In essence, all that life requires for the storage of electrochemical energy by proton gradients is a thin lipid membrane between two aqueous proton conductors (a biological capacitor, if you will). Important elements of the electron transfer package such as the cytochromes have not changed their winning ways for more than a billion years. That allows your very human *cytochrome oxidase* to coordinate perfectly with wheat germ *cytochrome C*— which again suggests that you and your distant wheat relatives both derive from the same long-ago-consummated mutually beneficial ancestral arrangement between bacteria or possibly between bacteria and early cells. And that man created God in his own image. But that maculate conception is no cause for concern, as we shall see.

CHAPTER 24

YOUR KIDNEYS

*The Value In Urine;... Countercurrent Extraction;... Cells Like It
Boring;... How Blood Filtrate Becomes Urine;... How You Urinate;...
Filtering Your Blood;... Salvaging That Filtrate;... ADH And All That;...
Renal (Kidney) Failure.*

The Value in Urine

Bird urine may undergo further extraction of salt and water within either the
distal intestine or the cloaca but birds usually release feces and urine upon arrival at
the common cloacal outlet. So the next time a bird craps on your hat, do not take
it personally or simply brush it off. Rather note how the dry central turd nestles
within that pure white halo of urate crystals. For birds and terrestial reptiles conserve water by eliminating their nitrogenous wastes as a moist paste of insoluble
urates (rather than in the form of urea as you do—although even you must dispose
of guanine and adenine as uric acid). Sea bird excretions sometimes accumulate to
many hundreds of feet in depth on arid offshore islands. That *guano* has great
commercial value as fertilizer (high in nitrogen as well as phosphate and potassium) or for making explosives. Indeed, human military forces have fought pitched
battles over ownership of such piles of bird shit. On the other hand, bird shit in
small amounts has little value. Indeed, after their exit from that same cloacal opening, bird eggs sell far better if rinsed before being packed into cartons. Now brush
off your hat.

Urea

When you boil a fresh beef or pork kidney, your home takes on an aroma
remini*scent* of the elephant house at the zoo. That same odor of urea breakdown
into ammonia (NH_3) is very obvious around dirty urinals. For elephants and cows
and careless little boys tend to eliminate most of their nitrogenous wastes as urea,

$$H_2N-\overset{\displaystyle O}{\overset{\|}{C}}-NH_2$$

471

which allows urea-splitting bacteria to enjoy a splendid life style around those barn-yards and urinals. Indeed, properly aged (hence microbe-infested) urine was formerly valued as an effective cleanser, just as household ammonia is today—that handy vat full of smelly urine quickly brought the gloss back to tired hair or fine manufac-tured woolens (e.g. tweeds). But despite its nutrient value, your urine usually remains bacteria-free until *voided* (released from your body). Although you seemingly pro-duce urea without effort, it requires a great deal of fossil fuel energy to produce urea fertilizer, since atmospheric nitrogen ($N \equiv N$) is very stable and thus quite resis-tant to being reduced (combined with hydrogen atoms) into ammonia (NH_3) and then urea. You are able to produce urea so efficiently because, rather than extract-ing energy from NH_3 (which is rather toxic), you generously eliminate it as biologically benign urea, allowing soil bacteria to flourish by its oxidation. In con-trast, camels and kangaroos on poor quality feed divert their own urea into the *rumen*—a large chamber preceding their stomach in which they ferment plant material. That added nitrogen encourages microbial fermenters and the protozoa that feed upon them to flourish and enrich the feed mix with their own high qual-ity protein before two thirds of them are swallowed and digested each day.

Your kidneys are the only tissues in your body with a high urea concentration. However, the ancient and honorable orders of cartilaginous sharks, rays and skates maintain a 2% blood urea level that helps their 1% salt blood stay in osmotic balance with the surrounding 3% salt seas. So when next you prepare home-cooked shark and fries, fake scallops punched out of skate flesh, or even steak and kidney pie, you will attract more favorable attention by first soaking out that urea. On the other hand, fresh water fish generally excrete NH_3 as it is produced during amino acid breakdown.

Countercurrent Extraction

Duck feet must be cold—otherwise those spindly shanks would lose too much body heat. Yet ducks have to provide warm arterial blood to their feet in order to keep them alive and kicking—and the venous blood returning from those uninsulated feet is often frigid. So how do ducks keep cool feet and warm bodies? The solution is not just ducky. Indeed, similar mechanisms cool your testicles (or those of your friends) and also your brain, at least to some degree. For duck leg arteries and veins travel so closely together to and from duck feet that warmer outward-bound arterial blood is continually exposed to cooler incoming venous blood. The resulting *countercurrent* transfer of heat (gradual cooling of arterial blood, progressive rewarming of venous blood) depends only upon warm arterial blood and cold venous blood flowing close by each other in opposite directions. More generally, the countercurrent flow of adjacent fluids bearing different gas (as in fish gills and swim-bladders—see Respiration), solute or water concentrations (your kidney tubules and vasa recta—see below) maximizes diffusion gradients. You depend upon such countercurrent exchanges of water and solute between tiny urine-containing tubules and also between extended blood capillaries to concentrate urea and salt within your *renal* medulla (the inner portion of your *kidney*). To make that countercurrent mechanism more effective, the million long narrow urine tubules of each kidney regularly reverse their course while sequential tubule segments display standardised but markedly different permeabilities to water, urea and various electrolytes—such arrangements greatly reduce the energy cost of concentrating or diluting your urine.

Cells Like It Boring

Your highly specialized cells function most effectively within an undisturbed extracellular fluid environment that has some resemblance to the ancient seas in which your unicellular ancestors prospered—to stabilize the volume and solute content of those internal seas, your fluid and solute intake and output must match. However, there are some fluid losses over which you have little control. Thus physical exertions (which increase your rate and depth of breathing), atmospheric temperature and humidity all affect the amount of water evaporated to humidify the air that you breathe. Activity and air temperature also help to determine your output of sweat. Furthermore, you may lose great quantities of fluid through vomiting or diarrhea if those huge GI tract surfaces that so faithfully absorb your fluids and solutes happen to malfunction because of infection, bacterial toxins or other cause. While they cannot fully compensate for all such disturbances, your kidneys generally manage to keep you moist enough in dry times, yet prevent excessive dilution during a college beer party or other near drowning. But how can those two small retroperitoneal collections of tubules and vessels possibly regulate all of your fluids and solutes? Well, we know that your kidneys need not refurbish your blood after each pass through your tissue capillaries since that would oblige you to have a *six*-chambered heart (see Circulation). However, your kidneys do drain off about a fifth of your resting cardiac output for treatment so your blood still averages one passage through kidney capillaries following every four or five passages through tissue and lung capillaries. And it is important that your urinary system is located entirely behind (as well as beneath) the peritoneal cavity, with one kidney lying on each side of your centrally located abdominal aorta and inferior vena cava—for this arrangement ensures the clean separation of your copious kidney circulation from the nutrient-rich flow of portal venous blood to the liver.

How Blood Filtrate Becomes Urine

One-tenth of the arterial blood entering each kidney is forced across glomerular filters into its one million urinary tubules. Ninety-nine percent of that filtrate soon reenters kidney capillaries to depart in your refreshed renal venous blood. The remaining one percent has all sorts of unwanted solutes added and useful solutes extracted before being discarded as urine. Thus its one million *nephrons* (filtration and processing units) provide each kidney with a relatively huge blood/urine interface that gradually converts blood filtrate into urine. By endlessly readjusting the fluid and solute content of that urine, you are able to maintain your blood pressure, blood volume and blood solute concentrations at comfortably boring levels. Certain marine fish achieve the same internal stabilization through *active secretion* (selectively tossing out unwanted molecules) but *blood filtration* allows you to explore many different fluid environments and strange foods while automatically excreting any new (and possibly dangerous) molecules that weigh less than about 70,000 daltons—unless you specifically try to recapture them. That gain in safety and convenience more than compensates for the extra effort required to recover useful solutes and water following their filtration.

How You Urinate

The operation of every nephron begins with fluid filtering out through the walls of leaky *glomerular* capillaries into a surrounding *Bowman's* capsule. Blood enters the high-pressure capillary tuft of each *glomerulus* through an *afferent* (incoming) *arteriole* and leaves by a smaller *efferent* (outgoing) *arteriole* (*not* a vein). Subsequent to this arteriole-capillary-arteriole arrangement, each efferent arteriole delivers its blood to a second set of lower pressure capillaries known as *vasa recta*. And it is through the walls of those vasa recta that filtered fluid and solutes reenter the blood stream. The vasa recta then channel their cleansed blood to small venules that lead into larger *renal* (kidney) *veins*. Note that the two sequential capillary beds of this high-pressure *arterial portal filtration and recovery system* are quite unlike the low-pressure double-capillary-bed *portal vein pick-up and delivery systems* bringing nutrients from gut capillaries to liver sinusoids or distributing hormonal instructions from hypothalamic capillaries to anterior pituitary capillaries.

Your high-pressure renal glomerular capillaries are lined by endothelial cells containing large pores. Those pores expose the fibrous meshwork of the underlying basement membrane that screens fluid leaking across into Bowman's capsule. The resulting blood filtrate then enters a long tortuous tubule which typically begins as a meandering *proximal convoluted tubule,* then becomes a *descending tubule* that dives deep into the *renal medulla* from the (outer) *renal cortex* (all glomeruli are located in the cortex)—there follows a thin-walled *loop of Henle* segment that turns into an *ascending tubule* rising back to the renal cortex—next is another meandering *distal convoluted tubule* that joins nearby tubules to empty markedly altered filtrate into a *collecting duct*—which then directs the remaining filtrate back down through the renal medulla for final modification before it reaches your hollow renal *pelvis* as *urine.*

The remainder of your urinary system simply provides for the sterile storage and intermittent elimination of that urine. It includes a tubular *ureter* (about the diameter of a pencil) that emerges from each renal *hilus* (the dented-in medial side of each kidney where renal blood vessels also enter). Newly formed urine moves intermittently through each ureter by peristalsis—passing quickly down behind your posterior peritoneum and across your pelvic floor to enter the less distensible base of your remarkably expansile *urinary bladder* close to where your bladder's central sphincter-controlled outlet opens into the *urethra.* The bladder wall smooth muscle contractions that force urine out through your urethra during voiding simultaneously squeeze those off-center ureteral orifices, thereby helping to maintain one-way outward flow as you urinate—this reduces the risk of *pyelonephritis* (kidney pelvis infection) from ureteral reflux of bacteria-laden bladder urine during any episode of *cystitis* (bladder infection).

We have seen how your respiratory tract, small intestine and liver sinusoid surfaces undergo repeated amplifications (those alveoli and villi/microvilli) that enhance the exposed area available for gas and nutrient exchanges. In contrast, your kidney pelvis, ureters and bladder serve best when they minimize the surface exposed to outgoing urine. It is particularly undesirable to have non-emptying folds or pockets in the bladder wall where infection-prone urine could stagnate. So its stretchy *transitional cell* epithelial lining thins markedly when your urine bladder becomes distended, then comes smoothly back together as the bladder empties. Of course, urinary tract infections are far more common in females since the short female urethra empties at a chronically moist and bacteria-laden surface (infected urine often has that same stale elephant house smell). The elimination of urine via ureters, bladder and urethra might seem cumbersome and indirect, but those hollow and intermittently empty structures keep you free of (predator attracting) odor while that rather forceful one-way urine flow also washes away urine-loving bacteria that might otherwise cause ascending pyelonephritis. The risk of ascending bacterial infection is further reduced by emptying that urine away from the body through a spout—rather than letting it flow directly from the kidney onto the continuously cold, wet, smelly and bacteria-laden skin of the flank, for example.

Filtering Your Blood

To prevent blowout of thin endothelial cell walls and their supportive basement membranes, each high-pressure capillary of the glomerular tuft is reinforced by wrappings from multiple many-armed *podocytes* (support cells) that leave only slit-like openings between their wraps. Basement membranes exposed at each endothelial cell pore bear a negative charge that tends to repulse blood proteins (which are mostly anions). Presumably your endothelial cells keep those basement membrane filters free of blood-borne debris by regularly cleaning (with phagocytosis) and also recycling those filters. Perhaps the usual ten to fifteen day delay before severely damaged nephrons recover their filtering and other functions (sometimes glomeruli and tubules do not recover) has some relationship to your usual schedule for filter maintenance and replacement. Glomerular filters are easily plugged by intraarterial debris released from badly damaged tissues or a mismatched blood transfusion, and exposed basement membranes are subject to attack by misdirected antibodies during certain autoimmune reactions. Many other sorts of filter damage also allow some of the small negatively charged serum albumin molecules to leak out into the urine—a condition known as *albuminuria*. Urinary tract infection generally is associated with *bacteruria* and *pyuria* (bacteria and white blood cells in urine samples) while red cells found during urinalysis may result from injury, tumor or other (possibly unimportant) cause. Of course, urine samples from healthy mature females may include red blood cells, white blood cells, bacteria and protein debris from the vagina (especially during menstruation) unless the urine specimen was obtained as a mid-stream sample (during the course of voiding) or through *bladder catheterization* (by insertion of a sterile flexible hollow tube up the urethra).

Any general increase in arterial blood pressure or dilation of its afferent arteriole boosts blood flow through a glomerulus—narrowing of the efferent arteriole also increases filtration into Bowman's capsule while simultaneously reducing pressures within the long vasa recta (second set of capillaries). Their low capillary blood pressures and high protein content (after donating that protein-free filtrate) help your vasa recta to recapture kidney tissue fluid by osmosis. However, any chronic high-glomerular-blood-pressure high-blood-flow condition (as in poorly controlled diabetes mellitus or following damage or removal of a majority of the nephrons) can progressively harm remaining nephrons (although blockage of *angiotensin converting enzymes* may reduce such damage—see below). But you are born with nephrons to spare and will only develop kidney failure if less than 10% of your nephrons retain normal function. That is why the loss or donation of one kidney is usually viewed as having no bearing upon life expectancy, provided the donor's remaining kidney is normal.

Salvaging That Filtrate

Warning: This section describes a continuous process in a step-by-step fashion. It is more easily understood if read through for an overview before concentrating on details.

The fluid filtered at the glomerulus is mostly salt water (almost 0.9% Na^+Cl^-). Over eighty percent of that fluid reenters your bloodstream while the filtrate is still passing through the proximal tubule and before it reaches the thin-walled loop of Henle deep within your renal medulla. This rapid recapture of fluid is accomplished by hard-working cuboidal cells topped with microvilli that allow easy Na^+ entry from the filtrate side, then dump that Na^+ out of their opposite side into the extracellular space in order to maintain their usual low intracellular Na^+ concentrations. That extracellular Na^+ (plus the HCO_3^- and Cl^- dragged along electrically and the water following those solutes osmotically) then diffuses into nearby vasa recta descending toward the renal medulla. Although your filtrate's Na^+ concentration is little changed by removing four-fifths of its Na^+Cl^- and water (along with any filtered glucose)—the filtrate urea and other solutes (to which descending tubules are not permeable) are thereby concentrated about five times. Equilibration of urea and other solutes across the freely permeable loop of Henle then raises renal medulla urea levels far above ordinary blood urea concentrations, with medulla Na^+Cl^- levels simultaneously being increased through countercurrent extraction of water from incoming vasa recta by solute-laden vasa recta heading back to the renal cortex.

The filtrate undergoes a further reduction of Na^+Cl^- content as it advances through the ascending tubule, but this time it is primarily Cl^- ion (now in higher concentration than Na^+ due to active removal of $Na^+HCO_3^-$ from the descending tubule) that is pumped out into surrounding tissues while Na^+ mostly follows along passively. The ascending tubule is impermeable to both water and urea so *total* filtrate solute concentrations decline as Cl^- and Na^+ are extracted—going from very *hypertonic* (about four times normal body solute levels) in the renal medulla to *isotonic* or even slightly *hypotonic* (below the normal blood solute concentration) in the renal cortex tissue fluid and distal convoluted tubule. By now the solutes within your advancing blood *filtrate* (soon-to-be-*urine*) have undergone major changes with urea and K^+ concentrations markedly elevated, Na^+ and Cl^- much reduced and varying concentrations of H^+ (or some buffered form of H^+ such as locally excreted ammonia [NH_3 plus $H^+ = NH_4^+$], with additional phosphate or bicarbonate added when urine pH falls significantly). In the course of stabilizing your blood pH and solute levels, your tubule and duct cells often interchange Na^+, K^+ and H^+ ions for each other—these exchanges of like-charged ions eliminate the need to move ions against an electrical gradient (and presumably prevent you from producing sparks when you urinate).

While 15 to 20% of the original filtrate volume reaches the loop of Henle, only about 1% of that original volume enters the renal pelvis. So 99% of all glomerular filtrate routinely reenters kidney tissue fluids to depart in your renal venous blood. Thus the urine concentration of creatinine (a substance neither secreted into nor removed from tubules and ducts beyond the glomerulus) will be about one hundred times higher than corresponding serum creatinine levels. On the other hand, filtrate urea concentrations often rise by much more than 100 times during this process of water recovery and filtrate modification (see below).

ADH And All That

To conserve water, your land-dwelling body releases *antidiuretic* (anti-piss-a-lot) *hormone* from the posterior pituitary. In addition to causing you to piss very little, ADH causes *vasoconstriction* (a temporary smooth-muscle-controlled narrowing of blood vessels) and affects neurons involved in regulating your circulation and body temperature—that is why antidiuretic hormone is also known as *vasopressin*. ADH-related molecules already had a water-conserving function in both your aquatic and your land-dwelling vertebrate ancestors. In amphibia, ADH improves hydration by increasing the permeability of skin and *urinary bladder* (where extra water is often stored) to water. Similarly, ADH acts upon the kidney collecting tubules of land vertebrates to enhance water passage (along an osmotic gradient) into extracellular tissues. ADH also acts as a neurotransmitter involved in the release of ACTH and it plays a role in the release of blood coagulation factors by blood vessel endothelial cells. Undoubtedly ADH inherited its many different jobs by having been part of an efficient signal/receptor complex for a very long time.

Anyhow, when antidiuretic hormone increases the permeability of your distal tubules and collecting ducts to water, some of the remaining filtrate water can be drawn out by the high concentrations of urea, Na^+Cl^- and other solutes in the surrounding kidney medulla. As mentioned, incoming (toward medulla) vasa recta lose water while outgoing (toward cortex) vasa recta pick up water and remove it from the kidney. Incoming vasa recta also gather up much of the Na^+Cl^- that has crossed tubule walls and, being permeable to urea, they develop the same high urea concentrations as the surrounding renal medulla. Outgoing vasa recta then tend to leave behind Na^+Cl^- and urea in countercurrent fashion. By allowing urea and water to escape into the renal medulla from those very-high-in-urea collecting tubules, ADH enhances urea concentrations in renal medulla tissue fluids adjacent to all those loops of Henle. Some of the urea that has just escaped from collecting tubules soon diffuses back into those very permeable loops of Henle to reenter the distal tubule and go around once again to the collecting duct. The urea in that filtrate may then be additionally concentrated by removal of useful electrolytes and a further ADH-controlled recapture of water and urea from the collecting ducts. Thus the impermeability of ascending and distal convoluted tubules to urea causes more urea molecules to collect within your renal medulla tissue fluids.

In summary, antidiuretic hormone helps you to conserve water when you are dehydrated by increasing the permeability of your kidney collecting ducts and distal convoluted tubules to water (while also reducing the output of your sweat glands). A lot of urea that goes through the distal convoluted tubule into the collecting duct then leaks back toward the loop of Henle to build up within your renal medulla. A lot of urea is excreted in the urine as well. Under maximal ADH influence, the collecting duct filtrate reaches the same osmotic concentration as surrounding renal medulla tissue fluids (four times the total solute concentration in your blood). Useful molecules or ions such as glucose or Na^+Cl^- only show up in your urine when blood glucose levels are very high (as in uncontrolled diabetes mellitus) or if you are eating lots of salty food. Thus *glycosuria* (sugar in the urine) occurs when the sugar grabber molecules of your proximal tubule cells are overloaded because of elevated blood (and therefore filtrate) sugars—excess sugar remaining in the urine osmotically retains water within the tubule so uncontrolled diabetes mellitus is associated with a larger urine output than normal. Drinking lots of fluid and ADH deficiency are among other causes of copious urine production—inadequate production and release of ADH minimizes water resorption beyond the loop of Henle so the urine becomes far more dilute than your original glomerular filtrate (because many useful solutes are still recaptured but a lot of water is not). So not surprisingly, ADH deficiency is known as diabetes *insipidus* which means "to pass flavorless urine".

Diabetes insipidus can be corrected by periodically blowing ADH up the nose (which allows rapid absorption). Otherwise a person with diabetes insipidus must drink large volumes of fluid rather continuously to avoid dangerous dehydration. In theory, since your urine solutes can be concentrated to about four times the level present in your blood plasma (which is equivalent to 0.9% saline), you should be able to drink seawater (3% saline) and still retain a bit of water after excreting most of that salt and some urea. Actually, it *may* be possible to drink a cup of seawater a day and gain a little water (in a serious emergency), but more than a cupful includes too much poorly absorbed magnesium sulphate. Since drinking more than a small amount of magnesium sulphate (also known as *"Epsom* salts" to honor the medicinal waters of a small English town) causes diarrhea, you will then lose more fluid through your bowels than you can possibly gain from drinking sea water and excreting maximally concentrated urine. You cannot increase your state of hydration by drinking whiskey either, since alcohol inhibits ADH secretion and thereby leads to a *diuresis* (a piss-a-lot condition) greater in volume than the whiskey water taken in. Among their many other effects, coffee and tea act as mild diuretics, while cigarettes increase your ADH production (so nicotine encourages fluid retention). Ocean mammals concentrate their urine sufficiently to get along without fresh water. Many sea birds, fish and reptiles utilize special salt-excreting glands to keep their blood salt concentrations safely below the salt levels of surrounding seas.

Other hormones that act upon your kidneys include *atrial naturetic factor* which is secreted by cardiac cells in response to cardiac distension. By blocking distal tubule and also collecting duct Na^+ channels, atrial naturetic factor reduces Na^+ and water resorption, thereby increasing urine volume and reducing blood volume. And when your glomerular filtration rate changes, the diameters of afferent and efferent arterioles can be decreased in varying degree by their sympathetic innervation as well as by increases in blood levels of *epinephrine* and *norepinephrine*. Blood pressure detectors in your aorta and carotid arteries, and the oxygen and CO_2 sensors in those same vessels and your brain stem, also guide the changes in your renal blood flow, glomerular filtration rate and renal secretion that help you to maintain a stable blood pressure, blood volume and arterial pH. Furthermore, if the circulatory delivery of oxygen to the kidney is chronically diminished by low blood hemoglobin levels (anemia with a hematocrit below 33 or so) or life at high altitudes, the kidney secretes more *renal erythropoietic factor* which encourages red cell maturation in the bone marrow, thereby increasing red cell output and raising the hematocrit.

Diminished delivery of Cl^- ion to your distal convoluted tubules is a signal that less glomerular filtrate is being produced because your afferent glomerular arterioles have narrowed or a low-blood-pressure low-blood-volume condition has developed that requires correction. Under such circumstances, your *macula densa* cells (located in the wall of the distal tubule where it swings back up near the glomerulus) dilate the afferent arteriole to increase glomerular blood inflow while your *juxtaglomerular cells* release *renin* into the circulating blood and constrict the efferent arteriole (renin is a hormone and quite unrelated to *rennin*, the enzyme that curdles milk in an infant mammal's stomach). Renin acts rather indirectly by converting circulating *angiotensinogen* (produced by liver cells) into *angiotensin I*, which is then converted by the *endothelial cell converting enzymes* of your pulmonary capillaries to *angiotensin II*, which both constricts peripheral arterioles and orders your adrenal cortex to secrete *aldosterone*. That aldosterone then enhances salt and water absorption from your distal convoluted tubules and collecting ducts (and usually increases urinary K^+ losses as well), thereby expanding your blood volume and raising your blood pressure back toward normal. A bit roundabout, perhaps, but many steps allow more controls. Many steps also hint at the fortuitous way that relationships have evolved between molecular causes and effects. As you might expect, tissue fluids soon become Na^+Cl^- depleted when the adrenals fail to secrete enough aldosterone. Adrenal cortical failure (*Addison's disease*) was not uncommonly associated with tuberculosis in past years, as tuberculosis *bacilli* (meaning rod-like bacteria—*cocci* are round bacteria) often flourished within and destroyed adrenal glands.

There are many other causes of adrenal failure, of course, but whatever the cause, adrenal failure markedly reduces an individual's ability to tolerate major stress such as a surgical procedure. Mildly Addisonian individuals may be the only ones aware of their own flimsy condition, since they frequently seem quite healthy. But by hard experience we learned to cancel non-emergency surgical operations if the patient seemed convinced that he or she would not survive it (surprisingly, patients were often willing to proceed despite such premonitions). Nowadays, the occasional patient who seems to be handling stress and recovery poorly can generally be helped by a short intensive course of synthetic adrenal steroids (which were neither understood nor routinely available as recently as the 1950's). To some extent, a diet high in salt (Na^+Cl^-) can compensate for abnormally low aldosterone levels under *non*-stressful conditions since increased salt intake or normal aldosterone production both enhance blood and tissue Na^+ levels and blood volume.

Renal (Kidney) Failure

Kidney failure may result from chronic diabetes mellitus, or from circulatory insufficiency, mismatched blood transfusion, severe trauma, glomerulonephritis (based upon autoimmunity), pyelonephritis (retrograde infection of kidneys through the urine), kidney stones, congenital renal cysts, overuse or adverse reactions to medications or other toxic chemicals and so on. Malfunctioning kidneys may shut down and become *anuric* (produce no urine) or *oliguric* (meaning very little urine output—less than 10 ml. per hour). Alternatively, damaged nephrons sometimes continue to filter adequate volumes of urine but without much tubule function (high-output renal failure).

A normal diet contains a preponderance of acidic foods. These include fatty acids, amino acids, citric acid (as in fruit and some carbonated beverages) and so on. A normal diet also contains a lot more plant and animal (K^+ containing) cytoplasm than (Na^+Cl^- containing) extracellular fluid—furthermore, the salty blood serum is usually drained from slaughtered animals to dry, preserve and enhance their meat. Therefore *renal acidosis* (low pH in blood and tissue fluids) and *hyperkalemia* (elevated serum potassium levels) are likely to result when poor kidney function compromises the ability to eliminate excess H^+ and K^+. And blood urea levels rise rapidly when urea no longer can be urinated away, especially if such anuric individuals are metabolizing much dietary protein or breaking down a lot of their own tissue proteins. So kidney failure patients do better if they oxidise mostly carbohydrates for reducing power, since carbohydrate combustion releases CO_2 and H_2O which can be lost through lungs and sweat glands. An appropriate diet for a person with non-functioning or mostly disabled kidneys would therefore be high in carbohydrate and low in K^+, protein, phosphate and sulphate. Thirst may be unrelated to fluid needs in such individuals—therefore, adults without kidney function are limited to about 1500 ml of water per day except when replacing unusual fluid losses (vomit, diarrhea or excessive perspiration).

Renal failure may be temporary or permanent. In either case, the patient must have access to some non-renal mechanism for eliminating K^+ since otherwise rising K^+ levels in blood serum soon interfere with cardiac electrical activity and cause cardiac arrest. The intravenous administration of *glucose* along with *insulin* can *temporarily* reduce circulating levels of K^+ during a hyperkalemic emergency, for insulin stimulates cellular uptake of glucose-plus-fluid and newly recruited intracellular fluid must match preexisting intracellular K^+ levels—which are always far higher than serum K^+ levels. With extracellular fluids the only available source for intracellular K^+, dangerously high blood serum K^+ levels are therefore effectively lowered (due to serum K^+ entering the cells) by intravenous fluids containing glucose and insulin.

Ion exchange resins given *by enema* (hosed up through the anus into the colon) will absorb K^+ from colon tissue fluids before the resins are then excreted. And artificial filtration devices can pull excess water and solutes from the blood to make room for appropriate replacement solutions. However, severe renal failure is usually best treated by intermittent renal *dialysis* utilizing an *artificial kidney* located outside of the body. During such treatments, a steady flow of the patient's blood is diverted via semipermeable tubing through a fluid-filled tank to allow equilibration by diffusion of his/her blood solutes with appropriately constituted (proper pH and osmotic strength, no K^+ or urea or phosphates or sulphates) dialysis tank fluids before that blood reenters the patient. In contrast, *peritoneal* dialysis utilizes the thin slippery peritoneal surfaces as a semipermeable dialysis membrane (instead of using colon wall or artificial dialysis tubing as discussed above). Here a liter or so of appropriate fluid is pumped into the peritoneal cavity where it equilibrates with extracellular tissue fluids before being drained out again. This technique removes K^+ and metabolic wastes at a more leisurely pace than blood dialysis. Although a lot of expertise, expensive equipment and dietary manipulation can stabilize a patient when those two fist-sized tubule-packed organs lying alongside of her/his spine are out of order, your kidneys clearly do it better, cheaper and with far less fuss—while also preventing anemia and helping to maintain your blood pH, skeletal integrity and brain function.

CHAPTER 25

WATER AND SALT

Your Inner Seas—A Quick Review;... Your Dynamic Fluid Balance;...
Fluid Losses May Be External Or Internal, Permanent Or Temporary;...
Blood Loss And Fluid Replacement.

Your Inner Seas—A Quick Review

Much of the water evaporated from Earth's oceans by solar energy inputs soon returns in rivers and streams bearing newly dissolved solutes from Earth's *crust* (which by weight is 46.6% oxygen, 27.7% silicon, 8.1% aluminum, 5% iron, 3.6% calcium, 2.8% sodium, 2.6% potassium, 2.1% magnesium and 1.5% everything else). On the other hand, when shallow landlocked seas evaporate completely, their salt content is soon buried beneath shifting sands. Furthermore, subducting sea floor plates carry huge quantities of Na and K silicates down into Earth's mantle. These incremental and decremental processes must more or less balance since the salinity of Earth's oceans appears to have remained relatively unchanged over the past three or four billion years. In any case, modern oceans average about 97% water, 3% Na^+Cl^- and 0.13% Mg^{++}, with much smaller percentages of SO_4^-, K^+, Ca^{++}, and $PO_4^=$. Many other atoms, ions, inorganic molecules and organic molecules are present in trace amounts that nonetheless represent huge quantities when entire oceanic volumes are considered. Your own body is two-thirds water by weight, or about 63% hydrogen, 25.5% oxygen, 9.5% carbon, 1.4% nitrogen, 0.3% calcium, 0.2% phosphorus, 0.06% potassium, 0.05% sulphur, 0.03% chloride, 0.03% sodium and 0.01% magnesium (plus many trace elements). While your inorganic ions are also those most common in the seas where life first prospered, life has markedly altered the concentrations of many oceanic solutes. For example, it is not surprising that the oceans contain far more Na^+ than K^+ (even though Na^+ and K^+ are distributed throughout Earth's minerals in similar amounts) since the larger K^+ ion fits less well between polar water molecules and is held more tightly within Earth's rocks. But another major cause of oceanic K^+ depletion is the way plant cells avidly extract K^+ for their cytoplasm from Earth's puddles, ponds, streams and seas. Earlier we noted how dissolved oceanic Ca^{++} and atmospheric CO_2 were biologically depleted during the formation of $CaCO_3$ (limestone and marble) and the way that bacterial processes have directly and indirectly created valuable mineral deposits by taking various metals and other elements out of solution. So life has gradually depleted the seas of many solutes ranging from iron to sulphur to gold, some quite toxic to life.

Phospholipid membranes have always separated life's organic molecules from the adjacent salty seas—and achieving osmotic balance with those waters required the expulsion of Na^+Cl^-. In addition, life's fatty *acids*, amino *acids* and nucleic *acids* all release H^+ into water at neutral pH—the resulting excess of organic anions further repulsed chloride (the dominant oceanic anion)—all of which produced ionic gradients across life's membranes from the outset. These electrochemical gradients have since been refined and expanded, and they continue to play critical roles (in the collection, retention and utilization of solar energy as well as for detecting and responding to changes in the environment), forcing every cell to pump ions continuously for both function and survival—so the first kidney did not invent anything new, it merely reorganized available skills in a slightly better fashion. Further flexibility came with the ability to import Mg^{++} as well as K^+ since Mg^{++} provides the same electrical effect with only half the osmotic effect. Of course, Ca^{++} has always been expelled by bacteria and cells because at higher concentrations it precipitates $PO_4^{\equiv}$, an ion essential for intracellular energy transfers and gene construction. And the onset of multicellularity and mobility made it increasingly essential to acquire extracellular fluids whose *electrolyte* (ion) concentrations had to differ markedly from those within the cell. Almost inevitably the extracellular fluids of your ancestors came to resemble ancient seas by being higher in Na^+, Ca^{++} and Cl^- than intracellular water while their intracellular solutes included higher levels of all the other common oceanic ions (K^+, Mg^{++}, $PO_4^{\equiv}$ and $SO_4^{=}$).

Your Dynamic Fluid Balance

Whether from your airway, a calm sea or an automobile battery, purely evaporative water losses do not include electrolytes. But frequently you are subject to other disturbances in your body fluid distribution and composition. And as you continue to sweat, bleed, urinate, vomit or have diarrhea, you may progressively deplete yourself of both fluid and electrolytes—which disturbs the stable extracellular fluid environment that your cells require in order to function competitively. Ordinarily, however, your GI tract absorbs enough of the water that you drink and the animal or vegetable parts that you eat so that your kidneys (and also sweat glands) can keep things even.

Dehydration is the simplest and most common disturbance of your extracellular fluid space. During a long hike or other heavy physical exertion (especially in warm weather) you may easily lose a kilogram or two (several pints) of fluid as sweat and urine (highly conditioned athletes exercising strenuously can lose up to two liters of sweat per hour). At such times it is simple for you to rehydrate—some soup, a sandwich, several glasses of water and you are on your way. On the other hand, a copiously perspiring midgame football player derives more immediate benefit from drinking a rapidly absorbed, *isotonic* (same osmotic pressure as blood) or slightly *hypo*tonic (lower osmotic pressure) electrolyte solution that contains *glucose* (which is utilized differently by your mouth bacteria and therefore less likely to cause dental decay than *sucrose*-containing solutions). That sports drink should also include Na^+, Cl^- and K^+, as well as some easily utilized anion such as lactate or citrate, so that his sweat glands, liver and kidneys can easily reestablish the appropriate electrolyte levels and acid-base balance while eliminating any excess.

If you lose four pints of sweat (or two liters, should you prefer to perspire in the metric), that fluid obviously came from somewhere. In this case, it was delivered by your sweat glands whose sole source supplier is the blood stream, so your circulating blood must have donated those four pints of sweat. However, as an average-sized male (unscientific research suggests that there are no average-size females), you only contain about five quarts of blood. And certainly you cannot have lost nearly half of your blood volume as sweat since that would include almost all of your blood plasma. Indeed, it turns out that dehydration concentrates your blood solutes, regardless of whether your dehydration results from vomiting, diarrhea, diuresis, perspiration, rapid respiration or whatever—and the resulting rise in blood protein concentrations draws fluid into your blood vessels until your tissues and circulating fluids again share the same osmotic pressure (except for nervous system fluids sequestered behind your blood-brain barrier).

Approximately two thirds of your entire body weight is fluid and two thirds of all that fluid is intracellular. Therefore only one third of the fluid lost as sweat depletes extracellular fluid stores while two-thirds of that sweat water is contributed by your cells. With more than three fourths of your extracellular fluid located outside of your blood vessels, only about one twelfth (¼ of ⅓) of that sweat represents lost plasma volume while three twelfths (¾ of ⅓) of that sweat fluid was donated by your other extracellular tissue fluids. In other words, each fluid compartment only contributes about 5% of its fluid volume when an average 70 kilogram male loses two liters of sweat. And as you have learned by experience, that sort of fluid loss is easily tolerated by a healthy young person whose vessels still can adjust their size to match the available blood volume.

Fluid Losses May Be External Or Internal, Permanent Or Temporary

In contrast, an older individual with less adjustable blood vessels and diminished cardiac reserve might find a four pint fluid loss quite disabling—perhaps enough to cause a serious or even fatal drop in cardiac output and blood pressure (it may seem silly that people become more concerned about health as they grow older with "less to lose"—but it also takes far less disturbance to incapacitate or kill the elderly so they no longer can ignore previously minor physiological derangements and symptoms). Furthermore, the increasing capillary fragility of older individuals allows more fluid and protein to escape from their circulation into surrounding tissues. While some of that surplus extracellular fluid soon returns to the circulation as lymph, other fluid (e.g. trapped in swollen ankles) may not reenter the circulation for hours or even days—perhaps not until the legs are persistently raised above heart level to improve venous and lymphatic drainage. Of course, such a legs-up position leaves the buttocks or entire back well below heart level so the *edema* (excess tissue) *fluid* may simply transfer to such newly dependent (hence higher venous pressure) places. Thus some internal fluids seem readily accessible to the circulation while other internally located fluids such as ankle edema may remain temporarily out of reach (not exchangeable with intracellular or extracellular fluid—although edema fluid solutes presumably remain in some sort of balance with the solutes of your other fluids—see also proteoglycans).

An elderly person who is seriously ill and lying flat on his/her back for several days may develop very boggy (edematous) subcutaneous tissues below heart level—ordinary flabby back-fat then feels like firm liver and the patient may gain a dozen pounds despite not eating. But no matter what or how much intravenous fluid is being administered, as long as that pool of extracellular fluid in those back tissues remains inaccessible and therefore irrelevant to the blood circulation, it also remains irrelevant to *current fluid requirements*—which *relate only to the* current status of that individual's *circulation*. It is worth reemphasizing that you cannot mobilize the edema from a swollen injured leg or edematous back or distended postoperative abdomen merely by fluid restriction—the need of a circulation for fluid does not make edema fluid suddenly available any more than the need of a homeless person for food will get them invited into a nearby restaurant.

However, as a bedridden individual becomes more mobile or the injury begins to heal, you can expect fluid from any swollen tissues to reenter the capillary circulation and lymphatic system. And that recovering patient who may previously have been physiologically *hypovolemic* (suffering from an abnormally low blood volume despite having sequestered a great deal of edema fluid) may even become *hypervolemic* from too much of that fluid reentering the circulation too quickly. While such a delayed postoperative mobilization of fluid and the resulting diuresis are favorable signs of impending recovery, that fluid excess can easily overload the lungs (cause *pulmonary edema*—as also seen with ordinary left heart failure), distend the liver and otherwise raise the right-sided venous blood pressures (as in right heart failure)—at such times it may pay to boost postoperative urine flows with a judicious use of *diuretic* medications.

Quite clearly, your life and function depend upon the effectiveness of your circulation, regardless of your total fluid content. So from the day you are born, you drink or administer fluids in response to circulatory changes. Of course, every newborn comes out a bit soggy and full of swallowed fluid after nine months of drifting (and recent dreaming) within the uterus. Therefore she/he has no need for fluid intake during the first day or two, when slight weight loss is normally anticipated. Indeed, zealously trying to maintain the birth weight of a hospitalized newborn receiving intravenous fluids can lead to dangerously wet lungs at a time when those lungs must suddenly adjust to above-water existence. In turn, wet lungs can stimulate unhealthy circulatory changes which adversely affect lung function and so on into a steepening spiral of dysfunction that could well end in death (but there are increasingly effective ways to intervene in such situations—see Respiration).

Gastrointestinal disturbances are a common cause of dehydration and electrolyte loss, especially in infants and young children with unrelenting diarrhea. Formerly it was believed that the GI tract should be put to rest during severe diarrhea, but it has repeatedly been shown that even severe diarrhea need not interfere markedly with intestinal absorption of orally administered isotonic or slightly hypotonic fluids (bearing an appropriate ratio of Na^+ ions to glucose molecules and including other electrolytes such as Cl^-, K^+, bicarbonate and citrate). A great many lives have been saved by World Health Organization (WHO) recommended oral salt solutions, especially in underdeveloped countries. Interestingly, the outcomes of intravenous fluid therapy given to compensate for acute diarrheal fluid losses are generally no better than the results from appropriate oral treatments, even though such intravenous treatments are far more costly and frightening to the sometimes unnecessarily hospitalized child. The goal of replacement water and solute therapy in this situation is to counter excessive losses and bring the patient back within a healthy range of fluid and electrolyte content so that the kidneys can once again sort things out. Adding complex carbohydrates such as rice starch to oral rehydration solutions improves nutrition while also reducing diarrheal stool volumes. Even solid high-carbohydrate foods such as rice, bananas or potatoes can be absorbed during severe diarrhea (of course, one should particularly avoid any foods likely to worsen the diarrhea).

Blood Loss And Fluid Replacement

Upon donating a pint of blood at a blood bank, you will probably show a slight increase in pulse rate as the minor reduction in venous blood returning to your heart causes a little less blood to be pumped forward by each heart beat. You may also feel thirsty or even dizzy if you sit up quickly before your circulation has readjusted. But if you just lie back and drink your juice, you should soon feel well enough to proceed with a regular day's light duties (assuming the blood bank only accepted your blood in view of your good general condition). Nonetheless, that one pint blood loss will reduce your hematocrit by a few points. And your hematocrit drops in this fashion because other fluids (those that you drank, as well as extravascular fluids from your intracellular and extracellular compartments) were drawn in to refill your circulation—this despite the fact that your blood vessels generally constrict as their fluid content declines. Those other fluids refill your circulation in response to decreased venous return, reduced atrial distension and lower arterial blood pressures. Lower arterial pressure results in less fluid escaping into the tissues at the arterial end of your capillaries and more fluid returning to the capillaries at their venous end. Furthermore, with reduced production of atrial naturetic factor and more ADH, aldosterone and other adrenal steroids being released, your kidneys respond to hypovolemic stress by recovering more of the fluid and salt still being filtered at each glomerulus. In addition, your spleen constricts to release some stored blood. And over the following days, your liver will replace albumin lost from the bloodstream while your kidneys may produce more erythropoietic factor if necessary to increase your hematocrit—which completes your rapid and uneventful return to normal.

But what if it turns out that your generously donated pint of blood cannot be used after all? Rather than waste it, you might as well have them pump it back into your own veins, perhaps the very next day. Right? Wrong. For even if given slowly, such an *autotransfusion* would likely overfill your already reexpanded (post-donation) circulation unless you receive a strong diuretic at the same time (to eliminate *several* pints of extracellular fluid as urine and thereby make space for that one returning pint of blood—see below). Otherwise the *slowly* dripping blood transfusion might cause wheezes or audible fine crackles at your lung bases posteriorly (best heard through a stethoscope during the brisk inspiration *subsequent to a sharp cough*—that cough encourages significant air entry into the lower lobes while you listen). Such an onset of expiratory wheezes and crackles at the lung bases suggests that *hypervolemia* (an overfull circulation) has caused "wet" lungs (the low-pressure capillary beds of lungs and liver easily accumulate fluid).

In general, unless you are "bleedin' bad", the bulk of any blood transfusion that you may receive will remain trapped within your circulation until those cells and proteins are routinely removed and recycled. Thus the distribution of your *whole blood* (whole blood is emphasized here because blood *components* such as *packed red cells*, blood *plasma* and *serum albumin* are often administered separately) is quite different from the distribution of your extracellular fluid. For extracellular fluid moves freely between your blood plasma (which is intravascular extracellular fluid) and your tissue fluid (extravascular extracellular) compartment. So the amount of whole blood required to refill the circulatory system of a person in *shock* (low blood pressure) due to blood loss is far less than the amount of intravenous salt solution (not containing proteins or red cells) it would take to provide the same beneficial effect. Furthermore, even when intravenously administered *Ringer's Lactate* solution restores the patient's blood volume and arterial blood pressure, that *extracellular fluid substitute* often has a more transient effect as it tends to enter already edematous tissues or be excreted by the kidneys.

Nonetheless, persons able to tolerate the anemia and reduction of serum proteins (hence more edema) that follow significant blood loss can often avoid the costs and risks associated with blood transfusion—and Ringer's Lactate solution is commonly used for circulatory enhancement along with or instead of whole blood transfusion. Since it closely resembles (and thus occupies the same space as) extracellular fluid, several liters of Ringer's Lactate will expand the circulating blood volume only as much as a single liter of transfused whole blood. More specifically, 3–5 pints of Ringer's Lactate should have about the same effect on circulatory dynamics (pulse and blood pressure) as one pint of whole blood. But that ratio is quite variable since it also depends upon the current blood volume, the plasma protein level and how much of the Ringer's solution leaks out to become temporarily useless or even dangerous edema fluid in the tissues or lungs. That is why frequently rechecking the pulse, blood pressure and lung bases, and occasionally reevaluating the hematocrit as well as serum Na^+ and K^+, is a necessary aspect of fluid replacement during circulatory stabilization.

Other commonly used intravenous electrolyte solutions are less effective for circulatory (intravascular) volume expansion during or after extensive blood or extracellular fluid losses. For example, giving large volumes of *isotonic saline* (also known as *normal* saline—an 0.9% NaCl solution) instead of Ringer's Lactate would provide too much Na^+ and way too much Cl^- since one liter of that saline solution contains 155 millimoles of each while your usual serum Na^+ ranges between 135–145 and your serum Cl^- should not rise much above 105 millimoles per liter. Consequently, the administration of normal saline in large amounts can cause *hyperchloremic acidosis* since many blood proteins and other solutes also are anions and that excess serum chloride may become electrically balanced in part by retained H^+ ions. Furthermore, 0.9% saline does not include other essential serum *electrolytes* (ions) such as K^+, Ca^{++} and Mg^{++}. So by being a simple isotonic electrolyte and glucose solution comparable to extracellular fluid, Ringer's Lactate solution expands the circulating blood volume most effectively (and its lactate ions are soon metabolized by the liver).

As you go through your usual cycles of dehydration and rehydration, your extracellular fluid remains the sole source supplier for new intracellular water and electrolytes. Of course, when tissue injury or bleeding cause a direct loss of intracellular fluid along with the cells that contain it, there is no need to restore fluid to those no longer present cells. More generally, when large volumes of fluid are lost, regardless of whether through burned skin or into injured tissues or by diarrhea or vomiting or bleeding, any non-blood replacement fluids given to expand your extracellular fluids should mimic extracellular fluids (your ordinary and only source of intracellular fluid). Which is just a roundabout way of reemphasizing the advantages of effective (circulation-expanding) volumes of Ringer's Lactate over other replacement solutions.

However, whether bleeding or not, you also have ordinary fluid and electrolyte requirements (daily fluid losses to replace). And your blood electrolytes could become markedly disturbed by some abnormal diet or because of gastrointestinal, renal, adrenal, pituitary or other metabolic derangement such as diabetes mellitus, or even by inappropriate fluid or electrolyte replacement. Therefore one must occasionally engage in a more individualized replacement of water or certain electrolytes as indicated by physical signs and blood tests. But such special fluid requirements ought not be delivered according to a set formula either, for just as one man's meat is another man's poison, every person also responds differently to so-called standard fluid replacement therapy—and since persons undergoing intensive fluid and electrolyte correction are probably quite ill already, it is best to make essential corrections in small and carefully evaluated steps rather than all at once.

A similar problem faces any revolutionary party that seizes power—gradual change is more likely to succeed. Cautious moderation with frequent feedback also prepares one for change for ordinary maintenance fluids may overload a patient who is actively mobilizing tissue edema—and a postoperative heart surgery patient with apparently adequate blood calcium levels may still benefit from supplemental intravenous Ca^{++} to boost circulating levels of *ionized* (versus protein-bound or otherwise combined) *calcium* for better cardiac contraction—and administering extra intravenous K^+ and Mg^{++} can bring about the high-normal serum levels of those ions that sometimes helps to reduce cardiac irritability (tendency to frequent extra beats). However, you should *always deliver any high-in-potassium solutions slowly* if through peripheral veins where they cause painful venous spasm, and *very slowly* through those long intravenous catheters that extend into the vena cava, for just a little high-potassium solution entering the coronary arteries can stop the heart (a little cardiopulmonary resuscitation will usually clear that solution from the coronary arteries but it is best to avoid such setbacks).

A person who is unable to drink fluids can generally be replenished with a dilute sugar and salt solution given intravenously (or even by enema). That solution should contain a little K^+ for reexpanding the intracellular fluid of shrunken cells, a little Na^+ and Cl^- for rebuilding extracellular fluids, some lactate anions to electrically balance these cations and more than enough glucose to prevent the solution from being hypotonic (note the similarity of this maintenance solution to currently popular sports drinks). Simple five percent *glucose* in water—also known as 5% *dextrose* in water (50 grams of glucose in a liter of water) is an isotonic intravenous solution that can safely deliver water at any reasonable flow rate without causing the *active tissue damage* or red cell hemolysis that *would result if solute-free water* was *given parenterally*—meaning by any route other than through the digestive tract—for example, by subcutaneous (S.C.—by clysis), intramuscular (I.M.) or intravenous (I.V.) administration.

If you wish to provide additional reducing power, you can increase the glucose content in any intravenous solution to 10% (or even higher, while checking occasional blood samples to make sure that blood glucose levels don't rise too far above normal). Indeed it is now possible to provide complete nutritional support and fluid replacement by vein, including appropriate fats and amino acids. This is referred to as *hyperalimentation* or *total parenteral nutrition* and requires significant expertise (although the normal GI tract does it far better without apparent effort or risk of blood stream infection or costly catheters or concerned consultants). Note again that appropriate fluids and solutes can be given intravenously with safety as long as the sterile solution administered is of isotonic or *hypertonic* (above-normal) osmotic pressure. Furthermore, most IV solutions are more easily tolerated if administered slowly—a concentrated glucose solution is generally given very slowly and carefully to avoid major disturbance of the blood chemistry unless that glucose is urgently needed for the correction of hypoglycemia (dangerously low blood glucose levels, perhaps from excessive insulin effect).

Your fluid balance tends to become disturbed in cold weather, especially if you camp out and become chilled and *vasoconstricted* (with your subcutaneous and skin vessels narrowed). Your body may then view the extra blood being forced centrally from your subcutaneous tissues as evidence of excessive blood volume and respond by restricting your ADH secretion and perhaps increasing atrial naturetic hormone release. Thus cold weather may inappropriately boost your urine output and make you *hypovolemic* (reduce your effective blood volume). Although you can still support your usual resting blood pressure despite being cold, miserable and vasoconstricted, your inadequate circulating blood volume rapidly becomes evident when you exercise. As your peripheral circulation then dilates and you begin to sweat or lose heat, you will also tire easily and perhaps become wobbly and faint. Exposure to cold weather therefore increases the risk of hypovolemia, which in turn makes you more susceptible to dangerous hypothermia. Your respiratory water and heat losses always increase under such circumstances since your airways must warm and humidify very cold (therefore very dry) air. So it is important to increase your fluid intake during prolonged exposure to the cold (under such circumstances you should continue to drink lots of warm fluids even though you are not especially thirsty and already urinating more than usual).

Hypovolemia can also result from various fad diets, including various herbal remedies and wrapping treatments (equivalent to cold weather—see above) that guarantee the loss of several ugly pounds in just hours or days. Of course, your extracellular and intracellular fluids or your colon stool content are the only ugly pounds you could possibly lose on such short notice without surgery. Herbal remedies often cause weight loss through persistent diarrhea—which can seriously deplete serum K^+ levels and lead to dangerous cardiac rhythm disturbances. So the diuresis and diarrhea caused by many quack "weight loss programs" may indeed bring about rapid weight loss. But that lost fluid and electrolyte will soon have to be replaced—unless you prefer to remain wrapped or fluid-restricted, feeling faint and potentially dangerously ill (due to the detrimental effects of those fluid and electrolyte disturbances on your heart, kidney, brain and other vital organs).

Furthermore, if you were constipated before those treatments, when that expensive medicine runs out and the diarrhea ends you will again be constipated, and back to your usual body weight as well (but with a lighter wallet).

CHAPTER 26

DOING WHAT COMES NATURALLY

Frequent Spousal Sex Strengthens Human Families As Well As Causing Pregnancies;... It Took A Lot Of Preparation;... The Egg Came First;... How It Works;... Your Inside-out Connections;... Avoiding Pregnancy And Avoiding Venereal Infections;... Menopause;... Testes Hang Out To Be Cool;... Sperm Are Not Very Smart;... Delivering Your Sperm;... Mitosis And Meiosis—Diploid And Haploid.

Frequent Spousal Sex Strengthens Human Families As Well As Causing Pregnancies

A glance. A touch. A kiss. That moment of bliss. No one ever felt like this before! It's magic! Well, not exactly magic, perhaps, but still rather impressive. For as billions of years of experimentation and bloody selection have confirmed, there are times when it becomes reproductively advantageous for your supposedly rational mind to see only six inches into the future—which is why temptation is more easily avoided than resisted. Were it otherwise, you might never have been conceived. Your parents should have gone bowling instead.

All human populations organize naturally into small family units because families are the most effective way to deal with the prolonged total dependency and continuing apprenticeship of human infancy and childhood. Even so, many adults are not loving, willing and able (strong, stable, healthy, tolerant and wealthy enough) to take on and successfully carry out such long-term commitments. But most humans and the females of other mammalian species inherit a strong inclination to help their own descendents at the expense of others—which is true of swans, some crocodiles and certain fish as well. We even feel sympathetic twinges when stricken parents rally around in support of their son, the mass murderer.

A natural mother can be quite confident that her new infant carries many of her genes but fathers regularly require reassurance. So it is not by chance that the usual first remark of a proud new mother to a proud new father is "He/She looks just like you, dear! (honest!)". And that is sure to please him, even though their generic-looking wrinkled and messy infant might be voted just plain ugly by any panel of objective observers. Not uncommonly, one reads of spousal murder or divorce when the husband suspects infidelity. Wives tend to be more tolerant of this sin as long as the husband continues to show interest and support the family. But clearly it is difficult to be a good adoptive parent—fortunately, such step-parents are often highly motivated. In contrast, a male lion taking over a pride (group of female lions) will usually kill all cubs. Since the hormonal effects of breast feeding tend to suppress *ovulation* (release of an egg for potential fertilization by a sperm) in all mammals, such instinctive (inherited) behavior by the male lion increases his likelihood for reproductive success before being displaced or killed by a younger, more vigorous male.

Even though it is not a completely reliable way to prevent ovulation and *conception* (fertilization of an egg), persistent *nursing* (breast feeding) maintains *lactation* (milk production) and is the principal method for spacing children in many third world countries—and that spacing is important, since the risk of death for the last infant rises if the mother gets pregnant soon again, especially when nutritional resources other than breast milk are limited and sanitation is poor. However, if the family system really has improved the odds of human survival to reproductive adulthood, there should be evidence of human adaptation to the constraints of family life. And such changes do appear in the reproductive habits that served our group-living ancestors so effectively. For example, unlike other mammals that advertise female ovulation by attention-seeking behavior as well as the odor and taste of vaginal secretions, human females strengthen family bonding and reduce family instability by remaining sexually receptive throughout most of the month-long ovulation cycle, as well as by not overtly signaling their monthly moment of optimum fertility to all interested males. Family stability is additionally enhanced by a relatively poor sense of smell that leaves human males less (consciously?) aware of any subtle signals being emitted by currently fertile females.

Furthermore, if adult human males engage in sexual intercourse more frequently as a result of not perceiving when their spouses are ovulating, they may develop greater confidence in the paternity of their children and thus be more supportive of the family. The rather incessant human male sexual drive and ongoing overproduction of sperm also increases the likelihood of pregnancy, but there are many other reasons to overproduce sperm, including the fact that they are cheap, and each one increases your chances in a competitive situation (as with an extra lottery ticket), and they formerly were broadcast at sea to all open receivers and so on. That human male sex drive certainly is well adapted to the slow maturation of children and the reproductive benefits that derive from a family structure for child care (especially in comparison to intercourse only during a mating season—often involving exhausting combat and nutritional depletion).

In reconfiguring human males to take better advantage of this family model, natural selection also appears to have reduced their aggressive tendencies—presumably by reducing the reproductive advantage of stronger, more aggressive males who might have dominated seasonal matings among group-living mammals. While human male aggression still varies markedly and at least somewhat in accordance with blood testosterone levels, it is increasingly diverted into more organized competition between males in business, sports and war. An inherently anti-female bias ("glass ceiling") might therefore be expected in those activities best suited for masculine displays of physical and mental prowess (to reassure a potential spouse about good genes, good health and the ability to provide ongoing support for the children). And not surprisingly, those actively seeking a sexual partner are likely to maintain themselves in a more attractive condition than those who already have caught one.

It Took A Lot Of Preparation

Early life grew and grew, then divided in two. If each half acquired its fair share of the parental assets and information, the population was doubled. Such an even distribution of assets was essential merely to keep up with competing life forms. As life gradually became more complex and *eucaryotic* cells (with nuclei) developed, an even distribution of information became the full-time responsibility of *centrosomes* (another intracellular organelle, possibly of endosymbiotic origin). That ancient and hereditary delegation to centrosomes of authority over intracellular probate was a good investment for all parties—it improved fairness in the distribution of assets and thus enhanced cell survival while also providing free room, board and replication for centrosome-producing information. Indeed this system has worked so well that all cells still rely upon these subcellular legal agents to ensure successful cell division.

But dividing in two is no longer a realistic option for a complex specialized multicellular creature that must detect, decipher, dream and digest as well as rip and run. Indeed, your only reasonable hope for reproductive success lies in producing new descendents more or less from scratch. That means each attempt at self-replication necessarily begins with tiny individual cells crammed full of the most intimate and intrusive inherited admonitions. Such a potential human must then be nurtured internally until its lungs, digestive system, kidneys, circulatory system and overall metabolism can function independently. Before being externalized, your latest descendent must also be capable of processing basic information and signaling loudly (if not accurately) regarding its needs for environmental change, more intake, removal of output and so on. However, that is just a start, for even a single reproductive success has a way of growing into a very long term project that consumes much of your remaining life. Not being able to produce useful descendents by subdivision therefore obliges you to multiply laboriously, more or less one descendent at a time—until sometimes it seems that your entire life has become invested in your pampered progeny.

Of course, the human race would long since have vanished from this dangerous world if the average adult only gave rise to a single potentially adult descendent. Yet multiple unit human pregnancies (twins, triplets, etc) have generally had a significantly reduced rate of success due to the additional long-term burden they place upon limited internal and external resources—so multiple human births never really caught on and human females (unlike dogs, cats and pigs that customarily produce litters) usually have only one set of breasts—nonetheless, an occasional human male or female is born with one or more *supernumerary* (extra) nipples located somewhere along the so-called ventral *milk line* (where nipples show up in other mammals). Especially when associated with functional breast tissue, such spare nipples may first attract attention (as other than a minor birthmark) by enlarging during pregnancy. Even so, there is increasing evidence that human pregnancies often begin with more than one embryo, and that post-fertilization deaths and dissolutions share some responsibility for the high percentage of single births. The natural reduction in embryo numbers during a multiple pregnancy may partially reflect the usual level of discards, since only about a fourth of all human embryos are carried to term—the rest undergo spontaneous abortion, often without being noticed.

In any case, all of the extra time and effort you have devoted to thinking of, talking about and experimenting with sex is soon dwarfed by the investment thereafter required to nurture each new human for whom you have parental responsibility. And as is true for art, stocks and diamonds, your dependents tend to take on whatever value you give them. So the greater the investment in time and appreciation, the greater the return. Since you and human society generally view the production of happy, competent and reproductively successful descendents as one of life's major payoffs, such an outcome alone will have to justify all of your investments and sacrifices—even though your health and other assets may be consumed in the process and those descendents do not seem nearly as appreciative of your efforts as you think they ought to be (does that sound like something your parents might have said about you?).

Speaking of being consumed in the process, one species of mite (a microscopic insect) exemplifies *total* parental investment since its one male and fourteen female embryos become sexually mature while still inside the mother mite. And as soon as the male embryo *copulates* (has sexual intercourse) with each of his 14 sisters, those pregnant siblings eat him (nothing wasted so far) before launching their own equally brief careers by consuming their mother from the inside. Although a just and loving God might not take pride in this particular design, it does provide reproductive advantage through optimal utilization of all parental resources. Interestingly, the more customary ratio of 7 males to 7 females would be wasteful under such sheltered circumstances since relying upon one good male and true to complete the entire project (on time and within budget) allows each pregnant female to double its output of egg-producing descendents. This inbred reproductive system may have come about when an unfortunate delay in egg-laying was associated with accelerated sexual maturation—of course, such incest among embryos fails to introduce new genetic material, which is what *sex* is all about. So that mitey male had no reason to be except lunch. But in the outside world, where sexual intercourse brings evolutionary advantage (and so is reinforced by great pleasure), the ratio of males to females usually remains close to even.

That ratio endures because any female predominance out there would reward each lusty male with more descendents than an individual female could hope to produce. The reproductive advantage in being male would soon be dissipated, however, as mostly male litters gained temporary reproductive advantage and came to predominate. Investing only in females would similarly maximize descendents in a world with a large male excess until producers of females again predominated. A possible exception to such an efficiently achieved equilibrium may occur in humans and other social mammals where inherited social status confers reproductive advantage on upper class males. Under such circumstances, the persistently wealthy and powerful ought eventually to have more sons than daughters. In contrast, lower class males might fail to attract any mate while lower class females can always produce descendents. So families of persistently lower social status should have disproportionately more daughters (as yet, there is insufficient evidence to allow an evaluation of this hypothesis with respect to humans).

Be that as it may, the descendents of ordinary male/female sexual relationships inevitably bear some recombined characteristics of each parent. Indeed sex is the only natural mechanism (excluding the infective horizontal transfers of inherited information from which sex may well have arisen) by which unrelated individuals bearing different beneficial (or detrimental) genetic changes are able to combine and recombine those traits within at least some of their joint descendents. In contrast, asexually reproducing individuals necessarily give rise to *clones* that differ from mother only as a result of random replication errors and other DNA mutations or through the horizontal introduction of infective DNA. The greater variability among products of sexual reproduction means some of them will eventually (as times change) outperform the previously well-adapted and more efficiently produced replicas resulting from *asexual* reproduction (such as the dandelions and whiptail lizards that have gained temporary reproductive advantage by abandoning sex). So males promote evolution by confronting natural selection with many more good and bad choices.

One commonly used definition of a species includes all variations capable of interchanging genetic material by sexual intercourse. Not only does such a definition leave clonally reproducing individuals without any species at all, it also fails to classify bacteria that have been caught illicitly exchanging genetic material with members of obviously different species. Apparently biological rules are made to be broken—but exceptions often provide valuable insights. So one asexually reproducing lizard still requires sperm from a related species in order to reproduce. But since that sperm is not utilized, one might reasonably ask why those males bother to help out when they could be contributing genes within their own species instead. It turns out that lizard females prefer experienced lovers, so donating those unused (but initiating) sperm brings reproductive advantage for both asexual female and sexual male. But regardless of all theory, sex is here to stay as far as you are concerned. And if you are a human female, you closely resemble all other female mammals (from tiny shrew to big blue whale) in the complex and magnificent hormonal and structural adaptions that not only enable you to nourish your embryo internally but also allow you to go on feeding your infant externally through the secretions of your mammary (modified sweat) glands. Since females dominate reproduction while males merely introduce additional recombined genetic information, we now review female reproductive anatomy and physiology.

The Egg Came First

The organs that produce *ova* (female germ cells or eggs) are known as *ovaries*. Your two tiny ovaries (or those of your friends) contained hundreds of thousands of eggs when you were born, so those eggs must have been many orders of magnitude smaller than the hard-shelled chicken eggs that cause such a cackle in the henhouse. But all chicken eggs and each of your own eggs originate as a small immobile cell—quite different from the tiny self-propelled sperm that might later be pumped into your genital tract. A chicken egg becomes large by enclosing all the food and water a cute but not terribly bright baby chick will need until it pecks out through that protective yet-sufficiently-porous-to-allow-gas-exchange egg shell. In contrast, your information-packed eggs require uninterrupted metabolic support from your ovaries, tubes and uterus.

How It Works

A hen has only a single external (cloacal) outlet while you have three—a separate small urethra that opens near the front of your vagina and an anus that empties behind. Not surprisingly, this arrangement corresponds with your adjacent internal anatomy. So your urinary bladder lies in front of your uterus and ovaries, up against that bony forward arch of your pelvis known as the symphysis pubis (which can be felt deep to the pubic hair at the bottom of your abdominal wall). And your rectum is located behind your uterus and ovaries, along the posterior pelvic wall and lowest vertebrae. The bony coccygeal tip of your vertebral column that once supported your useless embryonic tail can still be felt behind the anus in the midline. That tail protected your ancestor's perineal openings (as even the short moose-tail still does) from flies and other insects attracted to those sensitive, hairless and nourishing mucosal surfaces. The posterior placement of your rectum and anus ensures that feces will drop as far back as possible rather than smear across other body openings when you squat. That helped your ancestors to reduce fecal soiling of the vagina and avoid smelly fly-infested feet. The forward location of your bladder allows squatting or spreading the legs to direct urine away from the body and feet, but urine washing across the vaginal outlet or rectum is usually sterile in any case. Indeed, the ammonia released by urea-splitting vaginal bacteria may help to suppress more troublesome microorganisms. In contrast, a bladder outlet (urethra) behind the anus would be subject to regular fecal soiling.

As a central passage through the soft muscle-supported female *perineum*, your vagina undergoes tremendous stretching during childbirth. Were it located off-center, those surrounding tissues could not stretch so uniformly, resulting in a less distensible vagina more liable to tear. And as *perineal* (pelvic outlet) tissues bulged outward during childbirth, the baby might be caught under the overhang of the longer side (as in the hood of a sweatshirt). So the need to cushion the fetal head and direct it away from nearby bony prominences probably accounts for the posterior placement of the soft rectum as well as the anterior location of the soft bladder in both males and females. But there are many other constraints, such as the fact that your spine is best placed in back for protection/strength/body-support/ flexibility/movement, and the need to shelter your aorta and vena cava against a rigid spine. Furthermore, your kidneys must lie close to the aorta and cava yet outside of the peritoneal cavity in order to process lots of blood without disturbing the nutrient-laden portal venous drainage to the (necessarily intraperitoneal and moving-with-diaphragm-to-enhance-breathing) liver—which obliges your ureters to follow a lengthy retroperitoneal trail from behind your bowels to the bladder in front. Similarly, the entire circulation of your *female reproductive system* (ovaries, tubes, uterus and vagina) must remain extraperitoneal in order to isolate the copious venous return of your huge pregnant uterus from portal venous blood destined for your liver. As a sign of their descent from a higher retroperitoneal position during embryonic life, your two ovaries remain dependent upon long *ovarian artery* branches from your mid-abdominal aorta while your left and right *ovarian veins* drain into left renal vein and inferior vena cava respectively. In contrast, your uterine blood supply arrives through your internal iliac arteries and leaves via your internal iliac veins. However, except for its circulation, the female reproductive system is primarily intraperitoneal—indeed, each ovary is separately suspended inside the peritoneal cavity on its own short *mesentery* (fold of peritoneum that continues over the uterus, tubes and lateral peritoneal wall). Therefore every ovum that bursts through the ovarian surface during *ovulation* finds itself free within the peritoneal cavity.

That release of one or more eggs ordinarily occurs about once a month in human females of reproductive age until they reach *menopause*, when hypothalamic releases of *gonadotropin-releasing hormone* (*GNRH*) become less reliable or effective. In its active form, GNRH is a surprisingly short ten-amino-acid peptide message. Furthermore, only the first five amino acids are needed for release of *follicle stimulating hormone* (*FSH*) and *luteinizing hormone* (*LH*), while amino acids 6–10 apparently act directly upon your brain cells to alter your mating behavior. Clearly brevity of message need not imply lack of significance—sometimes a short message is less easily scrambled or misinterpreted. Menopause generally occurs in your mid to late 40's or early 50's and it usually represents cessation of the appropriate cyclic production of FSH and LH by your anterior pituitary. You enter menopause with your eggs mostly dead and gone—only a few hundred of those eggs will have been released during an ovulation into your peritoneal cavity (from either ovary), but a great many others start each cycle, then fail to compete and complete their development in one or another of your pituitary-driven cycles (well, hypothalamic-driven actually). Women in the third world may have only 10% as many ovulations since they are pregnant or lactating for so much of their reproductive lifetime. Eggs that fail to mature simply shrivel and disappear. Ovaries enlarge and become functional at puberty (as melatonin levels fall and the pineal gland calcifies—see Hormones) and tend to get smaller as the menopause approaches. Yet even elderly mouse ovaries can again become fertile if transplanted into young female mice. So the onset of puberty and menopause are determined by changes in pineal and hypothalamic function as well as altered ovarian responsiveness to the hormones being released.

Every successfully enlarging FSH-dependent egg-bearing follicle necessarily balloons toward the ovarian surface until it ruptures, releasing the egg along with estrogen-containing follicular fluid (which may include sperm attractants) and other cellular debris. Follicular rupture can be associated with small amounts of intraperitoneal bleeding that may cause mid-cycle peritoneal irritation (lower abdominal discomfort at the time of ovulation). Thereafter, LH stimulates the remaining follicular cells to form a yellow, highly cellular *corpus luteum* that produces both estrogen and progesterone until the pituitary-led cycle ends—at that point the corpus luteum atrophies and is replaced by scar tissue. Coordinated movements of its splayed-out fronds allow the open *fimbriated end* of a *fallopian* tube to draw in the nearby egg (along with any lysed blood clot and other intraperitoneal fluid). Smooth muscle peristalsis and ciliated lining cells then move that egg and fluid slowly through the narrow irregular fallopian channel until the smooth-muscle-enclosed endometrium-lined *endometrial* cavity of the uterus is reached. Fertilization usually occurs during those several days of leisurely travel if it is going to take place—perhaps the confining tube makes any sperm attractants more effective and the egg easier to find. But only a tiny fraction of the 400 or so eggs that could potentially follow this trail will ever be fertilized, since pregnancy terminates the monthly pituitary hormone cycle and ovulation until the child is born.

Thereafter the ongoing nipple stimulation of *nursing* (breast feeding) further suppresses GNRH and the pituitary-driven ovulation cycle while encouraging the pituitary release of *prolactin* (which enhances milk synthesis and secretion) and *oxytocin* (which brings about milk "letdown" or ejection by contraction of myoepithelial cells around the milk-producing mammary alveoli). Thus the risk of ovulation (and hence potential new pregnancy) remains low during the first six months of *full* (sole source of infant nourishment) breast feeding until the first menstrual period has occurred as a consequence of estrogen withdrawal. On the other hand, the first post-pregnancy menstrual period may be preceded by an ovulation if that period occurs more than six months after childbirth. In other words, the fully nursing mother is unlikely to ovulate during the first six months following childbirth so there is little risk of pregnancy during this time unless/until her menstrual cycle resumes.

Ordinarily your monthly FSH-stimulated competitive maturation of new follicle cells brings about production and release of ovarian estrogens which encourage the secretory-cell lining of your midline uterus to grow (thicken). That endometrial growth continues as LH-boosted corpus luteum cells also produce progesterone to enhance development of your endometrial glandular (secretory) structures. When the corpus luteum then regresses, its estrogen and progesterone production decline quite steeply, leaving the lush endometrial surface and its blood vessels bereft of hormonal support. The resulting closure of overgrown endometrial blood vessels causes superficial endometrial layers to separate and be shed along with blood escaping from the remaining raw surface. This more-or-less-monthly *menstrual discharge* passes out through the central canal of the *cervix* (uterine *neck*) into the vagina and may be associated with cramps as the uterus contracts to expel endometrial debris. The premenstrual peak of estrogen and progesterone production is often associated with noticeable breast swelling and tenderness, as well as fluid retention along with slight weight gain and mental irritability. Those signs and symptoms then subside as a new surge of FSH starts the next monthly cycle of reproductive organ stimulation and enthusiastic growth in preparation for possible fertilization and pregnancy.

Our Inside-out Connections

Your fallopian tubes, uterus, cervix and vagina provide an open channel between the bacteria-laden outer world and your sterile peritoneal cavity. Since your cervical canal normally is coated by bacteria-resistant mucus, the opportunity for bacterial or protozoal invasion (into the uterine cavity, tubes and peritoneal cavity) may be greatest when sperm or bloody menstrual discharge lead the way. Significant infection of the female genital tract is usually referred to as *pelvic inflammatory disease* (P.I.D.) and diseases primarily transferred by *genital* (penis to vagina) contact are considered *venereal diseases* (in honor of Venus, the Goddess of Love). With its moist opening located between urethra and anus, the *vagina* supports a variety of viral, bacterial, protozoal and fungal information—some of which may irritate vaginal surfaces or be potentially harmful. But the many glands that lubricate your thick (yet distensible) stratified squamous (but hairless) vaginal epithelium also help you to repulse most potential invaders. All fertile females (but not males) retain that open passageway between outside world and peritoneal cavity. Presumably the resulting risk of severe internal infection is offset by a significantly improved likelihood of becoming pregnant since otherwise some less troublesome arrangement would surely have ensued. For example, why not run the entire series of events from developing follicle to ovulation to menstrual period behind or beneath the peritoneal cavity. Surely your long-suffering ancestors could have evolved a more trouble-free reproductive organ to pass along. Indeed, with just pencil and paper and a bit of time you could probably design a better system yourself. But before you get too creative, consider the following limitations.

In the first place, were an egg to be released directly into a closed fallopian tube, the oft-associated blood clot might temporarily obstruct that tube which could prevent sperm from meeting egg in a timely and effective fashion. In comparison, your fallopian tube ordinarily draws in the egg along with dilute peritoneal fluid left over after clot lysis. Furthermore, various sections of the female genital tract may undergo directional peristaltic contractions, especially during *orgasm* (sexual climax). Those contractions, along with the piston-like effect of the erect penis repeatedly and forcefully sliding up the close-fitting vagina, can milk or pump sperm up through the nearly closed cervical opening into the uterus and on into the tubes. Under such circumstances, tubes without open fimbriated ends might be more likely to overfill and cause pain or even blow out. So as usual, there are a great many factors to consider since natural selection favors reproductively advantageous traits rather than seeking perfection. But a design that allows some of the powder applied to a baby girl's bottom to move into the peritoneal cavity by ordinary random genital tract contractions (for like penile erections, these systems must be tested and exercised regularly so they will be available when a reproductive opportunity knocks) seems a bit dysfunctional. And with some brands of talcum powder allegedly containing asbestos and some ovarian tumors possibly resulting from drawing in such asbestos, a clean and dry baby girl bottom probably is preferable to a powdered one anyhow.

Consider also *endometriosis*, an uncommon but troublesome condition characterized by ordinary endometrial tissue growing upon pelvic surfaces of the peritoneal cavity or at some other *extrauterine* (outside of uterus) site. That misplaced endometrial tissue still responds appropriately to the monthly hormone cycle with sequential growth, followed by sloughing and bleeding. But since the resulting blood and cellular debris cannot possibly be expelled from the peritoneal cavity through your long and thin fallopian tubes, much of that loose cellular material accumulates and grows upon local peritoneal surfaces instead, thereby enhancing the problem. Endometriosis can cause pain, inflammation, displacement and scarring of the tissues that surround the uterus, often to the point of *dyspareunia* (painful intercourse) or even *infertility*. Dyspareunia is not a specific sign of endometriosis, of course, since any irritated or inflamed part of the female (or male) genital tract or even a nearby disease such as acute appendicitis, can cause pain on intercourse. As for how endometriosis may originate, it seems likely that some episodes of inward-directed genital-tract peristalsis (perhaps during REM sleep or intercourse) delivered still alive but *sloughed* (discarded) endometrial cells onto those lower peritoneal surfaces—certain chemicals such as dioxin may also increase the likelihood that endometriosis will develop. However, uterine peristalsis (menstrual cramps) ordinarily discharges your shed endometrium via the vagina without subsequent difficulty.

Widespread peritonitis can accompany venereal infection of the uterus and tubes—intraperitoneal *adhesions* (bridging bands of scar tissue) associated with the healing of that peritonitis are sometimes found above the liver. Antibiotics are indicated early in the course of any non-viral venereal infection to reduce the severity and duration of that infection and thereby minimize inflammation and scarring of the fallopian tubes. Severely infected (abscessed) tubes may require endoscopic drainage or surgical removal. But even a mild infection that subsides quickly (with or without symptoms and antibiotic treatment) can partially or totally block one or both tubes—scar formation during healing may also bring about a *hydrosalpinx* (blocked tube distended by sterile fluid). A woman becomes infertile when eggs cannot pass through either tube (while there are many other causes for infertility, this particular problem can often be overcome by surgical extraction of ripe eggs for in vitro fertilization with subsequent embryo insertion into the hormonally prepared uterus—but the necessary hormonal manipulations may markedly increase the likelihood of developing an ovarian cancer). An *ectopic* (outside of the uterine cavity) pregnancy may result if sperm are able to get in and the fertilized ovum then continues its relentless development despite making minimal progress toward the uterus. Ectopic pregnancies usually are intra-tubal although intraperitoneal pregnancy also is possible (cigarette smoking interferes sufficiently with tubal ciliary function and peristalsis to increase the risk of tubal pregnancy). Unless detected and terminated, tubal ectopic pregnancies lead to tubal rupture and intraperitoneal bleeding—quite obviously, a tubal pregnancy cannot progress normally, nor can an embryo that implants and prospers elsewhere than within the uterus then be delivered through the vagina. So the exceedingly rare abdominal pregnancy that does go to term will require *caesarian* delivery (named in honor of Julius Caesar who supposedly was surgically delivered from his mother's uterus—presumably without benefit of anesthesia, but note who got all the credit).

Avoiding Pregnancy And Avoiding Venereal Infections

Venereal diseases include viral infections of the vagina and cervix, some of which can cause painful sores and inflammation or increase the risk of cancer on those surfaces (and on every penis exposed to them). There is still no good antibiotic or other certain remedy for viral genital sores (which often become recurrent). And the raw or inflamed surfaces that result from other venereal diseases may also increase opportunities for transfer of the AIDS virus through ordinary heterosexual vaginal intercourse—in contrast to the known higher risks of anorectal penile insertion which exposes single-cell-thick rectal mucosa to destructive seminal fluid enzymes as well as traumatic tears.

So how does one avoid nasty venereal diseases (that could endanger future reproductive success) and also prevent pregnancy? Well the only sure method is avoidance of sexual intercourse. Other than that, you had better protect yourself from direct penis-to-vagina contact as well as easy sperm access by covering the penis with a condom (or perhaps using the preplaced equivalent—the so-called female condom). *Oral contraceptives* (birth control pills that usually depend upon some combination of estrogen and progesterone to suppress ovulation) may reduce the likelihood of PID without reducing genital exposure to venereal infections, while spermicides seem to reduce the risk of venereal infections. Modern IUD's—simple intrauterine devices that usually prevent embryo implantation (perhaps by causing uterine irritability or loss of uterine milk through the cervix), may even increase the chance of contracting a serious uterine infection following exposure to causative organisms. To some degree, having sexual intercourse with an individual who has had intercourse with other persons carries a portion of the risk one would take by having intercourse with all of those other persons and their past partners, the previous partners of those partners and so on. Fortunately that potentially exponential risk of infection is reduced by the fact that many of those individuals happened to be relatively immune, careful or lucky or were successfully treated for their most recent venereal infection prior to the connecting intercourse in question.

Unfortunately, neither condoms nor birth control pills nor hormone implants nor IUD's nor spermicides placed in the vagina nor covers placed over the cervix within the vagina (such as diaphragms or cervical caps) can prevent pregnancy

100% of the time. But if properly used, all can markedly reduce the normal likelihood (risk) that pregnancy will occur within a few months of beginning "unprotected" intercourse. Some women undergo tubal ligation or hysterectomy to achieve reliable birth control. Surgical removal of the uterus (including cervix) results in closure of the upper end of the vagina. Because such a *hysterectomy* leaves the vagina in place, enjoyable intercourse should still be possible, but now without the uterine component of orgasm or any risk of pregnancy. The French *anti-progesterone* "morning after" *pill* (*RU486*) reliably prevents embryo implantation in the uterus by blocking essential progesterone effects. It is often used in combination with a *prostoglandin* that encourages uterine contraction for expulsion of the loose embryo. High doses of estrogen for 72 hours are quite effective in disrupting the progression by which a just-fertilized egg becomes an implanted embryo. RU486 with a prostoglandin as well as other more direct intrauterine interventions (depending upon the medical facility being used and the specific stage of pregnancy) also can dislodge an unwanted embryo even after implantation.

Menopause

Hysterectomy need not include the removal of normal tubes and ovaries but *total* hysterectomy with removal of tubes and ovaries or even bilateral simple *oophorectomy* (removal of both ovaries) will bring on *menopause* in an adult female if menopause has not already occurred. The various symptoms and signs of ordinary or surgically induced menopause are commonly treated with small doses of estrogens (sometimes progesterone is added). This hormone treatment usually prevents annoying *hot flashes* (sudden capillary flushes or blushing) and, along with supplemental calcium and Vitamin D_3, helps to minimize post-menopausal osteoporosis while maintaining still useful reproductive tissues such as the vagina in better condition for enjoyable sexual intercourse. Total hysterectomy during early adulthood may encourage the onset of arteriosclerosis (coronary artery disease, strokes and so forth) at an unusually young age. Substitute hormone therapy is generally recommended for young adults who must undergo removal of their ovaries. Even those ovaries left undisturbed during hysterectomy may have their often-variable blood supply sufficiently impaired to bring on an early menopause (so a woman in her forties who must undergo hysterectomy might as well eliminate the risk of ovarian cancer by simultaneous removal of those faltering structures. This situation is not analogous to the incidental removal of testes since those structures still function in many elderly and retain some decorative value and impact on self-esteem for the rest—much as your no-longer-functioning breasts do).

Being a distensible centrally located hole in the female perineum, the vagina becomes an area of potential weakness when aging tissues *atrophy* (thin out). Relaxation of supporting tissues may allow the bladder to press downward against the front wall of the vagina or even push loose anterior wall tissues out of the vagina as a *cystocele* (which becomes especially prominent when straining). The rectum can cause similar protrusion of the back vaginal wall as a *rectocele*, or the uterus itself may sag sufficiently to become visible as a *uterine prolapse*. If symptoms of difficulty in urinating or *incontinence* (loss of urine control) become sufficiently troublesome, a total *vaginal hysterectomy* can be performed from below (through the overstretched vagina) with subsequent repair of the trimmed and resupported vaginal tissues. Such a rebuilt vagina should allow satisfactory sexual intercourse and restore bladder control. A majority of older women in this country have undergone hysterectomy—which is a larger number than can be justified medically, even considering the high incidence of P.I.D., endometriosis, severely symptomatic cystoceles and genital tract malignancies. Surely humanity would not have endured if most women required hysterectomy or other major surgery upon their pelvic organs. In particular, a mild cystocele associated with minimal incontinence on coughing, or occasional difficulties in voiding, or irregular uterine bleeding as menopause approaches, are not in themselves indications for surgery.

However, most mature women should probably have a pelvic examination every year or two, especially those who have ever had intercourse (thus non-virgins) and those who *douche*—the French word for a "shower". While a regular shower is appropriate and non-controversial, a vaginal douche or irrigation brings little benefit and appears to increase the risk of developing cervical cancer, especially if used more than once a month. Pelvic examinations should include a PAP smear (honoring Dr. Papanicoulau who invented it) seeking cervical cancer cells. Unfortunately, cancers of breast, ovary and uterine cervix often develop insidiously (without early warning signals) so they continue to be common causes of death for women. While endometrial malignancies are associated with early bleeding and other signs that increase the likelihood of detection when cure is still possible, the initial spread of ovarian malignancies within the peritoneal cavity allows little opportunity for successful surgical removal once symptoms develop.

There are a great many causes for female infertility in addition to prior genital tract infection, purposeful tubal ligation or hysterectomy. Commonly blamed are ovarian unresponsiveness or hormonal imbalances or enzyme insufficiencies at the ovarian, pituitary or hypothalamic levels. Occasional causes of infrequent or irregular ovulations, or failure of the embryo to implant, are excessive physical or mental stress (as in highly conditioned female athletes) or simply not being fat enough. Malnutrition from any cause may delay puberty and leave the potential mother without adequate stores of fat (reducing power) to provide for herself plus a growing embryo, fetus and infant. And there is some evidence that female fertility decreases more rapidly in those that drink a lot of milk during adulthood (increased levels of galactose may be toxic to eggs or possibly even raise the likelihood of developing ovarian cancer). Pregnancy has an obvious impact upon the entire individual, including her skeleton, nervous system, hormones, circulatory system, gastrointestinal system and so on. Even mammalian lactation requires the interplay of at least GNRH, LH, FSH, estrogens, progesterone, thyroid hormone, insulin, prolactin, oxytocin, chorionic gonadotropin, chorionic somatomammotropin and growth hormone (see Chapter 27).

Testes Hang Out To Be Cool

Although ovaries work best when warmed and protected within the peritoneal cavity, your two *testes* (or those of your friends) are individually wrapped and externally suspended within a scrotal skin pouch where they can be held at less than body temperature. Indeed, male *testicles* (the site of production for your male germ cells) cannot function normally if kept at body temperature. Testes form in the same retroperitoneal area as ovaries but testes normally migrate out into the scrotum before birth. Sometimes one or both testes will be delayed in emergence or even remain trapped along that retroperitoneal or inguinal pathway. Such an *undescended testicle* often requires surgical mobilization in order to reach the scrotum and is commonly associated with an *inguinal hernia* (outpouching of peritoneum at the groin). Surgical mobilization often requires careful cord thinning and lengthening which may endanger the blood supply of the testicle, so not all testes remain functional postoperatively. But if retained high in the groin or within your retroperitoneal area, a warm and therefore non-functional testis has an increased likelihood of becoming cancerous. Your scrotal skin tends to loosen when warm, and constrict or shrivel upward when exposed to cold. The voluntary muscles surrounding each *spermatic cord* also draw the testicle upward toward warmth or out of harm's way (the *cremasteric reflex* causes an upward movement of the testicle when the inner side of your thigh is touched or stroked). Testicle cooling is probably enhanced by countercurrent heat exchange between testicular artery and veins—there can be many testicular veins within each spermatic cord. In addition to blood vessels, muscles and nerves, the other important component of the spermatic cord is the *vas deferens*, a small muscle-enclosed tube through which sperm are pumped from testicle to penis.

Vasectomy is a sterilization procedure that interrupts both of the vasa deferentia, thereby stopping sperm passage and release but not sperm production. Autoantibodies to sperm proteins are commonly detectable following vasectomy, for the lymphocytes and macrophages called in to deal with this blocked production would ordinarily never encounter sperm and therefore do not recognize them as self. Whether these autoantibodies have any impact at all on subsequent health is yet unclear—as with most contraceptive decisions there are many known and unknown costs and benefits to be considered. The intact vas deferens drains sperm from a multi-tube cap-like *epididymis* on the testicle. That epididymis is where mature sperm are stored for a few days until roused into action during sexual intercourse or otherwise discharged (e.g. during masturbation or a "wet dream") or engulfed by local phagocytes when outdated. As is true for ovaries, your testicular germinal and endocrine tissues are primarily controlled by the gonadotropins FSH and LH. FSH stimulates sperm production and maturation within your testicular ducts while LH stimulates *testosterone* secretion by special cells located adjacent to those sperm-producing ducts. In turn, testosterone brings about *secondary sex characteristics* such as male-pattern muscle and bone growth, enlargement of your male sex organs, production and maturation of sperm, enhanced growth of body hair and eventual male-pattern baldness as well as enlargement of your larynx (which leads to deepening of the voice). It seems likely that exposures to lead (Pb) can impair male fertility—perhaps by reducing circulating testosterone levels (at least lead decreases sperm effectiveness in rats).

Sperm Are Not Very Smart

Normal adult male testes produce sperm continuously. An individual sperm matures in about six weeks. During that time the developing sperm remains embedded in a large *sustenticular cell* that provides nutritional support. The DNA in each sperm is tightly coiled about a *protamine* core. Protamine is even more highly charged than those histone proteins that DNA coils around in other cells. Thus sperm DNA information is relatively inaccessible to the sperm. In addition, a sperm departs with little of the essential cytoplasmic machinery utilized by most cells for day-to-day living. But even if sperm were better endowed with cytoplasm and had easier access to their own DNA information, sustenticular cells would still have to nurse sperm along to maturity since sperm do not carry a full deck of chromosomes.

For it takes a *haploid* (half normal) chromosome count in each sperm and egg nucleus to provide the *zygote* (fertilized egg) with a full *diploid* set (23 *pairs*) of chromosomes, one of each pair being inherited from mother and one from father. Male-determining sperm carry twenty-two ordinary paternal chromosomes plus the small *Y* chromosome which lacks a lot of essential genetic information borne by the much larger *X* chromosome that is present within every female-producing sperm and every ovum. Thus male-producing sperm would require external life-support in any case. Furthermore, all sperm and eggs carry many detrimental *recessive* genes that might damage any sperm or egg relying upon inherited haploid DNA for advice. Your testicles maintain a fine inventory of mature sperm—in part as a response to the local release of *activin*. However, when your sperm production consistently exceeds your sperm discharge rate, your testicle cells will secrete *inhibin* (another testis-produced protein) to reduce pituitary FSH production and thereby decrease testicular sperm (and perhaps male hormone) production. Activin and inhibin are produced within the ovary as well—inhibin helps to limit or suppress excessive responses to FSH and other stimulants that if uncontrolled could lead to malignant tumors. And inhibin may be relevant to the "use it or lose it" experience of older males (so masturbation might help maintain male sexual function when acceptable alternatives for the regular discharge of sperm are not readily available).

Delivering Your Sperm

As sexual excitement mounts, sperm are transported through epididymal and vas deferens ducts by peristalsis of those tubes. Various nutrient, supportive and facilitating secretions are added en route before sperm finally are pumped from the urethral opening of your erect penis during orgasm. The urgent preparations for this potential reproductive event include activation of your two bulbo-urethral or *Cowpers glands* that put out small amounts of clear alkaline fluid to neutralize any acids left in the urethra by prior urination (the reflex closure of bladder sphincters prevents male or female urination during intercourse). *Seminal vesicles* and *prostate* (*not* prostrate!) *gland* add larger quantities of nutritious and suspensory *secretions* (your prostate surrounds the origin of your urethra at the bladder neck). Arousal with *erection* of the *penis* is a parasympathetic function, *ejaculation* (discharge of sperm and associated fluids during orgasm) involves sympathetic nerve discharges.

Penile erection is due to nitric-oxide (NO) induced smooth muscle relaxation that allows inflation of penile arteries and blood-containing chambers. The same sort of erection in the female *clitoris*—a comparable but far less prominent structure (except in the female spotted hyena, due to her high androgen levels), enhances female sexual sensitivity during intercourse. In both male and female, the areawide congestion of genital blood vessels that leads to erection and swelling of sexual structures continues only as long as arterial inflow markedly exceeds (and thereby also impedes) venous drainage—usually until male ejaculation and female orgasm (which are comparable events) or until the sexual excitement wears off and arterial inflows decrease. As you might expect, individuals vary greatly in their sexual responses, appetites and sensations. Thus one might simply define as "normal" (which everyone seemingly wishes to be, *only more so!*) any sexual activities or responses that satisfy sexual urges and could under some circumstances result in a pregnancy. Preliminary or exploratory sexual activities such as kissing or petting that do not involve full intercourse (penis to vagina contact) cannot cause pregnancy unless they lead to intercourse (or at least, sperm discharge into the vaginal opening).

Since your sexual partners may tolerate their venereal (viral, bacterial or protozoal) invasions with few or no symptoms, the source of your latest infection may not have been aware of his/her own "subclinically infected" status. Male sexual tubes, from the tip of the urethra back to the prostate and seminal vesicles and even back to the epididymis, can become infected by viral, bacterial or protozoal residents on a short-term or ongoing basis. *Urethritis* (*-itis* signifies inflammation) can lead to scarring and thus a *stricture* (narrowed urethral channel) that may interfere with urination. An infected prostate or epididymis may cause significant and persistent difficulties as well, but venereal infections rarely bring about male infertility directly—although prior or current infections can certainly contribute to sexual dysfunction in older males.

The male prostate (a secretory organ with many interlocking passageways) is easily felt anteriorly through the thin wall of your rectum by the physician's gloved finger. *Benign prostatic hypertrophy* (enlargement) in the middle-aged and elderly may expand that gland sufficiently to compress the urethra and interfere with urination. In that case, partial prostatectomy can restore free urine flow (when necessary, a sterile hollow flexible catheter inserted up the urethra into the bladder can maintain urine output prior to surgery). *Trans-urethral prostatectomy* is less disruptive to nearby autonomic nerves than a direct surgical attack from outside of the prostate so it is more likely to preserve penile erectile function. But the prostatic channels unroofed as the prostatic urethra is cored out from within may thereafter release ejaculate upward into the urinary bladder rather than toward the tip of the penis. As the second most common cancer of older American males (lung is first, in honor of the tobacco leaf), prostatic malignancy is somewhat more prevalent in blacks, perhaps due to their higher average testosterone levels. Prostatic cancer can be treated by surgery or radiation; hormonal manipulations are sometimes used as well. All such treatments tend to cause male *impotence* (sexual dysfunction with loss of erection).

As usual, there is much more to any intact and functioning person (or other complex mechanism or creation) than their parts and how these work and interact. Even the simple act of human sexual intercourse can be brutal and dehumanizing or a truly magical moment in a wonderful long-term relationship or more commonly, something in between. And the two partners involved in a particular sexual interaction may interpret all aspects of that event quite differently. For example, the average American male estimates the length of his erect penis at ten inches while the average American female views it as about four inches long. Perhaps "the old 6 inches" simply averages out such perceptual differences. But with every aspect of what we are or do so easily interpreted or misinterpreted in a thousand different ways, human relationships generally improve when there is an open and friendly communication of wants, needs, fears, pleasures and displeasures.

Mitosis And Meiosis—Diploid And Haploid

Each of your ordinary body or *somatic* cells carries complete copies of the 23 chromosomes contributed by your mother and 23 chromosomes contributed by your father. In a male, the small Y chromosome from father is paired with (but mostly not equivalent or *homologous* to) the X chromosome from mother. Therefore, in addition to its tiny male-determining portion and a few other genes, the Y chromosome carries numerous non-functional nucleotide sequences unrelated to (and thus unable to recombine with) many of the genes borne by its much longer X partner. Some researchers are investigating variations in nucleotide order on the Y chromosome of different races in an effort to determine ancestral relationships. Others are studying random *neutral* (causing no reproductive effect) changes within mitochondrial genes (inherited entirely from your mother in the egg cytoplasm). Such evidence should lead to more accurate lineages and better guesses at when and where the first human male and human female walked hand in hand under the stars. Of course, the unknown number of charming ancestral primate couples (often referred to as Adam and Eve) that gave rise to our common human line of descent so long ago can only have differed imperceptibly from their cousins whose descendents are the gorillas and chimpanzees.

Other than those *sex* (X and Y) *chromosomes*, your 22 chromosome pairs generally bear comparable information at matching sites on each chromosome of any pair. Here the term "comparable" implies that the gene may be identical, similar or different, but that the information each carries affects or is relevant to the same subjects or systems. Genes occupying matching sites on homologous chromosomes are referred to as *alleles*. Often one allele carries *dominant* information (such as eye color) that is always expressed. Sometimes both genes may be partially expressed— at other times either dominant or *recessive* (not ordinarily expressed when a dominant allele is present) genes may be randomly activated or suppressed in any particular cell or even in many cells throughout your body. Indeed, diploid female cells seem to randomly *imprint* (inactivate) one of their two X chromosomes—presumably this prevents those cells from being overloaded with X products, given that one X clearly suffices for all males. But whether or not a gene is expressed has nothing to do with how it is inherited—indeed, certain genes from father or mother may remain "turned off" in you (by methylation or whatever) until the time comes for them to be duplicated and passed onward.

Your unique genetic inheritance came through one plucky haploid sperm and a fine haploid egg, both too small to be seen without magnification. Except for your more or less unaltered Y chromosome (if you are a male), the chromosomes added by father's sperm and those present in mother's egg all delivered *recombinations* of the genes that your two parents inherited from each of their parents. Yet despite repeated recombinations, individual genes remained in proper sequence along each chromosome—just as the hundred-thousand recombined genes of your every sperm or egg continue to match up with those in the recombined chromosomes of your sexual partner's egg or sperm. In essence, the process of *meiosis* that gives rise to your germ cells causes matching chromosomes to *recombine* before *reducing* (halving) their final number. Such a *haploid* number of *recombined* chromosomes is in marked contrast to the *diploid* (paired) chromosomes that must be duplicated during every cell *mitosis* so that each *somatic* ("soma" means *body* or non-reproductive) daughter cell can receive a true copy of all your chromosomes just as they were inherited from mother and father.

More specifically, meiosis requires a *single duplication* of chromosomes, followed by *recombination* (an orderly but relatively random re-allocation of grandparent genes within each group of four homologous chromosomes), followed by *two cell divisions* without further duplication of chromosomal DNA in the cell nucleus. Thus the final product of one complete multi-stage meiosis in the male testicle is *four* motile sperm, each bearing a single (haploid) recombined set of chromosomes—the random nature of this recombination process ensures that no two sperm will carry exactly the same information. Each egg nucleus bears similarly unique recombined information in a single (haploid) set of chromosomes produced through the same reduction-division process (*one* chromosome duplication followed by recombination and then *two* nuclear divisions). But rather than giving rise to four little eggs, the outcome here is one bulky egg that eventually retains only a single haploid nucleus within its protoplasm while the other three tiny haploid membrane-enclosed egg nuclei (known as *polar bodies*) are discarded and soon disintegrate. The final meiotic division (the one that gives rise to a haploid egg nucleus plus polar body) is put off until a sperm nucleus has penetrated the egg cytoplasm—perhaps this reduces the likelihood of harmful recessive genes being expressed prior to fertilization.

The endless ways in which the maternal and paternal characteristics that you have inherited can recombine explains much of the great variability among your actual or potential offspring—some of your innumerable potential offspring would undoubtedly be more fit under current circumstances and others less so. And later tough times are far less likely to threaten all members of such a heterogeneous population than an equal number of identical or very similar individuals from a clone or an *inbred* group (consequent to many generations of sexual intercourse/intermarriage among closely related individuals). Even under normal circumstances, an inbred population of cotton, corn, humans or hogs may display weaknesses or deficiencies not prevalent in the wild strain. This comes about because different individuals vary in their susceptibilities to disease and all carry some frankly deleterious genes. Good or bad recessive genes are often masked by dissimilar dominant genes, and the owners of really harmful dominant genes soon die out. However, inbred populations of animals or humans often inherit and maintain matching *deleterious* recessive *genes* that contribute to embryo loss, mental retardation, deafness, bad hip joints or whatever. On the other hand, viable cheetah populations have become almost clonal in the similarity of chromosomal content between different individuals without displaying any deleterious genes under current conditions. Perhaps the closely related recent cheetah ancestors that successfully passed through some severe population bottle-neck just happened to carry few deleterious genes.

The opposite of inbreeding is the selection of two unrelated and very dissimilar parents. This usually results in the expression of few if any deleterious recessive genes—which accounts for the *hybrid vigor* (above-average growth rate and perhaps intelligence) of hybrid corn, *mongrel* dogs, cross-bred cows or people. As mentioned before, the whole point of sex is to maintain variability and prevent gradual deterioration of an entire species by the accumulation of damaged or detrimental genes—as might occur within a clone (very similar or clonal species are also more susceptible to elimination by viral and bacterial epidemics). And that is why men provide so much reproductive advantage. Since you are an outcome of *sexual* reproduction, you differ significantly from your siblings—undoubtedly some of you will achieve greater reproductive success than others. And a naturally selected few among the great variety of individuals resulting from sexual reproduction will probably be producing reproductively successful descendents long after the last asexual dandelion seed has flown.

CHAPTER 27

GROWTH AND DEVELOPMENT

You Used To Be A Good Egg;... The Egg That Would Be You;... Changing Patterns—How Living Matter And Energy Interact In The Fourth Dimension (Time);... Your Coming Out;... No Strings.

You Used To Be A Good Egg

You are the happy result of an unlikely encounter between a vigorous weeks-old sperm and the perfect egg that it roused from decades of magic slumber. Although barely mature, both were mere hours from apoptosis when their individually incomplete information packets cooperatively combined to escape that fate and possibly realize a prolonged and productive existence. While such an early post-release death sentence for all unmated sperm or eggs may seem unduly harsh, it favors those few sperm and eggs that remain strictly on schedule. At the same time, it reduces the risk that your many billion sperm (yearly output) or those hundreds of thousands of unused eggs (lifetime supply) might somehow become metabolically demanding or otherwise turn against you. After all, your every ejaculate may release as many sperm as there are people now living in the United States (at least this was true of 1940's sperm counts—current average sperm outputs have apparently fallen below half of that total in response to the prevalence of weakly estrogenic chemicals in the environment—those pesticide, plastic and petroleum residues may also contribute to the rising incidence of testicular and other reproductive system cancers). Even your many dozen trillion other cells would soon be hard pressed to maintain their competitive edge if they had such huge numbers of retired germ cells to feed.

Your sea-dwelling invertebrate and vertebrate ancestors usually released their sperm and eggs for external fertilization. The more successful of them came together or at least synchronized the discharge of their *gametes* (sex cells) in seasonal, often momentary, mass matings—that brief and glorious excess of gametes and larvae inundated all potential predators without providing them a reliable new dietary staple. Such synchronized underwater group sex also increased the likelihood that each egg emitted would meet some lucky sperm suitor of the same species—for death awaited any innocent egg seduced by some fast-swimming sperm of another species—a tragic waste of two fine (if misguided) young gametes. But even with perfect timing, the enormous numbers of sperm of so many different species being released into the sea during certain seasons made it important that each species develop specific barriers to prevent fertilization by inappropriate sperm—while still requiring arriving sperm of the right species to prove their mettle in a winner-take-all joust for that ovum's affections. The installation of an effective species-specific sperm-recognition-and-frustration system around each egg therefore brought great reproductive advantage. And that is just another reason why your mother never risked accidental fertilization despite the uncounted trillions of sperm that may have been released by corals and other sea creatures as she snorkeled beyond the reef. Being a mammal, she depended upon a penis for internal delivery of sperm anyhow—which further reduced opportunities for accidental fertilization by sperm of another species.

The Hebrew Bible (Old Testament) reports that God strongly disapproved of human sexual intercourse with animals. Recently discovered interspecies fertility barriers suggest that those biblical injunctions were not needed to reduce the prevalence of *centaurs* (half man, half wild horse) or *mermaids* (half woman, half cold fish)—so presumably interspecies *fornication* (sexual intercourse) was at least a perceived problem in the Days of the Prophets. More recent jokes about virgin wool coming from sheep that can run faster than sheep herders bear similar implications. It is well known that horse-donkey sex creates almost-always-sterile mules, although the progeny of wolf-coyote-dog or tiger-lion liaisons are known to enjoy reproductive success. In view of their minor genetic differences, the possibility of reproductively successful pairings between humans and chimpanzees ought to depend mostly upon any shared sperm recognition signals and the similarity of chromosomal gene sequences at crucial developmental sites (since surprisingly simple rearrangements of gene order and chromosome number may account for many quantitative differences between mammalian species—see syntenic sequences).

However, regardless of whether you came about as a result of careful planning or just a *coolie* (a quickie in the snow), it all began with what any unbiased observer would have viewed as a remarkably vigorous and less-than-rational interaction between your father and your mother. Undoubtedly, only the tiniest fraction of the several hundred million sperm swiftly pumped into your mother's vagina by your father's erect penis ever got anywhere close to the egg—which at that point was probably drifting within one of her fallopian tubes. Perhaps some of the first sperm to arrive were helped along by the plunger-like effect of penis in vagina as well as the peristaltic contractions of your mother's genital tract during and after her orgasm. But to enjoy any chance of success, each sperm approaching that egg had first to become *capacitated* (by removal of cholesterol that stiffened it) via exposure to glycoprotein-containing fluids in mother's genital tract. That capacitation prepared those sperm to release lytic enzymes from their acrosomal cap upon arrival at the egg-side. (The *acrosome* is derived from the spermatid Golgi apparatus. It carries enzymes to dissolve the proteins and complex sugars that surround the ovum. Thus the acrosome is equivalent to an externalized lysosome.) While capacitation involved some delay, it also prevented those sperm from ejaculating their introductory chemicals either prematurely or too late (in love as in war, timing is everything).

Still it took the combined chemical attack of many sperm suitors (each invigorated by the Ca^{++} influx following capacitation) to eventually break down the virginal integrity of cellular and acellular layers enclosing that shy ovum—until finally your tiny but tough founder sperm wriggled through into the splendidly provisioned cytoplasm of that huge 100 micron (0.1 mm) diameter egg. But first that sperm had to prove identity and suitability by establishing contact between its (formerly *sub*-acrosomal) *acrosomal process* and the firmly anchored microvillus recognition processes extending up from the egg surface. That contact secured enzymes exposed upon the sperm's forward tip to matching egg surface molecules, leaving sperm and egg locked together happily ever after by an enzyme unable to complete its usual reaction—yet another advantageous (recognition/docking) mechanism brought about by multiple minor modifications of previously available components. Your father's successful sperm dropped its tailpiece upon entering your ovum's chaste cytoplasm—its nucleus was then borne in triumph toward egg center while your mother's slumbering egg nucleus finally awoke to undergo its last meiotic division and expel the resulting polar body (spare haploid nucleus).

The Egg That Would Be You

Upon satisfactory proof of initial sperm entry, your egg became impenetrable to further sperm by electrical and enzymatically induced changes within its rapidly hardening surface layers. The outer cytoplasmic layers of your just-fertilized egg then rotated about the inner cytoplasm—a 30° twist that provided the fertilized cell with bilateral symmetry (left and right, front and back) where previously the ovum only had radial symmetry (head and foot). This post-fertilization twist may also have mixed previously segregated cytoplasmic ingredients together and thereby initiated all of those complex and wonderful molecular interactions that would eventually be you. Of course, the ovum already was well organized chemically, with its greatly enlarged cytoplasm packed full of essential nutrients and vast numbers of unexpressed messenger RNA molecules—mostly contributed by nurse cells of the surrounding egg follicle (a single egg nucleus could never have set up, organized and provisioned such a large egg by itself).

In any case, the fateful encounter between soon-to-die sperm and almost apoptotic egg that resulted in your fortunately long-lived *zygote* occurred some minutes to hours after intercourse, when mother and father were probably both asleep or else had gone on about their other activities. Only then did that sperm-induced metabolic awakening activate ancient ancestral advice that had been deposited within the egg several decades earlier while mother was still growing within grandmother's uterus. Finally stirred and stimulated by Ca^{++} influx and a rising pH, fueled and guided by the nutrient energy and informational endowment of the egg, that fertile seed began slowly to change. For half of that momentous day there was little obvious activity as sperm and egg pronuclei were brought closer together and their chromosomes duplicated. Then, when the fertilized ovum divided in two, sperm and egg nuclei finally met to match homologous chromosomes and compare strengths of alleles.

Even then, with its information combined, complete and appropriately activated or suppressed, the further development of this fertilized ovum still depended upon an endless number of past and future factors—ranging from a proper and undisturbed initial preparation of the egg while mother was still within grandmother's uterus, to the correct combination of a sperm and an egg carrying competitively useful information, to an appropriately supportive maternal genital tract. A lot of other special circumstances also had to be just right, a few of which are reviewed below. Not surprisingly, the large majority of fertilized zygotes never implant successfully, and that is reproductively advantageous since pregnancy not only depletes a mother but also terminates her lactation—often the principal source of nourishment for the previous child. In many nations where contraceptives are unavailable (or even prohibited!?), those little displaced ones then suffer from malnutrition and a high mortality rate. This problem has increased as "prolonged" breast feeding has gone out of style. Indeed, merely by keeping pregnancies more than two years apart, *contraception alone could save the lives of at least half a million young children every year.*[1]

Consider again the incredible odds. How many different ways could 100,000 diploid genes be recombined within the 23 haploid chromosomes of each parental

---

[1] Thapa, Shyam. Short, Roger V. and Potts, Malcolm. 1988. "Breast Feeding, Birthspacing and Their Effects on Child Survival". *Nature* 335. pp 679-82

sperm or egg? The number is almost incalculable. And how did one particular egg happen to outdo its hundreds of thousands of sister eggs (also present since mother's birth and seemingly just as susceptible to this particular hormonal cycle) in its rapid production of gonadotrophic hormone receptors? And how likely was it that everything else would work out just right this time? Judging from our overpopulated-by-humans world, it works very well, plenty often—indeed, on average, a healthy human female not using contraception is likely to become pregnant within about five months of regular unprotected intercourse (even though three out of four fertilized eggs are naturally discarded before or after the pregnancy is recognized). But a few months is not so long to wait when that pregnancy, birth, growth and development will probably consume the next several decades of your life— especially since a large percentage of parents eventually regret ever having reproduced. (There may also be days when a majority of those who are childless regret not having reproduced. But in this rapidly changing world, the teen-age parent with little education is probably destined for a lifetime of poverty and dissatisfaction).

During the first two days following fertilization, the standard progression of events proceeded flawlessly according to its ancient schedules, guided by prepositioned messenger RNA from the original ovum rather than fresh instructions manufactured within the nucleus of each newly formed cell. Although its external dimensions remained unchanged during this period, an ongoing series of cell divisions was subdividing the fertilized ovum within its clear flexible outer covering into an embryonic ball composed of many smaller cells. After a couple of post-fertilization days within the fallopian tube, that small tightly packed ball of new cells finally was propelled into your mother's uterine cavity by the cilia and peristaltic activity of her tube. Only at this point could the hardened outermost layer safely be discarded—any sooner and you might have implanted within the tube. Apparently your early cell bundle arrived at a time when endometrial growth and secretory activity were optimal, so you prospered. Within the uterus as within the tube, that nondescript and now hollowing clump of cells was continuously bathed in nourishing fluid (here known as "uterine milk").

Indeed, you had a really great start, with no bacterial or other cellular competitors unless a sibling parasite was developing nearby. Any non-identical sibling(s) originated in one or more simultaneously shed and fertilized egg(s)—identical siblings could only arise through a complete separation of your earliest embryonic cells—at a time when those initial cells still had access to all necessary information for producing an entire person. For as your cells increased in number, they also became more specifically destined to serve in certain embryonic regions or tissues.

That concentration upon particular informational subsets with their attendant limitations undoubtedly reflected the built-in maturational effects of multiple cell divisions and specific neighboring-cell-and-position effects—all responding to the way various diffusible cell organizer products influenced those sequential interactions designated by certain long-established *homeobox* (organizer gene) sequences that you share with the worm, the fly and the rat. Thus each member of your expanding cell population inevitably specialized as it followed long-established ancestral pathways. Many experiments on various animals have shown how additional cell divisions progressively limit embryonic cell options. For example, fossil evidence demonstrates that hen's teeth disappeared about 10 million years ago (along with honest politicians). Yet very early embryonic chick epithelial cells can still be induced to form teeth under the influence of adjacent mouse jaw mesothelium—an impact that the comparable chick mesothelium has lost.

If you had to share mother's uterus with siblings, it probably became too crowded for any of you to reach full normal infant size before birth. But at least you were not subjected to the more severe sibling rivalries encountered within some sharks that give live birth to a single descendent after it has consumed all of its younger weaker siblings. Instead, your now uncovered and enlarging mammalian cell bundle settled down and eroded into mother's rich and nourishing uterine lining for your embryo's first solid food (there are no intrauterine vegetarians). As your cells digested their way ever more deeply into that lush endometrium, they finally entered fresh puddles of mother's blood that were constantly being replenished by her uterine arteries and emptied by her uterine veins. And that is where "future you" took root, soaking within the plentiful oxygen and nutrients of your mother's bounteous blood, into which your wastes and instructions also diffused.

Your *chorion* (outermost embryonic cell layer) eventually combined with mother's endometrium to form the *placenta*. Thereafter your placenta served as GI tract, kidneys and lungs in addition to playing numerous other (endocrine and non-endocrine) roles until you were born. That partly-you/partly-mother placental tissue provided a relatively huge exchange surface between mother's blood passing close by on one side of a thin endothelial-like layer and your own embryonic and fetal blood being pumped past on the other. But long before your placenta was organized or you even had a semblance of a circulation, your chorionic tissues already were regulating mother's hormonal system on your behalf. The first and most urgent chorionic commandment released into mother's blood stream was *chorionic gonadotropin*. That signal (detectable as a positive pregnancy test on mother's blood or urine within eight days of fertilization) was primarily directed at the corpus luteum in mother's ovary. It ordered her corpus luteum to continue pumping out estrogen and progesterone rather than fade away with the usual monthly ebb of pituitary gonadotropins. For clearly you had to delay mother's next endometrial separation and menstrual discharge for nine months, since any menstrual period would have discarded your embryo as well.

Although your chorionic gonadotropin took charge of mother's ovary many days before she missed her first menstrual period, that was just a start—soon your placenta was producing estrogens, progesterone and placental lactogen (to prepare her breasts for giving you milk. Those increased estrogen levels stimulated renin and aldosterone secretion, causing her to retain Na⁺ and water) as well as somatomammotrophin (growth hormones) and relaxin—all helping to put you more directly in command of mother's reproductive physiology so you no longer had to rely upon her possibly overworked ovaries. Other placental products included corticotropin releasing factor (stimulating mother's ACTH release among its other effects) as well as 1,25(OH)$_2$D$_3$ (the activated form of Vitamin D) and an enzyme variously known as *oxytocinase* or *vasopressinase*. You circulated the latter peptide in your mother's blood to prevent any oxytocin effect from causing uterine contractions. It may also have reduced the effectiveness of her antidiuretic hormone (ADH or vasopressin) and thus caused her to drink more fluids to compensate for her increased urine output—pregnant women become thirsty at a lower blood solute level anyway (being slightly diluted makes water more readily available to the placental circulation).

As a half-foreign parasite invading your mother's tissues, your placenta had to deactivate mother's potential immune response to any paternal antigens expressed at the placental surface. While the entire mechanism by which your placenta avoided maternal rejection is as yet unclear, it apparently included suppression of mother's T-cell-mediated immune responses within the uterus by your own outermost cell layer—mother even remained tolerant of any ABO-compatible fetal cells that spilled into her circulation during the course of the pregnancy, although fetal cells remaining there after your birth were subject to immune attack (see Blood). Paternal genes dominate placental function because the placenta really represents father's interests more than mother's (especially in earlier times when placentas from different fathers had to compete for limited uterine resources. As yet it is unclear how one or another parent's chromosomes (those relevant to the placenta, for example) are *imprinted* (inactivated) under such circumstances until replicated for inheritance by the next generation—but such imprinting may explain the deaths and deformities that follow when an embryo receives a double set of chromosomes from one parent (as occasionally occurs due to some initial distribution error) rather than getting a matched pair—one from each parent.

Unlike your amphibian ancestors who necessarily returned to the water to breed, your mother produced only *amniote* eggs capable of carrying their own food and water supplies—the sort of late-model eggs that finally freed your reptilian relatives and their direct descendents (you and the birds) for sexual reproduction on dry land—arthropods independently evolved similar egg casings that made terrestrial invertebrates (such as the insects) possible. However, the tough yet porous egg shells within which all of these species protcted their eggs on dry land also made it necessary for those eggs to be fertilized within the female before those outer shell layers could be applied by the female oviduct. An oviduct already was essential for laying eggs. The penis only became an important tool for the internal delivery of sperm during reproduction on dry land. So even though you may treasure your penis or that of a friend, the female structure is the more basic—which explains why genetically male (XY) or female (XX) embryos that lack gonads always produce normal female structures (absence of gonads is a very rare problem that only attracts attention when there is no puberty). In view of the fact that males not only bear useless nipples but also must hang their testicles out to cool (below body temperature), plus all of the evidence for the female structure being more fundamental (what could be more fundamental to life than giving birth?) and durable (women generally outlive men under comparable conditions), those who are incurably devout might reasonably conclude that God created woman in Her own image—which would make women the righteous rulers of all God-based religions. Furthermore, since women cannot possibly make a worse mess of things than all of the Popes and Patriarchs, ayatollahs and rabbis, ought not those female impersonators now step apologetically aside?

Even such metaphysical ruminations become moot when you contemplate what a great majority of those *totipotent* (each capable of producing another complete and moral human being) early embryo cells were assigned to the four critically important egg membranes of your reptilian forebears. Thus despite the fact that *those membranes no longer served any valid function in mammals*, they still formed in the usual reptilian fashion before being modified for other purposes or merely recycled—as when your outermost chorion membrane fused with the underlying *amnion* and that amniotic membrane in turn soon secreted *amniotic fluid* to protect and float your soft and initially unsupported developing embryo until birth—by which time you had sufficient internal bracing to stay in shape when the amniotic sac (also known as mother's "bag of waters") emptied onto the seat of father's new car as they sped to the hospital. Although the mammalian placenta eliminated the need for egg yolk over 220 million years ago, your embryo still developed a fine *yolk sac* in continuity with its gut—you even extended your embryonic blood vessels onto that yolk sac surface in strict accordance with your inherited reptilian schedule. Naturally your functionless yolk sac and its blood vessels soon atrophied and disappeared—after contributing to your midgut and placental vessels respectively. But being useless and inactive at least made the yolk sac a tranquil temporary refuge (from the cellular hubbub and early flood of powerful activating and suppressing molecular messages) for the *germ cells* that later would give rise to your own eggs or sperm. Furthermore, to maintain a sanitary environment within its non-existent shell, your embryo formed yet another ancestrally important sac to collect the urates excreted by all birds and reptiles. That *allantois* showed up right on schedule (for reptiles) and eventually contributed additional blood vessel segments to and from the placenta. Of course, with every mammalian uterus (an enlarged oviduct) already ensuring easy elimination of embryonic wastes via placental surfaces and mother's circulation, your allantois' most important role may have been to remind you yet again of your humble origins.

But the tiny island of future-you cells that these four egg sacs enclosed or bordered upon was not idle—your first cell layers to differentiate (*ectoderm* and *endoderm*) soon were separated by invading *mesoderm* cells. In turn the mesoderm split to create your central *coelomic cavity.* Your *nervous system* originated as a central (mid-dorsal) infolding of your ectoderm and your germ cells migrated back from the yolk sac. Indeed, most of your organs were initiated during the eight busy weeks following fertilization. Luckily your cells were well protected from physical harm by the surrounding amniotic fluid and generously provisioned by your mother's metabolic and mineral reserves. For this period of rapid cell division with ritually choreographed streams and layers of cells growing over, around and across each other as they migrated about—separating and joining endlessly—was also a critical time when chemical injury (mother's use of medications, alcohol or illicit drugs) or illness (mother getting German measles, for example) could easily have caused you serious deformity or damage even unto death. Happily that thin placental barrier protected you from mother's many minor indiscretions or infections.

Those eight busy weeks brought you to a point (often referred to as the end of embryonic development and the start of *fetal* life) at which an experienced anatomist could finally begin to distinguish your own tiny embryo from that of a similar-stage fish, turtle, pig or chimpanzee embryo (see cover illustration). Certainly your future claim to the rights, privileges and duties of a fully accredited human being were less than obvious, especially in comparison with an apparently identical pig or chimpanzee embryo (the latter also displaying almost-human proteins—yet pigs and chimpanzees are commonly viewed as property). Evidently humans differ from other animals *quantitatively* but not *qualitatively* during their progressive growth and development until birth—which is why experiments with worms, fruit flies and mice can provide so much valuable information about our all-too-human selves. Furthermore, the fact that both plants and animals are basically bags of bugs (see Cell) also explains why our efforts to poison other life forms with herbicides and pesticides so often harm us as well (e.g. those declining sperm counts—though on average they still remain ten times higher than necessary).

Indeed, there is remarkably little chemical difference between the steroid hormones such as *estrogen* and *testosterone* that signal your entry into puberty and the *ecdysone* which orders insect larvae to molt and pupate. Nor ought you be amazed that *some plants* have acquired the DNA information necessary to produce *insect juvenile hormones*—thus preventing maturation and reproduction of the insect pests that feed upon them. And while prolactin promotes your breast cell milk production—among other things, by doubling the number of *casein* messenger RNA's and extending the functional duration those messenger RNA's by 25 times (casein is the principal phosphoprotein of your milk)—frogs produce prolactin to retain larval (tadpole) status and salamander prolactin signals the onset of a second metamorphosis in preparation for returning to life in the water and spawning. Note again how frugal evolution is, with its many appropriate but wildly different uses for the same signal modules, molecules and receptors. For example, your personal thyroid hormone could stimulate the metamorphosis of tadpoles into frogs. And among other interesting changes induced by that (or their own tadpole) T_3 is the transition from humble tadpole (prey and herbivore) demeanor to that of fierce frog (predator). As an herbivore, the tadpole is well served by eyes located laterally with the tadpole's right optic nerve going entirely to the left brain and vice versa. But a frog brain must continuously track moving visual images in addition to utilizing binocular vision for depth perception. So those progressively repositioned frog eyes send new connections to the brain until only half of the axons in each frog optic nerve cross the midline (just as in your eyes). And that midlife frog-eye-rewiring crisis is thyroid dependent. Clearly, in yourself as well as a frog, the T_3 signal activates some genes, suppresses others and leaves still others unchanged. Furthermore, there are threshold effects so that frog legs start to grow at a lower T_3 concentration than that which causes tadpole tail regression (thereby maintaining mobility throughout metamorphosis). A great many other signal proteins and their receptors interact in similar fashion during growth and development (many of those signal proteins belong to a single molecular superfamily that happens to fasten appropriately onto DNA—see Hormones).

Changing Patterns—How Living Matter And Energy Interact In The Fourth Dimension (Time)

Matter and energy tend to form patterns—ripples and sand dunes, rocks and galaxies, atmospheric clouds and life itself. Despite markedly different starting points, comparable conditions often bring about similar outcomes. Yet patterns tend to evolve with time, changing in all four dimensions (up and down, back and forth, side to side and past to future). The growth and development of every living thing fits seamlessly into the unbroken but ever-changing process of birth, maturation and reproduction that stretches back to life's very beginning. Life's most important molecules are distributed through all animals, plants, single-cell creatures and bacteria. The final messages that comparable molecules carry and the structures or even sequences that they designate may resemble one another or vary endlessly but the rules are simple—forgo useless complexity, avoid extravagance, recycle everything and above all, out-reproduce the competition. And so by chance and endless reiterations, through good times and bad, natural selection eventually happened upon exactly the right information to build the real you.

Change is inevitable. The Universe and its galaxies and their stars and planets continue to evolve and so does any life on or about them. Before your embryo lost its tail, it closely resembled the embryo of a rabbit or a bird or a turtle. For all vertebrate embryos still follow the same old step-by-step patterning process that gave rise to the embryos of your fishy ancestors—with each vertebrate embryo eventually deviating from that standard pattern in order to become yet another of its own kind. Obviously your own early embryo had no use for gill arches and 6 pairs of aortic arches—nor does any design principle require preplacement of a simple notochord before you can lay down your spinal cord—neither had you any reason to temporarily grow a pronephric kidney, or a cute tail, or skin like a fish embryo. Your early embryo simply resembled those other early embryos because your more recent ancestral embryos happened to pass through and develop from that generic, highly competitive, fish embryo pattern (which in turn came about by repeated minor modifications of what was before). You had to start somewhere. There was no need or way to reinvent every previous step each time your ancestors became slightly different. But had you been fully designed directly from scratch, your Creator could easily have come up with an endless number of far less complex, hence safer and more reliable routes by which to form the final you. Yet despite all of those incredibly circuitous interactions, Earth's greatest problem is the competitively developed, naturally selected and now suddenly excessive human reproductive potential. The fact that three out of four embryos are naturally discarded for design or developmental defects prior to completion of the pregnancy simply shows that even unnecessarily complex embryos are so cheap to construct that most can be tossed out after testing without risking reproductive disadvantage. Also that human evolution (which depends upon advances brought about by fierce competition between molecules as well as among organelles and cells and tissues and at every other level) has been advanced by selective survival at the zygote and embryo level and not just among humans approaching reproductive age (most of whom could once again cover Earth with their seed if things went badly for the rest).

Look at it this way. How does a fish become human? Verrry slowly. And in the process it must modify a whole lot. Some of those endless intermediate steps so lovingly constructed by your distant ancestors are still very easy for your embryo to follow since the trails are conscientiously maintained by natural selection and the necessary markers all remain in place. But quite clearly, each of the sequential modifications that took place during development of your own early fish-embryo-like structure represented a compulsive adherence to established ancestral embryo trails rather than any anticipation of your own design. While the temporary construction of 6 pairs of aortas might seem wasteful and unnecessary in retrospect, each of those aortas either contributed to your later vascular structure or it simply disappeared. Your stepwise embryonic construction through progressive modification of the embryos of your ancestors also explains why all of your embryonic cells and tissues had to move past, over, under, around and through each other in order to reach where they could best serve in you.

It would have been far simpler had your cells just marched out there with the band playing your song and formed you directly. Simpler for an embryologist to describe, that is. But for your embryonic cells, that direct approach presented insuperable organizational difficulties in comparison to simply repeating once again all of those long ago choreographed and endlessly perfected steps. When the half-time band at a football game already performs several sequential patterns perfectly, it is far simpler to make further modifications from where players currently are stationed than to reorganize everyone in the band back to their position on emergence from under the stands. Your cells continue those difficult rituals and complex marches because that is more or less how the equivalent embryonic cells did it in your earlier ancestors. "We've always done it that way" does not mean it is easier or even right. It simply states the obvious—that one more repetition is generally simpler than completely changing one's ways, even for good cause.

So while there is no fundamental reason for you to have been produced in this fashion, it just happens to be the way it all came about. Of course, had you been created directly, the nerves to your diaphragm could have emerged from your nearby lower thoracic spinal cord rather than stretching all the way from your cervical cord, and your single aortic arch would have developed in direct continuity with your carotids from the start, rather than being formed from some parts of paired arches #3 and 4 while pairs 1, 2, and 5 disappeared. And you certainly would not have bothered to grow a tail only to have it die later through programmed cell death (as did the tissues between your webbed embryonic toes). Nor would the recurrent laryngeal branches of your vagus extend way down from your neck to circle your intrathoracic aorta or subclavian artery en route to your throat. Yet that hardly seems inconvenient at all when you remember that the recurrent laryngeal nerves of a giraffe also loop down into the chest before returning to innervate the giraffe voice box. Those extra meters of axon between input and voice box make the impressive silence of giraffes less surprising. Surely no competent designer would ever build in such an unnecessary delay between thought and vocalization. While you have often regretted your words immediately, the giraffe has a far longer delay during which to regret messages still en route. But presumably that recurrent laryngeal nerve once followed a straight and narrow path across the short fat neck of your tiny ancestral mammal as it squeaked a quick warning of incoming pterodactyl. Furthermore your olfactory nerve would probably bypass your cerebrum entirely (and head off for the thalamus along with the rest of your sensory input) if that small initial group of cerebral nerve cells had not been dedicated to some long-ago ancestor's interpretation of odors—of course, were it otherwise, you might think quite differently as well.

A great deal of similar evidence strongly suggests that your uniquely human pattern is hardly the inevitable climax toward which all of Nature and life have been striving from the start. Indeed, it is impossible to show that human beings are anything other than the most effective information processors among the complex and specialized multicellular organisms currently alive. Of course, even that statement ignores the far larger brains of whales, dolphins and elephants as well as the embarrassingly greater brain-to-body-size ratio of monkeys, red squirrels and jumping mice (compared to your own). And let us not forget that all life is bacteria-based, bacteria-dependent and bacteria-recycled.

Your Coming Out

A pregnancy usually lasts about 280 days when measured from the start of the last menstrual period. Since successful fertilization ordinarily follows soon after ovulation and ovulation commonly occurs about 14 days after the start of that period, your internal parasitic existence probably lasted about 266 days (280 – 14), which is 38 weeks or 9 months. Quite appropriately, the internal portion of your existence ended in a flurry of activity and excitement. The events that led to your birth seem to have been initiated within your own brain as your hypothalamus notified its underlying pituitary to release ACTH which caused your adrenals to produce cortisone. In response to those changes, your mother's elevated blood progesterone levels (which had been inhibiting her uterine smooth muscle contraction) finally fell while her estrogens remained high. Her uterus then slowly began its rhythmic contractions, while at first providing long intervening periods of uterine smooth muscle relaxation during which your placental circulation could continue undisturbed. As the thick uterine *cervix* (neck) gradually shortened, thinned and dilated, it released bloody mucus and eventually that final flood of amniotic fluid (by now consisting of your own sterile urine) through mother's vagina.

Autonomic nerve reports of that reduction in uterine fluid distension then encouraged mother's posterior pituitary to release oxytocin while prostoglandins and oxytocin produced locally within the uterine wall also encouraged its contraction and thickening. As amniotic fluid emptied out, the increasing duration and strength of uterine contractions gradually compressed and molded you into an easier-to-expel torpedo shape. You responded to that tremendous pressure and stress by secreting both epinephrine and norepinephrine. Eventually, during a rapidly recurring series of uterine contractions, your head finally appeared through mother's maximally dilated cervix and painfully overstretched perineal tissues (even mother's pelvic ligaments were loosened as a result of your placenta's preparatory secretion of relaxin). Had you entered the cold dry world butt-first that would have been a *breech* delivery, which can be more difficult and even dangerous (since any hang-up or delay leaves the head inside, unable to breathe and therefore totally dependent upon the increasingly compromised or no longer available placental circulation— with the umbilical cord squeezed between baby and birth canal).

As it contracts and empties, the uterus eventually shears off its overgrown endometrial contribution to the placenta, which leaves a raw and bleeding surface—that separated *afterbirth* (placenta and ruptured membranes) is usually expelled soon after the baby is born. The blood loss associated with placental expulsion is generally impressive, temporary and limited. Since mother's blood volume was expanded by about 40% during the pregnancy, she can likely afford to lose more than a quart of blood at this point. In any case, significant and sustained post-delivery uterine muscle contractions compress those open endometrial vessels and help to bring about cessation of bleeding—those contractions can be encouraged by *pitocin* (oxytocin) supplements, as well as by gentle massage of the uterus through the now relaxed (because no longer over-stretched) belly wall. You were not ignored during this time, as you gasped your first breath of air and had your mouth cleared of mucus (and perhaps a tube passed to empty your stomach). Being very excited by the loss of flotation in your dry new world as well as by all of that squeezing through the birth canal and your own epinephrine, you probably cried a time or two. And that was a great relief to those giants who had gathered to serve you at such an inconvenient hour.

To enhance your safety and handling, your umbilical cord was soon tied and you were separated from the afterbirth (the Wharton's jelly inflating that umbilical cord would soon have dried and shriveled the cord circulation shut in any case). Had you been held way up in the air before your cord was tied, more of your blood might have remained in that placenta—but if your squatting mother expelled you onto a nice clean banana leaf, you probably acquired most of the blood from that afterbirth (thereby starting life with a higher hemoglobin content). The usual clamping of the cord occurs between these extremes—providing sufficient blood for a good start while minimizing the likelihood that an undesirably high blood bilirubin level might arise (consequent to ordinary red cell breakdown) and endanger brain function before the newborn liver could efficiently eliminate such potentially toxic materials.

While your birth may have seemed somewhat complicated and risky (indeed, it was!), it probably went very smoothly—which is not too surprising, since mammals have been working the kinks out of this system for more than 225 million years. After all, you are a direct descendent and inheritor of First Life and Early Life (who today might be nearing their four billionth birthday), as well as Old Eucaryote (now over two billion years old) and Multicellular Life (admitting to at least 750 million years). Furthermore, your vertebrate ancestors have been adapting to life on land for more than 400 million years. Even your most recent "first real human" ancestors (born between 250,000 and 1 million years ago) had time to modify and pass along a useful trait or two. But it is likely that the most important lesson you ever would learn was first received at mother's breast—as your persistent sucking stimulated her pituitary prolactin production, causing her breast tissue to secrete milk and release oxytocin which encouraged local smooth muscles to squeeze milk out of her breast alveoli so that you finally could take that warm, protective and nourishing *colostrum* (a dilute antibody-laden "first milk") from the numerous ducts serving her nipple.

No Strings

That lesson, of course, was that we all need love and help. For as you surely have learned, we only endow our own lives and the lives of others with significance to the extent that we value them. When we compliment a fellow worker, help a friend, bring food where there is illness or otherwise assist those in need, we add meaning to the lives of all involved. In our lifelong game of tit-for-tat and win-stay lose-shift, we are as good as we can be and as bad as we need to be to survive for a while and hopefully leave Earth a better place than we found it. So do something nice for someone—not to ingratiate yourself with some anecdotal vision of a frightful God—not because you might be judged either here or in the hereafter—but simply because it will aid both helper and helpee to move their information patterns more happily through this far-more-meaningful-because-not-preordained fourth dimension of life. Although no one gets out of that fourth dimension alive, time is part of life's essence. And greater love hath no human than to devote some of her/his remaining time to the happiness of others. Furthermore, you cannot help teaching any more than you can help learning, for your irresistible need to engage in meaningful brain-to-brain transfers of information during your own growth and development is naturally matched by your desire to contribute to the growth and development of your physical or intellectual descendents. Indeed, the joy in receiving and transmitting information is as "hard wired" into your brain as your interest in sex.

That it all came about by chance clearly increases the challenge, for then truly anything is possible. That you are not controlled like a puppet on strings merely means you must accept and work within circumstances that eventually could become partly of your own choosing. And not having a clue as to how it will all turn out simply enhances the interest, while also stimulating you to watch closely and get involved. For this land/air/water is your land/air/water as well as that of your peers and descendents, this information is your information and this life is your life. Chaos theory has shown how the most infinitesimal differences in starting conditions can quickly be amplified and distorted by non-linear and reiterated inputs until outcomes soon become unpredictable. That is what allows the gently beating wings of one small butterfly to alter those inherently unstable weather patterns (but not the climate) a month or two hence—for such initially tiny effects are endlessly modified, damped or amplified as they continue to influence all subsequent atmospheric states. Your life also depends upon countless interacting variables—and you too will just have to wait and see how it all turns out. So you remain inextricably entangled within life's marvelously dynamic, colorful and unpredictable four-dimensional tapestry of information. Like that butterfly, you cannot help but contribute to life's information as the uniquely tinted threads of your being permanently alter patterns yet to come through that tapestry. And whether you like it or not, you do make a difference. Indeed, you might even affect the climate.

INDEX

Bold entries indicate definition of terms.

ABOUT THE AUTHOR

Arndt von Hippel was born in Germany in 1932 and arrived in the United States in 1936. He comes from a multigenerational scientific family that recently was featured in *The Scientist.* Dr. von Hippel acquired a B.S. in Biology from M.I.T. and an M.D. from Harvard. In 1965, after eight years of surgical training, he moved to Anchorage with his pediatrician wife and their children. His broad and varied Alaskan medical experience included starting a very successful heart surgery program.

Following retirement, Dr. von Hippel taught a popular course in human anatomy and physiology at the University of Alaska—that course gradually evolved into this book.

STONE AGE PRESS ORDER FORM

Please send _____ copies of von Hippel's
Human Evolutionary Biology to:

_____

_____

_____

- ❑ **Check enclosed** (on prepaid orders we pay
  book rate shipping)
- ❑ **Bill me** (add $2.50 each to ship at book
  rate—may take 3–5 weeks: $3.50 each for
  priority mailing)

Single book price $29.95.

On prepaid orders for two or more books, send
only $23.95 per book.

Mail your order to Stone Age Press,
1649 Bannister Drive, Anchorage, AK 99508
(907) 279-3740
(907) 278-1475 msg (please speak clearly and slowly)

Satisfaction guaranteed or full refund.

STONE AGE PRESS
Anchorage, Alaska

STONE AGE PRESS ORDER FORM

Please send _____ copies of von Hippel's
Human Evolutionary Biology to:

_____

_____

_____

- ❑ **Check enclosed** (on prepaid orders we pa
  book rate shipping)
- ❑ **Bill me** (add $2.50 each to ship at book
  rate—may take 3–5 weeks: $3.50 each for
  priority mailing)

Single book price $29.95.

On prepaid orders for two or more books, sen
only $23.95 per book.

Mail your order to Stone Age Press,
1649 Bannister Drive, Anchorage, AK 9950
(907) 279-3740
(907) 278-1475 msg (please speak clearly and slowly

Satisfaction guaranteed or full refund.

STONE AGE PRESS
Anchorage, Alaska